Advances in
Applied Digital
Human Modeling

Advances in Human Factors and Ergonomics Series

Series Editors

Gavriel Salvendy
Professor Emeritus
Purdue University
West Lafayette, Indiana

Chair Professor & Head
Tsinghua University
Beijing, People's Republic of China

Waldemar Karwowski
Professor & Chair
University of Central Florida
Orlando, Florida, U.S.A.

Advances in
Applied Digital
Human Modeling

Edited by
Vincent G. Duffy

CRC Press
Taylor & Francis Group
Boca Raton London New York

CRC Press is an imprint of the
Taylor & Francis Group, an **informa** business

CRC Press
Taylor & Francis Group
6000 Broken Sound Parkway NW, Suite 300
Boca Raton, FL 33487-2742

© 2011 by Taylor and Francis Group, LLC
CRC Press is an imprint of Taylor & Francis Group, an Informa business

No claim to original U.S. Government works

Printed in the United States of America on acid-free paper
10 9 8 7 6 5 4 3 2 1

International Standard Book Number: 978-1-4398-3511-1 (Hardback)

Visit the Taylor & Francis Web site at
http://www.taylorandfrancis.com

and the CRC Press Web site at
http://www.crcpress.com

Table of Contents

Section IV: Product and Process Design

Section V: Motion Analysis

Section VI: Cognitive Aspects

Section VII: Human Response and Behavioral Aspects

Section VIII: Novel Systems Approaches

Preface

This book is concerned with digital human modeling. The utility of this area of research is to aid the design of systems that are benefitted from reducing the need for physical prototyping and incorporating ergonomics and human factors earlier in design processes. Digital human models are representations of some aspects of a human that can be inserted into simulations or virtual environments to facilitate prediction of safety, satisfaction, usability and performance. These representations may consider the physical, physiological, cognitive, behavioral or emotional aspects. They are typically represented by some visualization with the math and science computed in the background.

Each of the chapters of the book were either reviewed by the members of Scientific Advisory and Editorial Board or germinated by them. Our sincere thanks and appreciation goes to the Board members listed below for their contribution to the highest scientific standards maintained in developing this book.

K. Abdel-Malek, USA
T. Ahram, USA
T. Alexander, Germany
G. Andreoni, Italy
K. Bengler, Germany
R. Bhatt, USA
B. Boggess, USA
H. Bubb, Germany
J. Charland, Canada
Z. Cheng, USA
B. Corner, USA
M. Darby, USA
C. Du, USA
R. Dufour, USA
L. Fu, PR China
N. Fürstenau, Germany
R. Goonetilleke, Hong Kong
R. Green, USA
J. Grobelny, Poland
B. Gore, USA

D. Högberg, Sweden
H. Hsiao, USA
T. Inagaki, Japan
J. D. Lee, USA
K. Li, USA
Z. Li, PR China
R. Lind, USA
R. Marshall, UK
C. Moebus, Germany
A. Oudenhuijzen, The Netherlands
U. Raschke, USA
M. Reed, USA
K. Robinette, USA
M. Thomas, USA
L. Wang, PR China
X. Wang, France
J. Yang, USA
X. Yuan, PR China
W. Zhang, PR China

Explicitly, the book contains the following subject areas:

I. Applications
II. Mobility and Universal Access
III. Physical and Physiological Aspects
IV. Product and Process Design
V. Motion Analysis
VI. Cognitive Aspects

This book would be of special value to those researchers and practitioners involved in various aspects of product, process and system design worldwide. Engineers, ergonomists and human factors specialists will see a broad spectrum of applications for this research, especially in the automotive and manufacturing industries, military, aerospace and service industries such as healthcare.

April 2010

Vincent G. Duffy
School of Industrial Engineering
Purdue University
West Lafayette, Indiana, USA

Editor

CHAPTER 1

A Method for Positioning Digital Human Models in Airplane Passenger Seats

Rush F. Green[1], Jeffrey A. Hudson[2]

[1]The Boeing Company
Everett, WA, USA

[2]InfoSciTex
Dayton, OH, USA

ABSTRACT

In the airplane passenger cabin, there are several design features intended for use by the seated passenger. The design engineer needs to consider the extents of seated reach and vision for the passenger population. Digital human models (DHM) can provide this information, but for accurate results, the DHM must be positioned properly. Unlike some other common seated environments, there is no standard way to position a DHM in an airplane passenger seat. The method described herein is a proposal to create a standard passenger posture. Subjects representing the passenger population were scanned in representative airplane seats. The relationship of the subjects to the seats, along with their posture was measured. A method was then developed to place and position a DHM in a digital passenger seat.

INTRODUCTION

When designing for the seated airplane passenger, design engineers need to ensure reach and vision to several places in the passenger's environment. Passengers need to be able to reach to light switches, flight attendant call buttons and tray tables. Regulations require that passengers are able to view Fasten Seat Belt and Exit signs while seated. To evaluate the passengers' reach and vision in the digital design environment, digital human models (DHM) are commonly used. Accurate placement and posturing of the DHM is required to generate reliable DHM analyses.

In the airplane flight deck, the positioning of DHM is based on the eye reference point and in the automotive industry, SAE standards define both the location and posture of the driver and passengers. (Reed, et. al.) The currently is no standard method for the location and posture of DHM in the airplane passenger seat. Most commonly, the designer will "eyeball" the position of the DHM so that it looks reasonable, but there is no good way to validate this method. More sophisticated methods are available which may use the combination of a dynamic human model along with finite element definitions of the seat cushion and soft tissues of the DHM. This method may be physically accurate, but requires very sophisticated software not generally available to the design engineer. (Van Hoof, et. al.) The standard passenger posture defined in this paper is more accurate than the "eyeball" method and is simple to implement.

METHOD

Twenty six subjects were whole-body scanned with a Cyberware WB4 laser scanner and traditionally measured according to the CAESAR protocol (Harrison and Robinette 2002) prior to being scanned while seated in an economy class passenger seat (Koito Industries Limited, Model ARS-626). The subjects were chosen based on their anthropometric similarity to two idealized 95th percentile boundary manikin groups. (Zehner, et. al.) The groups were defined as obese (BMI >= 30) and non-obese (BMI < 30). This grouping was created to facilitate further design studies with these sub populations. Also, it was thought that by ensuring a large variability in the subject population, any relationships among sitting position, posture and anthropometry would be more readily apparent.

The subjects were scanned in various postures, from "seated erect" to "slightly slumped." The "seated erect" posture was used for the development of the standard passenger posture described below. Additional posture data was collected using a FARO 3-D digitizing arm. Twenty six anthropometric landmarks were digitized for each subject. The FARO data allowed for direct point-to-point measurement among the anthropometric landmarks and with landmarks on the seat.

The FARO data was used to determine the relationship of the subjects to the seat. The center point of the line defined by each subject's hip joint markers was measured relative to the seat reference point (SRP) of the seat. This SRP was defined as the center point of the intersection of the seat back and seat pan. The X, Y and Z distances for each subject-SRP combination were averaged and correlations for each distance to the anthropometric variables were also calculated.

To determine the definition of the subjects' postures, a DHM based on each subject's anthropometry was created using Dassault Systemes V5 Virtual Ergonomics Solution. Each DHM was then positioned and postured to match the "seated erect" scan for the respective subject. (Figure 1.) This alignment was done manually until the DHM and scan matched as closely as possible. Once this was done, the joint angles for each DHM were recorded. As with the FARO data, means and correlations with anthropometric variables were calculated.

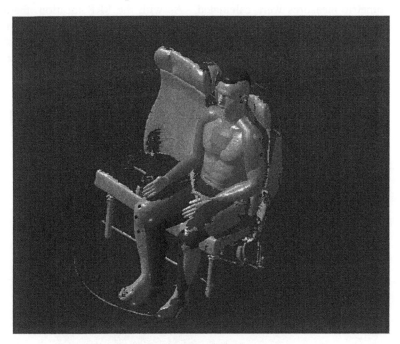

FIGURE 1. Example of DHM and subject scan alignment (Left side only).

RESULTS

Figures 2 and 3 show three different DHMs in the standard passenger posture superimposed on their respective scans. It can be seen that the standard passenger posture is a good match for some subjects and not as good for others. This was due to the great variability in size and posture among the subjects and also difficulties in matching the anthropometry and spinal posture of the subjects. In general, smaller DHMs ended up farther forward than the subjects and the larger DHMs ended up farther back.

The DHMs were positioned based on the mean distance of the DHM SRP and the seat SRP. Correlations between the SRP locations each of the anthropometric measures were also calculated. (Note that for protection of proprietary data, the calculated values cannot be shared here.) While there were reasonably strong correlations between some anthropometric measures and the DHM SRP location, it was decided that using the overall mean location was simpler and more practical for end users' applications.

For each degree of freedom for the lumbar spine, thoracic spine, neck, shoulder, elbow, hip, knee and ankle, the average posture was calculated. To match the pelvis orientation of the subjects, the DHM was rotated so that its hip segment aligned with the subject's pelvis. Correlations of all of these joint angles with the anthropometric measures were calculated. As with the SRP location, there were some strong correlations found. In this case, arm flexion was positively correlated with several of the weight related variables (e.g. chest depth and hip breadth). However, it was again decided to use the means for all joint angles as an easier to implement solution. Figure 2 shows an example of a DHM in the average posture along with the original seated scan.

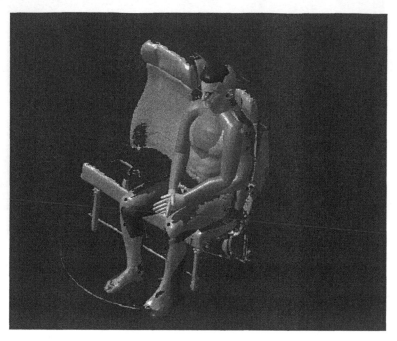

FIGURE 2. DHM in the standard passenger posture shows good alignment with original scan.

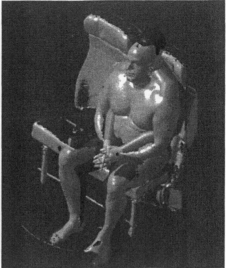

FIGURE 3. DHMs in the standard passenger posture appear reasonable but don't have good alignment with original scan.

DISCUSSION

A method for positioning and posturing DHMs in economy class seats was created. To apply this method, the designer first needs to determine the SRP of the seat. As stated above, this is done by defining the midpoint of the line defined by the intersection of the seat pan and seat back. If these parts of the seat don't intersect, which most often they wont, two planes can be defined that represent the seat pan and seat back. Variation among seats makes it difficult to come up with a definitive way to determine these planes, but it is suggested to create them such that their intersection would continue the primary seat pan and seat back angles. The intersection of these planes, along with a vertical plane dividing the seat in half, can be used to define the location of the SRP.

Once the SRP of the seat is defined, it may be necessary to create a corresponding SRP for the DHM. This is done simply by creating a point midway between the hip joint centers. Finally, the DHM can be positioned by placing the DHM SRP the correct distance from the seat SRP.

The next step is to rotate the DHM so that the hip segment matches the proper pelvic orientation. This rotation should be done on an axis defined by the line between the hip joint centers. The last step is to move all of the body segments to the joint angles determined in from the scan/DHM matching process. This final step can be greatly simplified by creating and saving a predefined posture. In V5 Virtual Ergonomics Solutions, a posture can be created once manually then saved in a library that can be applied to any other DHM.

Once the DHM is positioned in the standard passenger posture, a designer can begin analysis of reach and vision in the passenger cabin. By using this starting posture, the designer can have more confidence in the results and the consistency across analyses. Figure 4 provides a comparison of the standard passenger posture to the default DHM sitting posture.

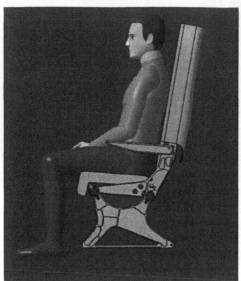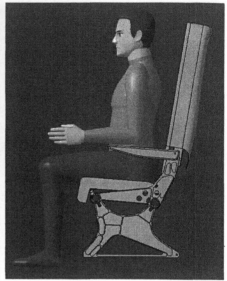

FIGURE 4. Standard passenger posture compared to the default sitting posture.

The standard passenger posture was created to provide a simple alternative to the "eyeball" method of placing DHMs in airplane passenger seats. There is much variability among airplane seats and how passengers sit in them. Also, there are limits to how accurately any DHM can represent human anthropometry and posture. If high accuracy is required for DHM placement and positioning, a replication of this study may be required for the specific seat, passenger population and DHM in question. It should also be noted that due to differences in the internal and surface definitions of commercially available DHMs, the SRP and joint angles described above may not be suitable.

CONCLUSIONS

A method was created to position and posture digital human models (DHM) in digital mockups of airplane passenger seats. The method is based on data collected on human subjects representing the extremes of the adult airplane passenger population. Anthropometry was collected using calipers, tape measures and laser scanning. The subjects were also scanned and measured with a 3D digitizing device while seated in an economy passenger seat. Average subject SRP location and posture were calculated and used to create the DHM positioning and posturing method.

The method was created to be easily implemented and applied by the general designer population. Some accuracy may have been sacrificed by only using averages for location and posture, however this method should be more accurate and repeatable than the "eyeball" method that is most commonly used. This starting posture can be used to provide consistency among results for reach and vision analyses in the airplane passenger cabin. This posture is not intended to be the most realistic in appearance. For instance, since the hip and knee angles are the same for each DHM, variation in leg length means that the feet may not initially match the floor level. (Figure 5.) Users who desire a more realistic appearing starting posture will have to make some minor adjustments. Designers and analysts still need to employ their professional expertise and judgment when using this new method.

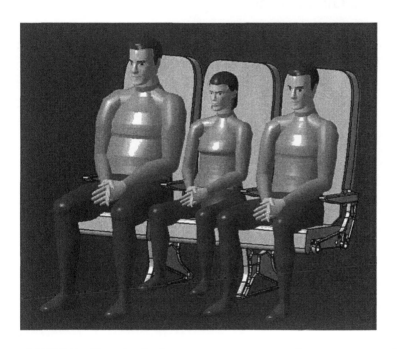

FIGURE 5. The standard passenger posture applied to large, small and average sized DHMs. Notice the variation in foot location.

REFERENCES

Harrison, C.R. and Robinette; K.M. (2002), *CAESAR: Summary Statistics for the Adult Population (Ages 18-65) of the United States of America*, Human Effectiveness Directorate, Crew Systems Interface Division, Wright-Patterson Air Force Base, Ohio, AFRL-HE-WP-TR-2002-0170

8

Reed, M.P., Manary, M.A., and Schneider, L.W. (1999), *Methods for Measuring and Representing Automobile Occupant Posture*. SAE Technical Paper Series Report no. 1999-01-0959. Society of Automotive Engineering, Warrendale, PA.

Van Hoof, J., Van Markwijk, R., Verver, M., Furtado, R. and Pewinski, W. (2004), *Numerical Prediction of Seating Position in Car Seats*, SAE Technical Paper Series Report no. 2004-01-2168. Society of Automotive Engineering, Warrendale, PA.

Zehner, G.F, Meindl, R.S., and Hudson, J.A. (1993) *A Multivariate Anthropometric Method for Crew Stations Design*: Abridged, (Publication No.: AL-TR-1992-0164), Armstrong Laboratory, Air Force Systems Command, Wright Patterson Air Force Base, OH

ACKNOWLEDGEMENTS

The authors would like to acknowledge the contributions of Gregory Zehner, Hyeg Joo Choi and Jim Fullerton.

CHAPTER 2

On the Creation of 3D Libraries for F-16 Pilots in their Crew Station: Method Development, Library Creation and Validation

Aernout J.K. Oudenhuijzen[1], Gregory F. Zehner[2],
Jeffrey A. Hudson[2], Hyeg Joo Choi[3]

[1]TNO Defence, Security and Safety
Soesterberg, The Netherlands

[2]USAF AFMC 711 HPW HP
Wright-Patterson AFB, USA

[3]Oak Ridge Institute for Science and Education
Wright-Patterson AFB, USA

ABSTRACT

In earlier verification and validation studies for digital Human Modeling Systems (HMSs) in an F-16 cockpit application, the initial positioning and posturing of the manikins were found to be the greatest source of error in calculations of manikin reach and clearance (Oudenhuijzen et al. 2002). The goal for this project was to develop a method to reduce these errors based on "training" the HMSs. In essence, this training enables the manikins to assume realistic postures by employing 3D body scans of real people in an actual F-16 ACES II ejection seat. This was the starting point for defining manikin initial position, and posture during reach, as well as to quantify the effects of the restraint system and the protective equipment in an F-16 cockpit environment. The Safework HMS was chosen as the modeling system to be "trained." Fourteen subjects with a considerable range in body dimensions were selected for the modeling activities in this project. Their scan data were collected under two conditions while seated in the F-16 ACES II ejection seat: 1) wearing stretch shorts (and sports bras for females) to serve as baseline data; and 2) wearing a full pilot cold water immersion ensemble (small subjects only). The resulting subject data were used to produce 15 reach posture libraries for the Safework HMS. These libraries can be considered as a kind of fidelity profile that quantified, and simultaneously accounted for, the effects of the restraint system, protective equipment, and tissue deformation in this seated cockpit environment. The average difference between the small subject reach envelopes and their corresponding manikin envelopes (compared at the radial styloid on the wrist for all 15 reach directions) had an error range of +/- 7 mm. Hence, the library is considered to be highly accurate and verified for anthropometric accommodation studies on the F-16 when using the HMS Safework and the resulting posture libraries. The positioning accuracy for accommodation tasks was also found to be accurate. Manikin eye location, during positioning in accommodation tasks, lies between +/- 6 mm in the vertical Z direction, and much smaller (+/- 2 to 7 mm) in the horizontal directions.

Keywords: Digital Human Modeling, F-16 Cockpit Accommodation, Anthropometry, 3D Scan, Reach Envelope, Safework.

INTRODUCTION

This project is the logical continuation from an earlier study in which Oudenhuijzen et al. (2002) concluded that "in their present form, Human Modeling Systems (HMSs) may yield inaccurate results when attempting to determine accommodation limits for an aircraft cockpit." However, the authors concluded that accommodation results offered by HMSs can be drastically improved if platform specific field studies are conducted to define the digital manikin's initial body position and posture.

The HMSs tested by Oudenhuijzen et al. (2002) revealed errors in modeling reach and clearance, all related to initial positioning and posturing of the manikin in the seat. For that study, the following HMSs were evaluated: CombiMan, Jack, BHMS, Safework, and RAMSIS. Results varied between the different HMSs. Specifically, their manikin reaches failed to reflect the subject reaches correctly with a mean error ranging from 20 to 80mm. The performance deviations between subjects and their corresponding manikins, with respect to an accommodation task, differed for small and tall pilots. The mean deviation for accommodation limits for tall subjects was 1.5 cm for maximum Sitting Height, while the maximum Leg Length deviation was 0.8 cm (Buttock Knee Length). The mean deviations for accommodation limits related to small subjects were much larger with 4.7 cm. for minimum sitting height and 9.5 cm. for minimum

"Comboleg," which is the sum of Buttock Knee Length and Sitting Knee Height measures.

The original aim of this project was to create a virtual environment to serve as an ergonomic test bed for the new F-35 (Joint Strike Fighter). However, this was not possible due to the fact that the F-35 ejection seat had not been finalized during our study. Instead, a virtual ergonomic test bed was created, verified and validated for the F-16. Still, we developed and validated methods that can be applied as soon as the F-35 ejection seat is ready.

METHOD

Fourteen subjects, representing both sexes and a wide range of anthropometric shape and proportions, participated in the creation of the F-16 Safework Posture Libraries. The "training" of the Safework manikins for the F-16 cockpit application followed these general steps:

1. Scanning of subjects in F-16 ACES II seat to record position and posture data;
2. Creation of Safework manikins based on subject anthropometry;
3. Aligning manikins to scans: creation of Safework Posture Library;

Each of these steps is discussed below. The Results section reports the validation of the reach posture libraries.

SCANNING OF SUBJECTS IN ACES II SEAT

After experimenting with different methods, a whole body 3D scanning technique to capture seated subject postures was chosen. All fourteen subjects (five men and nine women) were scanned according to the CAESAR protocol (Daanen & Robinette (2001), Figure 1). However, to achieve manikin alignment, ping pong balls (usually used for table tennis) were added to posture defining landmarks.

Two general accommodation issues were addressed with the scanning of subjects in the ejection seat. Larger male subjects were scanned with focus on clearance issues to knees, overhead, and shoulders, while the smaller female subjects were scanned with focus on reach issues. Hence, two different types of seated pilot postures were of interest when scanning the subjects in the F-16 ACES II ejection seat:

1. Resting or neutral posture - used for both male and female seated subjects in which hands are simulated to rest on stick and throttle controls (see Figure 2). The subjects were scanned three times in this posture.
2. Reach posture against locked inertial reels – used for female subjects, 15 different postures simulated reaching toward controls against locked

restraints. The 15 different postures are the result of 3 elevations with 5 directions. The smaller subjects (who were the focal point for reach postures) were scanned seated not only wearing CAESAR garments, but also while wearing cold water immersion suit, G Pants, G Vest, survival vest, and flight harness (see Figure 3). The subjects were scanned three times for each condition. Hence, the 15 reach postures had 45 scans associated with both with the full immersion suit, and with the CAESAR garments. (Two of the women could not return for this part of the study).

FIGURE 1 The CAESAR protocol standing scan posture (left and center) and seated posture on stool (right), with added ping pong markers.

FIGURE 2 A large subject seated in the ACES II ejection seat with added ping pong ball markers.

FIGURE 3 A small subject executing a low side reach while harnessed and belted into the ACES II ejection seat. She is wearing a full cold water pilot immersion suit. Small subjects were also scanned in the baseline CAESAR stretch shorts, as above in Figure 2, to produce baseline reach envelopes.

CREATION OF SAFEWORK MANIKINS

The subjects' digital twins in the Safework HMS were created from their traditional anthropometric measurements as described by Daanen & Robinette (2001). We verified the anthropometric dimensions on the manikin, using the method described by Oudenhuijzen et al. (2002), to ensure that the manikin had the same body dimensions as the subject (see Figure 4).

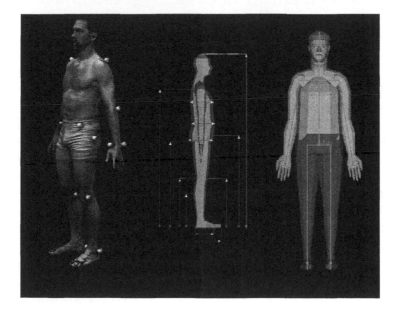

FIGURE 4 Creating a manikin (on the right) using the HMS software (in the middle) from the subject's traditional anthropometric data (on the left).

ALIGNING MANIKINS TO SCANS: CREATION OF THE SAFEWORK POSTURE LIBRARY

The digital manikin was superimposed on the scan of the subject to match the standard CAESAR postures. Forty millimeter spheres (constructed in the HMS software) were then created, positioned, and associated with the manikin based on the position of the subject ping pong balls marking the relevant bony landmarks (see Figure 5).

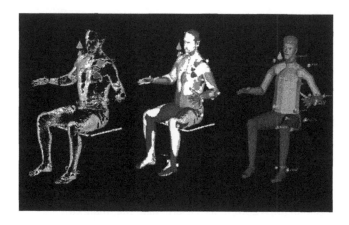

FIGURE 5 The subject's scan (on the left) was used to align the manikin (in the middle) and to position the 40 mm manikin spheres (representing the subject ping pong balls) on the manikin (on the right).

The next step was to visually align the manikin (now with reference balls) to the seated subject scans in the ACES II. First, the manikin's pelvis is rotated to match the angle with the subject's pelvis using the ball markers. Next, the manikin legs were aligned with the subject legs in the scan using the appropriate ball markers. The final step aligned the manikin's and subject's spine; now using the markers on the torso, neck, head and shoulders and arms resulting in the initial posture, or resting alignment, shown in Figure 6 (left) , below. This is the subject's initial posture when seated in his/her seat. In accommodation analysis this posture is used for clearance studies, which ask, for example, "what is the tallest or widest pilot that can be accommodated in the crew system?" or for small pilot issues, "what is the shortest eye height that can see over the nose of the aircraft?"

The initial posture also provides a starting point for generating the library reach postures. Any subsequent manikin reach posture begins with this initial seated manikin posture and is acquired by a series of body segment translations and rotations to match the position of the reaching scan. Below, in Figure 6 (right), is an example of the final alignment of a small subject manikin to one of her fifteen reaching scans, while wearing the full cold water immersion suit. This manikin position would then be captured and saved as part of the Safework Reach Library for the F-16 ACES II seat.

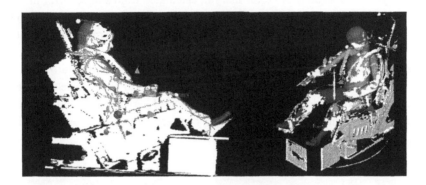

FIGURE 6 Left, the final alignment of the manikin's and subject's upper extremities using the appropriate markers resulting in an initial, resting posture. Right, the creation of a reach posture aligning the subject's and manikin's markers.

These resulting 15 reach postures for each manikin represent the small pilot area of interest with respect to modeling reaches to flight control and display locations in fighter aircraft like the F-16 and the F-35 (Joint Strike Fighter). Below, Figure 7 shows the three different heights (side view), and the five different directions (top view) which combined yield the 15 total reach postures.

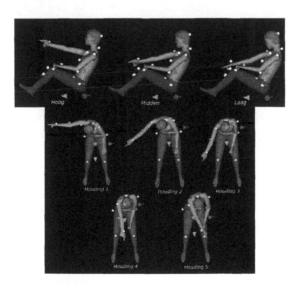

FIGURE 7 The fifteen resulting library postures for each small manikin are the combinations of the three heights (side view images) and the five directions (top view images).

RESULTS

To verify reach library accuracy, we used the same procedure that was used by Oudenhuijzen et al. (2002). An error margin, calculated from subject error in repositioning over the three trials for each scan posture, is used in this procedure. The reach envelope, shown in Figure 8, represents the average reach (over the three trials) for a subjects 15 reach directions. A manikin's posture, as defined by the position of its associated landmarks, is compared to this error margin around the reach envelope. Several landmark locations on the manikin were chosen for the verification and validation, however, the one reported here is the Radial Styloid, on the wrist, which combined with hand length, corresponds to the fingertip position and is directly related to reach.

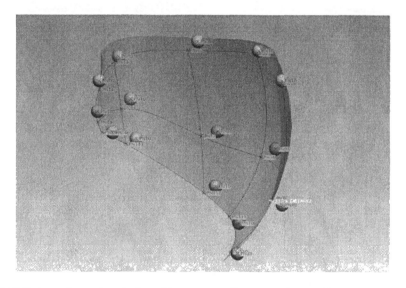

FIGURE 8 A reach envelope is the subject average (3 reaches) for reaching toward each of 15 directions. The distance from the spherical manikin markers to the reach envelope represent the deviation for a subject in a specific direction.

The difference, or deviation, between the subject's reach shell (defined by Radial Styloid locations) and their corresponding manikin's Radial Styloid, was calculated after posturing the manikin. These deviations were calculated in both a vertical direction (Z), as well as simultaneously in the horizontal directions (X and Y). Below, in Figures 9 and 10, the mean deviations between reaching manikin and subject are reported, while wearing the full immersion suit.

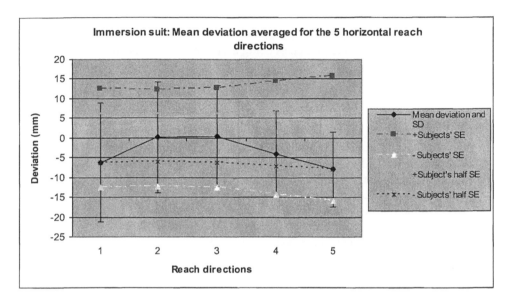

FIGURE 9 The box and whisker plot for the deviation between the subject's combined reach envelopes and the corresponding manikin's Radial Styloid for all 5 reach directions and the corresponding subject S.Es for full immersion clothing.

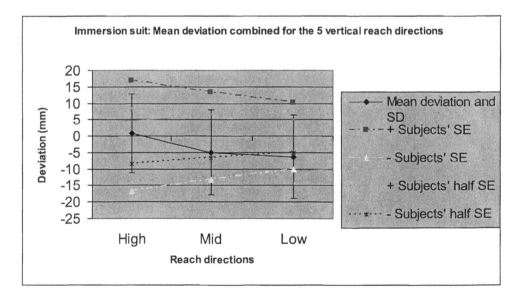

FIGURE 10 The box and whisker plot for the deviation between the subject's combined reach envelopes and the corresponding manikin's Radial Styloid for all 3 reach heights and the corresponding subject S.Es

The mean deviations, for all subjects' combined, was smaller in all reach directions than half the subject standard error. Hence, both libraries, one for baseline clothing (which had similar graphical results) and another for the full immersion suit, can be considered as being accurate for accommodation studies.

CONCLUSION

These 15 reach postures represent the area of accommodation interest with respect to flight control and display locations in fighter aircraft like the F-16 and the F-35 (Joint Strike Fighter). The reach postures are important in digital human modeling studies because the whole body subject scans of these postures incorporate the effects of the following issues, many of which the HMSs do not:

- Tissue deformation for the buttocks and thighs;
- Postural changes during reach compared to initial posture: e.g. translation and rotation of the pelvis and lower back;
- Effects from restraint systems; (the subject's were wearing the lockheed F-16 restraint system while being scanned);
- Effect from clothing. Two types of clothing were worn by the subjects for scanning: 1) CAESAR garment and flight harness to create a library representing a baseline with greatest freedom of movement ; 2) All available pilot gear, including a full immersion suit, G suit, etc., representing the most restrictive movement possible

With these posture libraries offering increased fidelity in manikin posture and position, it will be possible to correct or avoid the inaccuracies previously observed in our HMS accommodation studies (e.g. on Cougar, Apache and Chinook helicopter platforms) using the Safework HMS. Given that the original results for these airframes were conservative and on the safe side, a new DHM analysis would most likely yield broader accommodation limits, resulting in increased accommodation for smaller and taller pilots, and with fewer rejected pilot candidates.

REFERENCES

Daanen, H.A.M., Robinette, K.M. (2001) "CAESAR: The Dutch data set." Memorandum TM - 02 - M002, TNO Human Factors, Soesterberg, The Netherlands

Oudenhuijzen, A.J.K., Zehner, G.F., Hudson, J.A. (2002) "Verification and validation of human modeling systems." Report TM - 02 - A007 TNO Human Factors, Soesterberg, The Netherlands.

Chapter 3

Modeling Foot Trajectories for Heavy Truck Ingress Simulation

Matthew P. Reed, Sheila M. Ebert, Suzanne G. Hoffman

University of Michigan
Transportation Research Institute
USA

ABSTRACT

Digital human figure models are a useful tool for simulation of driver ingress and egress for passenger cars, light trucks, and heavy commercial trucks. Simulation allows evaluation of the suitability of steps and handholds as a system. Accurate simulation requires detailed, validated algorithms to predict driver motions. One critical component of such an algorithm is the accurate prediction of foot trajectories. This paper presents an approach to foot trajectory simulation based on statistical analysis of driver motions from a laboratory study. The movements of 20 truck drivers were recorded as they entered a reconfigurable truck mockup. The foot trajectories were parameterized using Bézier curves, which accurately represent the important characteristics of the trajectories with relatively few parameters. Statistical analysis of the fitted parameters showed that the shapes of the trajectories, after normalizing for overall displacement, were independent of truck step configuration but affected by driver characteristics. The resulting models are designed for use with the Human Motion Simulation Framework, a software system previously applied to simulating a wide range of task-oriented human behavior, including passenger car ingress and egress.

Keywords: Digital human modeling, ingress and egress, trucks

INTRODUCTION

Digital human figure models have been applied to simulation of driver ingress and egress for passenger cars, light trucks, and heavy commercial trucks (Andreoni et al. 2004; Cherednichenko et al. 2006; Dufour and Wang, 2005; Lestrelin and Trasbot, 2005; Pudio et al. 2006; Rasmussen and Christensen, 2005; Reed et al. 2008). The previous work on commercial truck ingress/egress (IE) has focused on cab-over-engine (COE) trucks, which are common in Europe and some other markets (Chameroy et al. 2008). In the U.S., the driver sits behind the engine in most heavy truck cabs (so-called conventional cabs) and the step and handhold configurations are substantially different from those of COE trucks.

Truck IE is a focus of ergonomic concern because of the relatively high likelihood of injury. Truck drivers are frequently injured due to slips and falls while entering and exiting the cab, and the design of the steps and handholds has been implicated in those injuries (Jones and Switzer-McIntyre, 2003). Lin and Cohen (1997), using data from a survey of trucking companies, reported that more than 25% of injuries, and more than 80% of the more severe injuries causing lost work days, were due to slips and falls.

As part of a larger study of truck driver IE with conventional cabs, a laboratory motion capture study was conducted. Among the goals of the study was the development of predictive methods for simulating driver IE with a wide range of step and handhold configurations. The overall simulation approach is based on the Human Motion Simulation (HUMOSIM) Framework, a hierarchical methodology for simulating complex, task-oriented human movement (Reed et al. 2006). The HUMOSIM Framework has previously been used to simulate a wide range of tasks, including passenger car IE (Reed et al. 2008). The previous IE work demonstrated the importance of flexible collision avoidance algorithms to guide the trajectories of the lower limbs when the simulated figure moves through a confined space.

The HUMOSIM Framework controls the kinematics of the limbs through a two-step process. The motion of the end-effector (hand or foot) is predicted, after which the joint angles necessary to achieve the prescribed end-effector movement are calculated using behavior-based inverse kinematics. This methodology exploits the empirical observation that end-effector trajectories are much more consistent across subjects and tasks than are joint angles, and hence are easier to model. This approach also supports generalization to alternative figure models with different joint definitions, because joint angles are not the primary control variables. End-effector translation trajectories are modeled using third-order Bézier curves, which provide a convenient and accurate parameterization of typical hand and foot motions. Previous studies have shown that hand trajectories in reaching tasks and foot translations in passenger car ingress and egress can be readily modeled using Bézier curves (Faraway et al. 2007; Reed et al. 2008). The current paper applies the Bézier parameterization to foot movements in truck ingress and develops a predictive model to assess the effects of step configuration and driver characteristics on foot movements.

METHODS

LABORATORY METHODS

A full-scale laboratory mockup (Figure 1) was constructed to simulate IE with conventional truck cabs having two steps and either interior or exterior handholds, a configuration found on more than 95% of US tractor/trailer truck cabs. A 13-camera VICON motion capture system was used to record movement data at 60 Hz. Figure 1 shows a truck driver entering the mockup wearing the passive optical markers used to track movements.

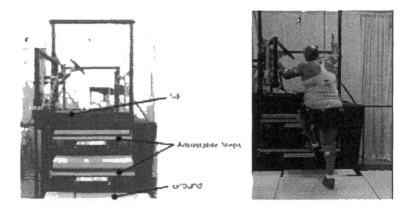

FIGURE 1. Laboratory mockup, showing adjustable steps; driver entering the mockup wearing motion-capture markers.

PARTICIPANTS

Data from 15 men and 5 women with at least two years of commercial truck driving experience (median 11 years) were analyzed for the current analysis (more than 95% of commercial truck drivers in the U.S. are men). The drivers ranged in stature from 1554 to 1862 mm and in body mass index (BMI) from 21.7 to 40.1 kg/m^2.

TEST CONDITIONS

Testing was conducted with 8 step conditions developed based on an analysis of 30 conventional truck cabs (Figure 2). The heights of the two steps and the cab floor were fixed, but the lateral positions of the two steps relative to the door sill were varied. Fifteen of the drivers were tested in all 8 conditions; five drivers were tested only in conditions one through four.

22

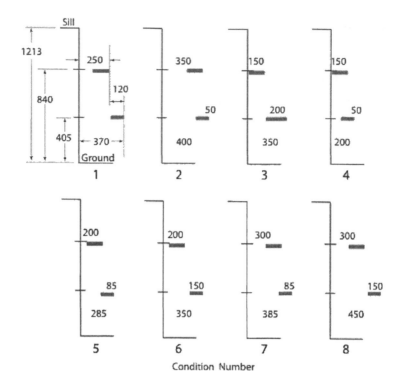

FIGURE 2. Step configurations. Dimensions in mm. Sill is at the height of the cab floor.

FOOT KINEMATICS

The analysis presented here focuses on the right and left foot moving from the ground to either the first or second step. The trajectory of each foot was modeled using the translation of single lateral ankle marker on each foot. Some participants placed their right foot on the first step from the ground, while others used the left foot. Left and right-foot trajectories were analyzed separately for analysis in each step condition.

Foot trajectories were modeled using a three-step process. The movement from the ground to the first or second step was extracted from the trial data by an automated algorithm that identified segments between stationary periods based on speed. The extracted motions, consisting of between 30 and 80 frames, were resampled using local linear interpolation to obtain 100 evenly spaced points.

Third-order Bézier curves were fit to each trajectory using a least-squares algorithm with adaptive knot assignment (Faraway et al. 2007). The third-order Bezier curve parameterizes the trajectory into two end points and two interior control points. Regression analyses were conducted to determine the effects of subject and step-location variables on the control point locations.

RESULTS

FOOT MOVEMENT PATTERNS

Out of a possible 280 foot motions (8 step conditions x 15 subjects x 2 feet + 4 step conditions x 5 subjects x 2 feet), 8 trials were excluded due to movement anomalies (e.g., slipping off of the step), leaving 272 trials for analysis. Table 1 shows the distribution of foot motions by the target step. The most common movement pattern (70% of the trials) was moving the right foot to the first step followed by moving the left foot to the second step. The left foot moved to the first step in only 30% of trials.

Table 1. Frequency of Movement with Each Foot to Each Step

Foot	Step 1	Step 2	Total
Right	105 (77%) *(70%)*	31 (23%) *(25%)*	136
Left	44 (33%) *(30%)*	92 (67%) *(75%)*	136
	149	*123*	272

BÉZIER FITTING

Figure 3 shows foot trajectories as marker data and Bézier curves for trials in step condition 1. The qualitative fit was excellent, with the Bézier curve readily capturing the basic character of the trajectories, including the departure angles from the ground and the approach angles to the cab steps. Over all trials, the 5th, 50th, and 95th percentile root mean square errors for the fits, evaluated at 100 evenly spaced points on each trajectory, were 3.8, 9.6, 37.8 mm, respectively. For comparison, the mean foot motion distance for the bottom and top steps were 689 and 1027 mm, respectively, meaning that the median root mean square error was usually less than one percent of the total movement distance.

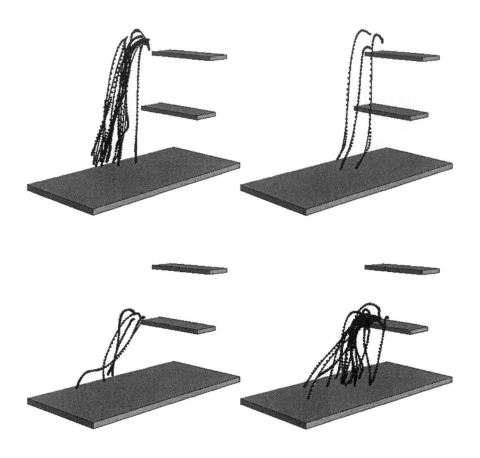

FIGURE 3. Sample trajectories fitted with third-order Bézier curves for left-foot trajectories (left) and right-foot trajectories (right). Trajectory points are shown as dots, Bézier curves as solid lines.

For digital human modeling applications, the foot placements on the ground and each step may be predicted using other statistical models (e.g., regression models), in which case a normalized trajectory analysis is more useful. Each splined trajectory was scaled to unit end-to-end horizontal and vertical displacement so that the starting point was at the origin and the endpoint was at $\{1,0,1\}$). In this representation, only the two interior control points remain to be predicted (six degrees of freedom) in the normalized coordinate system.

A multivariate statistical analysis was conducted to determine the effects of subject and test-condition factors on the shape of the normalized trajectories within each foot/step set. The interior control point coordinates were transformed using a principal component analysis and the principal component scores were predicted using linear regression. Potential predictors included driver stature, body mass index (body mass in kg divided by stature in meters squared), and three measures of the horizontal position of the two steps: step 1 relative to step 2, step 1 relative to sill, and step 2 relative to sill. Figure 4 shows the effects of varying the independent measures from 10% to 90% of the range of the independent variables in the dataset.

The analysis demonstrated that the step configuration variables did not have important effects on the trajectories after normalization. Stature and BMI affected the foot/step sets differently. Stature had a relatively large effect on the left foot trajectory when moving to the first step, with taller subjects moving the foot on a flatter trajectory. BMI had the largest effect on the movement of the right foot to the second step, with heavier drivers producing a steeper initial trajectory.

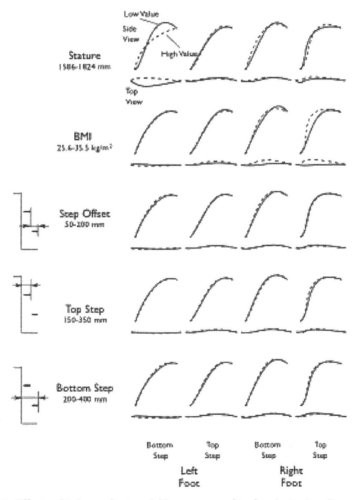

FIGURE 4. Effects of independent variables on normalized trajectories. Rows are foot/step sets. Columns are independent variables, each of which was exercised from the 10th to 90th percentiles of the values in the dataset. The normalized trajectories from all subjects and step conditions are shown in the background of each plot.

DISCUSSION

Foot motions during truck ingress demonstrate considerable regularity that is well represented by the third-order Bézier curve. A normalization approach was developed that removes both horizontal and vertical scale from the trajectories prior to statistical analysis, allowing the shape of the trajectory to be examined independent of the overall extent of the movement. The analysis demonstrated that the step configurations do not have a strong influence on the shape of the foot trajectory after accounting for scale. Relatively large anthropometric effects were seen, but only for the relatively infrequent left-foot movement to the first step and right-foot movement to the second step.

The results support a central tenet of the HUMOSIM Framework approach to human motion simulation, namely that movements of end-effectors, such as the hands and feet, exhibit smooth patterns that can be modeled with relatively simple empirical formulations. The parameters of the Bézier curve can be readily modified to impose obstacle avoidance on trajectory predictions (Reed et al. 2008) while maintaining the basic character of the motion.

The findings of the current analysis are limited by the relatively small subject pool. The trajectories show considerable residual variability that is not accounted for by overall measures of driver size. Accurate characterization of the range of driver behavior will require a larger and more diverse subject pool, particularly when multiple tactics are to be simulated. Driver behavior in the laboratory may not span the entire range of behaviors drivers exhibit in the field, and the laboratory conditions are necessarily a subset of the possible step and handhold configurations on trucks. Future work will include a more complete analysis of movement kinematics for both ingress and egress motions.

ACKNOWLEDGMENT

This research was supported in part by grant number 1-R01-OH009153-01 from the Centers for Disease Control and Prevention – National Institute for Occupational Safety and Health. Its contents are solely the responsibility of the authors and do not necessarily represent the official views of NIOSH. This research was also supported in part by the partners of the Human Motion Simulation Laboratory at the University of Michigan (www.humosim.org).

REFERENCES

Andreoni, G., Rabuffetti, M, and Pedotti, A. (2004). Kinematics of head-trunk movements while entering and exiting a car. *Ergonomics*, 47(3):343-359.

Chameroy, A, Monnier, G., and Roybin, C. (2008). Truck instep evaluation using a sample of manikins. Technical Paper 2008-01-1920. SAE International, Warrendale, PA.

Cherednichenko, A., Assmann, E., and Bubb, H. (2006). Computational approach for entry simulation. Technical Paper 2006-01-2358. SAE International, Warrendale, PA.

Dufour, F. and Wang, X. (2005). Discomfort assessment of car ingress/egress motions using the concept of neutral motion. Technical Paper 2005-01-2706. SAE International, Warrendale, PA.

Faraway, J.J., Reed, M.P., and Wang, J. (2007). Modeling three-dimensional trajectories by using Bézier curves with application to hand motion. *Applied Statistics*, 56(5):571585.

Jones, D. and Switzer-McIntyre, S. (2003). Falls from trucks: a descriptive study based on a workers compensation database. *Work*, 20:179-184.

Lestrelin, D. and Trasbot, J. (2005). The REAL MAN project: objectives, results, and possible follow-up. Technical Paper 2005-01-2682. SAE International, Warrendale, PA.

Lin, L-J, and Cohen, H.H. (1997). Accidents in the trucking industry. *International Journal of Industrial Ergonomics*, 20:287-300.

Pudio, P., Lempereur, M., and Lepoutre, F.-X., and Gorce, P. (2006). Prediction of the car entering motion. Technical Paper 2006-01-2357. SAE International, Warrendale, PA.

Rasmussen, J. and Christensen, S.T. (2005). Musculoskeletal modeling of egress with the AnyBody modeling system. Technical Paper 2005-01-2721. SAE International, Warrendale, PA.

Reed, M.P., Faraway, J., Chaffin, D.B., and Martin, B.J. (2006). The HUMOSIM Ergonomics Framework: A new approach to digital human simulation for ergonomic analysis. Technical Paper 2006-01-2365. SAE International, Warrendale, PA.

CHAPTER 4

Occupant Model Validation

Joseph Pellettiere, Ted Knox

711th Human Performance Wing
Air Force Research Laboratory
2800 Q Street
Wright-Patterson AFB, OH 45433, USA

ABSTRACT

Modeling and Simulation of crash scenarios can be a cost effective and useful tool when researching occupant motions and internal responses. The Articulated Total Body (ATB) Model is one such tool that is used to simulate the dynamic motion of a jointed system of rigid bodies. ATB is similar in function to other Multi-Body Solvers (MBS) in that it solves the equations of motion for a set of rigidly connected bodies. As with all Modeling and Simulation models, there are many limiting simplifying assumptions and parameters which must be determined. It is in the choices of simplifying assumptions and input parameters that the appropriateness and applicability of the particular simulation can be determined.

The ATB model was initially developed to complement experimental research in automobile crash environments and to provide a functional instrument for parametric investigations. It is primarily a research tool which is used to interpolate and extrapolate the results of full-scale tests. It has been used in research to assist in the understanding of a wide array of other events and environments such as explosions, floatation, and ejection. The finite element method originated as a tool for stress analysis. The structure of interest is discretized into a series of finite number of elements, each with simple geometry and associated material properties. These discretized elements are then assembled to form a structural stiffness matrix and prior to calculating displacement, strain, and stress. In recent years, new finite element methods have been developed to analyze stresses within a moving body while impacting another object or to assess the effects of heat flow in human tissue.

This paper will present an overview of some important considerations for model validation. Some metrics used to validate models as well as an overview process and where to look for guidance on conducting Verification, Validation, and Accreditation (VVA) of Modeling and Simulation (M&S) will be discussed. After this framework is laid, several different applications of where validation has been successfully used will be discussed. The concepts and limitations discussed will relate not just to MBS but also to more advanced modeling techniques such as Finite Element Modeling (FEM) when used in conjunction with occupant modeling for crash scenarios. Clarification of the definition of words such as 'prediction' in research will also be included.

Keywords: Model Validation, Occupant Simulation, Wavelet Analysis, Accreditation

INTRODUCTION

The steps necessary for conducting VVA can be separated into several process steps (Figure 1). Each one of these steps can have assigned roles and responsibilities (Pellettiere and Knox 2003). These could be steps such as the user defines the need for M&S thus starting the VVA process. It should also be mentioned that in the described framework, the VVA process can be a continuous loop as once the model is in use, its accreditation must be maintained whenever new uses or upgrades become available. It should be noted that the validation metrics are only one piece and appear in the implement V&V portion of the process.

The process begins when it is decided that M&S is needed to support or solve a particular need. This need could be the development of a new vehicle system and the M&S supports the vehicle safety design. To meet this support, it must be decide whether an existing model exists or if a new model must be developed. Once this decision is made, it must be determined if there is applicable VVA to continue, or if a new VVA process should begin. If a new process is begun, a detailed plan will be developed that will document the required steps to determine the VVA of the particular M&S. The plan can include which supporting test data, which simulation cases, and how the VVA will be determined. It is as part of this plan, that the validation metrics become important. The VVA plan will use the metrics as one of the measures to determine whether and how to move forward with the VVA process. This would occur during the implementation phase. If the validation of the model is accepted, then it can be recommended for accreditation and put into use. However, the process does not end here; its use will be monitored to determine if VVA will be needed for future applications.

As can be seen from this process, the metrics themselves are an important and integral part of the process. The metrics provide the objectivity and oftentimes, it is the result of applying the metrics that allows a validation determination to be made.

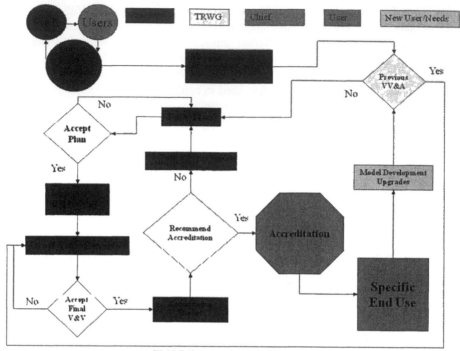

FIGURE 1. Model VVA Process

MODEL VERIFICATION, VALIDATION & ACCREDITATION

Commonly when discussing Modeling and Simulation (M&S), there are three terms that describe the status of the model: Verification, Validation & Accreditation (VVA). Recently accepted definitions have been adopted that help to clarify the meaning and intended purpose of any M&S:

1. Verification is the process of determining that a models implementation and its associated data accurately represent the developer's conceptual description and specifications (DoDD 5000.59, DoD 5000, DoDI 5000.61, DAR 5-11). Verification determines whether the Model has been properly programmed.

2. Validation is the process of determining the degree to which a model and its associated data are an accurate representation of the real world from the perspective of the intended uses of the model (DoD 5000, DoDI 5000.61, DAR 5-11). Validation checks if the model actually provides a realistic representation of an intended real world event. A model validated for a particular event, like a belted occupant in a frontal motor vehicle collision, may not be appropriate for modeling occupants in other impact scenarios (i.e., side impact? Rollover?)

3. Accreditation is the official certification that a model, simulation, or federation of models and simulations and its associated data are acceptable for use for a specific purpose (DoDD 5000.59). Accreditation is the determination whether the Model should be used for a particular application.

When considering these points, it is important to remember, that it is both the model and the data used in the development of the model that are important. Validation is normally accomplished by comparing a simulation with test data from a particular real world phenomena. The test data from the individual component or full scale tests may focus on only certain aspects of the real world phenomena and therefore the measured information may be incomplete. During the VVA procedure, many of the input parameters are not directly measurable. Validation and verification of any simulation model includes iterative adjustments and variations of some of the input parameters to permit a better correlation of the model with the component or full scale tests. The reference test or test series can be used as a baseline dataset and/or limiting event type for the particular simulation model verification and validation as appropriate. Also, if the tests themselves did not accurately represent the real world event, then can you consider the model to be validated?

Many times the purpose for simulation model development is to predict some response. The use of the word "predict" is often incorrectly misinterpreted to literally mean "to declare or indicate in advance; especially: foretell on the basis of observation, experience, or scientific reason" (DoDD 5000.59). When researchers use the word "predict," they are indicating that there is some measurable correlation of some observed phenomena between a given component or full scale test and the computer simulation model. Detailed occupant kinematics involves so many approximations, estimates, and assumptions that they cannot accurately "predict" injury or detailed movement. Whether a simulation model's "prediction" is to be used for product development or design, accident reconstruction, or just knowledge discovery will oftentimes determine the fidelity with which the model is developed and utilized. The VVA that is conducted with the model is then tailored to this specific use. For instance, a model that is developed to research trends on the capability of padding to reduce impact forces, cannot then be applied directly the development of the padding without further VVA. This is because the original VVA may have been geared towards demonstrating the correct relationship between the padding stiffness and the force outcome. There may not have been validation of the actual forces themselves.

REASONS FOR VALIDATION

Validation of a model gives future users of either the model itself or of the model's predictions a sense that the model accurately represents the process being modeled. Since models are often used in designing vehicles (cars or aircraft) and protection devices (seat belts, airbags, crushable vehicle structures or personal protection such as, helmets and pads for sports) it is important to know how accurate and reliable the model predictions are. Through the full VVA process, inherent flaws and limitations of the model can also be documented (Pellettiere et al. 2008). Also, when models are used in situations beyond their original intent then it is incumbent upon the user to revalidate the model for its new use. Having a catalog of previous validations will be instructive about the need and the type of validation needed.

VALIDATION METRICS

As part of the VVA process, the validation phase is sometimes the most difficult to accomplish. This is partly because of the subjectivity that can be involved and partly due to the inherent complexities in M&S. The subjectivity arises from the nonexistence of a single objective metric to state when a model is valid and to what extent is it valid. M&S can range from a simple single parameter model to one with thousands of degrees of freedom and responses, making it impossible to match every single combination. There are various techniques that can be used for validation and they can range from visual matching, parameter matching, to detailed time history matching. It is during this matching process that inherent model flaws are discovered. To determine the models validity, data can be obtained from either other validated models or tests that replicate the desired event. In either case, two sets of data (event and simulated) will be generated. It is from these data that the validation will occur and that metrics are needed to determine model validity.

Several different methods of conducting the model validation through metrics will be briefly discussed. These methods just provide the framework and it is up to modeler to tailor them to the specific simulations as necessary.

Correlation Analysis

If Data are available from two different signals, one being a test case and the other from a simulation of that test case it is possible to conduct a correlation analysis on these two signals. This correlation may range from the simplistic to more complex methods. On the simple end of the spectrum, pertinent response features may be chosen such as amplitude, pulse width, rise time, frequency response, total energy or some other discrete measurement that is a combination of these parameters. A simple mathematic ratio of the test and simulation parameters could then be calculated and a level of correlation determined. (Robbins 1989). Oftentimes this method is useful when building an initial model with only a few parameters of interest. Such a correlation lends itself directly into optimization schemes where the objective may be to minimize a specific response. However, this method can be extended to inspect the entire time history and not just discrete portions. If this is chosen, then the correlation becomes more of an error measurement over time. Here, care must be taken to synchronize the two signals as a small phase shift can create a large error. The relationship between two stationary signals can be determined using a correlation analysis (Bendat and Piersol 2000). However, there are times when the signals of interest occur over a very short time duration, such as for an automobile crash which is typical when conducting an occupant simulation. Automobile crash signals are not stationary but transient and strongly localized in the time domain. Consequently, conventional correlation analysis used for stationary signals may not be appropriate or efficient for the analysis of automobile crash signals.

Wavelets

Wavelets, as a new tool for signal analysis, have been introduced for the correlation analysis of automobile crash signals (Cheng and Pellettiere, 2005) and for the generalized cases of biodynamics including occupant modeling (Cheng et al. 2006). Wavelets are localized in both frequency and time domains, which matches with the major characteristics of automobile crash signals and the occupants within them. The basic tenets of a wavelet analysis are to break down the signal into its basic wavelet functions using approximations and details. The approximation can be thought of the basic rigid body motion of the signal of interest, while the details would correspond to higher order components. It should be noted that each successive level of detail contain less energy content, thus contribute less to the overall signal. At a certain point, the details could be considered noise and ignored. With the signal properly decomposed, correlations could then be performed on the individual approximation and details. This then yields information specific phenomena. In addition, a correlation of this type determines if two signals are linearly related and not just duplicates of one another. For model validation, it would be useful to know that one signal is a constant percentage of another, and not just to know that they disagree to some extent. Knowing this linear relationship would allow for correction factors, or changes to the model to be directly made.

Other Methods

The literature is full of other methods which can be applied to determine the validation of particular models. This includes variations of those discussed above, as well as other scalar and frequency domain methods. One such method that has been used more recently is the Sprague and Geers (S&G) metric (Sprague and Geers, 2004). The S&G metric is a simple method that accounts for both the amplitude and the phase errors, further it can be weighted to give preference to either the measured or the simulated signal. The amplitude portion of the S&G method can be thought of as a weighted scalar correlation analysis on the peak energy of the two signals. The phase error calculation gives more insight into the shape differences between the two signals. These two error calculations can then be further combined into a comprehensive metric.

Guidance Available

Many technical societies and government bodies provide guidance on the conduct of Validation. Oftentimes, these may be detailed requirements for specific uses or more generalized for informational purposes. One such is the Verification and Validation guide from the American Mechanical Engineers Society (ASME, 2006). While the guide itself provides a wealth of useful information, it is geared towards the validation of computational structural mechanics issues. There is no single source guide for validating occupant responses. Other guides of interest are the NASA STD 7009 (NASA 2008) and the DoD 5000. Each of which provides requirements for model validation for when they are used with specific programs for those organizations. The Federal Aviation Administration has been working to provide guidance to certify the safety of seats through computational means (FAA 2003). However, this advisory circular only provides the framework for acceptance, it is up to the individual engineer to determine the method whereby to validate the occupant model to be used for acceptance.

APPLICATIONS

The Air Force 711th Human Performance Wing has successfully developed a process based upon the available guidance and metrics described previously (Pellettiere and Knox 2003). The process was designed to be both rigorous and flexible to the needs of the developers and end users. This process has been successfully applied to a variety of applications including human motion simulation, burn simulation, and impact simulations (Pellettiere and Cheng 2007). This section includes a few applications to illustrate the both the range of models and the types of validations that have been useful.

Manikins are used as surrogates for cadavers and human volunteers. Human volunteers have the disadvantage that they must be exposed to non-injurious accelerations while cadavers have the disadvantage that they no longer have the same material properties as live humans. Each can help to bound the responses to high levels of acceleration and to give some estimate of tissue failure. In the last few years strides have been made in developing instrumentation to monitor people driving race cars, playing sports, and conducting military operations. The most promising of these is to monitor head accelerations using instrumented earplugs. New sensors were developed in 2008 (Knox et al 2008) that were tested in the laboratory and in studies of cadavers heads exposed to blasts from a shock tube (Salzar et al 2008) to simulate explosions. It was found that when the sensors were placed deep in the ear canal their response was within a few percent of sensors screwed to the skull. A transfer function to correct for the soft tissue between the skull bone and the sensor brings the ear canal sensor in very close agreement with the skull mounted sensor. The earplug was invented by Air Force Research Lab and the present version has been in use in the Indy Racing League and the

Championship Auto Racing Teams (CART) since 2003. There are now data from hundreds of crashes some of which resulted in concussions. The latest sensor (Yazdi et al 2008) is self contained and readable using RFID technology. These sensors have the ability to provide data that can be used during the validation phase of an occupant model. For instance, the chassis acceleration is available as well as the anthropometry and the injuries (if any) sustained by the occupant. These data can be compared with simulation results using the described validation techniques. In fact, both the standard correlation and wavelet techniques have been applied to these data (Knox 2004). It was shown that by using wavelets, a high level of correlation could be demonstrated by the sensors to other known metrics.

Vehicle crashes not only result in mechanical accelerations resulting in injury but on occasion post crash fires develop that can cause injury and death. Helicopter crashes in the Vietnam War were initial driver for the development of a model to predict when persons exposed to a post-crash fire would feel pain and receive a burn injury. BURNSIM 3.0.2 (Knox and Mosher, 2010) is the current version of this model that now includes all modes of heating and predicts six levels of pain, time to blister (second degree burn) as well as final burn depth. The Thermal Protective Performance index (TPP) is a number which quantifies the time it takes to develop a threshold blister when the skin is exposed to 2 cal/cm^2/sec and protected by a fabric. The higher the TPP value the more protection. The TPP index can also be used in BURNSIM 3.0.2.

BURNSIM is a mathematical model which calculates the heat flow to and through skin and the local temperature to specify a damage rate which can be integrated to give a local total damage. The transition between damaged and non damaged skin is the locus where total damage, omega = 1. The skin is considered as a number of 25 micron thick nodes and heat flow is calculated using finite difference mathematics.

BURNSIM has been validated using the developed process and metrics against 5 different data sets from experiments with pigs and humans. Uses include calculating the threat of burns or pain from a post-crash fire, from lasers, from explosions, touching hot surfaces, or hot air such as maintainers might be exposed to while working around an aircraft with the jet engine running. Recently BURNSIM successfully predicted the results observed in an industrial accident in which an explosion caused facial blisters and the protective garment prevented all other burns. Coupled with a fabric model, BURNSIM will aid in developing better protective clothing systems.

CONCLUSIONS

Validation of modeling and simulation is an important, if not the most important step in the full VVA process. Occupant modeling validation has the additional issues of dealing with data that is hard to collect and reproduce since human subjects react slightly differently to similar stimuli. Furthermore, it is difficult to expose them to necessary environments to run a full validation program. As such, great care must be taken on the limited validation available and the extrapolation of any data. A plan that follows a rigorous approach and uses suitable metrics for determining the level of validation can be successful.

DISCLAIMER

The findings and conclusions in this report/presentation are those of the authors have not been formally disseminated by the Air Force and should not be construed to represent any agency determination or policy.

REFERENCES

American Society of Mechanical Engineers. (2006). *Guide for Verification and Validation in Computational Solid Mechanics*. ASME V&V 10-2006; 2006

Bendat JS and Piersol AG (2000) Random Data-Analysis and Measurement Procedures, John Wiley & Sons, New York

Olvey, SE, Knox, T and Cohn KA (2004), The Development of a Method to Measure Head Acceleration and Motion in High-impact Crashes, Neurosurgery, 54:672-677, 2004.

Cheng, Z.Q., Pellettiere, J.A., & Wright, N.L. (2006). Wavelet-based test-simulation correlation analysis for the validation of biodynamical modeling. *Proceedings of XXIV International Modal Analysis Conference* February, 2006, MO.

Cheng, ZQ and Pellettiere, JA (2003). Correlation analysis of automobile crash response based on wavelet decompositions. *Mechanical Systems and Signal Processing, 17*, (6), 1237-1257.

Department of Defense Directive (DoDD) 5000.59: DoD Modeling and Simulation (M&S) Verification, Validation, and Accreditation, January 1994.

DoD 5000.59-M: DoD Modeling and Simulation (M&S) Glossary, December 1997

DoD Instruction (DoDI) 5000.61: DoD Modeling and Simulation (M&S) Verification, Validation, and Accreditation (VVA), April 1996.

Department of Army Regulation (AR) 5-11 (1997): Management of Army Models and Simulations.

Federal Aviation Administration Advisory Circular no. 20-146 (2003), Methodology for Dynamic Seat Certification by Analysis in Parts 23, 25, 27, and 29 Airplanes and Rotorcraft; May 2003.

Knox T and Mosher S, (2010). BURNSIM Model as a Predictor of Burn Injury Based on TPP Test Data. AFRL-RH-WP-TR-2010-0006.

Knox T, Pellettiere J, Perry C, Plaga J, and Bonfeld J (2008). New Sensors to Track Head Acceleration during Possible Injurious Events, SAE Motorsports Conference paper 2008-01-2976 published in Conference Proceeding and in SAE International Journal of Automobile Electronic and Electrical Systems, 2009.

Knox T (2004). Validation of Earplug Accelerometer as a Means of Measuring Head Motion, SAE Paper 2004-01-3538.

NASA-STD-7009, (2008), Standard for Models and Simulation, National Aeronautics and Space Administration.

Pellettiere, JA, McHenry, BG, Hu, J, and Yang, KH (2008). The appropriateness and applicability of Occupant Modeling and Simulation. *2008 Applied Human Factors Ergonomics International.*

Pellettiere, JA. and Cheng, ZQ (2007). Digital human modeling with Applications. *Proceedings of the Fifth Triennial International Aircraft Fire and Cabin Safety Research Conference,* Atlantic City, November 2007.

Pellettiere, JA and Knox, T (2003). Verification, validation, and accreditation (VV&A) of human models. *SAE Paper No. 2003-01-2204.*

Robbins, D (1989) Restraint Systems Computer Modeling and Simulation State of the Art and Correlation with Reality, SAE Paper 891976

Salzar, R.S., Bass, C.R., & Pellettiere, J.A. (2008). Improving Earpiece Accelerometer Coupling to the Head. *SAE Paper No 2008-01-2978.*

Self, B.P. Knox, T and Kairns C. Head Accelerations During Impact Events, Sept 2004 published in Hubbard M. , Mehta, R. D. and Pallos, J. M. (eds) (2004) *The Engineering of Sport* 5, Volum1 pp 616, Volume 2 pp 656 and presented by Dr. Brian Self.

Sprague MA and Geers TL (2004) *A Spectral-Element Method for Modeling Cavitation in Transient Fluid-Structure Interaction.* International Journal for Numerical Methods in Engineering. 60 (15), 2467-99; 2004.

Yazdi N, Knox T, Plaga J, Yafan Z, and Hower R (2008). Wireless Acceleration and Impact Recording Chips, SAE Motorsports Conference paper 2008-01-2979 published in Conference Proceeding and in SAE International Journal of Automobile Electronic and Electrical Systems, 2009.

CHAPTER 5

Experiments Under Water: Preliminary Findings on Individual Neutral Postures

Thomas Dirlich

Institute of Astronautics
Technische Universität München

ABSTRACT

The present paper describes a new method for the recording and analysis of human body postures under water in simulated microgravity. Experiments with a sample of 69 subjects were carried out aimed at exploring the motion behaviour and particularly the neutral postures spontaneously assumed by the body when specific motion tasks are absent. Three levels of data analysis are applied: visual analysis of the footage, a graphical overlay method for selected two-dimensional (2D) pictures and the detailed three-dimensional (3D) study of postures utilizing an individually adapted virtual man model. Examples of applications of these approaches are presented. The first results presented here seem to indicate the existence of an intra-individually stable, replicable and virtually constant neutral posture for each subject.

Keywords: Neutral Body Posture, underwater experiments, simulated microgravity, stereo-metric video recording, 3D posture analysis, Human Factors Engineering

INTRODUCTION

The detailed knowledge of human postural behaviour is of great importance for Human Factors Engineering (HFE). As a rule, the Neutral Body Posture (NBP) derived from space experiments and defined by NASA [NASA 1995] is referred to as a postural optimum. According to a widely accepted assumption the body „automatically" assumes the NBP when a person fully relaxes in a weightless environment. The NBP is conceived as a stable, replicable and nearly constant posture involuntarily and spontaneously assumed by the human body when specific motion tasks, internally triggered (voluntary movements) as well as externally triggered (e. g. to counteract gravity), are absent [Tengwall 1982].

The published body angles (Figure 1) are used in HFE as optimum values for joint angles in evaluating a given environment or man-machine-interfaces. Some digital discomfort models also use these values for their multi-body-simulation-based computations [Seitz 2003].

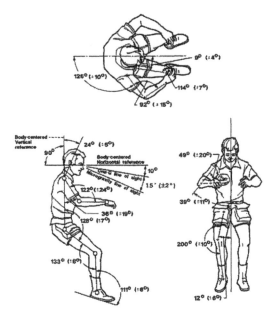

FIGURE 1. Neutral Body Posture drawing published in the NASA-STD-3000 [NASA 1995]

Follow-up experiments by NASA and ESA as well as re-evaluation of the original Skylab material raised doubts whether the NBP can be validly applied to specific individuals [Andreoni 2000], [Mount 2003], [Vergé-Dépré 1994]. Numerous observations indicate that the concept of differing individually typical neutral postures might be appropriate. Yet, even the concept of one constant and stable neutral posture for each individual needs to be studied with larger samples of subjects, including females. This may yield sound empirical evidence for understanding the variability of individual postures observed in prior experiments [Mount 2004] [Zimmermann 2002].

The experimental study reported herc leads in this direction. It is focused on the concept of the NBP. Can the motion activity of the body in the state of relaxation be interpreted as supportive of the concept of a stable, replicable and nearly constant NBP in microgravity? The experiments were carried out under water in a condition of simulated microgravity. Similarities of this condition with true microgravity are, however, counterbalanced by several essential differences. Hence, the preliminary results presented here need to be discussed carefully in the light of the similarities and dissimilarities between the two experimental conditions.

OBJECTIVE

The experiment is a series of simple movement tasks separated by distinct relaxation phases. The objective is to study the motor activity during these task-free breaks. The core questions are: does the subject's posture converge to a stable, replicable and nearly constant body posture during the relaxation breaks? If this can be empirically supported, is the assumed posture independent of the preceding motion task or not? How can the movements occurring during the relaxation breaks be characterized?

METHOD

Several test campaigns were conducted in pools at 3.00 - 3.70 m depth in chlorinated water at 26 - 28° C. A modularized setup was used consisting of an aluminum test rig, an ISS interior volume mockup, two camera stands with HD video cameras for stereo recordings in underwater-casings, a calibration body required for 3D posture analysis, a subject restraint to counteract buoyancy (CBR), an oxygen supply system with rig mounted pressure flasks and a breath apparatus. The subjects were continuously observed by safety divers and supported if necessary. The subjects could trigger an abort of the experiment at any time.

During the experiments the subjects were continuously monitored by the video cameras. The cameras were positioned on the test rig in side and frontal view positions of the subject enabling later stereo-metric (3D) interpretations of the recordings. The camera's fields of view were adjusted so that the fixed calibration body of 2.12 x 1.00 x 1.00 m built around the subject could clearly be distinguished. The cameras were synchronized with each other to time-coordinate the video data streams from the side and the front view cameras. Recordings were made in high definition quality (1920 x 1080 pixel) and stored on the cameras.

The study was conceived for a large group of subjects with each subject to be observed for a time interval of 8 to 20 minutes depending on personal capabilities and diving experience. Pilot experiments had shown that the duration of the experiment should be limited to not more than 25 minutes to avoid uncomfortable temperature sensations and feelings of boredom.

The schedule of the experiment was conceived as a sequence of 6 - 10 "effort cycles" to be performed by the subject. At the beginning of each cycle the investigator sent an acoustic signal to the subject to assume an "effort posture", i.e. one of the three postures "stretching", "crouching" or "asymmetric" (Figure 2).

FIGURE 2. Subject displays the three effort postures (stretching, crouching, asymmetric)

After keeping up the respective posture at maximum force for up to 10 s (Figure 3: effort phase) the subject relaxed (relaxation phase) and waited for the next instruction (neutral phase). This "non-effort" phase was terminated by the investigator with a signal to start the next effort. The cycle duration varied form 1:00 - 2:30 minutes. In each experiment the different efforts were arranged in differing pseudo random orders so that the subjects did not know in advance which of the effort tasks would be required in the next cycle.

Subjects were blindfolded during the experiment and fixed from both sides at the hip to minimize buoyancy effects. The CBR was optimized in order to minimize the bias of the displayed postures induced by the fixation by offering a rotational degree of freedom in the body axis.

69 subjects participated in the study. They had been recruited through announcements of the study in lectures at the university and by a website. The subjects were not paid for their participation. The majority of the subjects were 3rd to 4th year undergraduate students. There were 55 male and 14 female subjects in the sample. The age range was 20 to 38 years. All subjects were fit and healthy, most of them with a BMI values in the normal rage (19 - 25).

The video material of each experiment consists of two time-synchronized continuous HD video streams (25 pictures/s). First of all, these data were reduced to 1 picture per second automatically to diminish the workload for the first analysis. Trained analysts inspected this material and segmented it into effort cycles and these into three phases, effort, relaxation and neutral (the start of which was determined by the visual analysts as the end of apparent movements of the subject) yielding precise time protocols of every experiment (Figure 3). In addition, few failed experiments were excluded from further analysis.

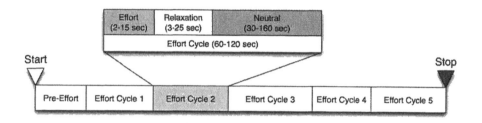

FIGURE 3. Graphical representation of one full experiment divided into effort task cycles

The images from every neutral phase were identified. A graphical approach for posture analysis was applied: using Adobe Photoshop side view images of a neutral phase were reduced to the outline of the subject's figure by cutting it free from the background and marking it in with signal colours. This procedure was applied to every third picture of the neutral phase. Image cut-outs after being rotated at the hip point to particularly compensate rotations of the body were stacked on top of each other using transparency effects (opacity 40%).

A man-model based measurement system, PCMAN (developed and provided by the Institute of Ergonomics at Technische Universität München (TUM) [Arlt 1999]), enabling marker-free 3D measurements of postures is used for detailed analyses of selected images. Picture pairs, i.e. side and front views, form the unit of analysis in PCMAN.

In a customizing session preceding to the analysis a standard 3D human volume model has to be adapted to the individual anthropometric features of the subject creating an individualized virtual 3D body model of the monitored subject. The customization requires images, taken outside the water, of the subject in two standard postures (standing upright first with arms hanging down then with one arm and one leg bent to a 90° angle).

The analysis of a picture pair requires (1) the reliable identification of the calibration body, which is part of the underwater setup and visible in each picture followed by (2) the adaptation of the individual body model to the postures exhibited in the pictures. Then, the 3D coordinates of a set of significant body parts respectively locations (the eyes, several anatomically well defined points on the spine and the major limb joints, in total 27 points in the 3-dimensional space) can be exported as a spreadsheet.

Interpretations and, moreover, intra-individual comparisons of different scenes require transformations of the PCMAN coordinate values which are related to axes of the calibration body into body related coordinate systems. The computation of 3D joint angles is an example for this approach. It is demonstrated here for the angles at the shoulder, elbow, hip and knee joints.

RESULTS

89 underwater experiments were carried out in 2009 and 2010 during which more than 1000 minutes of continuous video recordings were collected. On the one hand, the data represent processes in time allowing the study of the temporal microstructure of the movements and alterations of postures. On the other, the stereo-metric recording technique combined with the individually adapted body model opens up possibilities for quantitative studies of body postures in 3-dimensional space.

The analysis of this data material has recently been started. For some aspects an overview will be given, some results will demonstrate the breadth and depth of the analysis approach, however, not yet give a complete picture of the information contained in the data. However, at the moment no complete picture of the information contained in the data can be given.

A visual inspection of the entire footage was performed by two trained analysts to determine start and end time points of the effort task cycles and their three phases in each experiment (Table 1).

Table 1. Overview of all experiments up to January 2010

	Number of Subjects	Total Number Experiments	Experiment Duration (s) Mean, SD	Total Number of Cycles	Cycles per Experiment Mean, SD
Campaign 1	19	19	420	57	3
Campaign 2	34	45	534, 181	293	8, 1
Campaign 3	18	21	926, 190	141	8, 1
Total	69	89	-	491	-

At the end an effort phase subjects relaxed and their body began moving towards a neutral position. These spontaneous relaxation movements ended in most cases within not more than 20 s (Figure 4). In some cases longer durations of the relaxation movements were observed. In very few cases the movements were not terminated within 60 - 80 s. These experiments were aborted incomplete.

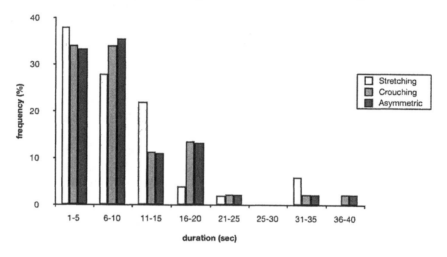

FIGURE 4. Distribution of the duration of relaxation phases for different preceding efforts: Stretching (n=50, light grey), Crouching (n=44, grey), Asymmetric (n=45, dark grey).

Figure 4 demonstrates that there is no difference between the distributions of the relaxation phases with respect to the three efforts. Obviously, the duration of the relaxation phases does not depend on the type of the preceding effort.

In order to explore how stable the postures are which are assumed at the end of the relaxation phase and throughout the neutral phase the graphical overlay technique was used. An example is shown in figure 5.

The series of 20 pictures starts with the first image of a neutral phase (picture 183) with subject in a stable posture. This phase lasts to picture 240, with picture 241 being the first showing effort task movements. Every third picture is computed as described above and placed over the preceding one. The images were rotated around the hip point and positioned for maximum overlay fit of buttock and upper thigh. After analyzing the variability over the complete set of images, the pictures are sorted into images taken while the subject was inhaling and while he was exhaling. These overlays labeled breath-in and breath-out where then investigated separately.

The analysis shows firstly the subject demonstrates a typical neutral posture, which seems even similar and constant under inspiration and expiration conditions. The lower limbs seem to be more stable during this cycle than the upper body. Secondly a constant postural relaxation can be observed over the complete neutral phase. The changes occur mainly in the upper body section and are far less evident (estimated 2 - 3° angular change per second) than the changes in the relaxation phase (estimated 10 - 20° angular change per second).

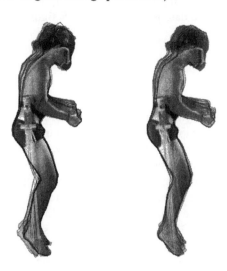

FIGURE 5. Posture variability using overlay (left: breath-in, right: breath-out)

An analysis of the intra-individual variability of posture features such as joint angles requires quantitative data obtained by the PCMAN method. Joint angles were computed for a series of nine images from one subject taken from four different consecutive effort cycles and from the pre-effort phase (Table 2). The results show that the magnitude of the variability of several joint angles was distinctly smaller than the values given in the NBP figure (Figure 1). In part this might originate from the fact that the NBP is based on mean values of data collected from three Skylab astronauts [Tengwall 1982] [NASA 1995]. Nevertheless the data listed in table 2 seems to indicate the existence of an intra-personally stable, replicable and nearly constant neutral posture.

Table 2. Variability of 3D joint angels collected from subject under water

		Shoulder (°)	Elbow (°)	Hip (°)	Knee (°)
Left side	Mean	27,2	101,7	147,6	145,1
	SD	8,1	5,3	3,8	6,9
Right side	Mean	26,6	108,3	146,2	142,2
	SD	8,4	2,7	4,5	5,1

These values are related to 8 joints (shoulders, elbows, hips and knees), which are also shown in the MSIS figure. In contrast to it, here the values describe angles in the 3D space while the data given in the MSIS figure describe angles in the 2D projections of the body.

CONCLUSION

The project reported here yielded two major results: firstly, the implementation of a reliable and feasible ergonomic method for observing movements and postures in simulated micro-gravity under water, and secondly, an experimental exploration of the inter- and intra-individual variability of body postures assumed during phases of relaxation when specific movement tasks are absent. A substantial data material was gathered. An extensive analysis has recently been started.

The first results indicate a large range of variability of postures between subjects and also within subjects. However, in almost all subjects distinct goal oriented movements were observed during the relaxation phases converging to neutral resting positions of all body parts. Visible movements ended after a few seconds and stable postures were maintained for time intervals of several seconds.

The data base allows an investigation of the central question: Is there sufficient empirical evidence to support the concept of a unique NBP which is inter-individually valid or do the collected data favour the concept of intra-individually valid neutral postures, yet, differing for individuals respectively groups of individuals? The answer to this question is important in the ergonomic conception and design of living and working environments in microgravity.

Manifold differences between simulated microgravity in an underwater environment and true microgravity on board a space vehicle exist and have to be considered when conclusions from this underwater experiment will be drawn. Yet, so far there is only limited empirical evidence on the postural behaviour when two of the most important factors determining the posture of the human body are absent: gravity and specific motion tasks. Results expected from the reported project are conceived as a contribution to this issue.

ACKNOWLEDGEMENTS

The Author would like to thank Olaf Sabbah and Heiner Bubb (Institute of Ergonomics, TUM) for their constant help and advice, Ulrich Walter (Institute of Astronuatics, TUM) for his guidance, Moritz Ellerbeck, Clemens Plank, Sebastian Diehl, the "Hydronauten" Team for their committed work and dedication, as well as all participants for their time and effort.

REFERENCES

Arlt, F. "Untersuchung zielgerichteter Bewegungen zur Simulation mit einem CAD-Menschmodell." *Dissertation, Lehrstuhl für Ergonomie der Technischen Universität München.* Herbert Utz Verlag. (1999).

Andreoni, G, et al. "Quantitative Analysis of Neutral Body Posture in Prolonged Microgravity." *Gait & posture* 12, no. 3 (2000): 235-42

Mount, F E, M Whitmore, and S L Stealey. "Evaluation of Neutral Body Posture on Shuttle Mission STS-57 (SPACEHAB-1)." *NASA Technical Memorandum* 104805 (2003).

Mount, F, and T Foley. "Section 6: Assessment of Human Factors." In: *Extended Duration Orbiter Medical Project.* December (2004).

National Aeronautics and Space Administration. „Man-Systems Integration Standards", *Rev. B, Vol. 1. NASA-STD-3000.* Houston, Texas. National Aeronautics and Space Administration. (1995). Figure 3.3.4.3.-1. downloaded from http://msis.jsc.nasa.gov/images/Section03/Image108.gif (01.03.2010)

Seitz, T. "Videobasierte Messung menschlicher Bewegungen konform zum Menschmodell RAMSIS." *Dissertation, Lehrstuhl für Ergonomie der Technischen Universität München.* (2003).

Tengwall, R, and J Jackson. "Human Posture in Zero Gravity." *Current Anthropology* 23, no. 6 (1982): 657-666.

Vergé-Dépré, K. "Body Posture in Microgravity." *ESA STM, Paris: European Space Agency. (1994).*

Zimmerman, J, and D L Akin. "Subject Effects Exhibited in Human Posture in Neutral Buoyancy and Parabolic Flight." *SAE International, Technical Paper* 2002-01-2538 (2002).

CHAPTER 6

The Analysis of Human Capabilities Through Human Capital Assessment for Covert, Clandestine and Denied Organizations

Marta S. Weber, William N. Reynolds

Least Squares Software, Inc.
12231 Academy Road, NE #301-192
Albuquerque, NM 87111
bill@leastsquares.com

ABSTRACT

Human Capital is defined as the collective knowledge; skills, talents capabilities, values and attitudes that people within an organization contribute operationally to its performance and productivity. Human Capital Assessment is increasingly important in the evaluation of any organization or human collective with a defined purpose. We contend that it is not only a high-value approach when used with open access to the subject organization and with the knowledge and cooperation of the subject organization, but might also be of significant value when utilized remotely for the evaluation of covert, clandestine and denied organizations, collectives and networks. A research design to establish that utility is proposed.

INTRODUCTION

The analysis of an organization's human capabilities is comprised of formal qualitative and/or quantitative valuation of an organization's human assets, often referred to as Human Capital, (HC). HC can be defined as *the collective knowledge; skills, talents capabilities, values and attitudes that people within an organization contribute operationally to its performance and productivity.* When the valuation is calculated in quantitative terms it is sometimes called a human capacity or human capital audit (HCA). As an emerging methodology, this assessment has not yet produced standardized terminology or procedures. It has attracted the greatest attention in business applications; where it is used most often in self-assessment to increase productivity, guide strategic and tactical planning and support succession planning. It is also employed in due diligence preceding Mergers and Acquisitions (M&A), in post-M&A integration and in competitive assessments of rival companies. Although human capabilities assessment is currently most recognized and valued in the corporate realm, it can be applied to any organization. Its application more broadly has produced highly insightful analysis capturing little-known strengths and weaknesses of various kinds of organizations, including political and social action and advocacy movements, groups and organizations, trade unions, professional practices, e.g. research groups, medical and legal consortia, NGOs of various types and sizes, both domestic and international and other non-state actors. One of its key value propositions lies in its ability to provide a standardized set of analytic units that allows comparison of organizational entities cross-culturally.

This presentation discusses techniques developed for conducting HCA of legitimate organizations and ways in which those approaches could be applied to conducting remote HCA on covert, or clandestine organizations and organizations in denied areas. In conventional HCA, available information can be used to make assessments of congruence between an organization's mission, structure, processes and human capabilities. Because some units of analysis – some elements of human capital – have been discovered to be correlated with other elements, it is possible in remote HCA, just as in remote profiling, to extrapolate from known characteristics of an organization to other, denied characteristics. (Weber, 2004, 2010) (Moore & Reynolds, 2006) This capability, if demonstrated in practical application, would have high utility when applied to secretive, covert and denied collectives and networks.

THE CONCEPT OF HUMAN CAPITAL

The concept of human capital originated with Adam Smith's eighteenth century exposition of the four kinds capital. (Smith, 1776) The use of the term in the modern neoclassical economic literature dates back to Jacob Mincer's pioneering article "Investment in Human Capital and Personal Income Distribution" in The Journal of Political Economy in 1958. (Mincer, 1958) The operational definition of human capital in contemporary economic theory is the contribution of U.S.

Economics Professor Gary Becker. In his 1964 landmark – and Nobel Prize-winning - study, Becker first defined the concept in practical terms: *Human capital is the collective knowledge, skills, talents capabilities, values and attitudes that people within an organization contribute operationally to its performance and productivity,* (Becker, 1964) .

The modern concept of human capital captures the recognition that people in all organizations are important and measurable assets who contribute to functionality, development and growth. Depending upon the extent to which an organization's output is knowledge-based or labor-intensive in its requirements, the collective skills, abilities and attitudes of people contribute more than any other factor to organizational performance and productivity.

The term 'human capital' has been criticized by some humanists as depersonalizing; certainly it is important to acknowledge the fact that human capital, unlike other assets, is not actually owned by the organizations to which it is attached, but rather, should be thought of as 'lent' or 'leased' by the individuals it represents. Of course, exceptions can be seen in those instances in which human participation is not by choice, as under totalitarian regimes or within organizational entities staffed through coercion.

Human capital assessment is increasingly used in business organizations, particularly large corporations where assessment within the organization may be directed toward maximizing personnel productivity, refining recruitment, staff development and management effectiveness, determining training needs and many other objectives. Internal HCA also supports strategic planning and in particular, succession planning. A more challenging application is found in the evaluation of other organizations for purposes of competitive assessment and other objectives, such as due diligence in advance of strategic alliances, merger and acquisition, where HCA is conducted remotely. It reduces uncertainty in the integration of personnel post- merger or acquisition in complex organizations.

ASSESSMENT OF HUMAN CAPITAL

Human capital (HC) is at once an established and an evolving concept as the role of human variables at all levels of an organization's life becomes both more important and more understood. Practitioners of human capital assessment employ a variety of concepts and measures. Assessments can be quantitative or qualitative. The former is most often used in the monetary valuation of a company. At the present time, quantitative assessment or audit measures remain largely proprietary, although a few basic metrics have been disseminated. (Fitz-Enz, 2009) In the age of information, it can be expected that methods of quantitative HCA will eventually come into the public domain. Tools of qualitative assessment measurement are more readily available. Metrics are also employed in qualitative reviews, e.g. the minimally functional ratio of succession pool candidates for mission-critical or leadership positions: 1:1.

Standardized terminology for human capital assessment has yet to be developed and widely accepted. Several core elements are common to the known approaches to HCA, however. Depending on the scope and objectives of the assessments, the following units of analysis of an organization might be addressed:

Strategic Planning - Aims of assessment: Identify and evaluate the organization's self-determined mission, vision of the future and its role in achieving it, core values, goals and objectives, and the strategies used to accomplish its mission. Identify planning processes, criteria for self-defined success mapped to externally determined measure of success. Evaluate strategic planning – does it advance the organization's mission? Does it include external and internal scans for key events and issues with potential impact on organization? Does it include periodic, objective planning and operational self-assessment? Does it assign implementation responsibilities? Does it include criteria for the assessment of success of the process itself?

Leadership - Aims of Assessment: Identify and evaluate leadership positions, persons occupying them, the responsibilities and powers associated with each, their styles and processes. Conduct SWOT (strengths, weaknesses, opportunities, threats) or similar analysis of key leadership, focused on knowledge, skills, abilities, technical expertise, management and leadership prowess, charisma, networks; underlying personality factors e.g. core world-view, values, motivational drivers, priorities; history of successes, failures, key influences and mentors, defining experiences; future vision, concept of legacy; selection and interaction with senior management team, likely tenure or longevity, variables influencing tenure or longevity; anomalies and areas of opacity and unknowns.

Talent - Aim of Assessment: Identify and evaluate the organization's available and necessary talent as represented by personnel. Identify and evaluate the technical and management skills of personnel, their current and projected intellectual property contributions. Identify and evaluate organization's processes for recruitment, hiring or equivalent engagement, training, development, and retention of talent. Identify and evaluate organization's plan, if any, for these processes. Is it linked to strategic planning? Identify and evaluate implementation.

Succession - Aim of Assessment: Identify and evaluate the succession pool in key areas of leadership, projected longevity, technical ability and advancement, other areas of organization's mission and operations; Identify succession plan if present and evaluate. Analyze contingencies and impact on succession. Assess impact of succession failure or gap on organization's productivity or other capabilities in accomplishing mission.

Organizational Alignment - Aim of Assessment: Evaluate alignment between organization's mission, strategy, leadership, human capital for current and future viability. Evaluate alignments between structure, processes and mission. Identify and evaluate organization's recognition of and approach to alignment issues. Assess organizational culture in the context of alignment issues and maximization of human capital. Identify key cultural themes within organization; assess impact of culture on organization's human capital and capabilities to achieve mission.

RELEVANT ISSUES IN HCA

GENERAL V. SPECIFIC HUMAN CAPITAL

Following Becker, the human capital literature often distinguishes between "specific" and "general" human capital. Specific human capital refers to skills or knowledge that is useful only to a single employer or industry, whereas general human capital (such as literacy) is useful to all employers. By extrapolation, the same issues apply to other organizations besides those engaged in business, as well.

SOCIAL AND EMOTIONAL CAPITAL AS ELEMENTS OF HUMAN CAPITAL

To Becker's original conception of Human Capital, business professors Bartlett (Harvard) and Ghosal (London Business School) have added the components of *social capital* - value from relationships and *emotional capital* - value from engagement and commitment. (Bartlett & Ghoshal 2002) A related measurement newly under consideration by some researchers is the level of synchronization between personal and company values among the leadership and other personnel. (Marrewijk and Timmers, 2003) The concept of human capital can be even more elastic, including variables such as personal character or prestige and other variables that add to any individual's value to the capabilities of the organization, collective or network.

"COMPETENCE" AS A FACTOR IN HUMAN CAPITAL

Competence is sometimes used interchangeably with skill or knowledge and is also sometimes defined as a combination of the later. Skill and knowledge tend to be narrow and domain-specific, where "competence" is broader and includes thinking ability ("intelligence") and further abilities like motoric and artistic abilities, facility with language and social interaction, cognitive mapping. From a macro perspective, it occupies a key position in HCA. Unlike labor, competence is:

- Expandable and self generating with use: as skilled professionals and other specialists such as doctors get more experience, their competence base will increase, as will their endowment of human capital. The economics of scarcity is replaced by the economics of self-generation.

- Transportable and shareable: competence, especially knowledge, can be moved and shared. This transfer does not prevent its use by the original holder. However, the transfer of knowledge may alter its scarcity-value to its original possessor.

THE PREDICTIVE OBJECTIVE OF HCA

Rarely is an assessment of human capital concerned only with the present. One of the significant challenges in HCA lies in its ability to forecast the human capital dimension in the subject organization over a future time frame. To do so, the assessment must include variables of longevity, fluctuations in the capabilities of key personnel and contingencies likely to affect the human capital landscape. Predictive measurement beyond an impressionistic qualitative   assessment will require additional algorithms.

HUMAN CAPITAL ASSESSMENT OF COVERT, CLANDESTINE AND DENIED ORGANIZATIONS

The specific requirements to assess human capital of covert organizations will necessarily differ   from those associated with denied organizations, but the two efforts confront several common challenges. Among these are the lack of desirable information and the need to attend to deception as well as denial in the organization's presentation of itself.   Many clandestine and denied organizations and collectives of interest make their presence known, intentionally or otherwise, only or most often through aggressive acts and strategic communications. It is possible to begin to analyze capabilities and other attributes of the entity through these acts, but so doing represents only a single window into them.   It is important to note that analysis of opaque organizations and groups has produced a significant body of work, particularly in the area of organizational decision-making, that is best characterized as theory that has been supported by evidence rather than theory that has been *proven*. (Davis and Cragin, 2009)   Nevertheless, with any available information on capabilities, often obtained through evidence of operations and intentional   exposed covert communications of various kinds about those operations, it should be possible to extrapolate to a decision matrix concerning the human capital sufficient for the operations and to develop basic forecasting hypotheses of future human capital needs and availability.

Various theories of organizational structure, process and dynamics might be brought to bear on an effort to discern human capital within a covert or denied organization. Because they are by definition, less traditionally structured and fluid, clandestine and denied networks represent additional challenges. All such entities exist in dynamic relationship to their environments, however, with mutual impact of some kind and scale. (Thomas, Kiser and Casebeer, 2005) For our purposes, the key concerns are issues of environmental acceptance or at least tolerance of the entity, resources, especially human capital, available to the entity from its environment and the responsiveness of the entity to demands or expectations of that environment, necessary for its own viability. The following examples are offered as ways HCA might be mapped from known elements of any organization through a process similar to reverse engineering,. Key questions proceeding from known elements – which, as noted are most often output involving actions of some kind, might include the following:

- What capabilities are required for this output?
- What technical, management and other capabilities are in evidence?
- Is there evidence suggesting the level of investment of available HC the entity has committed to the endeavor in question?
- What human capital is required to produce this capability? At what scale?

Further questions might include these:

- Which of these human capabilities must be internal, and which could be outsourced on an ad hoc basis?
- What are the implications of the existing and required identified human capital for the organization's structure, processes and succession needs?
- What human capital resources (HC pool) exist in the entity's environment?
- What must the entity do to procure them? Is there evidence of procurement/ recruitment?
- What development of HC must be undertaken to prepare available the HC resources for execution of the entity's mission?

RESEARCH DESIGN

We assert that a research design could be developed to implement an approach that would establish the utility of HCA in these contexts. The broad outlines of that research would look like the following:

PHASE I: Develop an HCA methodology for both open and denied organizations. The approach would be based on best available theories, practice and empirical evidence available. Identify required data sources. Define HCA capabilities on both open and denied targets. Develop an approach for validating the methodology. Prepare a final report, including system performance metrics and plans for Phase II. Phase II plans might include technological milestones and plans for at least one operational test and evaluation.

PHASE II: Develop a prototype software system implementing the Phase I methodology. This would consist of a modular package of assessment instruments and decision tree for application to the various organizations based on key intelligence topics, type of organization and available intelligence. All appropriate engineering testing would be performed, and a critical design review would be performed to finalize the design. Phase II deliverables would include: (1) a working prototype of the system, (2) specification for its development, and (3) test data on its performance collected in one or more operational settings.

PHASE III/DUAL USE APPLICATIONS: This technology would have broad application in military as well as commercial settings. HCA is a thorny issue for organizations, both military and civilian, conducting strategic planning, reorganization, succession and M&A. It is a key part of organizational capabilities assessment of both friendly and clandestine and denied adversary organizations.

CONCLUSIONS

Human Capital Assessment as an evolving yet established methodology has been applied to organizations, particularly corporate structures, with good results. The successes of this approach in evaluating present and future human capital needs of organizational entities argues for its utility in the assessment of covert, clandestine and denied organizations, collectives and networks, a s well. The research design proposed here offers the possibility of establishing the value of HCA in those endeavors. The results for intelligence applications could provide far-reaching benefits.

REFERENCES

Becker, Gary S. (1964, 1993, 3rd ed.). *Human Capital: A Theoretical and Empirical Analysis*, Chicago, University of Chicago Press. Gary S. Becker (1964, 1993, 3rd Ed

Bartlett, Christopher A. and Ghoshal, Sumantra, 1979, *The Individualized Corporation: A Fundamentally New Approach to Management*, Harvard Business School Press, 1979

_____, 2002, *Managing Across Borders: The Transnational Solution*, Harvard Business School Press, 3rd Edition.

Davis, Paul K., Cragin, Kim, (eds.) 2009, *Social Science for Counterterrorism*, Rand Corporation, Santa Monica, CA.

(Marrewijk, Marcel van and Timmers, Joanna, 2003, "Human Capital Management: New Possibilities in People Management," Journal of Business Ethics. Vol. 44, 171-184, Netherlands

Fitz-Ens, Jac, 2009, *The ROI of Human Capital: Measuring the Economic Value of Employee Performance,* New York, AMACOM

Mincer, Jacob, 1958, "Investment in Human Capital and Personal Income Distribution," The Journal of Political Economy. University of Chicago Press in its journal Volume (Year): 66 (1958)

Moore, David T. & Reynolds, William N. , 2006, "So Many Ways to Lie: Complexity of Denial and Deception". *Defense Intelligence Journal*, 15(2), 95-116.

Naroll, Raoul and Cohen, Ronald (eds.) 1973, Handbook of Method in Cultural Anthropology, Columbia University Press,

Smith, Adam, 1776, *An Inquiry into the Nature And Causes of the Wealth of Nations Book 2 - Of the Nature, Accumulation, and Employment of Stock*, Philadelphia

Thomas, Troy S., Kiser, Stephen D., Casebeer, William D, 2005, *Warlords Rising,* New York, Lexington Books

United States General Accounting Office, 2009, *HUMAN CAPITAL: A Self-Assessment Checklist for* U.S. Office of Personnel Management, *A Model of Strategic Human Capital Management.* Washington, D.C., GAO-00-0014, www.gao.gov

United States Office of Personnel Management, 2008, *Human Capital Assessment and Accountability Framework,* Washington, D.C. OPM-02-3738P www..opm.gov

Weber, Marta S., 2004, "Profiling for Leadership Analysis" in *Competitive Intelligence Magazine*, 7(4),
July-August 2004.

_____, 2010a. "Profiling: Behavioral Forecasting," *2010 Anthology of Competitive Intelligence*, Chapter 7, *passim*, Washington, D.C., SCIP

_____, 2010 b. "Human Capital Assessment in Competitive Intelligence", Society of Competitive Intelligence, Monograph, in process.

Chapter 7

Trust and Reliance in HSCB Models

Mike Farry, Jonathan Pfautz, Eric Carlson, David Koelle

Charles River Analytics, Inc.
625 Mount Auburn Street
Cambridge, MA 02138

ABSTRACT

Human socio-cultural behavior (HSCB) modeling technologies are gaining traction in operational communities as a means to analyze and predict human behavior at various scales. The National Research Council's study on behavior modeling and simulation has identified salient challenges to the acceptance and use of HSCB modeling technologies (Zacharias, MacMillan, & Van Hemel, 2008), and other efforts have identified specific factors that lead to distrust of models and their results (Pfautz, Carlson, & Koelle, 2010; Pfautz et al., 2009a). While these efforts directly address the issue of trust in HSCB modeling, we can achieve a more rigorous understanding of these complex design issues by studying the more mature body of work on trust from the human factors engineering community. The substantive and insightful literature in designing for appropriate trust and reliance in automated systems provides a promising opportunity for study.

In this paper, we review the leading principles of automation design to inform HSCB modeling efforts, extending and modifying principles to fit the unique challenges faced in the HSCB modeling domain. We present a framework and general guidelines for the design of HSCB models and modeling systems influenced by the principles of automation design (Parasuraman, Sheridan, & Wickens, 2000; Parasuraman & Riley, 1997), which have yielded effective and widely accepted automated systems for a wide variety of domains and user communities. Based on those factors, we hypothesize a set of five cardinal design guidelines that will engender appropriate trust and maximize the effectiveness of HSCB modeling systems.

Keywords: Behavior Modeling, Human Socio-Cultural Behavior Models, Modeling and Simulation, HSCB, Trust in Automation, Reliance, Human Factors

INTRODUCTION

As modern military operations have shifted from classic notions of conventional warfare, a premium has been placed on understanding and influencing those aspects of human behavior that contribute to phenomena such as terrorism and insurgency. While HSCB models have great potential in aiding decision makers as they engage adversaries in current and future operations, there are a number of gaps in understanding and development methodology that result in distrust and disuse of model results. These gaps frustrate operational users and model developers alike. The recent National Research Council study on Behavior Modeling and Simulation identified operationalization as a key challenge, and set a broad research agenda for the HSCB community (Zacharias et al., 2008). Modeling issues identified by that study center around the establishment of usefulness of model outputs to operational users, including lack of interoperability between models, support of data collection (specifically, the collection of *relevant* data), and the lack of standards for verification, validation, and accreditation (VV&A).. These gaps lead naturally into issues of trust (Pfautz et al., 2009a; Pfautz et al., 2009c). Relatively few HSCB modeling efforts have achieved significant operational approval for targeted user communities performing well-scoped tasks (For an example, see the paper that accompanies this one on the design of a system that uses the principles defined here for the benefit of a specific operational community (Carlson et al., 2010)). Clearly, there remains a large gulf between the number of operational problems where HSCB modeling *could* help and the number where it *does* help effectively.

In the face of a modern military environment where insight into social and cultural factors means the difference between success and failure, it is incumbent upon the research community to develop ways to rapidly accelerate the adoption of HSCB models. While continuing to focus research and development on specific operational applications may yield a string of one-off successes, the HSCB modeling community is likely to experience an inflection point in its rate of success upon the establishment of a set of rigorous guidelines to direct modeling efforts. A natural source for these guidelines is identifying patterns of success and failure among applications. While sound, this approach lacks alacrity; we simply cannot wait to be reflective of the products of our own community. Insight can be gleaned from other fields where the introduction of technology has posed issues of acceptance, trust, and reliance. Specifically, the ever-expanding body of research into designing for appropriate trust in automated systems, authored largely by the human factors community, can provide us with some knowledge that could be applied to HSCB modeling design efforts. This research area is relatively mature, and the acceptance of those guidelines has led providers to develop effective, appropriately trusted automated systems that, for many domains, would be unfathomable by its practitioners fifty years ago (e.g., advanced systems for automated piloting of aircraft).

We begin with a discussion of scoping. The most typical use case for HSCB models is to supply information for decision making and action. For the purposes of presenting focused definitions and guidelines in this paper, we do not consider "secondary" applications of HSCB modeling (e.g., using HSCB models for training). While the analysis of literature in automation design is likely to be fruitful in our effort to provide definitions and guidelines, there are many challenges in adapting those lessons to HSCB modeling arising from differences in the two areas of study. These differences, which receive more detailed discussion below, include the following:

- The increased difficulty in modeling complex human behavior as compared to physical systems, which comprise the most frequent applications of automated systems
- The fact that our selected use case for HSCB models is to supply information, while automated systems cover a wider spectrum from information gathering to full or supervised control of actions
- The operational tempo of the two use case classes. While automated systems are frequently employed for tasks that challenge or exceed the limits of human information processing and reaction, there are some differences in the applications of HSCB models
- The increased difficulty in VV&A for HSCB models; many automated systems can be tested in laboratory environments or scaled-down operational environments to empirically assess their effectiveness and limits, but this is not the case for HSCB

With an appreciation for these differences, we can now begin reviewing the issues surrounding design of automation for the benefit of HSCB modeling efforts.

BACKGROUND

HSCB MODELING: OPERATIONAL ISSUES

The 2008 NRC study surveys issues in use, details different types of behavior models that are currently employed, and sets a broader research agenda for behavior modeling researchers (Zacharias et al., 2008). Even when models and problems do align enough for operational use, those models that are anointed as "successful" cannot necessarily be reused for related use cases or situations. Similarly, models that use different terminology and come from different scientific disciplines are particularly challenging to merge. When a model fails to provide concrete information for an operational problem, a lack of trust in future model results is a natural consequence.

Applying the wealth of knowledge available from multiple scientific disciplines onto HSCB modeling problems may be beneficial from the standpoint of analytic rigor (Zelik, Patterson, & Woods, 2007), but it also increases the complexity and opacity of the model. That complexity is a key component of distrust, and distrust is compounded as complexity increases (Pfautz et al., 2009c). This insight has led to efforts that include the operational user as a key stakeholder in the model composition process, which has yielded success for specific applications (Pfautz et al., 2009a). However, while that approach engenders trust, it is untenable in many situations due to the required overhead. Trust in HSCB models is not a static quantity that depends on a single factor, but rather has several contributing meta-information factors that vary greatly across individuals, tasks, and domains (Guarino et al., 2009; Pfautz et al., 2007). A particularly thorough list of HSCB-relevant research communities and modeled factors is found in (Numrich & Tolk, 2010). These disciplines include anthropology, philology, linguistics, philosophy, psychology, theology, sociology, law and governmental study, criminal justice, political science, trade and finance research, and geopolitical studies, and the modeled factors include culture, ethnicity, language, religion, social norms, justice, rule of law, economics, and geography. If a common vocabulary could be established between all stakeholders for HSCB modeling, consistency in terminology would undoubtedly serve as a forebear for greater trust. Given the wide variety of communities and factors, however, the task of establishing a common vocabulary acceptable to all practitioners seems Sisyphean, indeed.

TRUST IN AUTOMATION: FOUNDATIONS AND DESIGN ISSUES

Modern automated technologies focus on a wide variety of domains and tasks. As a result, an oft-repeated theme in the literature is the establishment of *design guidelines* and *frameworks* to serve as bases and evaluative criteria, rather than over-describing the functionalities automated technologies should provide. Parasuraman et al. (2000) make an important and novel link between a four-stage model of human information processing and classes of automated technology that can assist with each stage. The stages are sensory processing, perception and working memory, decision making, and response selection. These, in turn, map to automation functions for information acquisition, information analysis, decision automation, and action automation—of which only the first two are truly relevant to HSCB modeling in its current envisioned use.

A deeper analysis of the term "trust" and its failures and associated concepts is also useful from the HSCB perspective (Parasuraman et al., 1997). Parasuraman and Riley define "misuse" as over-reliance on ineffective automation, "disuse" as the neglect of helpful automation, and "abuse" as the automation of functions—on part of designers—without due regard for its impacts on task performance. Trust, workload (including the cognitive overhead in deciding whether to use automation), risk, and communication of automation state all play into operators' decision to use automation, and from the designer's perspective, those quantities are hard to predict due to individual and situational variation (e.g., when antilock braking systems were first deployed, some drivers instantly saw value, while others were skeptical.) Once a decision is made to use automation, however, users frequently make a "premature cognitive commitment" and fail to monitor the automation's inputs, outputs, and actions (Langer, 1989).

Another useful perspective on trust comes from comparing the meaning of trust in social relationships to trust in automation (Lee et al., 2004). Since people tend to respond to technology in a social way, this comparison is apt. Guidelines presented in that work include designing for appropriate trust (not necessarily greater trust), showing past performance of the automation for selection decisions, increasing transparency, simplifying automation functionality, and training operators on expected reliability and intended uses. Lee and Moray (1992) identify universal bases of trust in automation as performance, process, and purpose. Performance is the "what" of an automated technology and refers to the actual operation of the automation and yields reliability, predictability, and ability. Process provides the "how" information, describing the constituent algorithms and their operation. Purpose provides information about "why" a particular technology was developed in the first place.

FOUNDATIONS FOR TRUST AND RELIANCE IN HSCB MODELS

Having reviewed the relevant literature, we now provide initial definitions and guidelines for the design of appropriate trust and reliance for HSCB models. The NRC study (Zacharias et al., 2008) lists five main research categories relevant to our discussion: modeling strategy (matching problems to the real world), VV&A, modeling tactics (the internal structure of a model), differences between modeling physical phenomena and human behavior, and combining components and federating models. While these are all fruitful avenues for research, they are also not merely questions that will yield concrete answers provided enough time and research dollars. However, if the HSCB research community establishes a set of proper definitions and guidelines for developing models, it can use these as a common vocabulary in designing and evaluating their efforts. This structure will result in an increased capacity to tailor HSCB solutions to the operational community for their specific domains, tasks, biases for and against particular modeling techniques and schools of thought, as well as other needs and preferences.

As mentioned in the introduction, insights from relevant literature cannot simply be imported wholesale for the benefit of the HSCB community due to the differences in the disciplines. Since HSCB is concerned with modeling human behavior, complexity grows and predictability decreases tremendously in comparison with physical systems. In light of this, the expected fidelity or accuracy of judgments for HSCB—whether those judgments are generated by a human or a model—are lower. Also, absent operational experience, automation can also prove itself through demonstrated performance in laboratory environments, again providing a starting point for users to assess its feasibility. This capability is also difficult to replicate for HSCB models; though models can be trained and tested on previously collected data sets, there is still likely to be unpredictable variability between previous sets and future operational environments.

On the positive side, however, the role of HSCB modeling efforts is envisioned to be more modest than that of automated technologies. HSCB model outputs are used to inform decisions about policy and operational strategy. This reduces the scope of trust in this domain: users do not have to approve of independently-made decisions or actions, but must instead inspect the fidelity of information provided to them in comparison to information from other sources and their own judgment.

Automated systems frequently consider linear processing-decision-action processes. Even if the supported processes themselves are not necessarily linear, the linear processing models often serve well enough for automation design efforts. This process is simplified by the exclusion of decision and action for HSCB, but it is clear that merely supporting "information acquisition" and "information analysis" for HSCB is insufficient. In defining information acquisition, it is important to recognize that there is a clear separation between data and information: *data* is anything that can be sensed or collected, while *information* has an effect on the decision-making process which uses HSCB modeling insight. We enumerate those impacts in a general way, and propose the following definition for information acquisition in the context of HSCB:

> **Information acquisition** for HSCB is the collection of information about human behavior that serves one of five purposes: (1) validating previous model outputs, (2) providing model parameters or inputs, (3) providing insight about how model structure should be changed, (4) suggesting the selection of a different model to be used, or (5) providing context that affects a human decision-maker's interpretation of model results.

Technologies or sources that support information acquisition for HSCB, then, are not models themselves, but are supporting technologies that provide information for those purposes. They can range in complexity from simple field reports, to more advanced trend analyses based on the content of open-source media (e.g., a Natural Language Processing-based statistical analysis of recent newspaper reports). The HSCB models themselves provide more of the information analysis functionality: once information has been acquired from sources, models will help to interpret, extrapolate, and identify patterns that may be observed in the future. Since the functionalities provided by modeling techniques and the tasks they support are rich and variable, we provide the following high-level definition:

> **Information analysis** for HSCB is the process of synthesizing the products of information acquisition to progress towards any of the following outcomes: (1) composing one or more decision alternatives, (2) providing values for the decision criteria that inform selection among a set of decision alternatives, (3) informing current or future information acquisition efforts, (4) providing input to a higher-level model or decision-making process for further information analysis, or (5) modifying the structure of a model.

Note that this definition includes both human and machine (i.e., HSCB model) actors. We can also characterize the functionalities performed by HSCB models under the classification of information analysis as follows:

- **Forecasting** or **extrapolation** is the analysis of past behavior and current observations to provide insight into possible and probable *future* actions of individuals, organizations, or societies.
- **Interpolation** is the analysis of past behavior and current observations to provide insight into possible and probable *current or past* actions or states.
- **Interpretation** is the transformation of acquired information into higher-order information that is relevant to decision criteria.
- **Need Identification** is a reflective process, where a model identifies areas of ambiguity in its analysis, and provides insight about the additional information that should be acquired to disambiguate the outcomes.
- **Mutation** is the process of realizing any change in modeling structure that results from acquired information.
- **Self-Explanation** is the communication of the model's presented results in any fashion, to provide greater transparency and engender trust. Examples of self-explanation include the presentation of intermediate results through an analysis process that takes multiple steps, or the representation of mathematical formulae used to calculate a specific factor.

It is unlikely that the above list of analysis functionalities is complete or exhaustive in light of the wide variety of decision making processes that are supported and modeling methodologies. However, it does represent a starting point and enables us to assess the interactions among these processes, as well as the decision-making processes they support. As mentioned above, in contrast to the automation literature's traditional linear modeling of these processes, a dynamic and iterative perspective is necessary for HSCB models.

Armed with a definition of the processes HSCB models and associated technologies can provide, we turn our attention to a useful definition of trust in HSCB and some design guidelines. From our analysis above, it appears likely that we have already identified the biggest difference between trust in automation and trust in HSCB modeling: the inability for a human user to consistently gauge the model's performance among the functionalities of information analysis, comparing both to the user's own performance, and the effects of any incorrect judgments on the eventual products of the decision-making or planning processes. Given the rich information landscape in which HSCB models are frequently applied, "trust" is inappropriate as a static concept intrinsic to any model. Still, we should be able to ascribe some analytical meaning to the term in a general way. To do that, our definition of trust must fully appreciate the greater context of other information sources. We propose the following definition:

Trust in an HSCB model, defined for any triad of HSCB model, planning process, and human user, is the human user's perception of the model's ability to provide relevant and accurate information, as compared to other available information sources.

The common theme in our definitions to this point has been an appreciation for the rich and complex information landscape of HSCB modeling problems. It is no surprise, then, that our HSCB design recommendations are largely centered on managing and carefully designing information flow. Our recommendations are also influenced by the excellent lists from (Lee et al., 2004).

1. **Consider all actors in the decision-making process: humans and machines alike.** As discussed above, not only should designers consider the information flow between human users and HSCB models, but also between models operating at different levels of fidelity or in parallel. Since the inclusion of technology in human processes changes rather than eliminates human tasks, the process flow and interactions should be considered and designed in great detail, and as soon as possible in the design process.
2. **Explicitly design information flow for the supported task.** What information is provided to the user as decision criteria? Where does this information come from? How will the user be able to select between conflicting information from different modeling sources? What about when the model disagrees with a human assessment? How will feedback be provided to information acquisition sources?
3. **Consider temporal effects and the asynchronicity of supported tasks.** As discussed above, information flows quickly among various levels of abstraction and complex temporal relationships result. In turn, that may lead to mistrust (due to the illusion of independent confirming information). In addition, (Mosier & Skitka, 1996) has shown that reliance on automated conclusions can make humans less attentive to contradictory sources of evidence.
4. **Design for openness about trust-related factors.** Related to Lee and See's concepts of designing for appropriate trust and showing past performance of automation, all meta-model taxonomic factors relating to trust and a model's potential for application should be clearly expressed (Guarino et al., 2009; Pfautz et al., 2009b).
5. **Make the user a stakeholder.** Automation designers cite training as a frequent need. However, for HSCB modeling, this is frequently unrealistic since those models are based on years of dedicated research to specific niches of science. Instead of expecting an understanding of the model's underpinnings, designers should include the targeted user community as much as possible during the process, achieving a higher level of understanding about how those insights may best be woven into their practices.

CONCLUSIONS AND FUTURE WORK

The establishment of those definitions and guidelines is intended to serve as just a starting point for designing for more appropriate trust in HSCB models. While the authors have sought to account for a variety of perspectives in composing them, it is unlikely that we have arrived at definitions that apply universally. Refining these definitions and guidelines requires the thoughtful and constructive criticism of the entire research community, and it is our hope that this work sets the foundation for those very important discussions. In addition to debate among the community, evaluation for these concepts must also come from practice. As HSCB modeling efforts progress, experience will be a driving force for the evolution of these ideas.

While the authors greatly value the concept of bringing multiple scientific communities together and encouraging a more diverse braintrust for HSCB models, this paper gently rejects the idea of composing a massive collaborative vocabulary of all HSCB-relevant terms to do so. We do this in the name of realism, and certainly do not intend to stifle creative efforts to bridge the gaps across multiple disciplines and communities. We hope that refined versions of our ideas can help bring these communities together and foster discussion in a more holistic way, encouraging all stakeholders to focus on the *benefits* of provided models, rather than the details of each model itself.

ACKNOWLEDGEMENTS

This work would not have been possible without the support of our Sponsors—principally Dr. James Frank, Ms. Shana Yakobi and Mr. Jareen Stubbs. The concepts and solutions presented here were funded across multiple efforts, including work performed, in some part, under Government Contract Nos. N41756-09-C-4558, FA8650-04-C-6403, FA8650-08-C-6921, FA8650-04-C-6403, and N41756-10-C-3317 by a number of scientists and engineers at Charles River Analytics, including Chester Tse and Geoffrey Catto. The authors would like to express their greatest appreciation for the willingness of the many dedicated and passionate (and necessarily anonymous) operational users that have participated in our analyses.

REFERENCES

Campbell, G. & Bolton, A. (2005). HBR Validation: Integrating Lessons Learned From Multiple Academic Disciplines, Applied Communities, and the AMBR Project. In R. Pew & K. Gluck (Eds.), *Modeling Human Behavior With Integrated Cognitive Architectures* (pp. 365-395). Mahweh, NJ: Lawrence Earlbaum Associates.

Gluck, K. & Pew, R. (2005). *Modeling Human Behavior With Integrated Cognitive Architectures*. Mahweh, NJ: Lawrence Erlbaum.

Guarino, S., Pfautz, J., Cox, Z., & Roth, E. (2009). Modeling Human Reasoning About Meta-Information. *International Journal of Approximate Reasoning, 50*437-449.

Langer, E. (1989). *Mindfulness*. Reading, MA: Addison-Wesley.

Lee, J. & Moray, N. (1992). Trust Control Strategies and Allocation of Function in Human-Machine Systems. *Ergonomics, 35*1243-1270.

Lee, J. D. & See, K. A. (2004). Trust in Automation: Designing for Appropriate

Reliance. *Human Factors,* 46(1), 50-80.

Mosier, K. & Skitka, L. J. (1996). Human Decision Makers and Automated Decision Aids: Made for Each Other? In R. Parasuraman & M. Mouloua (Eds.), *Automation and Human Performance: Theory and Applications* (pp. 201-220). Hillsdale, NJ: Erlsbaum.

Munson, L. & Hulin, C. (2000). Examining the Fit Between Empirical Data and Theoretical Simulations. In D. Ilgen & C. Hulin (Eds.), *Computational Modeling of Behavior in Organizations* (pp. 69-83). Washington, D.C.: Am. Psychological Association.

Numrich, S. K. & Tolk, A. (2010). Challenges for Human Social Cultural and Behavioral Modeling. *SCS M&S Magazine,* 2010(01).

Parasuraman, R. & Riley, V. (1997). Humans and Automation: Use, Misuse, Disuse, Abuse. *Human Factors,* 39(2), 230-253.

Parasuraman, R., Sheridan, T. B., & Wickens, C. D. (2000). A Model for Types and Levels of Human Interaction With Automation. *IEEE Transactions on Systems, Man, and Cybernetics,* 30(3), 286-297.

Pfautz, J., Carlson, E., Farry, M., & Koelle, D. (2009a). Enabling Operator/Analyst Trust in Complex Human Socio-Cultural Behavior Models. In *Proceedings of Human Behavior-Computational Intelligence Modeling Conference (HB-CMI) 2009.*

Carlson, E., Pfautz, J., & Koelle, D. (2010). Operator Trust in Human Socio-Cultural Behavior Models: The Design of a Tool for Reasoning About Information Propagation. In *Proceedings of 3rd International Conference on Applied Human Factors and Ergonomics.*

Pfautz, J., Carlson, E., Koelle, D., Potter, S., & Zacharias, G. (2009b). Using Meta-Information to Enable End-User Understanding and Application of HSCB Models. In *Proceedings of HSCB Focus 2010 Conference.*

Pfautz, J., Carlson, E., Koelle, D., & Roth, E. (2009c). User-Created and User-Adaptable Technosocial Modeling Methods. In *Proceedings of Technosocial Predictive Analytics: Papers From the 2009 AAAI Spring Symposium.*

Pfautz, J., Fouse, A., Farry, M., Bisantz, A., & Roth, E. (2007). Representing Meta-Information to Support C2 Decision Making. In *Proceedings of International Command and Control Research and Technology Symposium (ICCRTS).*

Pfautz, J., Roth, E., Bisantz, A., Llinas, J., & Fouse, A. (2006). The Role of Meta-Information in C2 Decision-Support Systems. In *Proceedings of In Proceedings of Command and Control Research and Technology Symposium.* San Diego, CA.

U.S.Department of Defense (2001). *VV&A Recommended Practices Guide Glossary.* Washington, D.C.: Defense Modeling and Simulation Office.

Zacharias, G. L., MacMillan, J., & Van Hemel, S. B. (2008). *Behavioral Modeling and Simulation: From Individuals to Societies.* Washington, D.C.: National Academies Press.

Zelik, D., Patterson, E., & Woods, D. (2007). Judging Sufficiency: How Professional Intelligence Analysts Assess Analytical Rigor. In *Proceedings of Human Factors and Ergonomics Society 51st Annual Meeting.*

Chapter 8

Are Force Dynamics Determined by Hand Posture During Hand Brake Manipulation?

Kang Li[1], Vincent Duffy[2,3]

[1] Department of Mechanical and Industrial Engineering
University of Illinois at Urbana-Champaign

[2] Schools of Industrial Engineering and Agricultural & Biological
Engineering, Purdue University

[3] The Regenstrief Center for Healthcare Engineering
Purdue University

ABSTRACT

Drive-by-wire technology enables braking and accelerating by hand. Operating the hand brake is essentially a multi-finger force synergy task. Few studies have been conducted on precision grip involving all the fingers of one hand or both hands when manipulating an object of irregular shape. This study examined the effects of target force level, numbers of hands used in the task, grip posture, grip duration, and training on hand-operated control device operation. Age and gender effects have also been investigated. The results showed that target force level, hand used in the task, grip posture and grip duration had the significant impact on the grip force coordination. With the development of digital human modeling and digital human simulation, analysis tools that facilitate prediction of variations in human performance and vehicle control based on modern design conditions can be useful for future designers before first physical mockups.

Keywords: Grip Force Coordination; Multi-Finger Synergy; Hand Brake; Manual Control

INTRODUCTION

While brakes and accelerators of most vehicles on the road today are mainly operated by foot with pedals, brakes and accelerators operated by hand have become promising options for next-generation vehicles. They are technically feasible in real vehicle (Szczerba, Duffy, Geisler, Rowland, & Kang, 2007). Brake-by-wire technology enables braking and accelerating by hand since much less force is required to be exerted on the control devices. These hand-operated control devices can also improve driving safety compared to the traditional foot-operated control devices. When operating the brakes and accelerators, hand could have less movement time than foot (Hoffmann, 1991), thus control by hand decreases the response time (Carter, 2000). As a result, the stopping time can be less and the stopping distance can be shorter (Szczerba et al., 2007). Therefore, the hand has the potential in future vehicle to play the role of the foot in current vehicles. Yet, although hand-operated control devices have been used for braking and accelerating for disable users (Curry & Southhall, 2002; Ostlund, 1999), few studies have been dedicated to explore the ergonomics issues with hand-operated brakes and accelerators. Without examining the ergonomics issues of the hand-operated brakes and accelerators and eliminating the incompatibility between design and hand characteristics, operating these devices by hand can lead to an uncomfortable user experience, hand physical injuries, loss of control, and fatality.

The aim of this study was to identify the contributing factors of the ability of human beings in precisely exerting grip force. We investigated the ability of grip force coordination under a variety of force production conditions. Our hypothesis was that the ability of grip force coordination were affected by the predetermined force need to be exerted (target force level), grip duration, hand posture, number of hands involved, and training as well as personal factors including age and gender. Subjects were required to grip the hand-operated control devices mounted on the steering wheel in a driving simulator and to exert predetermined force in the driving context. The ability of grip force coordination was then quantified based on the recorded grip force exerted on the control devices. The optimal force production conditions were identified.

METHODS

An experiment was conducted with sixty participants (30 males and 30 females). The participants were divided into two groups (younger and older) by age: the younger group aged 18-35; the older group aged 36-64. Each group had equal number of females and males. All participants had normal and/or corrected vision. Prospective participants were advised of the possibility of hazards such as motion sickness. Those who reported a history of simulator or motion sickness were discouraged from participating and those, who had epilepsy, heart problems, and carpal tunnel disorders (CTDs), or who had or thought they might be pregnant did not participate. The experimental protocol was approved by an Institutional Review

Board, and information consent was obtained from each participant. Seven anthropometric measurements were recorded including hand length, hand breadth, hand thickness, palm length, thumb length, index finger length, and middle finger length as defined in the Anthropometric Source Book Volume II: A Handbook of Anthropometric Data (Webb Associates, 1978).

The experimental environment contained a simulated driving environment in which a vehicle chair was installed on the wood platform . The chair was adjustable to adapt to different sized participants. The hand-operated control device consisted of a metal ring mounted behind the steering wheel, linear springs, and a force sensor (figure 1). When the ring was gripped, the distance between the center of the ring and the center of the steering wheel decreased and the linear springs was pushed. The force sensor measured the exerted force and transferred the information to the force measurement software developed by our group. The measured force was then displayed on a monitor with the sampling frequency 6 Hz and the resolution of 0.1N.

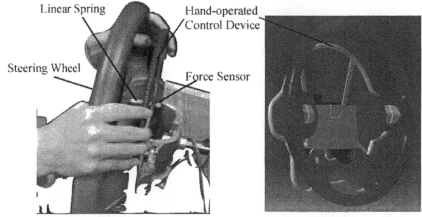

Figure 1. The hand-operated control device consisted of a medal ring mounted behind the steering wheel and a force sensor is mounted behind the steering wheel. The left one illustrates the real device and the right one shows a digital prototype of the device.

The experiment was composed of two sessions. In the first session, all the participants performed two simulated driving tasks in two different driving scenarios, then another eight simulated driving scenarios of 5-minutes each. The participants were divided into two groups: 40 participants used the hand-operated brake system in the driving ("trained") and 20 participants used the traditional foot brake system ("untrained") in the driving. In the second session, they were randomly assigned to grip the hand-operated control device and exert the force under 12 different production condition combinations. (3 levels of grip posture *2 levels of grip duration* 2 levels of target force=12 trials). Participants were asked to squeeze the hand control and hold for either 5 seconds or 10 seconds to reach two target force levels: 5N, 15N. They were also asked to exert the force with three squeeze types (Figure 2) which represented the possible postures that driver may adopt. The first one was to squeeze the ring on the top of the steering wheel with one hand (12 o'clock position). The second one is to put two hands at the 10 and 2 o'clock position of the steering wheel with two hands and then squeeze the ring. The third one is to put the two hands at the 10 and 2 o'clock position of the steering wheel with the finger touching the ring, then turn right to 12 and 4 o'clock position and squeeze the ring. The squeeze type in fact included two factors. One was one-hand task vs. two-hand task. The other was two-hand task in normal posture vs. two-hand task in awkward posture. In the following description, we used both, top and right to represent three squeeze types: two-hand squeeze at 10-2 o'clock, one-hand squeeze on the top of the steering wheel, and two hand squeeze at 10-2 o'clock, then turn right to the 12 and 4 o'clock positions. The trajectory of the grip force F(t) of each trial was recorded during the force exertion.

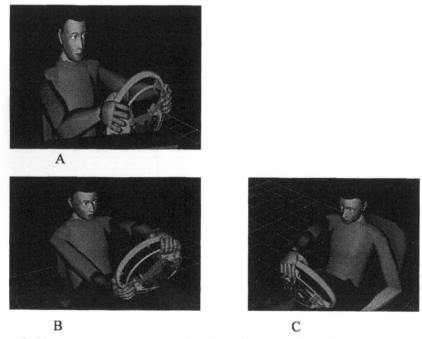

Figure 2. Three squeeze types were used to the subjects to grip the hand-operated control device: (A) two-hand squeeze at the 10-2 o'clock position (B) two-hand squeeze at the 10-2 o'clock position then turn right to 12-4 o'clock position (C) one-hand squeeze at the 10-2 o'clock position.

Four variables were then calculated and used to quantify the ability of grip force coordination including the peak force (P), dispersion (D), coefficient of variance (CV) and normalized overshoot (NO) of the force. These four variables were calculated as follows:

$$P = \max(F(t)) \tag{1}$$

$$D = \sigma(F(t)) \tag{2}$$

$$CV = \frac{\sigma(F(t))}{\overline{F}(t)} \tag{3}$$

$$NO = \frac{P - F_0}{F_0} \tag{4}$$

where $\sigma(F(t))$ is the standard deviation of $F(t)$, $\overline{F}(t)$ is the mean value of $F(t)$, and F_0 represents the target force.

Repeated-measures analysis of variance (ANOVA) tests were then conducted to examine the effects of age, gender, training, grip duration, hand posture and number of hands involved, target force level on each of the four variables quantifying the ability of grip force coordination. Of note is that the data of 1 untrained and 13 trained subjects were lost due to hardware problems. Since the hardware system could not measure forces exceeding 20N, there were 45 data points needed to be replaced with '20N' and were excluded from the analysis. In total 507 data points were used in the analysis. We also conducted a multivariate analysis of variance (MANOVA) using a Wilks' Lambda test to determine if the age, gender and training groups had significant difference in these hand anthropometric measures. We also examined the correlations between these anthropometric measures with the variables quantifying the ability of grip force coordination.

Only those significant main effects and interactions found in the ANOVAs and in the MANOVA were reported. In the MANOVA, the significant level were chosen at $\alpha=0.05$. In the individual ANOVAs, we considered the effects significant if $p < .005$ to avoid alpha-level inflation. Regressions were also performed to examine the relations between the peak value (P) and dispersion (D), coefficient of variance (CV) and normalized overshoot (NO) to gain insights on the optimal force range for controlling hand-operated devices.

RESULTS

No significant effect of training was observed on anthropometric measures, which suggested anthropometric characteristics were relatively consistent across trained and untrained subjects. Males exhibited greater anthropometric measures than females and the younger people exhibited greater anthropometric measures than the older people. However, neither personal factors including age and gender showed significant effects on the ability of grip force coordination. Training did not show impact on the ability of grip force coordination. No correlations were observed between the anthropometric measures and the variables quantifying the ability of grip force coordination.

The magnitude of the target force exhibited significant impact on three variables including the peak force (P), dispersion (D), and normalized overshoot (NO). Both the peak force and dispersion had smaller value at the lower target force . Yet the standard deviation of the two variables exhibited different patterns.

The mean (±standard deviation) of the peak value were 7.54 (±2.45) N at the target force of 5N and 16.61(±1.25) N at 15N. The mean (±standard deviation) of the dispersion were 1.27 (±0.45) N at 5N and 3.29 (±0.76) N at 15N. The normalized overshoot was much more affected by the target force level than the peak force and dispersion. The mean (±standard deviation) of the normalize overshoot were 50.79% (±49%) and 10.71% (±8.36%).

The squeeze type had significant effects on the peak value (P) and the normalized overshoot (NO). The mean (±standard deviation) of the peak value were 11.45 (±4.96) N for both, 12.18 (±5.01) N for top, and 11.48 (±4.84) for right. The mean (±standard deviation) of the normalize overshoot were 28.85% (±36.72%), 29.89% (±32.46%), and 41.14% (±38.68%), respectively.

Grip duration has significant impact on the coefficient of variance (CV) of the force. The mean (±standard deviation) of the CV were 26.70% (±8.00 %) and 23.30% (±6.63%).

The relations between the peak value (P) and the dispersion (D) of the grip force could be modeled using the linear, exponential, and power regressions with reasonable R-squares. The R-squares of the linear, exponential, and power regression were 0.7461, 0.8148 and 0.8204 respectively. However, no regression models have been found to be significant when examining the relations between the peak value (P) and the coefficient of variance (CV) and normalized overshoot (NO).

DISCUSSION

In the present study, we examined the ability of grip force coordination in a simulated driving context and contribute toward filling a research void concerning how accurately the drivers are able to control the force they exert on hand-operated device. Participants were asked to squeeze the hand brake ring to reach a certain force. Such a task emulated the speed tracking task in the real driving context. When people drive on the road, they usually adjust the vehicle speed using the brake and accelerator based on the information of the speedometer and external cues on the road. Although people have demonstrated their ability to precisely control the pushing force of the foot by using traditional foot-operated controls in daily driving, when the hand-operated control is in use, little information is available to determine whether the device could be operated smoothly and what kind of contributing factors should be taken into account in the device design.

Our experimental results suggest that the force magnitude drivers have to exert, the posture they have to choose, the duration of force exertion affect the ability of grip force coordination. It has been reported that the ability of grip force coordination should be better in the higher working force range. These results show specifically that drivers could control the hand-operated device better at 15N than at 5N in that the normalized overshoot and the standard deviation of the peak force were smaller at the greater target force. Thus, the higher working force range may help people achieve better performance in vehicle control. Therefore, the low sensitivity force range should be skipped when possible to avoid control requirements in this force range. This suggestion was consistent with the idea used in the conventional brake system as reported by Curry and Southhall (2002). They have shown that the device would not be activated below a threshold in the force effort curve of traditional foot brake system. Therefore, the working force range did not start from zero and was kept away from the low sensitivity force range.

The number of hands involved in force exertion also affected the ability of grip force coordination. The overshoot of the squeeze involving two hands was much smaller than that involving one hand while the mean peak force of the two tasks were similar. This result confirmed that force output in the bimanual task is more stable than that in unimanual task (Ohki & Johansson, 1999). Thus manipulation by two hands is shown to be better for vehicle control than manipulation by one hand for novice users. For the two-handed tasks with different posture, similar overshoots and peak force were observed across tasks. Thus, the posture for the two hand tasks seems not affect the ability of grip force coordination.

Therefore, when the vehicle is to be equipped with a hand-operated system, using two hands can decrease the risk of loss of control. How to avoid driving with one hand needs to be further studied in order to increase safety. Grip duration was also confirmed as a contributing factor of the ability of grip force coordination. Longer grip duration seemed to result in the lower coefficient of variance (CV) of the force. This is probably caused by that subjects have more time to make the exerted force stable.

Training did not show improvement on the ability of grip force coordination, which is in contradiction to our expectation. Usually training is effective on improving the task performance. Nowak et al. (2004) suggested that grip force control relied on accurate internal models of the dynamics of human motor system and the external objects we manipulate. Yet internal models are not static entities. They are dynamically updated by sensory experience. Sensory feedback signals the object properties and mechanical events to modify motor commands and update the internal representation automatically. Shadmehr and Brashers-Krug (1997) have observed that training was sufficient to allow the individual to form an accurate internal model and the memory of the consolidated skill lasts for half a year. Lazarus and Haynes (1997) have shown that subjects can improve their performance in isometric pinch force regulation significantly over the trials using the visuo-motor tracking task. We expected that users with different sensory experiences in the use of the hand brake may have different skill levels in controlling the devices. The result that no difference existed between trained subjects and untrained subjects may be a reflection of that numeric information was used as the input of human motor system rather than graphic information since Newell and McDonald (1994) showed that precision of the information affect the performance in the grip force task.

Personal factors seemed to have no effect on the ability of grip force coordination in this force range. The older and younger subjects showed no difference in controlling the grip force. This result did not agree with many studies on aging, which have shown that performance of the hand functions decreased with increased of age (Cole, 1991; Hackel, Wolfe, Bang, & Canfield, 1992; Kinoshita &

Francis, 1996; Lazarus & Haynes, 1997; Shim, Lay, Zatsiorsky, & Latash, 2004; Shinohara, Li, Kang, Zatsiorsky, & Latash, 2003; Shinohara, Scholz, Zatsiorsky, & Latash, 2004). Age generally has a significant impact on human behaviors such as reaction time, movement time, information transmission rates, grip strength and dexterity (Boucher, Denis, & Landriault, 1991; Chaput & Proteau, 1996; Francis & Spirduso, 2000; Hughes, Gibbs, Dunlop, Edelman, Singer, & Chang, 1997; Shim, Lay, Zatsiorsky, & Latash, 2004; Stelmach, Goggin, & Amrhein, 1988; Stelmach, Goggin, & Garcia-Colera, 1987). Yet, it should be recognized that younger subjects were selected from those aged from 18-36 and older subjects were selected from 36-64 in this experiment. The criterion was different than what a previous study used (Boucher et al., 1991). If older subjects aged from 65 to 74 were selected, the age effect on the grip force coordination capabilities could have been observed. The females and males also exhibited no difference in the ability of grip force coordination within this force range. Our results agree with Li (1998) and Li et al. (2001), which also did not show differences between men and women in indexes of finger interaction. However, this result was inconsistent with the results from Shinohara et al. (2003). They examined the effects of gender on finger coordination and suggested indexes of finger coordination scale with force-generating capabilities across gender and age groups. Therefore, whether age and gender are associated with the change in grip force coordination should be further examined before the hand-operated controls are available to public.

No simple relationships were shown between anthropometric data and the ability of grip force coordination, although anthropometric measures could affect the grip force coordination mode (Newell & Mcdonald, 1994). It was observed that a 5% female subject squeezed the hand control with only four fingers but a 95% male subject squeezed the hand control with five fingers. It was also shown that the involvement of the phalanges of the fingers in the task for these subjects were different. However, the low correlation between anthropometric data and the ability of grip force coordination did not support any clear effect of anthropometry on the ability of grip force coordination. That probably resulted from the dynamic nature of the gripping task. When a 5% female subject squeezed the ring, the coordination mode changed with the change of the grip span. An adjustable design could be considered in further commercial use.

These results have an important implication in designing hand-operated control devices. They showed that the task factors have more significant effects than personal factors on the ability of grip force coordination. When designing hand-operated devices, force magnitude drivers have to exert, the posture they have to choose, and the duration of force exertion should be paid more attend than the anthropometric data and gender. This study also provided an insight on the relations between the peak value and the dispersion. Since the differences between these R-square values were small, it was likely that the relationship between the target force and the dispersion of grip force was linear and the relationship between the target force and the CV was flat as suggested by Carlton and Newell (1993).

CONCLUSIONS

Drive-by-wire technology (Burns, Mccormick, & Borroni-Bird, 2002) has motivated this research regarding control by hand and the hand actuated brake system has been shown to reduce reaction time, reduce stopping distance and hence, may improve safety. Driving safety, in the context of using hand actuated braking system, requires an understanding of hand behavior and factors that influence control when driving by hand, since standards have not yet been developed. In the real driving context, people adjust the speed by the brake and accelerator based on the information of the speedometer and external cues on the road. How accurate the subjects can control the force they exert on the hand actuated system are important for the success in this kind of task. This study investigated grip force coordination capabilities and provided a direct application of grip force coordination capabilities to product design. To date, there is no existing analysis tool known to help people predict the human performance in such control tasks before the physical mockup for testing is made. With the development of digital human modeling and digital human simulation, this type of tool is needed to decrease the requirements of design variations and enable simulation testing before first physical mockups.

ACKNOWLEDGMENTS

Thanks to Zach Rowland, Ron Lewis and Will Jenkins for contributions to hand brake design as well as Vince Sanders for technical support. The authors also would like to thank Jinyan Du, Jihun Kang, Joe Szczerba, and Amanda McAlpin for their assistance throughout the project.

REFERENCES

Boucher, J. L., Denis, S., & Landriault, J. A. (1991). Sex differences and effects of aging on visuomotor coordination. *Perceptual and Motor Skills, 72*(2), 507-512.

Burns, L. D., McCormick, J. B., & Borroni-Bird, C. E. (2002). Vehicle of change. *Scientific American, 287*(4), 64-73.

Carlton, L. G., & Newell, K. M. (1993). Force variability and characteristics of force production. In K. M. Newell & D. M. Corcos (Eds.), *Variability and Motor Control* (pp. 15-36). Champaign, IL: Human Kinetics.

Carter, T. J. (2000). Motorscooter braking control response study. *SAE Technical Paper, 2000-01-0180.*

Chaput, S., & Proteau, L. (1996). Aging and motor control. *Journals of Gerontology Series B: Psychological Sciences and Social Sciences, 51*(6), 346-355.

Curry, E., & Southhall, D. (2002). *Disabled Driver's Braking Ability* (No. MIRA 02-211058): MIRA.

Francis, K. L., & Spirduso, W. W. (2000). Age differences in the expression of manual asymmetry. *Experimental Aging Research, 26*(2), 169-180.

76

Hoffmann, E. R. (1991). A Comparison of Hand and Foot Movement Times. *Ergonomics, 34*(4), 397-406.

Hughes, S., Gibbs, J., Dunlop, D., Edelman, P., Singer, R., & Chang, R. W. (1997). Predictors of decline in manual performance in older adults. *Journal of the American Geriatrics Society, 45*(8), 905-910.

Lazarus, J. A., & Haynes, J. M. (1997). Isometric pinch force control and learning in older adults. *Experimental Aging Research, 23*(2), 179-199.

Li, S., Danion, F., Latash, M. L., Li, Z. M., & Zatsiorsky, V. M. (2001). Bilateral deficit and symmetry in finger force production during two-hand multifinger tasks. *Experimental Brain Research, 141*(4), 530-540.

Li, Z. M., Latash, M. L., Newell, K. M., & Zatsiorsky, V. M. (1998). Motor redundancy during maximal voluntary contraction in four-finger tasks. *Experimental Brain Research, 122*(1), 71-78.

Newell, K. M., & McDonald, P. V. (1994). Information, coordination modes and control in a prehensile force task. *Human Movement Science, 13*, 375-391.

Nowak, D. A., Glasauer, S., & Hermsdorfer, J. (2004). How predictive is grip force control in the complete absence of somatosensory feedback? *Brain, 127*, 182-192.

Ohki, Y., & Johansson, R. S. (1999). Sensorimotor interactions between pairs of fingers in bimanual and unimanual manipulative tasks. *Experimental Brain Research, 127*(1), 43-53.

Ostlund, J. (1999). *Joystick-controlled cars for drivers with severe disabilities* (No. VTI rapport 441A -1999). Linkoping, Sweden: Swedish Road and Transport Research Institute (VTI).

Shadmehr, R., & Brashers-Krug, T. (1997). Functional stages in the formation of human long-term motor memory. *Journal of Neuroscience, 17*(1), 409-419.

Shim, J. K., Lay, B. S., Zatsiorsky, V. M., & Latash, M. L. (2004). Age-related changes in finger coordination in static prehension tasks. *Journal of Applied Physiology, 97*(1), 213-224.

Shinohara, M., Li, S., Kang, N., Zatsiorsky, V. M., & Latash, M. L. (2003). Effects of age and gender on finger coordination in MVC and submaximal force-matching tasks. *Journal of Applied Physiology, 94*(1), 259-270.

Stelmach, G. E., Goggin, N. L., & Amrhein, P. C. (1988). Aging and the restructuring of precued movements. *Psychological Aging, 3*(2), 151-157.

Stelmach, G. E., Goggin, N. L., & Garcia-Colera, A. (1987). Movement specification time with age. *Experimental Aging Research, 13*(1-2), 39-46.

Szczerba, J., Duffy, V. G., Geisler, S., Rowland, Z., & Kang, J. (2007). A study in driver performance: alternative human-vehicle interface for brake actuation. *SAE 2006 Transactions Journal of Engines, 2006-01-1060*, 605-610.

Webb Associates. (1978). *Anthropometric source book*. Washington,Springfield, Va.: National Aeronautics and Space Administration Scientific and Technical Information Office.

Chapter 9

Human Behaviour Analysis and Modelling: A Mixed Method Approach

Giuseppe Andreoni, Laura Anselmi, Fiammetta Costa,
Marco Mazzola, Ezio Preatoni,
Maximiliano Romero, Barbara Simionato

INDACO Dept.
Politecnico di Milano

ABSTRACT

Today the ergonomic assessment of objects and products is often treated with separation of the different components of the interaction: physical, sensory, subjective. This research aims to develop an integrated method and a protocol for the qualitative and quantitative study of motor functions for ergonomics. that is addressed to evaluation of strategies and efforts carried out by users in relation with products. In our case study these products are the home appliances and in particular the dishwater. Basic methodological approaches refer to biomechanics and product usability assessment techniques. The first one is based on the at measurement of angular excursions of the joints associated with the implementation of the human motion detectable in dedicated laboratory; the second methods rely on experiments with users and direct observations and questionnaires / interviews to quantify ease of use and user satisfaction by means of special scales of assessments carried out in usability laboratories. The integration of methods for a global and more comprehensive ergonomical assessment is the rationale of the research, that can then be used in proactive way in the early stages of development.

Keywords: Human behavior analysis, observational methods, quantitative movement analysis, HMI assessment.

INTRODUCTION

The idea of merging qualitative and quantitative methods has become increasingly popular, in particular in areas of applied research. HMI and ergonomics are multifaceted issues so it is important to approach the phenomenon under investigation from diverse sides and to combine data resulting from diverse methods. The proposed mixed method approach integrates different research methods into a research strategy [Brannen, 2005] increasing the quality of final results and to provide a more comprehensive understanding of analyzed phenomena.

For the development of quantitative indicators related to the usability of a given product, three methodologies could be adopted:

-   Pseudo-absolute.

In this case you can imagine to have an assessment based on the consideration that excursions of the anatomical angles considered in proper intervals - more or less preferential or ergonomic (for example, the greater the elbow flexion, the greater the muscular effort is and this corresponds to increasing values of discomfort). Therefore a very simple index given by the product of the joint angle for the time this posture is kept, can be adopted as ergonomical index related to the performed gesture or posture [Kee and Karwowski 2001].

-   Ordinal.

In this case we can structure a test in a protocol corresponding to with an expected value of comfort – in either ascending or descending order - (for example points to be reached gradually lower and lower is a worsening situation in terms of comfort). This approach is therefore based on a ergonomic intrinsic ranking according to the given protocol. This preliminary analysis can provide a classification by ranks (statistical approach for non-normal variables) of parameters of the movement more related to the act from the ergonomics point of view. Then by analyzing the several and different configurations of the same gesture with an experimental approach structure it may produce an ergonomic ranking that can then be adopted as an interpretive grid for similar movements in different conditions (as is the case with example of the movement of manual loading and unloading of food and products within two refrigerators which are different but similar being of the same typology thus giving rise to a relative ergonomic ranking between themselves) Optionally one could be assumed as reference for normalization or standardizations with respect to similar products or gestures.

-   Subjective.

In order to integrate the issues identified by physical measures and quantified using models with the subjective aspects related to quality of use of the products usability tests can be carried out [Jordan 1998]. The direct involvement of an adequate number of users, specially selected from the specific target audience, will identify the components of usability (according to the ISO 9241-11 norm): effectiveness (i.e. the accuracy and completeness with which users achieve their objectives of the test), efficiency (i.e. the accuracy and completeness of goals achieved in relation to resources spent), and satisfaction (comfort and acceptability in the use of the given product) - meaningful to the user. The technical standard ISO 9241-11 defines "usability" the 7 conditions with which a product can be used by specified users to achieve specific objectives, in a specific context.

Based on the above mentioned assumptions, this research was conducted by integrating the ordinal approach as regards physical ergonomics factors and the subjective ones related to aspects of quality of use in an innovative mixed approach. The case study where this new method was applied was the dishwasher appliance.

MATERIALS AND METHODS

The methodological approach here proposed follows three steps: a preliminary ethnographic analysis have been performed to design and drive the following phases consisting into two in-depth analyses regarding usability and biomechanics of the interaction with dishwashers.

In fact the object of investigation for this research are the dishwashers. Three commercial dishwashers have been identified as bestsellers, a competing model of the same type but higher market quality, an alternative type considered highly innovative. The given codes for identification were respectively H11, M33 and W46.

FIGURE 1   the three dishwashers in the case study where the proposed methodology was applied.

The ethnographic analysis was conducted by observing the interaction of people with your dishwasher in the natural home context to focus and better define the objectives and procedures of the two next analyses. In our case study we performed 4 ethnographic observations to detect user habits [Amit 2000]. On this basis we defined the usability analysis protocol.

The usability analysis about perceived and actual usability was conducted in the dedicated laboratory through tests with 10 users in two phases. In the first static session the user can only observe the products and while in the second dynamic session, the user interacts with the products [Rubin 1984]. The usability evaluation was performed in a lab where the dishwashers were placed paying attention to create a use situation the most similar to the real one. In detail the test was organized in two phases: a static one where the perceived usability was evaluated before interacting with the products and a dynamic one where the user were asked to fulfill several activities like opening the dishwasher, regulating the tray high, loading and unloading a complete set of vessels, simulating the filling of soap, salt and other operations. These tests with the dishwashers lead us to identify and isolate 6 main tasks to be considered in the next quantitative biomechanical study.

The biomechanical analysis was conducted at the instrumental laboratory of human motion analysis in two phases too. The first one allowed for defining an index for the measurement of comfort / discomfort of human movement involving 30 subjects, and the second session was carried out to compare between them dishwashers through the application of the index through the study of the motor behavior of other 15 subjects. We defined the biomechanical protocol and performed the acquisitions and evaluation with the MMGA index of discomfort [Andreoni et al. 2009]. Total body kinematics was recorded through a six cameras optoelectronic system while the subjects performed 6 specific motor tasks (i.e. opening the door and trays, placing 2 plates and 2 glasses, filling soap) with the three dishwashers. In a previous research [Romero et al 2008] we found some problems with the analysis of data acquired from natural users movements because of the excessive variability in behaviours that made inter-subject comparison very difficult. In this experience we consequently decided to constrain the sample people to some fixed points as the feet position. Constraining the people we partially loose naturality in the gesture, but using ethnographical observation and usability test results to define the fixed points, we found a acceptable compromise. The MMGA index, considering quantitative computation of the joints motion, a coefficient of discomfort, and the mass of the involved body district, has been calculated for each task in order to compare the score of the users interacting with the three different dishwashers.

Relevant differences between the three dishwasher have been verified through the application of Friedman's Test.

RESULTS

The preliminary ethnographic observations have provided useful information for the definition of laboratory setting (plan starting with sink, range and location of dishes, drawers with cutlery) and the protocol developed for testing usability (i.e. the identification of the most relevant tasks and their sequences in the user's experience; in thi case we selected: opening door, opening and filling the baskets in the preferred order interviewee, adjustment the top basket, detergent compartment filling simulation, simulation fill the rinse aid compartment, door closing, opening and emptying the baskets order preferred by the interviewer, simulation of filling up the compartment with a funnel specifically, simulation of cleaning filters).

Ethnographic observations and usability testing were also useful for identifying the most important movements in the human-product interaction (6 in our case study with dishwasher: opening door, positioning a glass in the corner nearest the basket superior position of another glass in the far corner of the truck higher placement of a plate in the bottom of the basket closer, positioning a second course in the far corner of the basket bottom, filling the compartment detergent) to be analyzed in terms of occupational biomechanics and physical ergonomics. In this case it was decided to omit the analysis phase of exhaust systems as well as the ethnographic evidence that the tests have usability highlighted how the strategies of discharge are subject to differences in the provision of kitchen.

The choice of dishes and position reflects the observations made in both context that, during the tests: it is in fact shown that users always ask glasses in the upper rack and large plates in the lower (for other objects the location instead was more differentiated). Between the filling of detergent, rinse, salt and cleaning filters, it was decided to concentrate on moving on to filling compartment detergent as it is the most frequently under normal conditions of use. Despite the criticality of use found in the tests could not be categorized by biomechanically filter cleaning and loading salt because their positioning in the dishwasher would make impossible shots optoelectronics. Usability tests have also highlighted several starting positions of users interaction with the dishwasher was therefore necessary to introduce even Protocol analysis of movements of a familiarization phase to define the position of departure feet for each subject.

As overall result we found interesting concordances between physical index of discomfort and user subjective appreciation in the usability test.

The total MMGA Index calculated for the whole interaction of the tasks highlighted a higher discomfort for one of the dishwashers. The same product received also a worst evaluation from the usability point of view. Also the biomechanical evaluation of the dishwasher with the lower MMGA Index has been confirmed by the usability test especially in the post use evaluation.

Despite a number of differences such as: a) the different nature, where qualitative and quantitative analysis of usability in the biomechanics; b) the order of interaction of users with the dishwasher (M33, H11, L46 in the first case and M33, W46, H11 in the second one); and c) the definition of tasks (many and complex in the first case, fewer and simpler in second one) it was considered interesting to investigate differences and similarities in the results obtained.

From methodological point of view, we found a positive conciliation to compare physical modelling movement quantitative data and usability test qualitative data and we reached the conclusion that it is not possible to arrive to an overall synthetic Index.

About the overall rating we can distinguish the subjective and quantitative contributions. In any case the comparison of two different methods of investigation has achieved results and more detailed information: verbalizations on the habits of use of the dishwaters have highlighted some distinctive features while the biomechanical analysis allowed for a more resolute dishwashers ranking in terms of effort/fatigue in all the tasks related to interaction with the product.

The main results of the usability and subjective analysis are related to human behavior about use habits that are also reflected in the mechanisms of choice and purchase. For example in the case study we found that during the assessment of the product you are looking for items that secures the closest possible as usual: standard programs, strategies for inclusion of dishes known use of cleaning products differ depending on the offers on outlets. At the time of purchasing the primary aspects of evaluation are represented by the relationship between guarantee of quality offered by the mark, understood both in time and duration of effectiveness washing, and reduction of costs in the short and long term, respectively acquisition cost and consumption.

Comparative analysis of the two surveys, has enabled a more comprehensive assessment of some elements and / or parts of the dishwashers: In the case of the entry system of the detergent, for example, if biomechanical analysis of L46 showed positive results compared to the positioning top drawer, the analysis of usability has shown also that the mechanism used in the case of this dishwasher does not receive favorable feedback, both for the orientation for the system Opening, reading together the data show that the positioning of the compartment detergent is correct, but is not resolved by the design point of view.

Further developments comprehend the evaluation of users intentions through eye-tracking technologies, the integration of overall movements recording placing few markers on the users while performing the usability tests to detect general motor strategies and the application of the same integrated approach to interface interaction analysis.

REFERENCES

Amit, V. (2000), Constructing the Field: Ethnographic Fieldwork in the Contemporary World, Routledge.

Andreoni, G., Mazzola, M., Ciani, O., Zambetti, M., Romero, M., Costa, F., Preatoni, E. (2009), *Method for Movement and Gesture Assessment (MMGA) in ergonomics.* In: Proceedings of the 2nd International Conference HCI/ICDHM, San Diego (USA), July 19-24, 2009, 591-598.

Brannen, J. (2005), "Mixing Methods: The Entry of Qualitative and Quantitative Approaches into the Research Process", *International Journal of Social Research Methodology*, 8(3), 173–184.

Jordan, P. W., (1998), An Introduction to Usability, Taylor and Francis, London

Kee, D., Karwowski, W. (2001), LUBA: an assessment technique for postural loading on the upper body based on joint motion discomfort and maximum holding time. *Applied Ergonomics.* 32(4), 357 –366.

Romero, M., Mazzola, M., Costa, F. and Andreoni, G. (2008), *An Integrated method for a qualitative and quantitative analysis for an ergonomic evaluation of home appliances*, In: A.J. Spink, M.R. Ballintijn, N.D. Bogers, F. Grieco, L.W.S. Loijens, L.P.J.J. Noldus, G. Smit, and P.H. Zimmerman (Eds.),Proceedings of Measuring Behavior 2008 (Maastricht, The Netherlands, August 26-29, 2008), 107-108.

Rubin, J. (1984), Handbook of usability testing: how to plan, design and conduct effective tests, John Wiley & sons, New York

AKNOWLEDGEMENTS

The authors want to thank INDESIT COMPANY S.p.A. for the financial support to the research.
The authors also thank all the subjects participating in the experimental trials.

CHAPTER 10

Vehicle Layout Conception Considering Trunk Loading and Unloading

Alexander Mueller, Thomas Maier

Research and Teaching Department Industrial Design Engineering
Institute for Engineering Design and Industrial Design
Universitaet Stuttgart
Pfaffenwaldring 9
70569 Stuttgart, Germany

ABSTRACT

In the field of Engineering Design and Industrial Design Engineering the fundamental question concerning the development of new products and the methodologies of product development is raised continuously. The present contribution gives an answer to the question how vehicles can be developed in a user-centered manner under particular consideration of trunk loading and unloading movement sequences.

Keywords: vehicle layout conception, digital human modeling, trunk loading and unloading, user-centered vehicle design

INTRODUCTION

In general, it can be observed that the extensive dimensioning of automobiles - the so-called vehicle layout concept - has gained considerable complexity in the last decades. This product complexity mainly results from the steadily increasing number of requirements imposed on modern vehicles (Braess 2005). Additionally, and regardless of high development costs, a strong increase of vehicle variants can be noticed in the market (Balzer 2002). Based on the outlined product complexity, vehicles are nowadays being developed in interdisciplinary teams which have to accomplish a multitude of conflicting goals (Eiletz 1999). In this context, ergonomic requirements do not necessarily play the most important role in the overall vehicle development and are not always sufficiently weighted.

In order to solve best the transport task, which can be described as the transport of occupants together with their luggage, vehicles should be configured in a user-centered way. This implies that vehicle development already starts in the concept phase with the user-centered and centrifugal vehicle layout conception.

From an ergonomic point of view, four fundamental aspects of the vehicle layout conception are of particular importance. The functional derivation of the interior space (Mueller et al. 2008) as well as vision requirements have a major effect on the systematic and user-centered vehicle layout conception (Mueller et al. 2009 A). Furthermore, the critical and complex ingress and egress movement tasks have a significant influence (Mueller et al. 2009 B).

Today trunk loading and unloading movement sequences are not systematically included in the vehicle layout conception. According to the current state of knowledge, a systematic and user-centered methodology, especially with regard to vehicle loading and unloading, has not been propagated in teaching and has not entered the relevant literature yet.

However, trunk packing is a critical issue (Aicher 1984), (Karwowski 1993), (Divivier 2008). Therefore the question about the integration of trunk loading and unloading movement sequences within the systematic and user-centered vehicle layout conception arises.

VEHICLE LAYOUT CONCEPTION

In this paragraph, a methodology is presented that provides a basis for the systematic and user-centered derivation of a vehicle layout concept that especially considers trunk loading and unloading movement sequences performed by vehicle passengers.

CURRENT STATE OF RESEARCH

The systematic and user-centered vehicle layout conception is based on main requirements.

Referring to previous publications e.g. (Mueller et al. 2009 B), and according to the current state of knowledge, the systematic and user-centered vehicle layout conception consists of 9 main steps that are listed below:

1. Definition of the main reference point.
2. Definition of the main reference planes.
3. Modeling of the drivers' workplace (includes drivers' vision as well as statically and dynamically required space).

4. Modeling of the passengers (includes statically and dynamically required space).
5. Definition of the vehicle interior.
6. Modeling of the luggage volume and modeling trunk loading and unloading movement sequences.
7. Picturing of the technical function units.
8. Definition of the vehicle exterior.
9. Picturing grid planes.

METHOD OF SOLUTION

Trunk loading and unloading corresponds to the 6th work step of the systematic and user-centered vehicle layout conception. Since the modeled luggage volume has an effect on the loading and unloading movement sequences, this geometry has to be defined first.

On the basis of analyzed movement sequences, trunk loading and unloading can be digitally modeled. As a consequence, the dynamically required space can be derived. This space forms a basis for the definition of certain vehicle exterior dimensions that are defined in the 9th work step. Dimensions derived from competing vehicle concepts that are deposited in a database, also serve as a guideline for the vehicle exterior definition.

Main requirements which are chosen exemplarily form the groundwork of the systematic and user-centered vehicle layout conception considering trunk loading and unloading. These requirements are briefly presented in the following. A short woman and a tall man defined as drivers and three tall men defined as passengers are specified as vehicle users. All vehicle passengers should be able to load up the vehicle with significant transport goods. The conception of a medium-sized passenger vehicle with a standard power-train (4-cylinder front transverse engine, with front wheel drive) is phrased as the design objective.

MODELING OF THE LUGGAGE VOLUME

Appropriate technical literature reports on conducted experiments that show clearly all relevant factors of influence for the design of a passenger car trunk (Karwowski 1993). It is reported that the ground to trunk lip height (H195) and the ground to trunk floor height (H252) have an essential impact on the design of a passenger car trunk. In the present paper, these two dimensions are regarded as equal in order to define the luggage volume, since a systematic and user-centered approach is followed for the vehicle layout conception.

In practice, vehicle trunks are mostly designed and evaluated either by application of ISO 3832 (ISO3832 2002) or by using a method described in SAE J1100 (SAEJ1100 2005). However, significant transport goods should be considered additionally and find their place within the luggage volume. The definition of significant transport goods (suitcase, golf bag, sports bag, beverage case and baby buggy) results from bibliographic references (Karwowski 1993), (J.D. Power 2009) and from surveys that were conducted with 65 vehicle users. The majority of these transport goods can be extracted from databases of modern CAD applications (Grabner 2006).

MODELING OF TRUNK LOADING AND UNLOADING MOVEMENT SEQUENCES AND DERIVATION OF THE DYNAMICALLY REQUIRED SPACE

In order to model the trunk loading and unloading movement sequence, this movement has to be analyzed first. Subsequently the dynamically required space can be derived.

Analysis

An experimental procedure for trunk loading and unloading with the significant transport goods defined above was conducted in order to identify applied movement strategies. One woman and four men served as test subjects. They were all healthy, between 25 and 28 years of age and had body heights of between 1630 mm and 1990 mm. Trunk loading and unloading movement sequences performed by the test subjects were recorded by means of a digital camera in front view and lateral view. A medium-sized (KBA 2008) station wagon served as test vehicle.

The analysis of the digitally documented movement sequences only show a minimal difference between trunk loading and unloading movement sequences. Since the dynamically required space of these movements is almost equivalent, loading and unloading movement sequences are regarded as equal. Consequently, the systematic and user-centered vehicle layout conception only considers trunk loading movement sequences.

As a result, movement sequences related to the loading of significant transport goods were analyzed and movement strategies were determined.

An exemplary investigation of the loading of a luggage volume with a 20 kg weighing suitcase (680 mm x 220 mm x 570 mm) is presented in the following. Two different movement strategies were observed and these movement strategies were divided into movement phases. Supporting postures characterizing the loading movement sequence are shown in Figure 1.

The so called "One-Hand"-Strategy was applied by all male test subjects. Test subjects using this strategy grasped the suitcase with the right hand and flexed the upper body. By straightening up the upper body the suitcase is lifted. Subsequently, the body of the test subject is oriented towards the luggage volume and the left hand is put on the suitcase's side. The suitcase is then set on the load floor and afterwards laid down. Finally, the suitcase is pushed into its final position in the luggage volume.

88

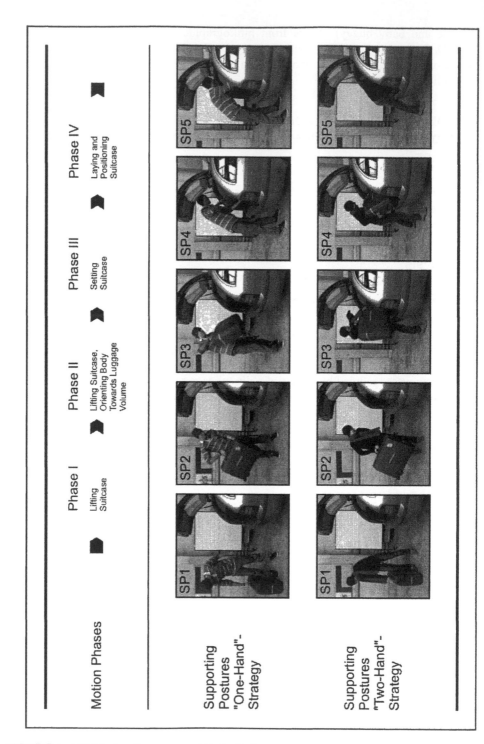

FIGURE 1: Vehicle loading movement strategies

While observing the female test subject loading the luggage volume with a suitcase of 20 kg weight, a different movement strategy was detected. This strategy was named "Two-Hand"-Strategy. The suitcase is also grasped with the right hand and is then lifted by erecting the subject's upper body. However, afterwards the suitcase is also grasped with the left hand. After orienting the body towards the luggage volume, the suitcase is set on the trunk lip with support of the subject's thigh and by flexing both arms. Finally, the suitcase is laid on the trunk floor and pushed into the final position.

Loading movement sequences with all transport goods defined as significant were analyzed in an analogous matter.

Digital Human Modeling

In order to limit the subject matter, loading movement sequences are modeled quasi-dynamically. This means that static supporting postures derived from the analysis serve as a basis for the modeling of the movement sequences.

The digital man model RAMSIS (Seidl 1994) within the CAD application CATIA (Dassault 2005) was used for the modeling of the test subjects' vehicle loading tasks. The movement tasks were modeled on basis of assessed joint angles and with respect to the observed movement strategies. The occupants' dynamically required space can be derived from the result of the digitally modeled trunk loading movement sequences. With respect to this space and by using predefined vehicle dimensions (SAE J1100 2005), a vehicle layout concept that especially considers trunk loading and unloading can be identified.

As the analysis of trunk loading with a suitcase is described above, the digital modeling of this movement sequence is exemplarily shown in Figure 2.

All transport goods defined as significant can be digitally modeled on the basis of the above described exemplary analysis.

DERIVATION OF VEHICLE EXTERIOR DIMENSIONS

Among others, the overall body shape of the vehicle can also be defined by taking into account the vehicle users' dynamically required space for trunk loading by means of digital human modeling with RAMSIS and by using a certain vehicle interior layout concept.

Furthermore, vehicle dimensions taken from competing vehicle layout concepts (autograph 2010) can be evaluated on the basis of the dynamically required space. In order to identify decisive competitors, vehicle registration statistics can be helpful (KBA 2008).

Especially the functional influence of the dynamically required space on the window slope angle of the backlight (A121-2) becomes apparent. Additionally, the body opening dimensions of the tailgate, in particular the minimum loading width of the rear opening (W208), the body opening width (W_{BO-1}) and the rear opening height (H202) can functionally be derived and evaluated against competitors (see

Figure 3).

The systematically generated and user-centered vehicle layout concept, which is based on the above and exemplarily defined main requirements, is also shown in Figure 3.

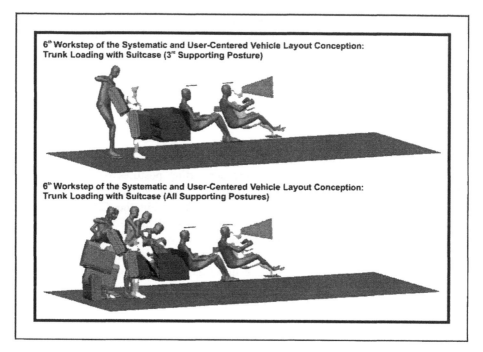

FIGURE 2: Digital human modeling of trunk loading with a suitcase

DISCUSSION

From an ergonomic point of view the systematic and user-centered vehicle layout conception would be ideal. However, in practice a vehicle layout conception is not necessarily user-centered.

The analysis of trunk loading and unloading movement sequences under consideration of transport goods that were defined as significant was performed with five test subjects. In order to support the findings described in the present paper, an increase of the number of test subjects could be reasonable.

When loading and unloading movement sequences were analyzed, it became apparent that the ground to trunk lip height is especially critical for short passengers loading the vehicle (cp. Figure 1). Hence, this dimension was reduced to 370mm. It is well established that the selection of this dimension also depends on the loading movement of a tall passenger. In addition, this dimension has a major influence on the positioning and orientation of technical function units (e.g. rear axle, tank, exhaust system).

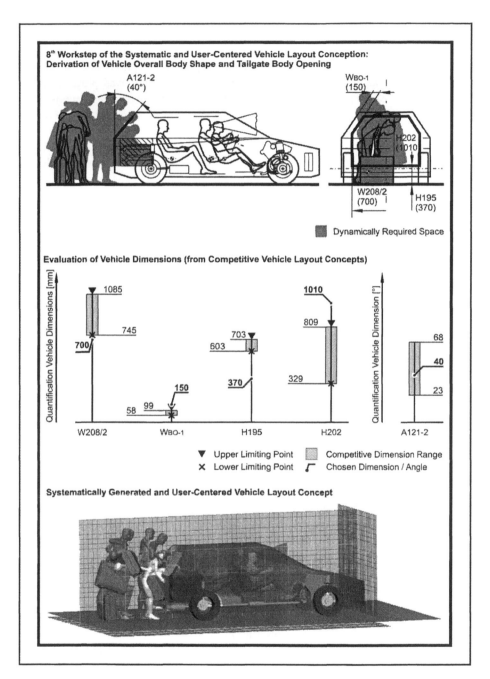

FIGURE 3: Derivation of vehicle exterior dimensions and vehicle layout concept

CONCLUSION

The systematic and user-centered vehicle layout conception that especially considers trunk loading and unloading movement sequences was phrased as the main objective of the present paper. The results show that this is generally possible.

For ergonomists quickly generated and user-centered vehicle layout concept variants could serve as an argumentation aid in the interdisciplinary vehicle layout conception. Especially the implementation of the described methods in digital human modeling software could be reasonable, as user-centered vehicle layout concepts could then be generated systematically and quickly (Mueller et al. 2009 C).

As cost intensive loops based on an insufficient dimensioning of the occupants' dynamically required space can be avoided, the application of the new methodology additionally offers an interesting potential for economization.

ACKNOWLEDGMENTS

At the former Institute of Machine and Gear Unit Design, Research and Teaching Department Industrial Design Engineering, the idea to configure vehicles systematically and user-centered was first conducted in 2005 in collaboration with Prof. Dipl.-Ing. i.R. Hartmut Seeger (Seeger 2007). The systematic and user-centered vehicle layout conception considering trunk loading and unloading in particular was first carried out in a research study supervised by the authors (Breyer 2009). The methodology benefited from the implementation of data from competitive vehicles provided by Mrs. A. Peters (autograph 2010). Sincere thanks are given to all persons involved.

LITERATURE

Aicher, O. (1984). Kritik am Auto: Schwierige Verteidigung des Autos gegen seine Anbeter. Eine Analyse. Muenchen, Germany, Callwey.
autograph dimensions, (2010). Top-level vehicle measurement. http://www.autograph.de, lastly checked: Feb. 2010
Balzer, R. (2002). Modellierung der Außengestalt von Personenkraftwagen zur Ermittlung eines Gestaltwertes. Stuttgart, Germany, Universitaet Stuttgart, IMK - Technisches Design, Dissertation.
Braess, H.-H. (2005). Vorwort. In: Vieweg Handbuch Kraftfahrzeugtechnik. Wiesbaden, Germany, ATZ/MTZ-Fachbuch.
Breyer, S. (2009). Untersuchung zur nutzerzentrierten Fahrzeugbe- und -entladung. Student Research Project. Stuttgart, Germany, Universitaet Stuttgart, IKTD.
Dassault Systèmes, (2005). CATIA. Version V5 R16. Velizy-Villacoublay.
Divivier, A. (2008). Beladen und Servicetaetigkeiten rund um das Fahrzeug. In: RAMSIS USER CONFERENCE 2008. Kleine Autos innen ganz groß.

Kaiserslautern, Germany.

Eiletz, R. (1999). Zielkonfliktmanagement bei der Entwicklung komplexer Produkte - am Beispiel PKW-Entwicklung. Bd. 32: Reihe Konstruktionstechnik Muenchen. Aachen, Germany, Shaker, Dissertation.

Grabner, J.; Nothhaft R. (2006). Konstruieren von Pkw-Karosserien. Berlin, Heidelberg, Germany, Springer Verlag,

ISO3832 (2002). Standard: ISO 3832: Passenger cars - Luggage compartments - Method of measuring reference volume.

J.D. Power (2009). J.D. Power and Associates: Vehicle Interior Needs Study (VINS).http://www.jdpower.com/corporate/automotive/download/2009_VINS .pdf, lastly checked: Nov. 2009.

Karwowski, W.; Yates, J. W.; Pongpatana, N. (1993). Ergonomic guidelines for design of a passenger car trunk. In: Automotive Ergonomics. London, GB. pp. 117-139.

KBA, Kraftfahrtbundesamt (2010). http://www.kba.de, Flensburg, Germany, lastly checked: Feb. 2010.

Mueller, A.; Maier, T. (2008). Ganzheitliche Methodik zur systematischen Auslegung des Fahrzeuginnenraumes. In: Produkt- und Produktions-Ergonomie. Aufgaben fuer Entwickler und Planer. Dortmund, Germany, Gesellschaft fuer Arbeitswissenschaft, pp. 403-406.

Mueller, A.; Maier, T. (2009 A). Vehicle Layout Conception Considering Vision Requirements: SAE Technical Paper 2009-01-2296. http://www.sae.org/technical/papers/2009-01-2296. Warrendale, PA, USA, Society of Automotive Engineers, Inc.

Mueller, A.; Maier, T. (2009 B) Systematic Integration of Complex Movement Sequences in the Vehicle Layout Conception. In: 9. Internationales Stuttgarter Symposium. Automobil- und Motorentechnik. Dokumentation. Wiesbaden, Germany, P. 267-279.

Mueller, A.; Maier, T.; Wirsching, H.-J. (2009 C). Workshop: Focusing on the Human Being in the Vehicle - Vehicle Design, Centered on the User. In: RAMSIS USER CONFERENCE 2009. Zukunft Marktposition - Mit Prozessinnovationen Wettbewerbsvorteile sichern. Kaiserslautern, Germany 2009.

SAE J1100 (2005). Standard SAE J1100, Motor Vehicle Dimensions. Warrendale, PA, USA, Society of Automotive Engineers, Inc.

Seeger, H. (2007). Generierung des Maßkonzepts und des Karosseriegrundtyps von Personenkraftwagen fuer deren Formentwicklung und Design. Stuttgart, Germany, Universitaet Stuttgart, IKTD.

Seidl, A. (1994). Das Menschmodell RAMSIS, Analyse, Synthese und Simulation dreidimensionaler Koerperhaltungen des Menschen. Muenchen, Germany, Technische Universitaet Muenchen, Lehrstuhl fuer Ergonomie (LfE), Dissertation.

Chapter 11

Modeling and Simulation of Under-Knee Amputee Climbing Task

Fu Yan, Li Shiqi, Bian Yueqing

Industrial Engineering Department
Huazhong University of Science & Technology
Wuhan, 430073, P.R.C

ABSTRACT

In the non-obstacle construction for the handicapped life and work, benefits of the prosthesis wearers should never be neglected. Various disabled functions on the lower limb lead to a wide variety of prosthesis type and thus the prosthesis design should be a personalized design issue. A digital design and analysis platform would be very helpful to the personalized design of the prosthesis. It is the goal of this study to build up a biomechanical model of the prosthesis wearer mobility which can provide basic data model for the digital design. Based on the present on-site investigations, the research is mainly focused on modeling the climbing of the under-knee prosthesis wearers based on the experiment observation. The model would be validated by the simulation based on Poser and the results would be compared to the statistical data from the observed experiment, which validate the model itself. By transferring the model into the mathematical model combined with the finite element analysis, a digital design analysis platform for the limb prosthesis design suitable for more complicated motions can be established for the comfort design of prosthesis.

Keywords: Modeling, simulation, under-knee amputee, ladder climbing

INTRODUCTION

China Census of the Handicapped (2006) shows that 6.34% of the population is the handicapped, among which the limb handicapped has the highest percentage of 29.07%, excluding the multi-handicapped with limb disabled. More than 70% of those limb handicapped were after-born disabled due to accidents and pathological changes. As for the return-to-work possibility and life convenience, a great urgency on the rehabilitation and aiding facility design such as wheelchair and prosthesis is prosperous. There are several approaches to evaluate the design and mainly focused on 3 criteria: function, comfort and appearance. For the function analysis, the physical therapists still rely less quantitative tools and observation and provide large samples of statistical data for the designers to improve their design (Radcliffe, 1998). Digital data and tools developed in OpenGL for CAD/CAM have been introduced into the design and manufacturing of the prosthesis socket to follow the principles: accurate measurement of the stump geometry, perfect close fitting of the prosthesis to the stump, good response to forces and mechanical stress, safety, and that each single area of the socket must have a tight connection to the stump anatomy without affecting blood circulation. No matter how digitalized the process, the starting point of customer-fit design largely depends on the discomfort/comfort analysis of the patients.

On the other hand, with strong self-statue, the handicapped expect more mobility other than walking and standing with the assistance of the prosthesis. At present researches are mainly focused on the knee load and flexion as well as the ankle joint when people walk and stand (Spinelli, 2009; Fuchs 2005). In our 1-year on-site investigations in the workplace of the limb handicapped, 50 out of 104 limb prosthesis wearers would do more mobility than walking. It is the frequent occasion that the limb prosthesis is to climb in order to finish job tasks.

Ladder Climbing activities are performed as part of many occupational tasks, such as pollard worker, firemen. There are many researches on ladder climbing activities from observe analysis to quantitative analysis. Donald R. McIntyre (1983) studied the gait pattern of climbing of twenty two male subjects, found little evidence to suggest a preferred climbing gait and the two most commonly utilized methods of ascent for all trials were the lateral and four-beat lateral gaits. Hammer and Schmalz (1992) observed that three-point contact occurred 37 to 52 percent of the total climbing time on ladders tilted 60° and vertical ladders respectively. Marco J.M. Hoozemans and Michiel P. de Looze et al (2005) found no significant differences between 30 and 35 cm rung separation were observed for the energetic workload and results concerning the perceived exertion, discomfort, and safety indicate that 35 cm rung separation is preferred.

Dewar(1977) indicate that at the steeper ladder angle, the hands play a greater part in maintaining the balance of the body and there are greater differences between the movement patterns of tall and short subjects by thirty-five male subjects climbing a ladder set at two angles: 70.4 degree and 75.2 degree with the horizontal. Wen-Ruey Chang, Chien-Chi Chang et al (2004, 2005) indicated that the ladder inclined angle and the climbing speed were the most and second most critical factors, respectively, in friction requirement among the factors evaluated. The required friction coefficient increased by 77% on average when the ladder inclined angle was decreased from 75 degree to 65 degree. They also indicated that the contact at the top of the ladder and ladder type had minor effects on the required friction at the ladder base. Bloswick and Chaffin (1990) conducted the most comprehensive biomechanical study to date, reporting climbing forces for both feet and hands on several different vertical ladder slants and rung separations.

Nevertheless, few researches have been focused on the climbing activities of those prosthesis wearers who can never be neglected in the climbing worksites and life sites. It is quite meaningful to study the bio-mechanism of ladder-climbing for the limb prosthesis when we are to create a more friendly work and life environment for those handicapped.

METHODOLOGY

The simulation modeling is mainly based on the experiments. The independent and dependent variables included in the study are summarized on Fig. 1.

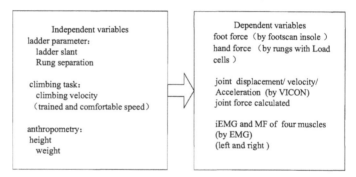

Fig.1. Independent and dependent variables used in the study. This paper do not analyze data gained by EMG and APAS.

The levels of the independent variables included in the main study were:
Ladder slant: 10deg , 20deg (with vertical)
Rung separation: 28cm, 32cm
Height and weight: 2 groups were divided according to anthropometric data (Table 1).
The selection of ladder slant was based on the previous study of Dewar (1977) and Bloswick, Chaffin (1990), and the choice of rung separation relied on national criterion of the people republic of china (GB/T 17889.1-1999) in which the ladder's functional sizes were decided.

Table1
Ladder climbing subject statistics (n = 10)

Subject#	Gender	Height (level)	Weight (level)
1	M	166cm (1)	51kg (1)
2	M	177cm (2)	72kg (2)
3	M	172cm (2)	72kg (2)
4	M	176cm (2)	57kg (1)
5	M	164cm (1)	61kg (1)
6	M	166cm (1)	59kg (1)
7	M	167cm (1)	56kg (1)
8	M	178cm (2)	62kg (1)
9	M	166cm (1)	70kg (2)
10	M	174cm (2)	95kg (2)

On a force/dynamic measurement platform (See Fig.2) developed for the experiment, the biomechanics data of the limbs and torsos of those subjects are collected.

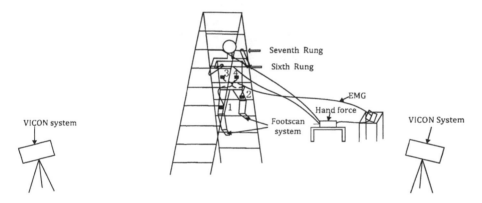

Fig.2. An instrumented ladder for measuring selected rung forces (sixth rung and seventh rung) was designed. The two rungs instrumented with strain gage load cells to measure hand forces. Also two sides of the ladder have different rung separation (28cm and 32cm) and ladder slant can be easily adjusted from vertical (0 deg slant) to more than 30 deg slant. Foot scan System instrumented with strain gage load cells in insole was used to measure foot resultant forces. In this paper, hand and foot forces were primary analysis objects. The experiment also has EMG system and VICON system. EMG system was used to measure four muscle information and VICON was used to record the kinematics in climbing process.

RESULTS

As Figure 3 shows, the hand average peak forces of the samples are average 29.5% body weight. By ANOVA, groups (height and weight subjects) (p<0.001, n=180) and ladder slant (10deg and 20deg) (p<0.001, n(normal)=93 and p<0.001, n(handicapped)=87) have a significant effect on the peak force observed, and rung separation (32cm and 28cm) doesn't have a significant effect on the peak hand force. Considering the standards for the rung separation, it can be proved that present standard ladders fit the handicapped well. But it is alerted that the placement of the ladder during the life and work is quite important factor for the safety and capability of the handicapped during the climbing tasks.

Height and weight also should be considered as important variables to hand force. Figure.4 shows the averages and standard deviations of the peak hand force based on two height levels. Figure.5 shows the averages and standard deviations of the peak hand force based on two weight levels.

Figure 6 and Figure 7 show the averages and standard deviations of the peak foot force based on two levels of height and weight. The height of the handicapped subjects ($p<0.001$, $n=88$) have significant effects on the peak foot force and the weight has significant effects on the peak foot force.

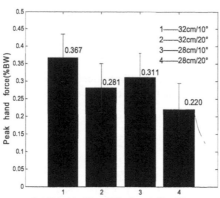

Fig.3. The peak hand force of 10 handicapped subjects(fake curs)

Fig.5. The peak hand force of two levels of weight

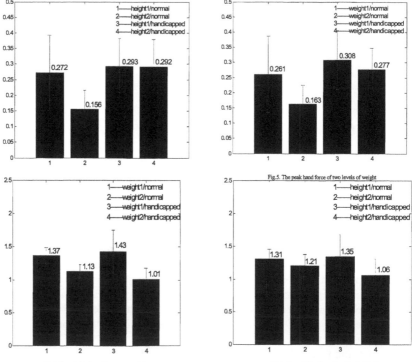

Fig.6. The peak foot force of two levels of weight

Fig.7. The peak foot force of two levels of height .

By statistical analysis, a co-relational model can be set up to explain the relationship between task paremeter and hand/foot force while the force parameter is quite important factor to explain the coupling relationship between human body and the ladder.

$$F_{foot} = F(W, H, SD) \tag{1}$$

$$F_{hand} = F(W, H, SD) \tag{2}$$

Where F stands for foot or hand force while climbing. W is weight, H is height and SD is slanted degree of the ladder.

More subjects are required to build up the definite weighting for the calculations of the hand and foot forces. Meanwhile the correlation result of this experiment provided a reference for further validation.

CLIMBING MODELING & SIMULATIONS

To validate the possibility of the above the relationship, the foot force is calculated as an example based on the following function:

$$F_i = W \times P_i \tag{3}$$

Where F_i is the foot force and W is the weight and P_i is the matrix of the moment of the ankle. The moment can be easily calculated based on VICON captured kinematic data. A digital man with under-knee prosthesis is built by Poser, which is very friendly to CAD category of the prosthesis design. One subject among the 10 people whose anthropometric data is among the average is chosen as the prototype of the digital man. Climbing motion data from VICON is applied to drive the digital man to climb. A graphic user interface is established to visualize the coupling model of hand/foot with the ladder and the coordination model of the hand and foot. The coupling visualized model is mainly based on the collision graphic calculation of hand and foot and deformation of muscle as well. The posture model, mainly originated from the coordination model is obtained directly by the interface between VICON and the graphic platform.

The difference between the modeling result and the experiment result for the foot force is 0.012% of Body Weight, which is acceptable in this experiment. But more subjects motion data can be simulated to find whether the same acceptable difference exist as well as the accumulated difference should be the important factor in the comparison.

Fig 8 Simulation of ladder-climbing scenario

DISCUSSIONS AND CONCLUSIONS

The ergonomic design of prosthesis and related rehabilitation treatment should consider different task scenarios highly related to the handicapped. Ladder climbing modeling can be a very good starting point for a more complicated motion simulation than gait modeling because it considers the coordination of hand and foot when the torso functions against the gravity. And the coordination model between hand and foot together with the coupling model of hand/foot with the ladder provides deeper insight into the biomechanics of the under-knee amputees. In the future, more samples of the handicapped are expected to participate the experiment and provide enough data for build up a quantitative model calculating the hand force and foot force while climbing. Deeper exploration can be done to validate the relationship between kinematic data and the weight in the domain of larger ladder-climbing samples. In all, It is quite meaningful to build up a digital model of biomechanics during the climbing present a more accurate and focused discomfort/comfort analysis for the under-knee prosthesis patients.

REFERENCE

M Spinelli, S Affatato, etc. *Bi-unicondylar Knee Prosthesis Functional Assessment Utilizing Force-control Wear Testing.* Proc. IMechE Vol. 224 Part H: J. Engineering in Medicine., 2009, 11

Charles W. Radcliffe. *Functional Considerations in the Fitting of Prostheses. Selected Articles from Artificial Limbs.* Robert E. Krieger Publishing Co. Inc, Huntington, N.Y., 2001.

Fuchs, S., Tibesku, C. O., Frisse, D., Genkinger, M., Laass, H., and Rosenbaum, D. *Clinical and functional comparison of uni- and bicondylar sledge prostheses.* Knee Surg. Sports Traumatol. Arthrosc., 2005, 13(3), 197–202.

Donald R. McIntyre, 1983. *Gait patterns during free choice ladder ascents.* Human Movement Science 2(3), 187-195

Hammer W., & Schmalz U, 1992. *Human behavior when climbing ladders with varying inclinations.* Safety Science15, 21-38.

Marco J.M. Hoozemans, Michiel P. de Loozeb, Idsart Kingma, Karen C.N. Reijneveld, Elsbeth M. de Korte, Maarten P. van der Grinten, Jaap H. van Dien, 2005. *Workload of window cleaners using ladders differing in rung separation.* Applied Ergonomics 36, 275-282

Dewar, M.E. 1977. Body movements in climbing a ladder. Ergonomics, 20 (1), 67-86.

Wen-Ruey Chang, Chien-Chi Chang, Simon Matz, Dal Ho Son, 2004. *Friction requirements for different climbing conditions in straight ladder ascending.* Safety Science 42, 791 – 805

Chien-Chi Chang, Wen-Ruey Chang, Simon Matz, 2005. *The eVects of straight ladder setup and usage on ground reaction forces and friction requirements during ascending and descending.* Safety Science 43, 469 – 483

Donald S. Bloswick, Don B. Chaffin, 1990. *An ergonomic analysis of the ladder climbing activity.* International Journal of Industrial Ergonomics 6, 17-27

National criterion of the people republic of china, GB/T 17889.1-1999, ICS 13.110

CHAPTER 12

Assessment of Dynamic Reaching Tasks Through a Novel Ergonomical Index

Ezio Preatoni, Marco Mazzola, Giuseppe Andreoni

Dipartimento di Industrial Design
Arti, Comunicazione e Moda (INDACO)
Politecnico di Milano, Milan, Italy

ABSTRACT

The aim of this study was to provide a descriptive analysis of a novel ergonomic index, i.e. the Method for Movement and Gesture Assessment (MMGA), which has been designed to exploit the potentialities of modern motion analysis technologies and to be applied to both static and dynamic movements. The 3D total-body kinematics of 15 male and 15 females subjects that performed reaching tasks was recorded by an optoelectronic system. Joint angles patterns and segmental mass distribution were used to estimate the comfort/discomfort scores, in a movement domain that was defined by 21 target points, on a 7-by-3 vertical grid placed at 2 different distances from the subject. The agreement between the MMGA and the corresponding LUBA indexes was analyzed and appeared consistent. Furthermore the MMGA may provide a more sensitive ranking over a wider range of dynamic motor strategies. Therefore, it may be integrated into motion simulation software and used for the quantitative ergonomic assessment of working environments/activities.

Keywords: ergonomic index, joint discomfort, movement and posture assessment

INTRODUCTION

The analysis of the man-product and man-environment interaction is a key ergonomic issue. Inappropriate movement and positions, and excessive articular loads may result in physical and psychological discomfort, may induce fatigue, and may cause the insurgence of musculo-skeletal disorders (David, 2005). Many authors have tried to score comfort/discomfort during working tasks (e.g. Karhu et al., 1977; McAtamney and Corlett, 1993; Kee and Karwowski, 2001). However, to our knowledge, there is still lack of literature concerning the definition of quantitative indexes that can exploit the potentialities of modern motion analysis technologies and rate movement comfort/discomfort during dynamic tasks.

Therefore, the aim of this work was to define and assess a novel ergonomic index that could overcome the aforementioned limits and that could be applied for quantifying movement comfort/discomfort in a wide range of man-product or man-environment interaction contexts.

METHODS

POPULATION

Thirty adults volunteered for participation in this study. The population consisted of 15 males and 15 females, whose age, height (H), weight (W) and forearm length (L_{fa}) were (mean and std): 22.0 (2.5) years; 181.5 (13.5) cm; 72.5 (13.0) kg; 48.0 (4.0) cm for the male group; 26.0 (14.5) years, 166.5 (6.0) cm, 56.0 (8.8) kg, 43.8 (2.4) cm for the female group. All the participants were right-handed and were free from any neuro-musculo-skeletal disorder.

Every subject was properly informed about experimental procedures, personal data treating and possibility to withdraw at anytime, and signed written informed consent before participation.

EXPERIMENTAL PROTOCOL

33 retro-reflective spherical markers were glued onto the subject's anatomical landmarks according to the *Vicon Plug-in Gait* protocol. Each subject was asked to align their lateral malleoli to a reference line, to stand comfortably in front of a vertical grid and to subsequently touch the 21 intersections formed by the 7 rows by 3 columns. Points to be touched (P1-P21) were named so that P1 was the upper-left corner and P21 the bottom-right one. Intra-row and intra-column distances were 30 cm (Figure 1). The reference line was set at 2 different distances from the target plane: (i) the length of the subject's forearm (A); (ii) L_{fa} + 40 cm (B). The idea behind this mock-up was to reproduce and explore the most common 3D movement domain that may be experimented by the man during actual interactions with daily

environments and appliances. The range of movements at the different levels of height and depth corresponded to different stages of difficulty in terms of ergonomics. Each subject repeated the reaching sequence three times, without being previously instructed about the motor strategy to be followed. For each pointing, they were asked to start from the reference orthostatic position (t_i), touch the target with the forefinger of the dominant hand, and to return to the same posture at the end of the task (t_f). Therefore, the analyzed movement was defined as the interval (Δt) between the subsequent standing positions, i.e. between t_i and t_f. The best trial in terms of quality of data (all markers always visible and correctly reconstructed in the 3D space) was selected for the analysis.

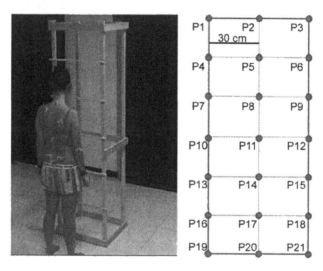

Figure 1 The experimental set-up (left), with the subject standing in front of the target grid. The 21 reaching point that formed a 7 rows by 3 column vertical grid (right)

INSTRUMENTATION

Total-body kinematics was recorded through a six-cameras (TVCs) optoelectronic system (Vicon M460, Vicon Motion System Ltd, Oxford Metrics, Oxford, UK) working at 120 Hz. TVCs were placed so that a volume of about 3 x 2 x 2 m was covered. Calibration procedures were carried out before each experimental session and a maximum mean error of 0.6 mm concerning markers placed on a rigid wand was obtained.

DATA PROCESSING

The subject's anthropometric measures, markers 3D coordinates and the *Vicon Plug-in gait* model were used for the estimation of joint angles patterns. The following variables were considered for this study: wrist, elbow, knee and ankle flex-extension; shoulder and hip flex-extension, intra-extra rotation and abd-adduction; trunk flex-extension, rotation and lateral bending. The LUBA (Kee and Karwowski, 2001) and MMGA (Andreoni et al., 2009) discomfort indexes were calculated for each subject, pointing target and reference line.

DEFINITION OF THE MMGA INDEX

The Method for Movement and Gesture Assessment (MMGA) index was defined and calculated as follows:

  i. angle-variables, $\alpha(t)$, were normalized to 100 points ($\alpha(p)$, p=1-100), independently from actual movement duration. $\alpha(1)$ corresponded to t_i and $\alpha(100)$ to t_f;

  ii. a coefficient of discomfort $\varphi_j(p)$ for each joint, j, at time-percentage, p, was computed through a spline fitting of the discomfort ranks proposed by Kee and Karwowski (2001) for varying joint motions;

  iii. the contribution of each joint was weighted (δ_j) on the mass of the j-th distal body district participating to the movement; the mass of body segments was estimated referring to the anthropometric tables proposed by Zatsiorsky and Seluyanov (1983).

Therefore:

$$MMGA = \sum_{p=1}^{100} \sum_{j=1}^{N} \varphi_j(\alpha(p)) \cdot \delta_j$$

where N is the number of functional joint movements considered.

STATISTICS

The normality of data samples was tested through the Lilliefors test (P=0.05).

The comparison between LUBA and MMGA results was carried out by applying a Bland-Altmann analysis (1995).

RESULTS AND DISCUSSION

The anthropometric characteristics of the analyzed sample appeared well distributed over the 1st-99th reference percentiles (Dreyfuss, 1993) and thus suitable for the purposes of this research. Subject's height spanned between 1.55 and 1.93 m, weight between 49 and 90 kg, forearm length between 39 and 55 cm. H, W and L_{fa} were all normally distributed (respectively, P=0.27, P=0.08, P=0.45). The age of participants ranged from 21 and 59 years, but its distribution was positively skewed, and concentrated in the 20-30 years interval. This choice came from the need of analyzing subjects who did have full articular mobility and did not present any neuro-musculo-skeletal disorder and/or limitation.

LUBA and MMGA scores were estimated for each subject (s1-s30), pointing task (P1-P21) and reference line, so that 30 x 21 x 2 conditions were collected for both the indexes. This allowed to map the man-pointing space interaction and to describe discomfort ranges depending on target, anthropometric characteristics, gender and motor strategies in accomplishing the task. LUBA results were distributed over the 1165-3089 range (1st-99th percentiles, in arbitrary units), while MMGA ones spanned between 159 and 1162 (1st-99th percentiles, in arbitrary units). The most comfortable reaching areas were the ones corresponding to the 2nd, 3rd and 4th row (P4-P12). In particular, the target points on the right side of that part of the grid were the ones with the lowest range of values. P9 showed the following measures (median, 25th-75th percentiles): 1301 (1252-1337) for the LUBA; 245 (207-281) for the MMGA. In contrast, the lowest rows (and points on the left in particular) evidenced the most increased discomfort indexes. P19 reported 2354 (2105-2749) for the LUBA and 690 (613-812) for the MMGA. These results were consistent with the subjects' motor strategies and right-handedness. In fact, subjects just needed to move their upper limb for the most comfortable points, while they had to pass through more demanding joint patterns and sometimes assume awkward postures for the least accessible ones. Very tall individuals experienced the highest discomfort in reaching the lowest targets in condition A. Short subjects had difficulties in reaching the upper-left points in condition B.

Iso-comfort state planes were drawn starting from discrete data points, and were used to visually evidence critical areas for pointing targets. These 2D+ maps (2D at 2 depth) are a visual representation of reaching ergonomics (Figure 2).

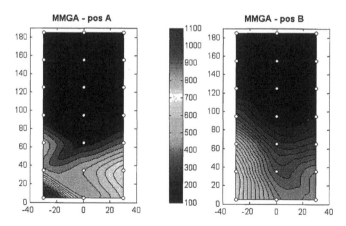

Figure 2 Example of iso-comfort state planes concerning one of the subject in condition A (left) and B (right). Different colors correspond to different levels of the MMGA index, where red areas represent increased values. The reported subject was one of the tallest within the analysed population. The difficulty in reaching the lowest targets (in particular point P19) from position A (L_{fa} distance from the vertical grid) may be observed

MMGA and LUBA measures presented an evident difference in magnitude, even though MMGA was derived from LUBA. This difference was the result of two main factors: the step-like definition of LUBA scores (Kee and Karwowski, 2001); and the multiplication of the coefficient of discomfort φ by the percentage ergonomic contribution δ within the MMGA index. Therefore, both LUBA and MMGA were normalized to their 1st-99th percentiles ranges in order to be compared. Results of the limits of agreement analysis is reported in Figure 3.

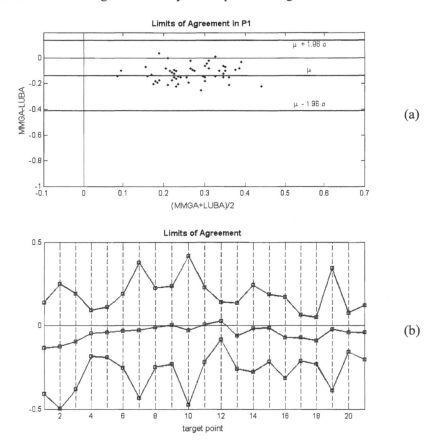

Figure 3 (a) An example of Bland-Altman plot of the normalized LUBA and MMGA scores. Scores of the whole population concerning target point P1 are reported (black dots) together with the corresponding Limits of Agreement (mean±1.96std band, in red). (b) Summary of the Limits of Agreement over the whole reaching domain. Targets P1-P21 are in abscissa, mean±1.96std in ordinate

LUBA and MMGA showed a rather good concordance, and appeared consistent in discriminating between comfort levels. However, they manifested different sensitivity to discomfort in correspondence of points P1-P3, P7, P10 and P19, where Limits-of-Agreement bands were larger. Those targets were among the ones that had been previously identified as critical, because they induced very variable motor solutions within the population. The MMGA scoring takes into account the contribution of both the upper and the lower body segments, and weights their intervention according to the mass involved. Hence it covers a wider spectrum of motor strategies and thus may provide a more complete and robust analysis of the observed movement.

CONCLUSION

A novel Method for Movement and Gesture Assessment (MMGA) was presented and compared to the LUBA index, from which it was derived. The MMGA can exploit the potentialities of the modern motion analysis systems and evaluate both static and dynamic tasks by assigning instantaneous discomfort scores to the recorded total-body movements. The index may be integrated into motion simulation software and used in the ergonomic assessment of working environments and activities in order to provide a first reference for the future definition of discomfort-warning scores.

REFERENCES

Andreoni, G., Mazzola, M., Ciani, O., Zambetti, M., Romero, M., Costa, F., Preatoni, E. (2009), "Method for Movemente and Gesture Assessment (MMGA) in ergonomics". *Proceedings of the 2nd International Conference HCI/ICDHM*, San Diego (USA), July 19-24, 2009, 591–598

Bland, J.M., Altman, D.J. (1995), "Comparing methods of measurement: why plotting difference against standard method is misleading". *The Lancet*, 346, 1085–1087.

David, G.C. (2005), "Correcting ergonomic methods for assessing exposure to risk factors for work-related musculoskeletal disorders". *Occupational Medicine*, 55, 190–199

Karhu, O., Kansi, P., Kuorinka, I. (1977), "Correcting working postures in industry: a practical method for analysis". *Applied Ergonomic*, 8(4), 199–201

Kee, D., Karwowski, W. (2001), "LUBA: an assessment technique for postural loading on the upper body based on joint motion discomfort and maximum holding time". *Applied Ergonomics*, 32(4), 357–366.

McAtamney, L., Corlett, E.N. (1993), "RULA: a survey method for the investigation of work-related upper limb disorders". *Applied Ergonomics*, 24 (2), 91–99

Dreyfuss, H. (1993), "The measure of man and Woman". New York: Whitney Library of Design

Zatsiorsky,V., Seluyanov, V. (1983), "The mass and inertia characteristics of the main segments of the human body". In: H. Matsui, K. Kobayashi, eds. *Biomechanics VIII-B*, International series of Biomechanics, 4B, 1152–1159, Champaign (IL): Human Kinetics

Chapter 13

Fast and Frugal Heuristics conquer the airport

Christoph Möhlenbrink[1], Eckehard Schnieder[2]

[1]Institute of flight guidance
German Aerospace Center (DLR)

[2]Institute for Traffic Safety and Automation Engineering
Braunschweig, Germany

ABSTRACT

This paper is concerned with a human-machine-model for the tower controller working position to investigate human decision making in dynamical environments. The authors assume that human decision making in the highly dynamical task environment of tower controllers is guided by heuristics. Tower controllers don't have the time to look for all information in their work environment before making a decision. They have to use heuristics to interact successfully with their task environment.

Colored petrinets are used (CPN-Tools) for modeling the human-machine-system. The three-step-principle of fast and frugal heuristics is implemented into the controller model. The strength of petrinets is their mathematical background that allows for calculating state spaces. Analyzing fast and frugal heuristics in dynamical environments, using state space analysis, is introduced here as a methodological approach.

This approach can not only be used to analyze heuristics implemented in formal models, but also to analyze empirical data of human decision making. The application of this approach for the analysis of heuristics used by air traffic controllers in the field is evident.

Keywords: Bounded rationality, air traffic controller, decision making, heuristics, microworld, petrinets,

INTRODUCTION

TOWER CONTROLLER WORKING POSITION

Control tasks at an airport are divided in different functional areas and executed by different working positions. The air traffic control service at the aerodrome is executed by tower- and apron controllers. The tower controller is responsible for controlling the runway and the airspace of the whole aerodrome. Tower controllers take over the control of arriving traffic from and hand over starting aircraft after take off to the approach control. At large airports, the taxiing from the runway to the gate and back to the runway is controlled by apron controllers. At small-sized airports there is one tower controller and one coordinator in the control tower.

This paper will focus on the tower controller (executive) working position. The executive is in contact with the pilots via radio and has to coordinate and control all movements on the aerodrome. A pilot is just allowed to land on the runway, or roll on a specific taxiway, or to be pushed from the gate, after the executive delivered the right clearance. When several pilots wait for clearances at the same time, the executive has to solve this multiple task situation. He has to decide in which order the clearances are given. The here presented work is motivated to better understand decision making of the executive, when confronted with multiple task situations.

FAST AND FRUGAL HEURISTICS

For research about human decision making, it has to be clear, which assumptions are made about the human as an information processing system. Here the assumption is made that humans do not aim for optimal solutions, but adapt to the environmental constraints. This assumption goes in line with the vision of bounded rationality propagated by Herbert Simon (Simon, 1982). Simon pointed out the limitation of the human mind, but also its adaptability to the task environment.

Gigerenzer introduces the term of fast and frugal heuristics to consider human decision making (Gigerenzer et al., 1999). In general, a heuristic can be understood as a "rule of thumb" that describes an easy way to come to a decision. Fast and frugal heuristics are defined by Gigerenzer as heuristics that "employ a minimum of time, knowledge and computation to make adaptive choices in real environments." (Gigerenzer et al., 1999, p.15) For the approach of fast and frugal heuristics a three-step-principle is defined, how humans come to their decisions. The three steps are characterized by:

(1) a search rule, that defines the order to look for information
(2) a stopping rule, that defines when to stop the search
(3) a decision rule, that directs how the information found determines the decision

Important for the understanding of fast and frugal heuristics is the terminus "cue". A cue can be defined as an attribute or characteristic, that includes the relevant information. Therefore it can be used to simplify the decision problem. For the executive controllers fast and frugal heuristics are of interest as they permanently have to make their decisions under time pressure, with limited knowledge and with the limited information processing capacity of the mind.

MODELING HUMAN-MACHINE-SYSTEMS

In engineering science it is common practice to use models and simulations to evaluate new technologies. There are established models to simulate pure technical systems.

More demanding than modelling technical systems is modelling of holistic human-machine-systems. For holistic human-machine-models see e.g. (Weingarten and Levis, 1989; Cacciabue, 1998; Corker, 2005). There are a lot of questions that arise, when human behaviour has to be integrated into such models. Issues relevant for modelling human interactive behaviour can be found in (Kirlik, 2007). For modelling of the controller in his task environment the following four points are of main interest:

- what aspects of human information processing have to be considered
- which information is relevant for the operators task
- which cues are present, so that the operator can successfully interact with the task environment
- what about timing issues of (1) the task environment, (2) human information processing and (3) the interaction of both

Especially for a safety critical domain, like the air traffic control domain, it is desirable to have holistic human-machine-models for the evaluation of new operational concepts.

HUMAN-MACHINE-MODEL REALIZED WITH COLORED PETRINETS

In this paper colored petrinets are used to model the executive controller and the airport processes on a small-sized airport. This modeling approach was suggested and initialized by Werther (Werther, 2004, 2006; Werther et al., 2007).

In the first part of the method section it will be shown, why CPN-Tools was used for modeling the whole human-machine-system of a executive controller working at a small-sized airport (Jensen, 1992). In the second part of the method section, a microworld approach is introduced to collect empirical data for validation of controller heuristics (Möhlenbrink et al., 2008). The implementation of fast and frugal heuristics within the petrinet model is explained in the third part.

However, the main advantage of petrinets for modeling controllers' behavior at the airport is the mathematical basis that allows for formal analysis of the whole system (Werther 2006). Hence the method section will end with a discussion of the difficulties to generate the full state space for complex systems. A solution for an untimed, causal petrinet model is depicted that is able to analyze the decision heuristics for the complex human-machine-system. In the result section the generation of the state space will be demonstrated, for two heuristics. Furthermore it will be discussed, that the state space analysis is able to reveal the decision space for each single human participant of the microworld study.

METHOD

MODELING TOOL

For modeling the executive controller and the processes on the airport CPN-Tools is used. There are four positive properties (color, hierarchy, time and openness) that supported the decision for CPN as a petrinet tool in this context. First of all, it is a tool using colored petrinet, what enables to model realistic, complex systems in a compact manner. Second, the tool allows for a hierarchical representation of the human-machine-model. Modeling all system states on one layer would result in a confusing representation of the whole model. Especially for the process description of the infrastructure of an airport, it is helpful to divide the whole net into subnets. The third point is the timing issue. Within CPN-Tools it is possible to add timing information. For the understanding of controllers' heuristics, considering timing is important, as it is a crucial factor for the interactive behavior between controller and the processes on the airport. Openness, as the last positive property discussed here, allows for the communication of CPN-Tools with other software tools via TCP/IP protocols. As it will be outlined in the next section, the petrinet model will be used to drive a microworld simulation as an experimental platform to collect data by human-in-the-loop studies.

MICROWORLD STUDY

The microworld simulation represents the top view of a small-sized airport with one runway. The layout is depicted in figure 1. On the bottom of figure 1, there is an arrival list (left side) and a departure list (right side). They include the flightstrips with information about the time, the callsigns and the requests of aircraft. In the middle part there are six interaction buttons that can be activated by mouse clicks. The task for the participants is to control the airport and to give clearances to the aircraft. There are the (1) land, (2) taxi-in, (3) taxi-in-apron, (4) pushback, (5) taxi-out and (6) take off clearance. The participants have to select the aircraft in the lists by a mouse click and have to deliver the right clearance afterwards via the interaction buttons. The traffic scenario is driven by the CPN-Tools petrinet model. The microworld was designed to investigate how human participants solve the multiple task situations. The assumption is, that participants show different behavior guided by different heuristics they use to control the processes on the airport.

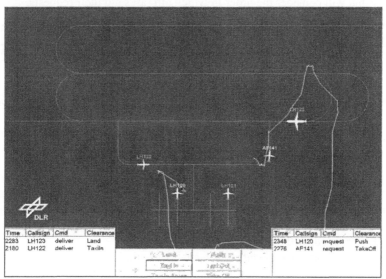

Figure 1: Microworld FAirControl to study decision making in dynamic multiple task situations.

IMPLEMENTATION OF FAST AND FRUGAL HEURISTICS

As suggested by Cacciabue, the human-machine-model for a controller at a small-sized airport consists of three submodels (Cacciabue, 1998). In the petrinet model a controller model, an interaction model and an airport process model are realized. For research about fast and frugal heuristics, the three-step-principle, (1) the information search rule, (2) the stop rule and (3) the decision rule have to be implemented into the controller model. Before defining the heuristic it has to be decided what is considered as cue, the controller is looking for. Within the controller model it is defined that each aircraft waiting for a clearance is considered as cue.

So far the implementation of the heuristic is fulfilled for (3) the decision rule. In the future the scan path of human participants, recorded with an eye tracker, can be used to evaluate (1) the information search and (2) the stop rule used. For the decision rule two different heuristics are given as an example. The first fast and frugal heuristic (heuristic 1) delivers clearances in a strictly hierarchical manner.

Heuristic 1:   TaxiIn > Land > TakeOff > TaxiOut > TaxiInApron > Pushback

For the heuristic it is assumed, that the controllers will prioritize the taxi-in clearance to all other clearances. Second the landing clearance is prioritized to all other clearances, all but the taxi-in clearance.

Another heuristic (heuristic 2) is called "first come, first serve". In this case the cue is more demanding. The controllers are not just looking for the aircraft waiting for a clearance, but will consider the order in which aircraft send their requests.

Heuristic 2: [TaxiIn, TakeOff, Pushback, Land, TaxiIn, Pushback]

The heuristic is implemented as a list of requests. The first request in the list will be the next clearance. New requests will be added to the end of the list. In the example the controller will deliver the taxi-in clearance followed by the takeoff clearance and the pushback clearance (etc.).

STATE SPACE ANALYSIS

The strength of petrinets as a modeling tool is the ability to calculate state spaces containing all reachable system states and their interconnections. For a predefined scenario, the state space is able to ensure with 100% certainty, that no other system state than calculated, will occur in any simulation.

The main disadvantage of state space analysis is the fact that they easily become very complex, if all system states of a human-machine-system have to be analyzed (Hasselberg et al., 2009). This problem is known as the state space explosion problem (Jensen 2009). The state space in the case of the timed colored petrinet of the controller interacting with an airport becomes very complex due to:

- parameters (callsign, stands, etc,)
- timing issues
- non-determinism of parallel processes (airport, controller)

For the analysis of fast and frugal heuristics in dynamical environments it is of interest to identify, how human participants deal with the multiple task situations. This consideration is motivated by the simple idea, that applying different heuristics in the same dynamical environment will result in different decision situations at a later time point. Based on the insight that the state space generation of the whole human-machine model is too complex, preconditions were derived that will allow generating the decision space for the behavior of one participant.

First, it is suggested to eliminate parameters that increase the number of system states, but are not crucial for the decision itself. This action helps to avoid the state space explosion problem. One example is the callsign for each aircraft. If a participant has to decide to deliver an aircraft a landing clearance or a takeoff clearance, the callsign is not necessary to fulfill this task. However, if the callsign is represented in the petrinet, each decision situation with one aircraft asking for a landing clearance and one asking for a takeoff clearance will be represented by a different node in the state space.

Second, time has to be eliminated from the petrinet. Generating the state space for a timed petrinet will always result in a state space with dead markings. The same decision situation at different time points will result in different nodes in the state space. The state-space will explode and the larger the traffic scenario, the larger the state space.

As the last point it must be understood, how to handle the nondeterminism to restrict the state space to only those decision situations caused by a heuristic, or the behavior of a human participant of the microworld study. The solution for the nondeterminism is the most demanding precondition for generating the decision space. For all decision situations the next decision situation must be already known. Hence the decision space for a human participant can only be generated afterwards. Therefore all possible system states that are generated by a heuristic, or human participant, have to be initialized as input scenario. Figure 2 depicts the petrinet model that is used to generate the decision space caused by a heuristic.

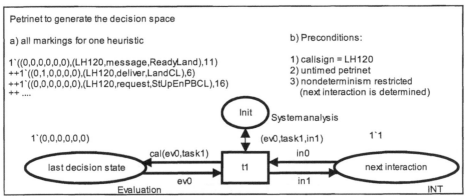

FIGURE 2 Petrinet model to generate the decision space of the controller

For this petrinet the decision space will be generated for a traffic scenario of five aircraft for heuristic one, heuristic two and for both heuristics together.

RESULTS

STATE SPACE

The state spaces are represented in figure 3.1-3.3. It can be seen that there are different decision situations in figure 3.1 and 3.2. Even though heuristic one generates 24 different decision situations (equates to the number of nodes in the graph) quite similar to heuristic two with 23 the decision spaces are very different. There are decision situations that come up when using heuristic one, but do not come up when using heuristic two. The black nodes and arcs in figure 3.1 represent situations and decisions that occur only for heuristic one. Accordingly dark grey nodes and arcs in figure 3.2 are just apparent for heuristic two. The decision space

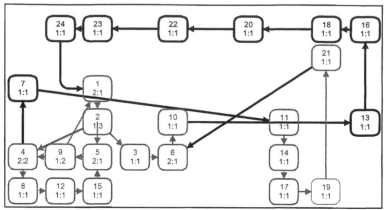

FIGURE 3.1 State space for heuristic 1 (CPN-tools)

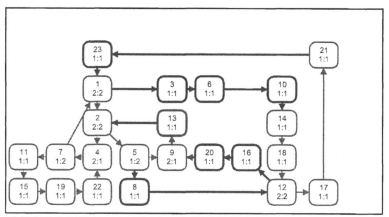

FIGURE 3.2 State space for heuristic 2 (CPN-tools)

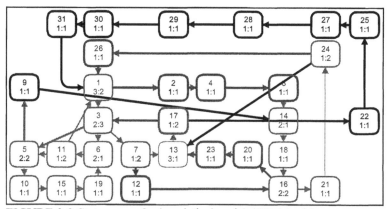

FIGURE 3.3 State space for heuristic 1 and 2 (CPN-tools)

for all possible situations, regardless if heuristic one or two was used, is finally depicted in figure 3.3. It becomes clear, that each heuristic spans a subset of a decision space considering all possible decision situations.

DISCUSSION

In the result section it was made explicit for a traffic szenario of 5 aircraft, how the advantage of petrinets to calculate the state space can be applied to span the decision space for the working position of a tower controller. Two heuristics were simulated within this simplified example to make the methodological approach explicit. However the petrinet model (figure 2) is not only able to analyze data of simulated heuristics. The formal approach can be appropriated to calculate the decision space for human participants e.g. of the introduced microworld study. It is possible to calculate the decision space for all participant of the study, or for a single human participant. In the latter case the state space will include all decision situations the participant was dealing with. For each single decision situation the state space demonstrates which decisions were made by the participant. If a participant is using an efficient heuristic, the decision space will be less complex, compared to a participant with an inefficient strategy.

The aim of the formal state space approach is the identification of fast and frugal heuristics in dynamical environments. It offers the possibility to depict interindividual differences in behavior based on the decision space. Although the petrinet suggested for spanning the decision space is an untimed, causal petrinet, it is important to note, that the heuristics were applied on a timed human-machine-model with a certain airport infrastructure (compare microworld).

To the knowledge of the authors, this is the first paper using state spaces to analyze heuristic behavior in dynamical environments. It was pointed out by Simon that for research about human decision making the adaptability of the human mind to the task environment is important to consider (Simon, 1982). This approach nicely respects both constraints, the limitation of the human mind and the constraints of the task environment.

CONCLUSION

Starting with the motivation to learn about human decision making behavior in dynamical environments, it was argued to use human-machine-models, so that the restrictions of the task environment are adequately incorporated when modeling the decision making of human operators. On the basis of the implementation of two heuristics it was shown how state space analysis can be used to research about fast and frugal heuristics in dynamical environments.

It became evident, how formal state space analyses can be applied to empirical data to better understand human decision making in interactive dynamical task environments. The adaptability of the method for tower control research is given.

REFERENCES

Cacciabue, P.C. (1998). Modelling and simulation of human behaviour in system control. (1 Ed). London: Springer.

Corker, K.M., Smith, B.R. (1993). "An Architecture and Model for Cognitive Engineering Simulation Analysis: Application to Advanced Aviation Automation", paper presented at the AIAA Conference on Computing in Aerospace, San Diego, CA

Gigerenzer, G., Todd, P.M. and the ABC Research Group (1999). Simple Heuristics that make us smart: Oxford University Press.

Hasselberg, A., Oberheid, H., Söffker, D. (2009). „State-Space-Based Analysis of Human Decision Making in Air Traffic Control", paper presented at the 7th Workshop on Advanced Control and Diagnosis, Zielona Giora, Poland.

Jensen, K. (1992). Coloured Petri Nets: Basic Concepts, Analysis Methods and Practical Use. Berlin, Springer Verlag.

Jensen, K., Kristensen, L.M. (2009). Coloured Petri Nets Modelling and Validation of Concurrent Systems. Berlin: Springer.

Kirlik, A. (2007). "Ecological Resources for Modeling Interactive Behavior and Embedded Cognition", in: Integrated Models of Cognitive Systems, Gray Wayne (Ed.), pp. 194-210, New York: Oxford University Press.

Möhlenbrink, C., Oberheid H., Werther, B. (2008). "A Model Based Approach to Cognitive Work Analysis and Work Process Design in Air Traffic Control" in: Human Factors for assistance and automation. pp. 401-412.

Simon, H. A. (1982). Models of bounded rationality, Cambridge, MA: MIT Press.

Werther, B. (2004). "Modellbasierte Bewertung menschlicher Informations-verarbeitung mit höheren Petrinetzen", paper presented at Entwerfen und Gestalten. 5. Berliner Werkstatt Mensch-Maschine-Systeme.

Werther, B. (2006). "Colored Petri net based modeling of airport control processes", paper presented at Computational Intelligence for Modelling, Control and Automation, Sydney, Australia.

Werther, B., Möhlenbrink, C., Rudolph, M. (2007). Colored Petri net based Formal Airport Control Model for simulation and analysis of airport control processes. HCI, Beijing, China, Springer.

CHAPTER 14

Capability Test for a Digital Cognitive Flight Crew Model

Andreas Lüdtke[1], Jan-Patrick Osterloh[1], Tina Mioch[2], Joris Janssen[2]

[1] OFFIS e.V., Escherweg 2
26127 Oldenburg, Germany

[2] TNO Human Factors, Kampweg 5
3769 DE Soesterberg
The Netherlands

ABSTRACT

The objective of the HUMAN project is to develop virtual test pilots for cockpit systems, in order to improve the human error analysis during cockpit system design. Virtual test pilots should supplement simulator-based testing of new cockpit systems with human pilots, by providing the possibility to simulate human behavior in early design phases. In this paper we will present how we modeled and tested four relevant basic capabilities, necessary for the simulation of pilot behavior, namely crew coordination, sophisticated perception, reactive behavior and multitasking.

Keywords: Cognitive flight crew model, cognitive architecture, rule-based modeling

INTRODUCTION

In the project HUMAN (funded by the European Commission under the 7th Framework Programme) a digital cognitive pilot crew model is developed which will be used as a virtual tester of complex cockpit system designs in realistic flight scenarios. The paper describes our modeling approach, and shows that the model is able to simulate some relevant basic capabilities to cope with highly dynamic flight scenarios. To demonstrate the capabilities our model has been connected to a generic cockpit simulator within a simulated air traffic environment, called simulation platform throughout the paper. We simulated flight scenarios in which the model has to interact with an Advanced Human Machine Interface (AHMI) in order to manipulate the Advanced Flight Management System (AFMS; both developed by the German Aerospace Center, Braunschweig, Germany) and further cockpit systems (e.g. the autopilot) to navigate the aircraft according to flight procedures. The basic capabilities investigated in this paper are: (1) Crew Coordination, to simulate a flight crew (sharing of task execution and communication); (2) Perception, to simulate sophisticated perception in terms of a restricted visual field with a visual focus including attention according to current visual focus and realistic times for eye-movements; (3) Reactive Behavior, to react to events (e.g. ATC calls, warning lights) by initiating adequate flight procedures (e.g. an altitude change maneuver); (4) Multitasking, in terms of maintaining multiple goals at the same time and switching between them, as well as interrupting a goal to deal with reactive goals.

These four capabilities are achieved with the cognitive architecture CASCaS (Cognitive Architecture for Safety Critical Task Simulation), whose key concept is the theory of behavior levels (cf. Anderson (2000) or Rasmussen (1983)). A flight crew is modeled by using two instances of the model: a pilot flying (PF) model and a pilot monitoring (PM) model. To test the model capabilities we devised and simulated two flight scenarios and defined expected plausible behavior. The resulting model behavior has then been tested against these expectations.

The following section presents more details on the four basic capabilities and describes the modeling approaches within the cognitive architecture CASCaS. Afterward the process for testing the capabilities and the test results are described.

MODELLING THE BASIC CAPABILITIES WITHIN A COGNITIVE ARCHITECTURE

The first basic capability we modeled is to simulate a flight crew. The crew is a very important safety net and consists of two pilots, a pilot flying (PF) and a pilot non-flying (PNF). While the PF has control of the aircraft, the PNF monitors the actions of the PF and the aircraft, and has other duties like communication with the air traffic controller (ATC). Main duty of the PNF is monitoring, thus this pilot is therefore referred to as the pilot monitoring (PM) in this paper. Monitoring the PFs actions includes sharing of task execution, e.g. procedures that foresee cross-checks at certain points. This implies communication between PF and PM.

The second important basic capability concerns perception. This capability is needed to simulate realistic visual attention allocation with regard to what a pilot can perceive at a certain point in time of the simulation. It is based on visual fixations and transitions between these fixations (saccades). Therefore it was necessary to create a restricted visual field with a visual focus and modeling realistic times for eye-movements.

The third basic capability of our model is to react to events. These events can be for example ATC calls (clearances for altitude) and warning lights. When perceiving an event, the model should initiate the associated flight procedure (e.g. an altitude change procedure).

The fourth and final basic capability which has been modeled is to simulate multitasking. The model is capable of maintaining multiple goals at the same time, switching between them, as well as interrupting a goal to deal with reactive goals.

The basic capabilities have been modeled within the cognitive architecture CASCaS, which was developed by OFFIS in the EU project ISAAC (see Lüdtke et al. (2006) and Lüdtke et al. (2009) for further information on CASCaS). The architecture is based on a flight procedure formalization in the form of "if-then" rules. The rules formally describe a mental representation of flight procedures. In order to perform a simulation, the flight procedure rules are uploaded to the cognitive architecture. A cognitive architecture with uploaded procedure rules is what we call a pilot model. The cognitive architecture can be understood as an interpreter or executor of formal flight procedure rules. Within a simulation platform the pilot model interacts with a system under investigation (e.g. the AHMI/AFMS) and a simulated environment (including the aircraft). A simulation kernel synchronizes the different models and organizes the dataflow.

CASCaS is based on Rasmussen's (1983) three behavior levels in which cognitive processing takes place: skill-based, rule-based and knowledge-based behavior. The levels of processing differ with regard to their demands on attention control dependent on prior experience: skill-based behavior is acting without thinking in daily operations, rule-based behavior is selecting stored plans in familiar situations, and knowledge-based behavior is coming up with new plans in unfamiliar situations. Anderson (2000) distinguishes very similar levels, but uses the terminology of autonomous, associative, and cognitive level, which will be used throughout the paper. In HUMAN CASCaS has been adapted to integrate the basic capabilities. The modeling effort focuses on the associative and cognitive behavior level. Figure 1 gives an overview on the components of CASCaS. These components form the following control loop: The "Perception" component retrieves the current situation from the "Simulation Environment", and stores the information in the "Memory" component. The "Processing" component contains components for the behavior layers. These layers can retrieve information from the memory and process this information according to their cognitive cycle (rule-based or knowledge-based). The layers may store new information in the memory, or start motor actions in the "Motor" component.

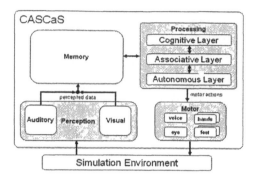

Figure 1. CASCaS Architecture

CREW COORDINATION

In order to simulate a crew, two instances of CASCaS are running in parallel. Each instance has its own procedure set, one for the PF and one for the PM. In addition, CASCaS has been extended with two new components, a vocal motor and an auditory percept component. The rule-based language has been extended with vocal actions. A vocal action consists of a text that is spoken, a variable instantiation which represents the meaning of the text and a receiver. See Figure 2 for an example of a rule with a vocal action. The time for the vocal action is determined by the text "confirm target altitude", estimated with 190 words/min. The vocal action is transferred to its receiver, in this example the modeled Pilot Monitoring (PM), via the simulation platform. The receiver perceives the text and stores the instantiated variable ("cross_check_altitude_request = 1") in the memory component. These memory entries can then trigger reactive rules, see the section on "Reactive Behavior" below.

`Rule 103005, type=regular`	rule header
`Goal (al_ap_change_altitude_cc)`	active goal
`Condition (target_altitude == fcu_altitude)`	condition
→	
`Vocal ('confirm target altitude',`	spoken text
`        cross_check_altitude_request=1,`	instantiated variable → memory entry for receiver
`        PM)`	receiver
`Goal (al_ap_change_altitude_cc_wait)`	subgoal

Figure 2. Rule activating a vocal action

The vocal action allows the two instances of CASCaS to interact with each other verbally.

PERCEPTION

In order to model eye-movements and a visual field, information on the topology of the environment is needed. For example, information is needed about the locations of instruments, their size and colors. We defined an XML schema that allows us to describe the instruments in terms of areas of interest (AOI), together with several parameters, like location and size. This is a new input for the percept component of CASCaS. Based on this information, eye- and head movement times can be calculated. We use a 2 degree angle of focus, and a visual field angle of 170 degree horizontal and 110 degree vertical. The visual field is used in order to model selective attention (also called bottom-up attention). Selective attention refers to the phenomenon of automatic shifts of attention due to external events, like a flashing signal, or a moving object. A more detailed description on the percept component can be found in (Osterloh & Lüdtke 2008).

REACTIVE BEHAVIOR

There are two types of reactivity: The first type is a non-cognitive, almost automatic reaction to an event. In this case, no reasoning needs to be done, since there is a standard procedure to handle the event. A reactive rule on the associative layer is activated. Reactive rules have the same structure and semantic like regular rules (cf. Figure 2), with the exception that they are not associated to a goal (on the left-hand side of the rule). As soon as the condition of a reactive rule evaluates to true, the rule is fired, independently from the current active goals.

The second type of reactivity is the reaction to events that involve the cognitive layer, e.g. if a decision needs to be made whether a standard procedure is appropriate at a particular moment. In that case, a goal to handle the event becomes active on the cognitive layer, which reasons about appropriate reaction. The cognitive layer may decide to activate a goal on the associative layer if standard procedures are appropriate, or decide to handle the event itself.

MULTITASKING

Multitasking is modeled in three ways in CASCaS. First, on the associative layer, several goals can be interleaved. The mechanisms for interleaving are based on the general executive for multitasking of Salvucci (2005). Furthermore, the reactive capabilities of the associative layer, which are described in the previous section, allow interrupting an ongoing goal by adding a new goal onto the agenda that is processed immediately.

Second, goals on the cognitive layer are executed based on their priority. It is possible that a currently executed goal is stopped if another goal has a higher priority. However, the first goal is not automatically resumed after finishing the second goal; it is first evaluated whether the first goal is still relevant to the situation by checking its activation conditions again.

A third way of multitasking is realized because the associative and cognitive layer run in parallel and interact in the following ways: (1) The cognitive layer can start (and thus delegate), monitor, temporarily halt, resume and stop activities on the associative layer by manipulating the associative layer's goal agenda. Monitoring of the associative layer is realized through determining whether the appropriate goals are placed in the goal agenda. (2) The associative layer can inform the cognitive layer about the status of rule execution. For example, the current execution is stuck because no rules are available in long-term memory for the chosen goal or execution of a perceived event cannot be started for the very same reason. In these cases the cognitive layer starts to perform the goal or event. (3) The cognitive layer can take over control at any time. Currently this is initiated by setting the parameter "Consciousness". If the value is "associative" then every event will first be processed if possible and the cognitive layer becomes only active if no rules are available. If the value is "cognitive" then the cognitive layer processes each event independent of the availability of rules (more details can be found in Lüdtke et al. 2009).

CAPABILITY TEST

We devised and simulated two main scenarios in order to test the basic capabilities. For each scenario expected plausible behavior has been defined, and resulting model behavior has then been tested against these expectations. The plausibility of the model behavior has been analyzed based on time series data recorded during the virtual simulation. The results show that the model is able to perform basic behavior in flight scenarios in a way as it is expected based on human pilot behavior.

SCENARIO DESIGN

In order to prove the basic capabilities, two scenarios have been designed, in which the basic capabilities are needed to fly the aircraft safely. The basis for all scenarios is the same flight plan on flight level 90 (9000 feet) at 200 knots. The flight is controlled via a scenario controller, which induces events, like ATC requests, based on the current position of the aircraft.

SCENARIO 1

The first scenario tests three capabilities: crew coordination, multitasking and reactive behavior. The aircraft is flying in cruise mode on flight level 90. The basic duty of the PF and the PM is (under normal conditions) to monitor the flight. In the scenario, an ATC call requests the pilots to change the altitude. The crew has to *react to this event* (reactive behavior), has to *interrupt the monitoring task* (multitasking), and then it has to follow the procedure for an altitude change via autopilot, which includes *cross-checks* (crew coordination). When the altitude is changed, both pilots have to *resume their monitoring task* (multitasking). In order to change the altitude, the PF has to dial the new flight level into the autopilot using the altitude selector. After the PF has selected the new altitude the PM should compare the flight level cleared by the ATC with the flight level dialed in by the PF. The PM and PF models should behave in the following way:

1. If the value dialed by the PF is correct, the PM should acknowledge verbally and the PF should answer the acknowledgement verbally and activate the new target altitude by pulling the altitude selector.
2. If the value dialed by the PF in incorrect, PM should reject which should trigger the PF to dial in the correct target altitude. The PM should check again and acknowledge verbally. PF should answer the acknowledgement verbally and activate the new target altitude.

In the scenario the second case can be forced by the scenario controller by overwriting the value of the altitude selector. Figure 3 shows the behavior for the first alternative.

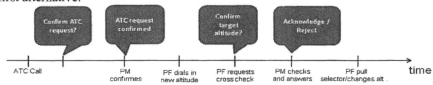

Figure 3. Expected behavior for correct crew coordination

SCENARIO 2

The second scenario is dedicated to test the basic capability "Perception". This mainly concerns the test of the visual component, particularly the focus and the visual field. The scenario has been divided into three steps:

1) The PF model has to perform procedures that involve reading autopilot flight mode annunciations on the Primary Flight Display (PFD) and values from the engine display. In order to read the values the model has to simulate eye-movements.
2) While the visual focus is on the engine display, a flight mode is changed by the scenario controller. This is displayed on the PFD with a flashing box for 10 seconds around the flight mode annunciation. The flashing box should attract the visual attention of the PF model, because it appears in the visual field. If the PF model detects the mode change, a vocal output should be generated.
3) Finally, the PF should shift the visual focus to the track ball. While the PF focuses the trackball, the scenario controller changes the flight mode. This time the PFD and thus the mode change is outside the visual field, because the trackball is located near the knees of the pilot. In this case the mode change should not be detected.

RESULTS OF THE CAPABILITY TEST

Scenario 1

As described above, scenario 1 tests the basic capabilities crew coordination, multitasking, and reactive behavior.

Crew coordination: As can be seen in Figure 4, for the cross-check, the two cognitive models communicate with each other, the light gray bar representing the PF, the dark gray one representing the PM. It can also be seen that the communication depends on the status of the environment. The PF's actions depend on the vocal input he receives from the PM. In Figure 4, the dialed-in altitude is accepted by the PM, and thus executed by the PF, whereas in Figure 5, the dialed-in altitude is not accepted, and therefore the PF needs to correct the altitude.

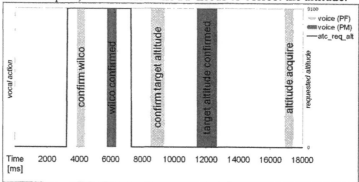

Figure 4. Diagram depicting verbal protocols from PF and PM, taken from logfile. The flight level dialed by the PF corresponds to the flight level requested by ATC.

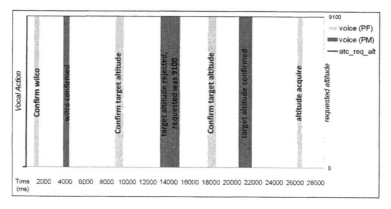

Figure 5. Diagram depicting verbal protocols from PF and PM, taken from logfile. The flight level dialed by the PF does not correspond to the flight level requested by ATC.

Multitasking: As can be seen in Figure 6, the monitoring task is interrupted when the ATC call occurs (at time 82300). After handling the ATC call (around time 112300), the monitoring task is resumed at the same point as it was interrupted before.

Reactive behavior: When receiving the ATC call, the monitoring task is interrupted and it is reacted to the ATC call by choosing the appropriate procedure to handle the ATC call (i.e. change altitude).

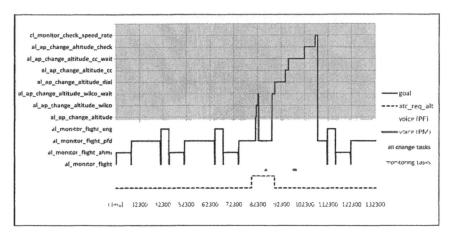

Figure 6. Diagram depicting switching between monitoring tasks and altitude change task

SCENARIO 2

Scenario 2 tests the perception, particularly the focus and the visual field. As can be seen in Figure 7, the percept component is able to focus on different parts of the interface.. The first mode change is recognized by the pilot, as it is in the visual field. The second mode change, even though the same event occurs (flashing box on screen), is not perceived by the PF, as his eyes are on the mouse, which means that the flight mode annunciation is outside the visual field. As can be seen, no verbal action is taken by the pilot.

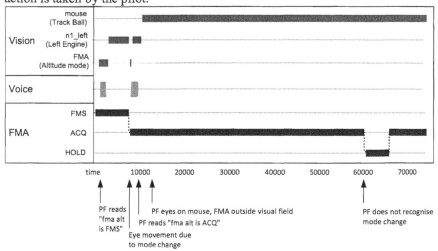

Figure 7. Diagram that displays the location of the eye gaze, and recognition of events in the visual field.

CONCLUSION AND NEXT STEPS

In this paper, we presented how we modeled basic capabilities, relevant for simulation of pilot behavior. These basic capabilities where tested within two scenarios against expected plausible behavior. The next validation step will be to compare the model data with human pilot data.

It has to be pointed out that our model abstracts from several features of human behavior. Especially, we have not modeled the influence of performance shaping factors like fatigue or emotions. For example it is well known that under fatigue, pilot performance degrades significantly. Our model represents pilots that are impacted neither by fatigue nor emotions, thus generated predictions have to be interpreted taking this into account.

The work described in this paper is funded by the European Commission in the 7th Framework Programme, Transportation under the number FP7 – 211988.

REFERENCES

Anderson, J. R. (2000), Learning and Memory, John Wiley & Sons, Inc.

Anderson, J. R., Bothell, D., Byrne, M. D., Douglass, S., Lebiere, C. & Qin, Y. (2004). An Integrated Theory of the Mind. Carnegie Mellon University, Pittsburgh and Rice University, Houston.

Lüdtke, A., Cavallo, A., Christophe, L., Cifaldi, M., Fabbri, M., Javaux, D.: Human Error Analysis based on a Cognitive Architecture. In: Reuzeau, F., Corker, K., Boy, G. (eds) Proceedings of HCI-Aero, pp. 40-47, Cépaduès-Editions, France (2006)

Lüdtke, A., Osterloh, J.-P., Mioch, T., Rister, F., Looije, R. Cognitive Modelling of Pilot Errors and Error Recovery in Flight Management Tasks. Proceedings of the 7th Int. Working Conference on Human Error, Safety, and System Development, Brussels, Belgium, Sept. 23-25, 2009

Rasmussen, J. (1983), "Skills, Rules, Knowledge: Signals, Signs and Symbols and other Distinctions in Human Performance Models" IEEE Transactions: Systems, Man and Cybernetics, SMC-13, pp.257-267

Salvucci, D. D. (2005). A multitasking general executive for compound continuous tasks. Cognitive Science 29, S. 457-492.

CHAPTER 15

Development and Evaluation of Task Based Digital Human Modeling for Inclusive Design

Russell Marshall[a], Steve Summerskill[a], Keith Case[b], Diane Gyi[c] and Ruth Sims[a]

Departments of: [a]Design & Technology
[b]Mechanical and Manufacturing Engineering, [c]Ergonomics
Loughborough University
Loughborough, LE11 3TU, UK

ABSTRACT

HADRIAN is a digital human modeling (DHM) system that is currently under development as part of an EPSRC funded project in the UK looking at accessible transport. The system is a partner tool to the long established SAMMIE DHM system and aims to address issues with the lack of applicability of DHM tools to inclusive or universal design problems. HADRIAN includes a database of 102 manikins based directly upon data taken from real people, many of whom are older or with disabilities and who span a broad range of anthropometry, age, and joint mobility. This database is combined with a task analysis tool that provides an automated means to investigate the accessibility of a workstation or environment. This paper discusses the issues and subsequent refinement of the tool that resulted from validation using an ATM design case study. In addition the results from a second validation are presented. This second study examines the accessibility of a Docklands Light Railway station in London. The results highlight that whilst physical simulations can be made with a generally good degree of accuracy there are still many opportunities to be explored in the cognitive and emotional areas that can be used to inform designers of accessibility issues during virtual assessments.

Keywords: Human modeling, task analysis, inclusive design, ergonomics

INTRODUCTION

Digital Human Modeling systems (DHM) have been in development for more than 40 years (Porter et al., 2009) and during that time have seen increasing use for the assessment of ergonomics issues within design. As design has become ever more CAD driven DHM tools have been exploited earlier in the design process. In addition, as DHM systems themselves have increased in sophistication and ease of use they have been used to evaluate a greater range of issues, and potentially, with a greater level of confidence. However, much of the capability of DHM systems are data driven and if the data are unapplicable or unavailable then the accuracy of the simulation can be severely flawed. Even if the simulation is accurate any concerns with the data are likely to lead to a lack of confidence in the findings. Whilst there are numerous anthropometric databases available for DHM system use (e.g. Adultdata, 1998; CAESAR, 2010 etc.), there are many limitations including the age of the data, the lack of data on all national populations, and whether all the relevant measures are available. Potentially, the two most fundamental issues are: the lack of data on older people and people with disabilities; and data that reflect real world task capability and not scientifically standardized measures that, whilst consistent, are generally unapplicable to design problems.

Summerskill et al. (2009) reported on the initial validation of a DHM based tool called HADRIAN with an ATM based case study. The HADRIAN system (Marshall et al., 2004) has been developed to address concerns with the availability and applicability of data particularly for those who are older or who have disabilities. This paper discusses the subsequent modifications made to the tool in light of the findings from the initial validation work and details a second validation aimed at addressing a broader scope of capability and assessing the effectiveness of the modifications.

THE HADRIAN SYSTEM

The Design Ergonomics Group (DERG) at Loughborough University in the UK recognized the need for a new approach that addressed the need to improve the data available to designers, ergonomists and engineers on older and disabled people together with a means to employ these data for design evaluations. Research funded by the EPSRCs Extending Quality Life programme (EQUAL) and later the Sustainable Urban Environment programme (SUE) resulted in HADRIAN.

The HADRIAN system consists of two core elements, the first is a database of physical and behavioural data on over 100 individuals, aged 18 to 89 years. Body size and shape is captured through a comprehensive list of anthropometric measures spanning from 1st to 99th %ile with at least one person in each decile. In addition, the database provides joint range of motion data and covers over 20 recognised impairments including: cerebral palsy, epilepsy, multiple sclerosis, limb loss, arthritis and Parkinson's disease, with many individuals having more than one impairment. The full scope of the technical data associated with each individual includes: anthropometric body measurements, joint angle ranges of motion, reach range in the form of a reach envelope or volume, grip strength, manual dexterity and a range of more general information on age, occupation, nationality, work history, a range of activities of daily living and transport use. These data are also accompanied by photographs and video clips of the individuals.

The data contained within HADRIAN are conveyed to designers by means of numerical data outputs, personal information, through videos of tasks being performed and through video clips of participants describing their experiences. In this way the individuals in the HADRIAN database are initially presented much like personas (Cooper, 1999). This extension and personalization of ergonomics data is a deliberate attempt to engender a greater degree of empathy and understanding within designers and other practitioners.

One particular issue that HADRIAN intended to address by this methodology was the use of anthropometric percentiles for establishing design limits and multivariate accommodation. Designing for 5th to 95th percentile has become a de-facto standard and yet these univariate percentile manikins are poor representations of real people and convey none of the actual impact of designing out the bottom or top 5% of the population for any single measure. By personalising these data it was hoped that establishing design limits becomes a process of understanding individual needs and striving to accommodate as many people as possible rather than just making an arbitrary decision based upon a percentile.

AUTOMATED TASK ANALYSIS

The 102 people within the HADRIAN database can be used to generate 102 digital human manikins with their appropriate size, shape, joint range of motion, and task based capability. Working directly with the long established SAMMIE DHM system (Porter et al., 2004) these manikins can be employed in an automated virtual fitting trial. The HADRIAN system provides the ability to define a task using a relatively simple element breakdown requiring the need for the system user to specify an activity (reach, view, move etc.), a target (coin slot, on button, door handle etc.), and a number of optional parameters (hand or foot reach, grip type, acceptable viewing distances etc.) (Marshall et al. 2002). Through the use of a CAD model of the design to be evaluated, the system can automatically create manikins and sequentially get them to perform the task as described. The analysis results in a percentage excluded giving a representation of the number of people who would be unable to perform the task with the given design. In addition to highlighting who failed the task, the system also indicates why failure occurred and allows the designer to adjust their design, or to try alternative concepts with the same task, thus examining the inclusiveness of any given design.

VALIDATION OF HADRIAN'S TASK ANALYSIS

To investigate the accuracy and representativeness of HADRIAN's ability to replicate a task analysis it was considered critical to perform a series of validation studies. The studies were designed to compare the use of human modelling in the form of SAMMIE and HADRIAN firstly, to one another, and secondly to a benchmark established by performing a user trial using a real product and real people.

THE FIRST VALIDATION STUDY

The first validation study focused on assessing the inclusiveness of a commonly used and widely available product in the form of an ATM provided by a project collaborator NCR. The study consisted of three assessments to provide a full picture of HADRIAN's capability to represent real people and its ability to highlight the same issues that would be found by an experienced ergonomist using a more conventional DHM system. The three assessments included:

1. A validation assessment of a CAD model of the ATM in SAMMIE by an experienced ergonomist and DHM system user. 10 manikins created from data in the HADRIAN database were used for the assessment. The positioning of the participants in relation to the ATM was considered to be a crucial aspect of the analysis and the experience of the ergonomist was heavily relied upon to position the human models appropriately.
2. A validation assessment of a CAD model of the ATM using HADRIAN by a designer with a small amount of training in the HADRIAN system. The same 10 subjects from assessment 1 were reused.
3. A benchmark user trial with a physical model of the ATM conducted by an experienced ergonomist. NCR provided an ATM fascia that was then mounted on a rig that allowed the ATM to be adjusted in height through a range of 250mm. The height of the highest interaction point (statement printer output) was adjustable from 1200mm to 1450mm in line with international variability in mounting height from the floor. The 10 real people from which the HADRIAN and SAMMIE manikins had been created were used for the user trial.

In each of the three assessments the tasks used for the analysis were the same and resulted in 160 tasks being performed in each assessment: ten participants performing eight tasks at two ATM mounting heights.

The sample of participants was carefully selected to cover a number of issues that posed a challenge for both forms of virtual analysis and to provide a basis for potentially interesting results. The sample included people with a range of age, size and ability. The 10 people used in the study included an ambulant disabled female with cerebral palsy who uses a wheeled walking frame, an ambulant disabled male who uses crutches due to leg injuries sustained during a car crash, a crutch user with balance and coordination issues, a powered wheelchair user with limited strength in the right arm due to a stroke, a powered wheelchair user with mobility issues due to a broken back. two wheelchair users with good upper body mobility, and a mobility scooter user with balance and coordination issues. Two non-disabled members of the HADRIAN sample were included as a control. These were a UK male with 99th%ile stature, and a UK female with 1st%ile stature.

The Results of the First Validation

Comparing assessments 1 and 3, (the SAMMIE assessment and the user trial) showed a very close match in task successes and failures, and the postures adopted in the various stages of the tasks (Summerskill et al., 2009). These results demonstrated that given the appropriate experience and accurate and applicable data, a user of a DHM system can perform a valid assessment of a design in a virtual environment that can be confidently used to predict the issues likely to be encountered by real people.

The results comparing assessments 2 and 3, (the HADRIAN assessment and the user trial) showed that a much more conservative approach was taken by the HADRIAN system. As a result there were more incorrect predictions of task failures by HADRIAN when the subject in the user trail had successfully performed the task. Fundamentally, HADRIAN also made a number of incorrect predictions based upon wheelchair orientation to a task interaction point. Wheelchair orientation in the three assessments were categorised to be one of: facing, oblique and lateral (see Figure 1). HADRIAN correctly predicted the facing and the lateral orientations but failed to use the oblique orientation. Whilst this error resulted in incorrect posture prediction it did not always result in incorrect task success/failure prediction. As HADRIAN was designed to assess tasks in a representative manner incorrect posture prediction was a more serious issue than the total predictions of task successes and failures. It is likely that under different circumstances incorrect posture prediction of this kind would lead to much more substantial errors in task prediction.

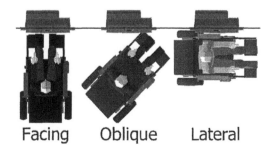

Facing   Oblique   Lateral

FIGURE 1 The classification of wheelchair user orientation in relation the ATM.

FURTHER DEVELOPMENTS

The results of the first validation study highlighted four main areas in which the HADRIAN assessment protocol could be improved (Summerskill et al., 2009). These recommendations identify that HADRIAN could make improved use of existing data available on each of the individuals in the database. In addition, they also suggest a change of approach to the assessment strategy coded into the HADRIAN system for wheelchair users.

The failure of HADRIAN to correctly use the oblique wheelchair orientation results from the 'adaptation processes' that HADRIAN employs to manage task failures (Marshall et al., 2009). In this case HADRIAN employed a number of strategies to complete the task, starting from a default facing orientation. After attempting a large number of variations in posture for the reach task HADRIAN would still be faced with a task failure. HADRIAN was then moving immediately to the perceived optimal positioning of the manikin that essentially located the shoulder belonging to the dominant arm to be used for the reach task as close to the target reach point as possible. In most cases this resulted in a lateral orientation being used to minimise the reach distance.

From this observation an amendment to the HADRIAN assessment protocol was proposed. The core of the issue was that whilst the protocol rigorously evaluated the potential solutions to a task failure for a particular location and orientation of a manikin it did not apply the same level of rigour to evaluating solutions that might result from a different location and orientation. Thus the proposal resulted in the following strategy being implemented:

1. If the reference to target distance is greater than those accommodated by a posture change one or more of the following will be applied:
   - If possible the human model is turned to an oblique orientation with respect to the target.
   - If possible the human model is moved closer to the target.
   - If possible the human model is oriented with respect to the target to minimize the shoulder target distance.

THE SECOND VALIDATION STUDY

The second validation study focused on pushing the assessment capability of the HADRIAN system beyond a single discrete interaction area. The study aimed to assess the inclusiveness of a Docklands Light Railway (DLR) station at Greenwich in London. Greenwich DLR station is a modern station with ramps as well as stairs, lifts, and level access. However, it also has a significant underpass as the track is raised well above ground level. For this study there were two assessments performed. A key aim of the assessments was to evaluate the representativeness of the HADRIAN data in being able to identify issues highlighted by new participants who do not feature in the database. The assessments included:

1. A validation assessment using HADRIAN of three CAD models of areas of interaction within the station namely: the ticket machine, the lift and the train. All 102 manikins created from the data in the database were used in the assessment.

2. A benchmark user trial at the station in Greenwich with 9 real people who are not part of the sample of 102 people in the HADRIAN database. The participants were tasked with making a journey through the station including the purchase of a ticket from the ticket machine at the station, through navigation to the appropriate platform, and finally getting on the appropriate train.

In both of the assessments the tasks used for the analysis were the same. The ticket machine required a total of nine reach and vision tasks, the navigation to the platform required between two and six reach and vision tasks depending on if a subject decided to use the lift or not. There were also four movement tasks involved in navigating the station that were all on level ground unless a subject chose to take either the stairs or ramp to the platform. This resulted in a maximum of 19 tasks being performed by each participant. Table 1 shows the full range of tasks.

Table 1 Second validation task breakdown.

Major task element	Vision task sub-elements	Reach task sub-elements	Movement task sub-elements
Buy ticket from ticket machine	Look at screen Look at control dial	Reach to dial Twist dial to select Push centre of dial Reach to coin slot Reach to note slot Reach to retrieve change tray Reach to retrieve ticket tray	Move to ticket machine
Navigate to platform	Look at direction signs Look at lift call button Look at platform selection button Look at platform direction signs	Reach to lift call button Reach to platform select button	Move to lift / or climb stairs / or ascend ramp Move along platform
Board train			Move onto train

The sample of participants was again carefully selected to cover a number of issues that posed a challenge for both forms of virtual analysis and to provide a basis for potentially interesting results. Thus the sample included: a male powered wheelchair user with cerebral palsy, a female powered wheelchair user with familial spastic paraparesis, a female manual wheelchair user with spina bifida, two ambulant disabled females who both use a walking stick, an ambulant disabled female who uses a wheeled frame and three non disabled people. These included a UK male with 99th%ile stature, and an Indonesian female with 1st%ile stature. The third person without a disability was a 96th%ile stature UK female who performed the task with a pushchair and two small children.

The Results of the Second Validation

Assessment 2 (real people) resulted in 10 task failures, all of which were associated with the ticket machine. The most common failure (3 out of 10 participants: subjects 1,2 and 6) was associated with the coin slot (to insert coins for payment) as it was the highest interaction point. By contrast the note slot (to insert notes for payment) was in a much more accessible location but was still a problem for two subjects (subjects 1 and 6) due to the complexity of the process of having to manipulate a flap within the slot and carefully feed the notes in for the machine to accept them. A similar issue arose for one participant (subject 1) with the ticket and change retrieval tray. The security flap installed over the tray that had to be pushed inwards to retrieve the contents actually prevented access altogether.

Assessment 1 (HADRIAN) resulted in a total of 68 failures across all 102 participants. HADRIAN successfully identified the same task failures highlighted by assessment 1 but also identified failures in reach to the lift controls.

DISCUSSION

After assessing the various physical elements of the journey in HADRIAN with the people in the database it was clear that most of the physical interaction issues with the ticket machine were accurately predicted. Whilst there were few failures across the 9 real people HADRIAN predicted many more. However, a feature of this trial was that the participants were not matched. One of the aims of the trial was to see if the 102 people in the HADRIAN database were suitably representative to identify the likely issues experienced by any subset of real people. Thus, whilst the results are not comparable on a like for like basis it is possible to see that the same issues experienced by the real people were indeed flagged up by at least one of the individuals in HADRIAN. Clearly this requires more from the user of the HADRIAN system to understand, evaluate, and then make some decision about the results. However, it also indicates that the HADRIAN database may be sufficiently representative to capture the majority of concerns that would need to be looked at to improve accessibility and inclusiveness.

In addition to the overall task successes and failures, HADRIAN's ability to predict appropriate postures was evaluated in light of the changes made in response to the first validation. From the results it was clear that wheelchair users were now postured much more realistically, due to the introduction of the oblique orientation in many more cases (see Figure 2). However, HADRIAN still achieved a number of successes in the facing orientation that is used by very few wheelchair users in reality. This does suggest that it may be more representative to always start in an oblique orientation to the reach target for a wheelchair user.

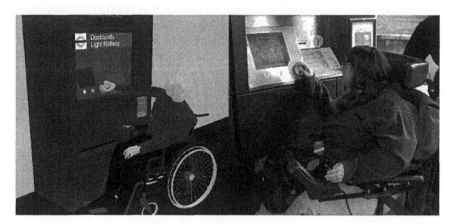

FIGURE 2   A predicted posture in HADRIAN and a comparable real subject's posture in attempting to use the DLR ticket machine.

In addition to the physical barriers, Greenwich also poses psychological challenges due to the nature of the environment. Whilst the study with real people was conducted during the day and good lighting is present during the evening it is still not the most inviting environment. This perception was noted by the participants who did say that they may be put off making a journey via this station due to issues to do with their perception of personal security. Whilst HADRIAN does contain information about each person in the database concerning their thoughts about environments and their likelihood to avoid areas with a lot of graffiti or that are poorly lit, this data is as yet not built into the task analysis tool. Thus it was not possible to assess if HADRIAN was capable of identifying the same issues raised by the real people automatically. However, as the data is included in the database, it is possible for designers and other practitioners to merely browse the database and be alerted to the possibility that the nature of this particular environment might be an issue. The automation of this feature is the focus of future work.

CONCLUSIONS

The HADRIAN automated inclusive design DHM system has undergone a two stage validation process designed to verify and improve the accuracy and representativeness of the system's ability to analyse products. The results from the first validation highlighted a need to use the data on the individuals in the database more thoroughly and to more realistically orient wheelchair users to an interaction target. A series of changes were implemented to the HADRIAN system and a second validation study performed. This study addressed the need for HADRIAN to be able to identify issues that may be found with a design by people who are not explicitly represented in the database. It also investigated the assessment of an environment with multiple interaction areas, as opposed to one discrete interaction point. The results of the second validation confirmed that HADRIAN was able to accurately predict issues of accessibility for a new set of test subjects. In addition, posture prediction was improved for wheelchair users. However, the validation also showed that further modifications to the behavioural coding of HADRIAN may need to be fine tuned to avoid postures that are very rarely adopted (even if possible). The validation also confirmed the need for further research into the area of cognitive and emotional behavioural predictions to compliment the physical analysis.

REFERENCES

Adultdata. (1998). The handbook of adult anthropometry and strength measurements – data for design safety. L. Peebles and B. Norris, eds. Department of Trade and Industry.

CAESAR. (2010). Civilian American and European Surface Anthropometry Resource. http://store.sae.org/caesar/ [Accessed 28/02/2010].

Cooper, A. (1999). The Inmates Are Running the Asylum: why hi-tech products drive us crazy and how to restore the sanity, Macmillan Publishing Co., Inc. Indianapolis, IN, USA.

Marshall, R., Case, K., Summerskill, S.J., Sims, R.E., Gyi, D.E., and Davis, P.M. (2009). Virtual Task Simulation for Inclusive Design. Lecture Notes in Computer Science: Proceedings of the Second International Conference, ICDHM 2009, Held as Part of HCI International 2009, San Diego, CA, USA, July 19-24. 700-709. Springer: Berlin.

Marshall, R., Case, K., Porter, J.M., Sims, R.E. and Gyi, D.E. (2004). Using HADRIAN for Eliciting Virtual User Feedback in 'Design for All', Journal of Engineering Manufacture; Proceedings of the Institution of Mechanical Engineers, Part B, 218(9), 1st September 2004, 1203-1210.

Marshall, R., Case, K., Oliver, R.E., Gyi, D.E. and Porter, J.M. (2002). A task based 'design for all' support tool, Robotics and Computer-Integrated Manufacturing, 18(3-4), 1st June 2002, 297-303.

Porter, S,. Case. K., Summerskill, S., and Marshall, R. (2009). Four decades of SAMMIE. In Ergonomics at 60: a celebration. 12-13. The Ergonomics Society: Loughborough, UK.

Porter, J.M., Case, K., Marshall, R., Gyi, D.E. and Sims, R.E. (2004). Beyond Jack and Jill: designing for individuals using HADRIAN, International Journal of Industrial Ergonomics, 333: 249-264.

Summerskill, S.J., Marshall, R., Case, K., Gyi, D.E., Sims, R.E. and Davis, P. (2009). Validation of the HADRIAN System using an ATM evaluation case study. Lecture Notes in Computer Science: Proceedings of the Second International Conference, ICDHM 2009, Held as Part of HCI International 2009, San Diego, CA, USA, July 19-24. 727-736. Springer: Berlin.

CHAPTER 16

Anthropometrics and Ergonomics Assessment in the IMMA Manikin

Erik Svensson[1,2], Erik Bertilsson[1,2], Dan Högberg[1], Lars Hanson[2,3]

[1] School of Technology and Society
University of Skövde
Skövde, Sweden

[2] Department of Product and Production Development
Chalmers University of Technology
Göteborg, Sweden

[3] Industrial Development
Scania CV, Södertälje, Sweden

ABSTRACT

Digital Human Modeling (DHM) tools are useful for simulating human work and proactively evaluating ergonomic conditions. IMMA (Intelligently Moving Manikin) is a project that aims to develop software that combines digital human modeling and path planning. The work in the IMMA project is divided into a number of work packages that gradually increases the complexity of the problem. This poster paper regards both the functionality for ergonomics assessment and consideration of anthropometric diversity in the DHM tool being developed. Reviews of current DHM tools and interviews with DHM users and ergonomics specialists were done to clarify problems, needs and opportunities when working with anthropometrics and ergonomics evaluations. Interviews showed that simulations and following evaluations are almost solely based on static postures and with few human models. The main reason for this is claimed to be complex and time consuming processes when creating and evaluating simulations. Both the review of current DHM tools and the interviews confirmed that there is an evident need for more time-dependant evaluation methods and a better coverage of the intended users' diversity. Attained knowledge from the analysis of current DHM tools and interviews are used to create work processes and two specific modules intended to be implemented in the new IMMA DHM tool. Key issues for the modules are ease of use and flexibility. The overall objective with the IMMA DHM system is to offer a tool that support faster and more correct ergonomics analyses.

Keywords: Digital Human Modeling, Ergonomics, Evaluation, Simulation, Work Process, Anthropometry

INTRODUCTION

In product and production development it is often necessary to study how a product, workplace or task will affect a potential user, both related to physical and cognitive ergonomics. Human-machine interaction has traditionally been evaluated relatively late in the development phase (Porter et al., 1993), and this has often been done by physical mock-ups which have been expensive and time demanding (Helander, 1999). Ergonomics evaluation of productions systems focuses on identifying and eliminating risks for musculoskeletal disorders (MSDs) among the workers. Software tools for simulation of the human work and ergonomics evaluation, known as Digital Human Modeling (Chaffin, 2001) have been introduced to the industry to support a proactive and efficient consideration of ergonomics in the design process. In the DHM tool it is possible to manipulate the manikin manually (manipulating individual segments in each degree of freedom), use predefined postures, import motion capture data or use motions generators to generate manikin postures or movements. IMMA (Intelligently Moving Manikin) is a project that aims to develop a software tool that combines digital human modeling and path planning. The purpose is to consider anthropometric and other relevant diversity, minimize biomechanical loads, increase assembly quality and to reduce simulation and analysis time.

The objective of the study performed is to evaluate some of the current DHM simulation systems and clarify problems, opportunities and solutions when working with human diversity and ergonomics assessment in DHM systems. This information then leads to a suggestion and overall description of two state of the art DHM modules and work processes for considering anthropometric diversity and ergonomics assessment.

METHODS

To review current DMH tools and their capabilities within ergonomics assessment and anthropometric diversity, an analysis of Catia V5 Human, UGS Jack and Pro/Engineer Manikin were conducted.

Additionally, five qualitative interviews were conducted with actors within the Swedish automotive industry in order to get an insight into how they work with anthropometric diversity and ergonomics assessment. A total of eleven persons were interviewed. Their work positions varied from simulation engineers to ergonomics specialists and people with more managerial positions within virtual manufacturing. The interviews were carried out as focused (semi-structured) interviews, where the interrogator asked predominately open questions regarding the focus area. The interviews were audio recorded with accompanying notes taken.

Attained knowledge from the analysis of current DHM tools and the interviews were combined with informal brainstorming sessions (together with colleagues within the IMMA project) to create a general plan and flowcharts for two new modules regarding the anthropometrics and ergonomics assessment in the IMMA manikin software.

RESULTS

According to the interviews, a DHM tool needs to be fast and easy to use. Using a simulation system should lead to better quality with the same work effort and the results need to be trustworthy. An advantage of using DHM tools is the possibility to solve problems in an early phase and being able to make better decisions based on the simulations. Today, working with DHM systems is frequently combined with qualified guessing of the final result due to manual positioning of the manikin and estimation of a person's balance. DHM systems are seen as expert tools which few people within the originations are qualified to use. The interviewed DHM users believe that it is too hard and expensive to customize their DHM tools to fit their specific needs and requirements. Also, they feel that their requests on upgrades and functionality development do not get hearing at the large software companies.

Simulations are often done with static postures and only one or two manikins. The results for the rest of the population are produced with guesswork based on simulation results and self-knowledge. The reason for these simplified solutions is the time-consuming processes when working with several manikins. The study among DHM users gave that the ergonomics evaluation methods integrated in the DHM tools are rarely utilized. Instead, they often have customized their DHM tools and integrated their own specific ergonomics standards and evaluation methods. Also, the study gave that simulations and following evaluations are almost solely based on static postures aimed to represent "worst case" scenarios. This is partly driven by the lack of evaluation tools for dynamic motions and the time required creating simulations representing full work cycles.

DISCUSSION

Current DHM tools are very advanced, but the interviews gave that many of the features are not used in industry because of lack of time and that several of the methods are not suited for the user's specific activity. Furthermore, there are several user requirements that are not met by today's DHM systems. For example, a

142

consequence of the subjective evaluations and sometimes guesswork make it difficult to get evaluation results to produce credibility when communicating with product developers, designers and managers.

Results from the study were the basis for brainstorming sessions and conceptual generation of two new DHM modules dealing with anthropometry and ergonomics assessment. This is very much a work in progress and the resulting work methods and modules will most likely evolve and transform during the development process. Both the review of current DHM tools and the interviews confirmed that there is an evident need for more time-dependant evaluation methods and better possibilities to cover all intended users of a population.

Key issues for the modules are ease of use and flexibility. The overall objective with the IMMA DHM system is to offer a tool that supports faster and more reliable ergonomics analyses. It should work as a fast and easy to use screening tool, but also provide more in depth analysis. An important factor for the success of the IMMA manikin is the close collaboration with the Swedish automotive industry. Working closely together will produce a tool that fulfills the needs of today's users and their call for high level of customization.

A concept for a module considering anthropometry is illustrated in Figure 1. Depending on the situation it is possible to perform a number of different operations. One part of the module is where the user can add own anthropometric data which can be processed and saved for later use in simulation. It will also be possible to choose data from within the system from which one or several manikins can be created. Depending on the simulated situation, different methods for creating manikin families can be used. The choices and data are processed in the anthropometric module and the result can either be used within the IMMA DHM tools or inserted into another DHM tool.

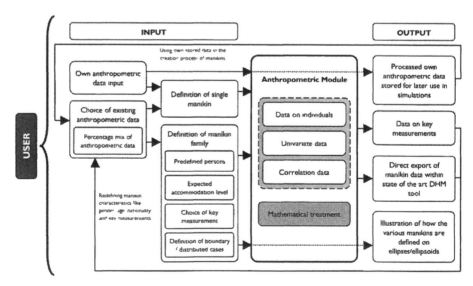

FIGURE 1 Flowchart depicting the anthropometric module and work process.

Figure 2 shows a proposed module for ergonomics assessment. One key objective is that the module will work together with the IMMA motion generation functionality. This means that the integrated ergonomics evaluation methods will support the path planning functionality to minimize ergonomic loads for predicted motions. After the motions are determined, the module will enable evaluation of full work cycles by considering time related factors and aggregation of ergonomic loads. The aim is to move away from pure static posture analyses and instead evaluate motions patterns over time, i.e. assess dynamic work tasks rather than snapshots of human work. This approach would make it possible to discern motions patterns and show which parts of the movements that generate high ergonomic loads and the total accumulated load/strain. The module will facilitate customization for different users' requirements and ergonomics standards in an easy and intuitive way.

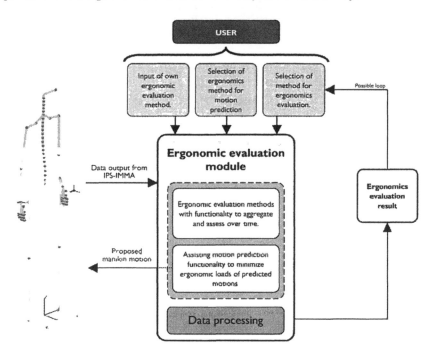

FIGURE 2 Flowchart depicting the ergonomics assessment module and work process.

Further work will require a deeper examination of existing knowledge combined with analysis of anthropometric measurements, correlation data, biomechanics and current evaluation methods. Focus will be on covering the target population with several manikins and including the time factor to find possible and biomechanically acceptable ways for the simulated human interacting with a workplace or product.

ACKNOWLEDGEMENTS

This work has been made possible with the support from the Swedish Foundation for Strategic Research/ProViking and by the participating organisations. This support is gratefully acknowledged.

REFERENCES

Chaffin, D.B. (2001) Digital Human Modelling for Vehicle and Workplace Design. SAE International, Warrendale, PA.

Helander, M.G. (1999) Seven common reasons to not implement ergonomics. International Journal of Industrial Ergonomics, Vol. 25, No. 1, pp. 97-101.

Porter, J.M., Case, K., Freer, M.T., Bonney, M.C. (1993) Computer aided ergonomics design of automobiles. In: Peacock, B., Karwowski, W. (Eds.), Automotive Ergonomics. Taylor & Francis, London, pp. 43-77.

CHAPTER 17

Dynamic Foot Scanning. Prospects and Limitations of Using Synchronized 3D Scanners to Capture Complete Human Foot Shape While Walking

Timo Schmeltzpfenning[a], Clemens Plank[a],
Inga Krauss[a], Petra Aswendt[b], Stefan Grau[a]

[a]Medical Clinic; Department of Sport Medicine
University of Tuebingen
Germany

[b]ViALUX GmbH
Germany

ABSTRACT

Investigation of dynamic foot shape was performed using a multiple scanner assembly (ViALUX, Germany) based on fringe projection technology. Five synchronized camera – projector systems were used to capture the entire foot during walking (4.5 km/h) of 144 subjects. With this measurement technique, the foot could be captured at a 3D – frame rate of 21 to 49 fps under natural walking conditions. Significant and practical relevant changes during the stance phase of gait were determined whereas differences between static and dynamic values were not significant.

Keywords: Foot morphology, dynamic foot measurement, foot deformation.

INTRODUCTION

BACKGROUND OF THE STUDY

The human foot shape has been the focus of ergonomic research for the last decades. Understanding individual variations of foot structure and function is essential knowledge which last designers need in order to provide a perfect fit of shoes and insoles. Footwear fit is not only important for comfort but particularly for health reasons (Kleindienst et al., 2006). Previous studies postulate that inadequately fitting shoes influence the development of foot deformities like hammer toe and bunions (Frey 2000; Janisse 1992; Kleindienst et al. 2006). However, it is well known that shoe shape does not necessarily match foot shape (Channa 2004) and that a foot changes shape depending on the load imposed on it (Tsung et al. 2003; Xiong et al. 2009). Previous studies investigated foot deformation with regard to different weight bearing conditions. They found that the foot becomes longer, wider and reduced in height with increased weight imposed on it (Tsung et al., 2003; Xiong et al., 2009). Even though these static results help shoe designers understand foot deformations, we need to capture and analyze foot shape in a dynamic situation to get more precise and realistic data on foot changes during the gait cycle. Therefore, the basic consideration of this study was to examine whether the exact measurement of the human foot under natural walking conditions could give additional information for optimizing the fit of shoes and insoles.

CURRENT MEASUREMENT TECHNIQUES - POSSIBILITIES &LIMITATIONS

There has been a tremendous technical development in foot measurement devices (Branock, ink foot print, 2D- and 3D- laser scanner) leading to a major improvement in the objective analyses of human foot shape. However, standard measurement devices are still limited to static conditions. In the past, investigators mainly used plantar pressure plates, three-dimensional motion capture systems and two-dimensional video-based analysis systems to characterize the changes in foot shape, e.g. arch height, during walking and running. (Williams and McClay, 2000). However, all three techniques have major limitations with regard to detecting the shape of an object. Using plantar pressure analysis systems, the measurable information is limited to the plantar surface of the foot and cannot be used to determine reliable vertical height information, for instance of arch structure (McPoil and Cornwall, 2006)). Motion analysis systems are strongly limited to the number and density of markers (Kimura et al. 2009). Two dimensional video based systems can only measure data points in one projection plane and are therefore affected by perspective distortions (Grau et al. 2000). Generally, these systems are only useful for single point measurements and not suitable for capturing the surface of an object.

To capture the 3D - information of a moving object with an inconsistent shape like the human foot, there are currently two approaches conceivable: the stereo matching method and the structured light method.

The **stereo matching method**, as a passive technique without texture projection, is based on triangulation of multiple synchronized cameras. This technique has been used in studies investigating dynamic foot shape (Coudert et al. 2006, Kouchi et al. 2009, Wang et al. 2006). The main disadvantage of this method, however, is the intricate correspondence between images from multiple views. Investigators in a recent study used painted cross-section lines on the foot to investigate its shape during the walking process. Only predefined discrete values were obtainable instead of the entire three-dimensional foot shape (Kouchi et al., 2009).

The **structured light method** uses a projector–camera system as an active device with defined light patterns projected on the moving object. The correspondence between images is therefore easily solved. Furthermore, information on depth is more precise, especially by applying the phase-shift technique, and noise is reduced compared to a camera–camera system. This method was used by Kimura et al (2009), who investigated fractional dynamic food shape using a synchronized 3 camera – projectors system based on structured light methodology. To avoid pattern switching they applied one single color coded pattern per 3D - frame (Kimura et al., 2009). However, up to now, synchronous measurement of the entire foot using multiple pairs of projector-camera systems has not been reported. This is essential for capturing lateral and medial anatomical landmarks simultaneously to analyze foot width, length, circumferences and arch changes during the roll over process (ROP).

In conclusion, there is still only very limited knowledge about foot anthropometry during walking or running which is mainly due to limited measurement techniques. There is currently no measurement system which is capable of capturing the entire foot shape during walking with sufficient accuracy and measurement frequency. In addition, most studies were performed to determine the feasibility of a new measurement system without actually investigating the human foot shape using large sample sizes (Coudert et al. 2006, Kimura et al. 2009, Wank et al. 2006).

148

Based on these considerations, the aim of this project was to develop a 3D-scanner-system to capture the dynamic foot shape during the roll over process of walking. Furthermore, differences in foot measures within the dynamic gait cycle and between dynamic and static foot measures were investigated using a large number of healthy subjects.

MATERIALS AND METHODS

PROPOSED MEASUREMENT TECHNIQUE

The proposed measurement system is based on the structured light method combined with the phase-shift technique, where a series of fringes is incrementally shifted by a constant phase angle. This sinusoidal pattern is projected and recorded at ultra-high speed using digital micro mirror device technology (DMD™ by Texas Instruments, USA). The Accessory Light modulator Package (ALP) from ViALUX (Germany) enables a pixel wise high speed control of the micro mirror array, consisting of 1024x768 mirrors. Each pixel (micro mirror) generates a binary light switch (± 12°), which either does or does not direct the incident light (blue LED light) into the aperture of the projection optics (Frankowski et al., 2003). Unlike other pattern switching scanner devices, the described technique provides the ability to combine high resolution with a short shutter time (<1.0 ms) and short recording time which results in a maximum 3D video frame rate of 49 fps by running the cameras in a 4 by 4 binning mode.

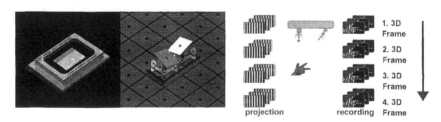

Figure1: Digital micro mirror device by Texas Instruments and series of 3D Frames. (ViALUX 2007)

DYNASCAN4D FOOT SCANNER

Using the proposed measurement technique as a synchronized scanner ensemble, containing five scanner units each consisting of one CCD-camera (CCD = Charge-coupled Device) with VGA resolution (640 x 480) and one projection unit, it is possible to capture the entire foot shape including height, circumferences, width and length measures during walking (see Figure 4). A safety glass platform (0.6 x 0.4 m) is integrated into a walkway that is 4.6 m long and 0.8 m high which enables plantar data capture (see Figure 2). During the current study, the cameras were run in a 2 by 2 binning mode, whereby four pixels were combined to one pixel. With this setup, the foot can be captured at a 3D frame rate of 21 fps under natural walking conditions by running the cameras at its highest frequency.

Figure2: Picture of walkway (left) and scanner ensemble (right).

MEASUREMENT PROCEDURE AND DATA ANALYSIS

One hundred and forty-four subjects ($♀$ 90 $♂$ 54, age 44 ± 13.5, BMI 24.6 ± 4.4) were scanned in a static and dynamic situation. Static scans were performed in different weight bearing conditions.

1. full weight bearing: standing on one foot
2. semi weight bearing: remaining in bipedal position with body weight equally distributed on both feet
3. non weight bearing: subjects were sitting on a chair

Each static measurement procedure was repeated twice. Dynamic measurements were taken during the ROP of walking. All subjects were asked to walk at a predefined velocity of 4.5 km/h (±5%), monitored by light barriers. Measurements started only when subjects felt comfortable and normal walking conditions could be assured. Measurements were taken unilaterally, randomized by coin toss. Five dynamic scans were performed for each subject, three of which were analyzed afterwards.

Thirteen measurement values for foot length, width, angle, circumferences and the structure of the medial longitudinal arch were analyzed in five different stance phases: First Foot Contact, First Metatarsal Contact, Forefoot Flat, Heel off and Terminal Standing Phase (Blanc Y. et al., 1999). In each stance phase the foot was aligned to a medial foot axis, defined as a tangent line to the most medial point of the first metatarsal head and the most medial edge of the heel.

Differences between static and dynamic data as well as comparisons of foot measurement values during the five stance phase were tested against the null hypothesis (H_o: $\alpha < 0.05$; i.e. no difference between static and dynamic conditions) with a paired t - test. Normality of each dimension was controlled using quantile - quantile plots (Q-Q Plot). All statistical analyses were run in JMP® 7.0 (SAS Institute Inc, USA), and Microsoft® Excel 2002 (Microsoft Cooperation, USA).

Table 1: Description of foot measures

Measurement categories	Foot measurement	Description of foot measures
Foot length measures (a-c)	Foot length (a)	Heel to foot tip
	Ball length medial (b)	Heel to MTH1
	Ball length lateral (c)	Heel to MTH5
Foot width measures (d-f)	Ball width (d)	MTH1 to MTH5
	Plantar arch width (e)	Medial to lateral border of loaded foot part within the midfoot intersection plane
	Heel width orthogonal (f)	Widest part of the heel orthogonal to medial foot axis
	Midfoot width	Most medial and lateral part of the midfoot within the midfoot intersection plane
Foot angle measures (g - h)	Arch angle (g)	Angle between medial foot axis and apex of the medial longitudinal arch
	Ball angle (h)	Angle between line MTH1 orthogonal to medial foot axis and diagonal ball line MTH1 to MTH5
Foot arch measures	Arch height	Vertical distance between highest soft tissue margin of medial longitudinal arch and ground
Foot circumferences	Circumference measures	At ball line, 50% of foot length and 65% of foot length

RESULTS

Foot circumferences

Figure3: Measurement values plantar (right) and foot circumferences (left)

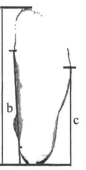

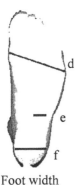

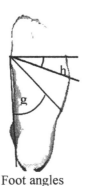

Foot length   Foot width   Foot angles

DIFFERENCES DURING THE ROLL OVER PROCESS OF WALKING

The results showed significant foot deformities in length, width and height during the roll over process. Especially arch height, arch width and arch angle showed significant changes during walking. Arch variables indicate a decrease of the medial arch structure during ground contact of 23% for arch angle and 17% for arch height. Plantar arch width, which is also an indicator for changes in arch structure, increases during ROP by 8.3 mm. All three arch variables suggest that the most flattened arch structure occurs during midstance.

Foot width measures show lowest changes in midfoot width compared to heel width and ball width. In 89% of the population, the widest heel width measures occurred during stance phase of first metatarsal contact (FMC). In contrast, ball width exhibited maximum value in heel off in 78% of the population and minimum width in FMC in 98% of all subjects. All changes of foot dimensions during the ROP were statistically significant.

DIFFERENCES BETWEEN STATIC AND DYNAMIC FOOT MEASURES

Comparing dynamic and static situations showed only small differences in the measured foot values. An increase of dynamic foot dimensions compared to the static situation was observed in FMC in heel width as well as in midstance in medial ball length and ball angle. Ball width increased in the terminal stance phase. However, in most dynamic situations, the foot dimensions were decreased compared to static conditions. All arch measurements indicated higher values during the ROP than in a resting position. 75% of all subjects showed a more flattened medial arch structure in static condition (half weight bearing) compared to dynamic situation. In midstance, all width measures were slightly decreased compared to the static situation, in which 70% of the population exhibited a larger midfoot width.

Figure4: Point clouds of the entire 3D - foot shape captured during walking (4.5km/h)

DISCUSSION

The main purpose of this study was to investigate three – dimensional foot structure during ground contact while walking by using a novel measurement technique. The development of this synchronized scanner ensemble allowed the investigation of foot structure within the dynamic gait cycle and between static and dynamic conditions. Even though the proposed measurement technique requires a series of pictures to generate three-dimensional information on dynamic foot structure, which is a time-consuming process, it still achieves a recording time of 21 - 49 fps, which is suitable for walking conditions.

Present results show the greatest changes in foot structure during walking for longitudinal arch, which reaches a minimum arch height and arch angle during the midstance phase. Additionally, arch values show the highest standard deviation of all foot dimensions, which indicates high individual differences in arch flattening during walking. These findings of absolute changes in arch structure are comparable to previous studies investigating foot shape under different weight bearing conditions (Carloo and Wetzstein 1968; Tsung et al 2003; Xiong et al. 2009). Tsung et al. (2003) showed significant changes in foot shape between non-weight and full-weight bearing within a comparable range of motion in decreasing arch height (Δ -20% ± 9.2). A current study by Xiong et al. (2009) demonstrated that the foot becomes significantly reduced in height while everting with increased loading on the foot. However, present results show a higher arch structure during walking compared to static conditions. This indicates an influence of foot muscle strength on soft tissue and arch structure behavior which is only activated in dynamic situation rather than in a relaxed, inactive resting position.

Present results of foot width measures show changes during ROP which corresponds to more than one shoe size in width. Previous studies show comparable differences within dynamic width measures (Coudert et al. 2006; Gefen et al. 2001). Ball and heel width increased the most in phases of the ROP where the contact area of the foot is reduced and body weight has to be borne by a smaller area. These results could be essential for last designers to decide whether last geometry has to be wider or narrower than static foot shape. Further dynamic foot width measures including different walking and running conditions are required to optimize the allowance in last construction.

CONCLUSIONS AND FUTURE

Even though differences in static and dynamic foot measures are relatively small, dynamic foot scanning can provide important information on foot geometry during active movement. Especially knowledge about the individual behavior of dynamic arch flattening can provide a further step to dynamic last construction and manufacturing of orthopedic insoles. Previous studies on static foot shape, which were based on a large number of subjects, demonstrated the necessity of considering different foot types (Mauch et al. 2009; Krauss et al. 2008). Therefore, a cluster analysis considering different influencing factors is required for further research to provide specific dynamic foot type classification. Overall, the present study showed that the proposed measurement system is a suitable technology for capturing the human foot shape during the roll over process with high potential in different application areas.

REFERENCES

Carlsoo S., and Wetzenstein H. (1968), "Change of form of the foot and the foot skeleton upon momentary weight-bearing." Acta Orthop Scand; 39: 413-423

Cashmere T, Smith R, Hunt A. (1999), "Medial longitudinal arch of the foot: stationary versus walking measures." Foot Ankle Int; 20: 112-118

Channa P. (2004), "Dimensional differences for evaluating the quality of footwear fit." Ergonomics; 47: 1301-1317

Cheng JC, Leung SS, Leung AK, Guo X, Sher A, Mak AF (1997), "Change of foot size with weightbearing. A study of 2829 children 3 to 18 years of age." Clin Orthop Relat Res; 123-131

Coudert, T, Vacher, P, Smits, C, and van der Zande, M. (2006), "A method to obtain 3D foot shape deformation during the gait cycle." Ninth International Symposium on the 3D Analysis of Human Movement

D'Apuzzo (2006) "Overview of 3D surface digitization technologies in Europe." Corner, B. D. and Tocherie M. SPIE-IS&T Electronic Imaging; Three-Dimensional Image Capture and Applications VI. SPIE-IS&T Electronic imaging 6056.

De Cock A, De CD, Willems T, Witvrouw E (2005),Temporal characteristics of foot roll-over during barefoot jogging: reference data for young adults. Gait Posture; 21: 432-439

Frankowski, G, Chen, M, and Huth, T. (2000), "Real-time 3D Shape Measurement with Digital Stripe Projection by Texas Instruments Micromirror Devices DMD." Proceedings SPIE 3958[90(2000)], 90-106. 7-7-2003. 24-1-2000. Ref Type: Conference Proceeding

Frey C. (2000), "Foot health and shoewear for women." Clin Orthop Relat Res: 32-44

Gefen A., Megido-Ravid M., Itzchak Y. (2001), "In vivo biomechanical behavior of the human heel pad during the stance phase of gait." J Biomech; 34: 1661-1665

Grau S., Mueller O., Baurle W., Beck M., Krauss I., Maiwald C., Baur H., Mayer F. (2000), [Limits and possibilities of 2D video analysis in evaluating physiological and pathological foot rolling motion in runners]. Sportverletz Sportschaden; 14: 107-14

Janisse DJ (1992), "The art and science of fitting shoes." Foot Ankle, 13: 257-262

Kimura, M, Mochimaru, M, and Kanade, T. (2009), "Measurement of 3D Foot Shape Deformation in Motion." International Conference on Computer Graphics and Interactive Techniques. Proceedings of the 5th ACM/IEEE International Workshop on Projector camera systems. New York, NY, USA, ACM.

Kleindienst FI, Krabbe B, Walther M, Bruggemann GP (2006). Grading of the functional sport shoe parameter "cushioning" and "forefoot flexibility" on running shoes. Sportverletz Sportschaden, 20: 19-24

Kouchi M., Kimura M., Mochimaru M. (2009), "Deformation of foot cross-section shapes during walking." Gait Posture; 30: 482-486

McPoil TG. and Cornwall MW. (2006), "Use of plantar contact area to predict medial longitudinal arch height during walking." J Am Podiatr Med Assoc, 96: 489-494

Krauss I., Grau S., Mauch M., Maiwald C., Horstmann T (2008). Sex-related differencess in foot shape. Ergonomics, 51:11,1693 - 1709

Mauch M., Grau S., Krauss I., Maiwald C., Horstmann T. (2009) A new approach to childrens's footwear based on foot type classification. Ergonomics 52:8, 999-1008

Nachbauer W and Nigg BM (1992), "Effects of arch height of the foot on ground reaction forces in running." Med Sci Sports Exerc; 24: 1264-1269

Rossi W. and Tennant R. (1984), "Professional shoe fitting." Pedorthic Footwear Association, National Shoe Retailers Association: 90-100

Tsung BY, Zhang M., Fan YB, Boone DA (2003), "Quantitative comparison of plantar foot shapes under different weight-bearing conditions." J Rehabil Res Dev; 40: 517-526

Wang J, Saito H., Kimura M., Mochimaru M., Kanade T. (2006), "Human Foot Reconstruction from Multiple Camera Images with Foot Shape Database." IEICE Trans Inf & Syst; E89-D: 1732-1742

Xiong S., Goonetilleke RS, Zhao J., Wenyan L., Witana CP (2009), Foot deformations under different load-bearing conditions and their relationsships to stature and body weight. ANTHROPOLOGICAL SCIENCE; 117: 77-88

Chapter 18

The Effects of Extravehicular Activity (EVA) Glove Pressure on Hand Strength

Miranda Mesloh[1], Scott England[2], Elizabeth Benson[2],
Shelby Thompson[1], Sudhakar Rajulu[3]

[1]Lockheed Martin
Anthropometry and Biomechanics Facility
NASA Johnson Space Center
Houston, TX 77058, USA

[2]MEI Technologies, Inc.
Anthropometry and Biomechanics Facility
NASA Johnson Space Center
Houston, TX 77058, USA

[3]National Aeronautics and Space Administration
Anthropometry and Biomechanics Facility
NASA Johnson Space Center
Houston, TX 77058, USA

ABSTRACT

The purpose of this study was to characterize hand strength while a Phase VI extravehicular activity (EVA) glove was worn in an Extravehicular Mobility Unit (EMU) suit. Three types of data were collected: hand grip, lateral pinch, and pulp-2 pinch, under 3 different conditions: bare-handed, gloved with no Thermal

Micrometeoroid Garment (TMG), and gloved with a TMG. In addition, during the gloved conditions, subjects were tested when unpressurized and pressurized (4.3 psi).

The TMG reduced gloved grip strength to 55% of bare-hand strength in the unpressurized condition and 46% in the pressurized condition. Without the TMG, gloved grip strength increased to 66% of bare-hand strength in the unpressurized condition and 58% in the pressurized condition. For lateral pinch strength, the reduction in strength was the same for the 2 pressure conditions and with and without the TMG, about 85% of bare-hand strength. Pulp-2 pinch strength with no TMG showed an increase to 122% of bare-hand strength in the unpressurized condition and 115% in the pressurized condition. While the TMG was worn, pulp-2 pinch strength was 115% of bare-hand strength for both pressure conditions.

Keywords: EVA, Spacesuit, Hand Strength, Glove, Pressurized Suit, Phase VI Glove, NASA, Human Systems Integration, Grip, Pinch

INTRODUCTION

Proper human-system integration is essential to mitigate risks to mission and crew while promoting human health and performance. This study focused on EVA glove performance, in terms of hand and finger strength, to generate requirements for the next-generation glove and the design of hardware such as tools, controls, mobility aids, and hatch operation. In fact, a vital step in human-centered design is the application of these results to designing better human-machine interfaces.

The results reported here for hand and finger strength are part of a large set of data collected to characterize the Phase VI EVA glove. All data was collected by the Habitability and Human Factors Branch at NASA's Johnson Space Center. Other data collected were hand mobility, tactility, dexterity, and functional task performance using EVA tools and a cursor control device. Because an immense amount of data was collected, this publication will focus only on the hand strength results.

The specific hand strength postures (grip, lateral pinch, and pulp-2 pinch) were chosen because of the availability of values in existing literature for comparison and the popularity of their use by astronauts manipulating EVA tools. Current NASA requirements state that the next-generation EVA glove must perform as well as, or better than, the current Phase VI glove. Therefore, it was important that values for quantifiable and replicable measures be collected to characterize performance. To that end, a set of conditions that generalized known operations was developed.

METHODS

Three types of hand strength data were collected: hand grip, lateral pinch, and pulp-2 pinch. The included results were compiled from 8 subjects (4 men and 4 women)

who represented a range of hand anthropometry. Subjects were instructed to exert and sustain a maximum grip or pinch force for a total of 3 seconds during each trial. The force data was recorded using a data acquisition system from which the maximum produced force was extracted. This data acquisition was repeated until a total of 3 trials were completed. A rest period of 2 minutes was given between trials to prevent fatigue (Rajulu, 1993). The test conductor then verified that the maximum strength values for each trial were within 10% of each other. This was done to ensure the accuracy of results. If a trial was outside of this 10% interval, the value was excluded and an additional trial was performed. The 3 trials were then averaged to produce a final strength value for that particular subject and condition.

This procedure was implemented for all types of hand strength and conditions involved in the test. A Phase VI EVA glove was worn in an Extravehicular Mobility Unit (EMU) suit. Hand strength was tested under 3 different conditions: bare-handed, gloved with no Thermal Micrometeoroid Garment (TMG), and gloved with a TMG. In addition, during the gloved conditions, subjects were tested when the glove was unpressurized and pressurized (4.3 psi). Figure 1 shows the glove without a TMG. Figure 2 shows the test setup used for each subject in the EMU supported by a suit stand.

Figure 1. Phase VI EVA glove with no TMG, shown unpressurized (left) and pressurized (right).

Figure 2. Subject test setup in Extravehicular Mobility Unit (EMU) supported by a suit stand with a TMG.

Results will be discussed in terms of actual force applied (pound-force, lbf) and as a percentage of bare-hand strength. The latter is important in characterizing the glove itself instead of the strength capabilities of the test subjects.

Grip strength was measured using a JAMAR hand dynamometer. Pinch strength was measured using a calibrated load cell. When the strength data was collected, subjects stood using an arm posture similar to the one seen when the suit is worn. This was done in an effort to eliminate as many variables as possible.

RESULTS

HAND GRIP STRENGTH

The hand posture used while gripping the dynamometer is seen in Figure 3 and the average, minimum, and maximum force are shown in

Table 1. The TMG condition reduced grip strength to 55% of bare-hand strength in the unpressurized condition and 46% in the pressurized condition. Without the TMG, gloved grip strength increased to 66% of bare-hand strength in the unpressurized condition and 58% in the pressurized condition. Gloved grip strength was reduced the most when subjects wore a TMG at 4.3 psi and reduced the least with no TMG and unpressurized. The percentage grip strength for no TMG at 4.3 psi was similar to that found with a TMG and no pressure (see Figure 4).

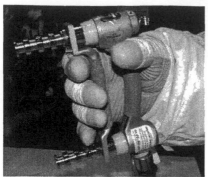

Figure 3. Posture used to measure grip strength, shown with a TMG (left) and without a TMG (right).

Table 1: Grip strength results

	Bare-Hand	Gloved (No TMG)		Gloved (TMG)	
		0 psi	4.3 psi	0 psi	4.3 psi
Minimum	60	43	35	33	24
Maximum	135	80	66	73	66
Mean	93	59	52	50	43

All values are reported as pound-force (lbf).

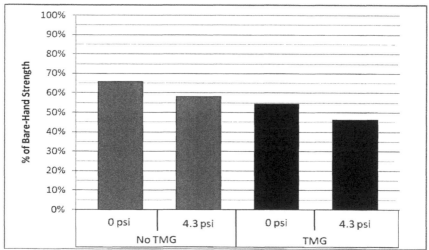

Figure 4. Grip strength of subjects wearing EVA gloves as a percentage of bare-hand strength.

These results suggest that overall the TMG at both pressure conditions lowered grip strength more than no TMG, as shown by the lower percentages of bare-hand strength.

PINCH STRENGTH

The hand posture subjects used while gripping the load cell can be seen in Figure 5. Lateral pinch posture is also commonly referred to as a key pinch as it resembles the hand configuration used when starting a car or unlocking a door (Mathiowetz, 1984). The load cell was positioned between the pad of the thumb and the second knuckle of the bent index finger. The pulp-2 pinch hand posture, also seen in Figure 5, utilizes the thumb and index finger. While keeping the fingers as straight as possible, the subject used the pads of the thumb and index finger to apply a force to the load cell.

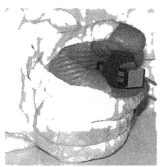 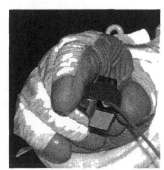

Figure 5. Posture used to measure lateral pinch strength (left) and pulp-2 pinch strength (right).

Lateral Pinch

Table 2 shows the average, minimum, and maximum lateral pinch strength values. The TMG condition reduced lateral pinch strength to 84% of bare-hand strength in the unpressurized and 86% in the pressurized condition. Without the TMG, lateral pinch strength was 83% of bare-hand strength in the unpressurized and 84% in the pressurized condition.

Table 2: Lateral pinch strength results

	Bare-Hand	Gloved (No TMG)		Gloved (TMG)	
		0 psi	4.3 psi	0 psi	4.3 psi
Minimum	17	14	13	13	8
Maximum	28	24	23	26	22
Mean	21	20	16	19	16

All values are reported as pound-force (lbf).

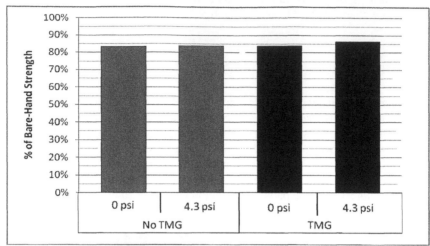

Figure 6. Lateral pinch strength results for subjects wearing EVA gloves as a percentage of bare-hand strength.

Overall, lateral pinch strength was about 15% less for gloved hands than bare hands. Pressurizing the glove actually provided a slight increase in performance for both TMG and non-TMG cases. Further explanation will be provided in the discussion.

Pulp-2 Pinch

Table 3 shows the average, minimum, and maximum pulp-2 pinch strength values. The TMG condition increased pulp-2 pinch strength to 115% of bare-hand strength in both unpressurized and pressurized conditions. Without the TMG, pulp-2 pinch strength when EVA gloves were worn was 122% of bare-hand strength in the unpressurized condition and 115% in the pressurized condition.

Table 3: Pulp-2 pinch strength results

	Bare-Hand	Gloved (No TMG)		Gloved (TMG)	
		0 psi	4.3 psi	0 psi	4.3 psi
Minimum	7	9	9	7	9
Maximum	18	19	19	18	17
Mean	12	12	12	13	12

All values are reported as pound-force (lbf).

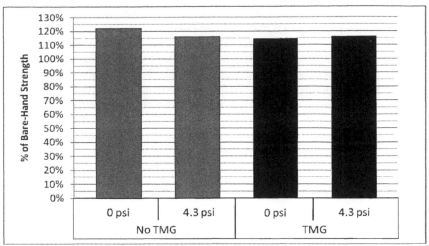

Figure 7. Pulp-2 pinch strength of subjects wearing EVA gloves as a percentage of bare-hand strength.

Unlike the other measures of hand and finger strength, pulp-2 pinch strength performance increased while subjects wore the Phase VI glove. In all gloved configurations, subjects had greater than 110% of bare-hand strength. Whereas with a TMG pressurization slightly increased the strength ratio, without a TMG pressurizing the glove decreased the percentage of bare-hand strength by about 5%.

DISCUSSION

To verify that the subjects represented the population well, a comparison was made with existing literature. With the bare-hand grip strength ranging between 60 lbf and 135 lbf, the expected astronaut population was represented well, as an earlier study of astronaut grip strength reported a range of 65 lbf – 150 lbf (Rajulu, 1993). For bare-hand lateral pinch strength, a previous study (Mathiowetz, 1984) reported an average range for the general population of 16 lbf – 25 lbf, which compares well with the range of this study, from 17 lbf to 28 lbf. Pulp-2 pinch was also on point with a literature range of 11 lbf – 17 lbf (Mathiowetz, 1984) compared to the bare-hand range of 7 lbf – 18 lbf found in this study.

Reduced applied hand strength due solely to wearing an unpressurized glove can significantly reduce performance. This was evident mostly in the grip strength evaluation. On average, subjects wearing an unpressurized glove without a TMG could produce only 55% of the force seen when they were bare-handed. Therefore, understanding how the glove can reduce certain hand strengths is of utmost importance when designing interfaces with a gloved crewmember, whether the suit is pressurized or unpressurized.

Pressurizing the Phase VI glove noticeably decreased hand grip strength, but had little to no effect on the lateral and pulp-2 pinch strengths. The cause of this difference may reside in the mechanics of the glove. When the glove is pressurized, hand postures like grip may require more force to overcome the natural shape of the glove, whereas the glove design may be more conducive to lateral and pulp-2 pinch postures. In other words, the subject is applying the same force as when bare-handed, but the motion may be resisted or encouraged by the pressurized glove.

The TMG decreased the hand grip and pulp-2 pinch strengths, but had no effect on the lateral pinch strength. With a decrease of greater than 10% seen in bare-hand grip strength, designers of the next-generation glove could greatly improve performance by looking for a novel solution in glove thermal and micrometeoroid protection. This could decrease fatigue and also prevent injuries associated with repeated tool manipulation.

CONCLUSIONS

Although quantifying the characteristics of glove performance can be challenging at times, future designs will benefit by baselining existing hardware as this study did with the Phase VI glove. Measurement of the fluctuations in applied hand strength associated with pressurization and the TMG will also aid in the design of tools and other human-machine interfaces. These types of studies are also useful for other industries that require a gloved hand to manipulate tools and interface with other hardware. Careful consideration should be taken of how hand strength is affected by a glove before assuming the hand will perform the same way as it would without a glove.

Future work on gloved strength performance could involve glove box testing, fatigue studies, improved methods to quantify glove fit, and incorporation of new technologies for data acquisition to get a better understanding of what is happening inside the glove (Hinman-Sweeney, 1994). The current EMU does not accommodate humans at the extremes of the anthropometric spectrum. For this reason, additional testing could be performed in a differential pressure glove box using the same methods as this grip and pinch strength evaluation. Smaller women and very large men who cannot be properly fitted with an EMU suit would be used as subjects. The results could be compared to the results of this study. Fatigue studies can be useful in assessing what types of motions and tools are best for long-duration EVAs or highly repetitive tasks. Glove fit is an integral part of task performance and very difficult to quantify because of the natural subjectivity in human preferences. Although they are normally derived using questionnaires, metrics of glove fit would be more accurate if hand position inside the glove could be measured. New technologies such as internal pressure sensors could help in quantifying glove fit.

REFERENCES

Hinman-Sweeney, E.M. (1994) Extra-Vehicular Activity (EVA) Glove Evaluation Test Protocol. NASA Technical Memorandum, NASA TM 108442, Lyndon B. Johnson Space Center, Houston, Texas.

Mathiowetz, V., Kashman, N., Volland, G., Weber, K., Dowe, M., Rogers, S. (1984) Grip and Pinch Strength: Normative Data for Adults. Milwaukee, Wisconsin: Occupational Therapy Program, University of Wisconsin-Milwaukee.

Rajulu, S.L., Bishu, R.R. (1993) A Comparison of Hand Grasp Breakaway Strengths and Bare-Handed Grip Strengths of the Astronauts, SML III Test Subjects, and the Subjects from the General Population. NASA Technical Paper 3286, Lyndon B. Johnson Space Center, Houston, Texas.

CHAPTER 19

Physically-Based Grasp Posture Generation for Virtual Ergonomic Assessment Using Digital Hand

*Yui Endo[1], Natsuki Miyata[1], Makiko Kouchi[1],
Masaaki Mochimaru[1], Satoshi Kanai[2]*

[1]Digital Human Research Center
National Institute of Advanced Industrial Science and Technology
Tokyo, 135-0064, JAPAN

[2]Graduate School of Information Science and Technology
Hokkaido University
Hokkaido, 064-0867, JAPAN

ABSTRACT

Our research purpose is to develop a system for ergonomics design, which enables an ergonomic evaluation for handheld information appliances without "real" subjects and physical mockups by integrating a digital hand with a product model. In this paper, we propose a new method for physically-based grasp posture generation by using a commercial realtime physics engine.

Keywords: Digital hand, digital human, grasp synthesis, physics simulation

INTRODUCTION

Recently, as handheld information appliances, such as mousse devices, mobile phones, digital cameras, have widely spread, the development of these appliances should pay more attention in their ergonomic design. However, the user tests for developing the "ergonomic" appliances are still done with many real subjects and a variety of these physical mockups, which require the expensive cost and take a long time.

So we propose a software system of a virtual ergonomic assessment system for designing handheld information appliances by integrating the digital hand model with the 3D product model of the appliance. The goal of our system is to provide the following feature functions for the ergonomic assessment: 1) Generation of kinematically and geometrically accurate digital hand models with rich dimensional variation, 2) automatic grasp posture generation and evaluation of the ergonomics for product models, 3) automatic evaluation of ease of the finger motion in operating the user interfaces, 4) aiding the designers to re-design the housing shapes and user-interfaces in the product model.

So far, we proposed three quantitative indices for the virtual ergonomic assessment (Endo 2008, Endo2009): 1) grasp stability for the product geometry and 2) ease of grasping from aspect of finger joint angle configuration 3) grasp fitness between the hand and the product model. In addition, we proposed several kinematically-based grasp posture generation methods. These methods find the optimized grasp posture to the ergonomic assessment indices by searching finger joint angle configurations, starting with a user-specified initial posture of the hand. As a result of several experiments using real subjects and commercial products, we found that these methods could generate humanlike grasp postures.

However, in these methods, there is a possibility that slight difference of the user-specification for creating an initial hand posture may cause a large difference of the final grasp posture, so users sometimes have to try the user-specification many times to acquire a desirable grasp posture. In addition, these methods take large computational costs and take much time because the search space for a number of control variables in the optimization is extensive. Moreover, in these methods, the system could not simulate the hand surface deformation when the hand surface contacts the surface of the product model, so dynamically-based virtual ergonomic assessment (e.g. evaluation of the contact forces) cannot be performed accurately.

In this paper, we proposed a new physically-based approach for the grasp posture generation. Physics simulation by a commercial game physics engine provides stable and fast control of the posture of the digital hand, so we can stably and rapidly generate the appropriate grasp postures where the result is not affected by the slight differences of any user-specifications. We used PhysX (NVIDIA PhysX) for our physics simulation because its performance of the physics simulation is relatively fast and stable and soft bodies using tetrahedral meshes can be simulated.

DIGITAL HAND MODEL

Figure 1 shows the digital hand model used in our physically-based grasp posture generation method. The digital hand consists of three models: A triangular mesh of a hand surface (called "surface mesh"), a link structure model which approximates to the rotational motion of bones in the hand, and a tetrahedral mesh. We use a part of Dhaiba-hand (Kouchi 2005) with respect to the surface mesh and the link

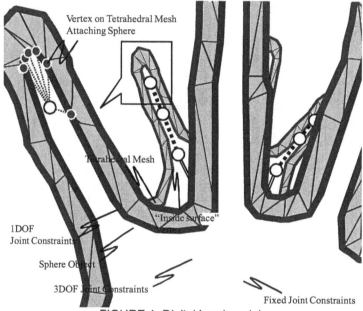

FIGURE 1 Digital hand model.

structure model of our digital hand. The digital hand is constructed as follows:

1. The number of faces of the surface mesh is reduced and a closed shell mesh is generated. The newly added vertices when making the shell mesh are tagged as "inside surface".
2. A tetrahedral mesh is automatically generated from the shell mesh and soft body object (Müller 2006) is created from the tetrahedral mesh.
3. Sphere objects are created at the position of the joint center of each link based on the link structure model. These spheres are used to control the hand posture and ignored in collision detection.
4. Joint constraints of 0-3 DOF between the sphere of each link and the one of the parent link are added, as shown in Figure 1.
5. The weight of each link for each vertex on the shell mesh tagged as "inside surface" is calculated. We use the skin attachment method based on heat equilibrium (Baran 2007) to the weight calculation. Each "inside surface" vertex is attached to the sphere of the link which has the largest weight for

the vertex. These vertices follow the movement and the rotation of the attached sphere object.

The number of faces was reduced from 59904 to 5602, and the thickness of shell is set to 3[mm]. We used "PhysXViewer" (NVIDIA PhysX) to generate the tetrahedral mesh from the triangle mesh.

While physics simulation is running, each vertex position on the triangular mesh is updated by linear interpolation of each vertex position of the associated tetrahedron on the tetrahedral mesh based on barycentric coordinates.

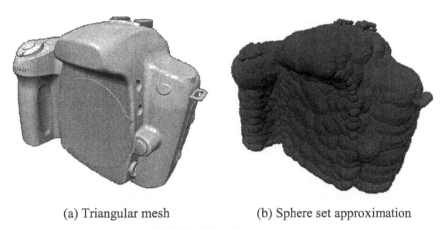

(a) Triangular mesh   (b) Sphere set approximation

FIGURE 2 Product model.

PRODUCT MODEL

The product model is located as a static object, which is fixed in the world reference frame and is not affected by any external forces or torques. Using a triangular mesh (Figure 2(a)) as the shape of the product model in collision detection may greatly increases the calculation cost. Instead, we approximate the triangular mesh of the product model by a set of spheres (Figure 2(b)). We construct a sphere-tree by the adaptive medial axis approximation (Bradshaw 2004) and use the highest level of the tree as the shape of the product model in collision detection. Except in collision detection, we use a triangular mesh as the shape of the product model.

PHYSICALLY-BASED GRASP POSTURE GENERATION

Figure 3 shows the overview of our grasp posture generation method based on the physics simulation. The algorithm of the method is as follows:

1. As shown in Figure 4, the user specifies one sphere object related to a link of the digital hand and six vertices on the triangular mesh of the product model: one sphere object s_p is around the palm, and six vertices v_p, v_i ($i =$ 1,...,5) are expected to contact to the region of the palm near the s_p and five tips of fingers f_i ($i = 1,...,5$) respectively when the product is grasped.

2. The system locates the digital hand so that $\mathbf{v}_{sp}$ is identical to $\mathbf{v}_{vp} + d\,\mathbf{n}_{vp}$ and $-\mathbf{n}_{spz}$ is identical to $\mathbf{n}_{vp}$, where $\mathbf{v}_{sp}$ defines the position vector of the center of the s_p, $\mathbf{v}_{vp}$ and $\mathbf{n}_{vp}$ is a position vector and a normal vector of the v_p respectively, $\mathbf{n}_{spz}$ is a unit vector directed to the z-axis (adduction / abduction) of the joint local frame with respect to the link where the s_p belongs, and d is defined by the user. We set $d = 50$[mm].

3. The system starts to run the physics simulation. At each time step, external forces and torques (see the next subsection) are applied to the sphere objects of the links of the digital hand.

4. If the sum of the velocity of each sphere object is less than a user-specified tolerance τ_{sim} in a few time steps, the system stops the simulation and output the grasp posture.

EXTERNAL FORCES AND TORQUES APPLIED TO DIGITAL HAND

At each time step in the simulation, the system applies the following external forces and torques to the sphere objects of the links of the digital hand.

Forces and Torques for Applying the Joint Constraints of Human Hand

*   The dependency between the joint angles of the DIP and PIP joints of each finger: for each finger f_i ($i=2,...,5$), if $r_{iDIP} > \tau_{DIP}r_{iPIP}$ (We defined that flexion of a joint decreases the joint angle value) and $r_{iDIP}r_{iPIP} > 1.1\tau_{DIP}$,the system applies torque to the s_{iDIP}, where r_{iDIP} and r_{iPIP} are the DIP and PIP joint angles of the f_i respectively, τ_{DIP} is usually set to 2/3. In case that the product needs to be grasped tightly (like heavy SLR cameras), τ_{DIP} is set to 4/3. The direction of the torque vector is the x-axis (flexion / extension) of the local frame of the DIP joint of the f_i.

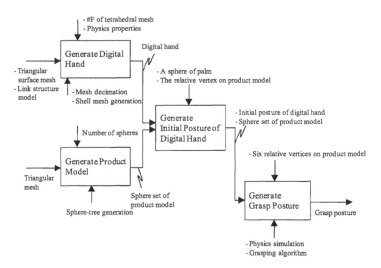

FIGURE 3 Overview of our grasp posture generation method.

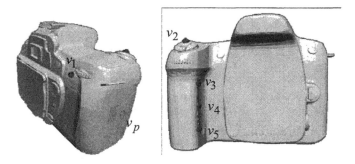

FIGURE 4 User-specified vertices on a product model.

- The dependency between the joint angles of the MP joints: for each finger f_i ($i=2,3,4$), if $| r_{iMPx} - r_{(i+1)MPx} | > \tau_{MPx}$, the system applies a torque to the s_{iMPx}, where r_{iMPx} is the MP joint angle of the the f_i with respect to the x-axis, τ_{MPx} is user-specified value. We set τ_{MPx} to 40[deg]. The direction of the torque vector is x-axis (if $r_{iMPx} < r_{(i+1)MPx}$) / the inverse of the x-axis (otherwise) of the local frame of the MP joint of the f_i.

- Convergence of fingers when the fist is clenched: for each finger f_i (i=2,3,4), the system applies a force to the sphere objects related to the DIP joints of the f_i and the $f_{(i+1)}$. The applied position of each force is the center of the link where the force is applied. The direction of the forces are $\mathbf{v}_{(i+1)DIP} - \mathbf{v}_{iDIP}$ (for the f_i) or $\mathbf{v}_{iDIP} - \mathbf{v}_{(i+1)DIP}$ (for the $f_{(i+1)}$), where $\mathbf{v}_{iDIP}$ is the position vector of the DIP joint rotation center of the f_i. The magnitude of the forces are proportional to $r_{iMP} + r_{(i+1)MP} + r_{iPIP} + r_{(i+1)PIP}$.

These properties of the joint constraints of human hands were referred to the research of Lee and Kunii (Lee 1995).

Forces for Grasping the Product Model

The system applies forces to the sphere object s_{iDIP} related to the each DIP joint of f_i (i=1,...,5). The applied position of each force is the center of the link where the force is applied. The direction of each force is $\mathbf{v}_i - \mathbf{v}_{iDIP}$, where $\mathbf{v}_i$ is the position vector of user-specified vertex v_i. The magnitude of each force is proportional to $|\mathbf{v}_i - \mathbf{v}_{iDIP}|$. As same as these forces applied to the fingers, the system applies forces to the s_p so that $\mathbf{v}_{sp}$ is identical to $\mathbf{v}_{vp}$.

RESULTS

Figure 4 shows the user-specified vertices v_p and v_i (i=1,...,5). We selected the sphere on the MP joint of the Index finger as s_p. The number of the spheres of the product model was set to 2000. The coefficient of the restitution and the static / dynamic friction were set to 0.1, 0.5 and 0.5 respectively. The volume stiffness and stretching stiffness of the soft body of the hand were set to 0.5 and 0.8 respectively.

Figure 5 and Figure 6 shows the simulation process and the result of our grasp posture generation method for the product model of a commercial SLR camera. The total processing time was 41s. We could find that the system stably and rapidly generates the humanlike grasp posture.

Figure 7 shows another result where the initial posture was different from the previous result. We could find that these two grasp postures are almost same postures even though each simulation started with the different initial hand posture. The digital hand used in these tests has the representative adult average size of Japanese.

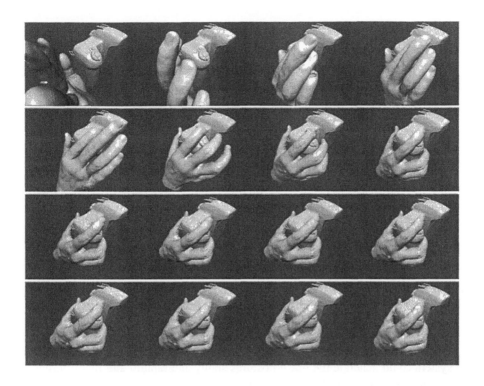

FIGURE 5 The simulation process of grasp posture generation (top left to bottom right).

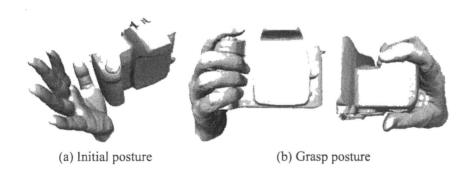

(a) Initial posture　　　　　(b) Grasp posture

FIGURE 6 The result of grasp posture generation.

174

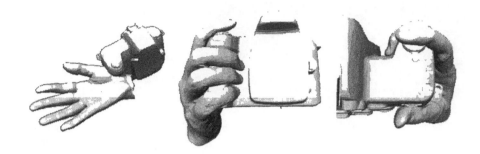

FIGURE 7 Another result of grasp posture generation.

CONCLUSIONS

We proposed a physically-based grasp posture generation method of the digital hand by using a commercial game physics engine. From results, we could find that the system could generate the grasp postures stably and rapidly and the differences of the initial hand postures hardly affect the result grasp postures. As future works, we have the following plans:

- The system automatically detects contact regions on a product model for the fingertips of the hand.
- The system sequentially generates grasp postures for various sizes of product models and the digital hands.
- The system evaluates the dynamically-based ergonomic assessment indices such as contact forces and ease of operating.

ACKNOWLEDGEMENT

This work was supported by Grant-in-Aid for Scientific Research (08J02105). All product models of SLR cameras used in this paper were provided by Nikon Corporation.

REFERENCES

Baran, et al. (2007). "Automatic rigging and animation of 3D characters", ACM Transactions on Graphics Volume 26 No. 3.

Bradshaw, G. et al. (2004). "Adaptive Medial-Axis Approximation for Sphere-Tree Construction", ACM Transaction on Graphics Volume 23 No. 1. pp1-26.

Endo, Y. et al. (2008). "Virtual Ergonomic Assessment on Handheld Products based on Virtual Grasping by Digital Hand", SAE 2007 Transactions Journal of Passenger Cars – Electronic and Electrical Systems, 2007-01-2511.

Endo, Y. et al. (2009). "Optimization-based Grasp Posture Generation Method of Digital Hand for Virtual Ergonomic Assessment", SAE International Journal of Passenger Cars-Electronic and Electrical Systems Volume 1 No.1, pp590-598.

Kouchi, M. et al. (2005). "An analysis of hand measurements for obtaining representative Japanese hand models", proceedings of the 8th Annual Digital Human Modeling for Design and Engineering Symposium, 2005-01-2734.

Lee, J. and Kunii, T. L. (1995). "Model-Based Analysis of Hand Posture", IEEE Computer Graphics and Applications Volume 15 No. 5. pp77-86.

Müller, M. et al. (2006). "Position Based Dynamics", proceedings of 3rd Workshop in Virtual Reality Interactions and Physical Simulation 2006. pp71-80.

NVIDIA PhysX. Website: http://www.nvidia.com/object/physx_new.html.

CHAPTER 20

The Geriatric 3D Foot Shape

Ameersing Luximon[1], Yan Luximon[2],
Duo Wai-Chi Wong[1], Ming Zhang[3]

[1] Institute of Textiles and Clothing

[2] School of Design, Polytechnic University

[3] Department of Health Technology and Informatics
The Hong Kong Polytechnic University
Hung Hom, Kowloon, Hong Kong

ABSTRACT

Inappropriate footwear is a major cause of foot illness and injuries among the elderly. Although, proper fitting footwear improves fit and comfort; balance; and reduces illnesses and injuries, foot shape data needed to design footwear is still lacking. Very few studies have characterized the geriatric foot shape. This study provides an in-depth study of foot anthropometry and foot surface model. The foot shapes of 100 (50 males and 50 females) elderly were measured using a 3D foot scanner. Foot shapes and 18 foot parameters together with stature, body weight, and age were collected. Results showed no significant paired differences between the left and right foot for both genders, despite absolute individual differences between left and right foot. As for gender, female foot parameters were significantly smaller compared to male foot parameters. In order to see shape differences, the foot was normalized by foot length. There were significant differences for ball girth, foot width, lateral sphyrion fibulare height, medial sphyrion fibulare height and medial malleolus height; and marginally significant differences for the lateral sphyrion height and instep girth between genders. This indicates differences in shaped between genders. The average foot shape was then computed and showed differences between genders. Normalized foot shape was compared to visually see differences in genders. A look at individual feet indicates that the geriatric toes were generally more deformed and hence may require more toe space during footwear design. Results of this study can be used for the footwear development for the Hong Kong elderly Chinese.

Keywords: Elderly, foot shape, foot dimension, geriatric, footwear design

INTRODUCTION

As age advances, the foot undergoes various changes from tissue level, such as skin ulceration and nail disorder to structural level, such as foot deformity (Helfand, 1995). These aging problems are further exaggerated by wearing inappropriate footwear and walking habits. Hammer toes, callus, bunions, corns, hallux valgus and lesser digital deformity are profound in elderly, with pain and lost of mobility in consequences. While it is believed that risk of falling is attributed to foot pain (Chaiwanichsiri et al., 2009), proper footwear design could maximize fit and comfort, so as to reduce the risk of falling.

Large-scale foot anthropometric studies have been performed on populations, while that of elderly is scarce (Manna et al., 2001; Wunderlich & Cavanagh, 2001). In fact, these reports only made use of few foot dimensions. Information for characterization of footwear geometry was inadequate. Chaiwanichsiri et al. (2008) have conducted an anthropometric research on the Indian population to evaluate proper shoe size. However, local anthropometric data is essential due to reported ethnical differences in foot geometry (Ashizawa et al., 1997).

Research on design of footwear has been performed to evaluate footwear features on stability (Li et al., 2006, Menant et al., 2008a, 2008b, 2009). Better design of footwear could reduce the risk of falling among elderly, whilst foot anthropometry pays an important role in the design factors. This study aims at providing 3D foot shape information in addition to anthropometric measures such as foot length and width. Furthermore weight, height, Body Mass Index (BMI), gender and age have been considered in the calculation for the Chinese geriatric population.

METHODS

One hundred Hong Kong elderly (50 males and 50 females) aged 60 and above, were recruited to participate in the study. There were able to ambulate with or without walking aids. They also could achieve equal weight-bearing with bare-feet. Participants with trauma, severe foot deformity, and weight-shifting problems were excluded in the study. The study was approved and all participants were required to

178

sign a consent form before the experiment.

A 3D foot laser scanner (INFOOT, Japan) was used to acquire the foot geometry. Eleven markers of 4mm in diameter were attached to selected anatomical landmarks (Figure 1). Subjects were asked to stand with equal weight-bearing when one foot was inside the scanner and the other on a leveled platform. Eighteen foot parameters were computed by the software accompanied with the scanner. The details of the parameters are listed in Table 1. Average foot shape was calculated using foot alignment (Goonetilleke & Luximon, 1999; Luximon et al., 2003), uniform sectioning, and averaging methods (Luximon, 2001; Luximon et al., 2005; Luximon & Goonetilleke, 2004).

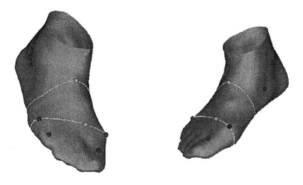

Figure 1 Foot scanning (INFOOT, Japan) with eleven surface landmarks

Table 1 Definition and notation of different foot parameters used in this study.

Foot Parameters	Definition	Notation
Foot length	The distance between pternion and the midpoint of the maximum breadth of the ball cross-section. The ball cross section is the vertical cross section passing through metatarsale tibiale(MT) and metatarsale fibulare(MF)	FL
Foot circumference	Circumference of the ball cross-section. Ball cross-section is the vertical cross section passing through MT and MF.	BG
Foot breadth	Maximum breadth of ball cross-section. Ball cross-section is the vertical cross section passing through MT and MF.	FW
Instep circumference	Circumference of vertical cross-section of the foot perpendicular to the sagittal plane containing the foot axis at the 59% of foot length.	IG
Heel breadth	The breadth of the heel measured perpendicular to the foot axis at the 16% of foot length from pternion.	HW
Instep length	The length from pternion to MT projected on the foot axis.	MPJ_m
Fibular instep length	The length from pternion to MF projected on the foot axis.	MPJ_l
Ball height	The maximum height of the ball cross-section	H_{MPJ}
Instep height	The maximum height of the instep cross-section. Instep cross section is the vertical cross section of the foot perpendicular to the sagittal plane containing the foot axis at the 50% of foot length.	H_l
Toe 1 angle	Acute angle made by the following two lines drawn on the foot outline. Line 1 is passing through the medial measuring point of the heel breadth and the most medial point of the ball cross section, and line 2 is the tangent line of the medial side of the 1st toe passing through the most medial point of the ball cross section.	$T1_{Ang}$
Toe 5 angle	Acute angle made by the following two lines drawn on the foot outline. Line 1 is passing through the lateral measuring point of the heel breadth and the most lateral point of the ball cross section, and line 2 is the tangent line of the lateral side of the 5th toe passing through the most lateral point of the ball cross section.	$T5_{Ang}$
Great toe height	The height of the 1st toe at the interphalangeal joint.	H_{T1}
5th toe height	The height of the 5th toe at the distal interphalangeal joint.	H_{T5}
Navicular height	The height of the most medial point at the navicular bone.	H_{nav}
Sphyrion fibulare height	The height of the lowest point of lateral malleolus of fibula.	HS_l
Sphyrion height	The height of the lowest point of medial malleolus of fibula.	HS_m
Lateral malleolus height	The height of the most laterally protruding point of the lateral malleolus of fibula.	$Hmal_l$
Medial malleolus height	The height of the most medially protruding point of the medial malleolus of fibula.	$Hmal_m$

RESULTS AND RESULTS

Table 2 Descriptive statistics of foot parameters.

| | Female (n=50) | | | | Male (n=50) | | | |
| | Right | | Left | | Right | | Left | |
Measure	Mean	SD	Mean	SD	Mean	SD	Mean	SD
FL	228.72	10.08	229.38	10.68	240.82	12.99	241.73	13.21
BG	222.58	10.28	221.53	10.06	238.66	16.12	235.95	14.79
FW	94.11	5.11	93.60	4.84	100.79	6.71	99.86	6.38
IG	222.27	10.36	221.93	11.58	245.21	54.69	244.96	56.43
HW	65.24	3.53	65.13	3.43	67.92	4.26	68.58	5.51
MPJ_m	168.73	8.15	169.08	8.90	176.88	9.84	178.26	11.46
MPJ_l	148.55	7.82	149.32	7.90	156.42	8.95	157.59	9.17
H_{MPJ}	34.08	2.76	34.30	2.97	36.72	3.40	36.56	3.18
H_l	60.10	3.95	59.69	4.48	65.13	11.38	64.76	11.26
$T1_{Ang}$	11.73	9.66	13.62	10.29	12.49	7.74	11.85	7.07
$T5_{Ang}$	11.99	6.01	10.90	5.84	13.40	3.54	11.58	4.59
H_{T1}	20.29	1.49	19.90	1.66	21.14	2.83	21.28	1.63
H_{T5}	13.96	1.77	13.93	1.69	14.90	2.43	14.99	2.30
H_{nav}	40.87	4.13	38.29	5.08	42.52	6.87	41.39	6.53
HS_l	45.16	3.99	45.58	4.80	45.23	9.08	47.29	5.53
HS_m	55.21	5.02	54.14	4.85	60.19	7.57	60.33	6.24
$Hmal_l$	60.95	5.27	60.16	5.03	63.98	5.76	63.29	6.23
$Hmal_m$	70.77	4.60	70.31	4.95	76.03	6.53	76.75	6.52

The mean ages of male and female were 74.08±7.40 and 74.42±7.42 respectively; the mean height was 159.24±6.69cm and 153.00±5.87cm respectively; and the mean weight was 60.44±8.96kg and 53.92±9.12kg respectively. In terms of BMI, though male and female shows significant differences in stature and weight with $p<0.001$, there was marginal difference in the BMI (p=0.07). Males (mean of BMI = 23.91, Stdev of BMI = 3.83) had slightly higher BMI than females (mean of BMI = 22.99, Stdev of BMI = 3.41). A basic descriptive statistics on the measured parameters, separated by gender, is presented in Table 2.

Table 3 Differences between left and right foot

	Male			Female		
	Mean	SD	Max	Mean	SD	Max
		Absolute difference			Absolute difference	
FL	2.64	2.14	8.3	2.54	2.24	8.9
BG	3.99	2.68	10.1	3.02	2.63	11.7
FW	2.04	1.51	6.6	1.95	1.52	7.3
IG	3.04	2.36	11.5	2.85	2.23	8.8
HW	1.53	1.18	6.2	1.39	1.44	8.4
MPJ_m	3.65	3.15	11.8	3.11	2.53	14.1
MPJ_l	3.72	3.22	13.5	2.83	2.44	10.7
H_{MPJ}	1.55	1.27	4.6	1.52	1.32	6.5
H_l	2.43	1.73	6.9	2.24	1.60	7.3
$T1_{Ang}$	4.66	3.57	17.6	4.54	3.45	16.5
$T5_{Ang}$	3.50	2.46	8	3.15	2.14	8
H_{T1}	0.94	0.72	3.4	0.94	0.92	3.5
H_{T5}	1.74	1.50	7.1	1.48	1.18	5.3
H_{nav}	4.97	3.59	16.5	4.41	3.66	19.5
HS_l	4.85	6.44	29.9	4.15	3.40	15.3
HS_m	4.54	3.51	15.5	3.07	2.34	9.4
$Hmal_l$	3.36	3.00	16.9	4.13	3.24	13.5
$Hmal_m$	3.95	2.83	12.6	3.03	2.40	8.8

A paired comparison was completed between the left and right foot. There were no significant difference between left and right foot, except for Toe 5 angle (p <0.03) for male and navicular height (p= 0.006) for females. The Toe 5 angle for the elderly male right foot (13.40°) was bigger compared to the left foot (11.58°). The navicular height for the elderly female right foot (40.87mm) was higher than the left foot (38.29mm). Even though not significant, there were differences between left and right foot. For example, as shown in table 3, the maximum difference between left and right foot length was 8.3mm and the average absolute difference was 2.638mm. These differences need to be considered when making footwear, especially customized footwear.

Since there were no significant differences between left foot and right foot, the data was combined and further analysis was carried out. There were significant differences between male and female for all foot parameters (p<0.001) except Toe 1 angle and Toe 5 angle. The mean value of all foot dimensions of females were small than those of males.

The value of the foot parameters was then normalized with foot length to see study differences in foot shape. After normalization, there were significant difference between males and females for ball girth (p=0.04), foot width (p=0.04), lateral sphyrion fibulare height (p<0.001), medial sphyrion fibulare height (p<0.001), and medial maleolus height (p=0.01). In addition there were marginally significant differences for lateral sphyrion height (p =0.08) and instep girth (p=0.07). After foot length normalization there are still gender differences in most foot parameters indicating shape differences between male and female feet.

THREE DIMENSIONAL FOOT SURFACE

In order to see the differences between the left and right foot shape, the point cloud data of the participants were merged to create an average foot shape as discussed in Luximon (2001). First the feet were aligned (Goonetilleke & Luximon, 1999; Luximon et al., 2003). Then the feet are sampled to create sections at regular interval based on percentage of foot length (Luximon & Goonetilleke, 2003). Ninety nine sections at 1% interval were created. For each section, 360 points were extracted based on polar coordinate system (Luximon et al., 2005) at 1° interval. In order to calculate the average foot shape, simple mean for each points (99x360) were computed (Luximon, 2001). A program was written in Matlab to process the data. Figure 2 shows the average geriatric male and female left and right foot. Figure 2c shows the comparison of the male and female feet. The smaller ones are the female feet. Except for size, the feet seems similar, hence a normalized average foot shapes are shown in Figures 3 and 4. It can be visually seen there are differences between left and right feet.

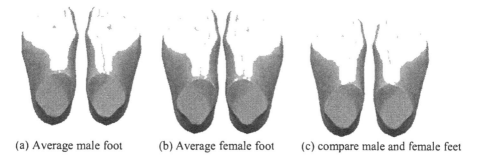

(a) Average male foot   (b) Average female foot   (c) compare male and female feet

Figure 2 Average male and female feet

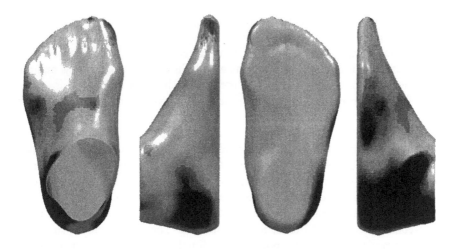

Figure 3 Comparison of normalized male and female left feet (darker color means larger error)

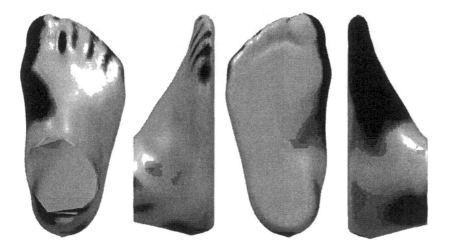

Figure 4 Comparison of normalized male and female right feet (darker color means larger error)

CONCLUSIONS

The risk of fall fall-related injuries has been associated with inappropriate footwear (Menz et al., 2001;Koepsell et al., 2004). Mass-customization could provide a mean to improve shoe design in terms of anthropometry and foot shape, while maintaining manufacturability and feasibility. However, studies on foot geometry of elderly for shoe design are inadequate (Chantelau & Gede, 2002). The study could provide a database of foot geometry for the elderly in Chinese community, and thus could further enhance the development of prevention shoes for elderly.

Though current studies was limited to cross-sectional populations, the subject temporal factors and various foot parameters showed clear trends and relationship. Further studies should increase the number of participants. Further investigation upon non-weight bearing condition could give us more information about the tolerance during shoe development for the elderly. In addition, instead of visual differences between normalized foot shape, new techniques for error comparison, alignment and shape averaging need to be developed.

Current studies had characterized the geometry of the elderly feet in the Chinese community with different foot parameters. The differences in most foot parameters among gender after foot length normalization showed that male and female feet are quite different. Information from this study will be useful for footwear and accessories design for the elderly in the Chinese community.

ACKNOWLEDGEMENTS

The geriatric data was jointly collected with Prof KS Leung and his team (PPY Lui, PS Lam and HW Mok) from the Chinese University of Hong Kong. The study was partially funded by the Hong Kong Polytechnic university grants No: A-PD1V.

REFERENCES

Ashizawa, K., Kumakura, C., Kusumoto, A., Narasaki, S. (1997). Relative foot size and shape to general body size in Javanese, Filipinas and Japanese with special reference to habitual footwear types. Ann. Hum. Biol. 24: 117-129.

Chaiwanichsiri, D., Janchai, S., Tantisiriwat, N. (2009). Foot Disorders and Falls in Older Persons. Gerontology 55: 296-302.

Chaiwanichsiri, D., Tantisiriwat, N., Janchai, S. (2008). Proper Shoe Sizes for Thai Elderly. The Foot, 18: 186-191.

Chantelau, E., Gede, A. (2002). Foot dimensions of elderly people with and without Diabetes mellitus – a Data Basis for shoe design. Gerontology 48: 241-244.

Goonetilleke, R.S. and Luximon, A. (1999). Foot flare and foot axis. Human Factors, 41: 596-607.

Helfand, A.E. (1995). The foot – a geriatric overview (part 2). The Foot 5: 19-23.

Koepsell, T.D., Wolf, M.E., Buchner, D.M., Kukull, W.A., LaCroix, A.Z., Tencer, A.F., Frankenfeld, C.L., Tautvydas, M., Larson, E.B. (2004). Footwear style and risk of falls in older adults. JAGS 52: 1495-1501.

Li, K.W., Wu, H.H., Lin, Y.C. (2006). The effect of shoe sole tread groove depth on the friction coefficient with different trad groove widths, floors and contaminants. Applied Ergonomics 37(6): 743-748.

Lord, S.R., Bashford, G. (1996). Shoe characteristics and balance in older women. J Am Geriatr Soc 44: 429-433.

Luximon, A. and Goonetilleke, R.S. (2004). Foot Shape Prediction. Human Factors, 46: 304-315.

Luximon, A., and Goonetilleke, R. S. and Zhang, M. (2005) 3D foot shape generation from 2D information. Ergonomics, 48(6): 625-641.

Luximon, A., Goonetilleke, R.S. and Tsui, K.L. (2003). Foot landmarking for footwear customization, Ergonomics, 46: 364-383.

Manna, I., Pradhan, D., Ghosh, S., Kar, S.K., Dhara, P. (2001). A comparative study of foot dimension between adult male and female and evaluation of foot hazards due to using of footwear. J. Physiol. Anthropol. 20(4): 241-246.

Menant, J.C., Perry, S.D., Stelle, J.R., Menz, H.B., Munro, B.J., Lord, S.R. (2008a). Effects of Shoe Characteristics on Dynamic Stability When Walking on Even and Uneven Surfaces in Young and Older People. Arch. Phys. Med. Rehabil. 89:1970-1976.

Menant, J.C., Steele, J.R., Menz, H.B., Munro, B.J., Lord, S.R (2009). Effects of walking surfaces and footwear on temporo-spatial gait parameters in young and older people. Gait and Posture 29: 392-397.

Menant, J.C., Stelle, J.R., Menz, H.B., Munro, B.J., Lord, S.R. (2008b). Effects of Footwear Features on Balance and Stepping in Older People. Gerontology 54: 18-23.

Menz, H.B., Lord, S.R., McIntosh, A.S. (2001). Slip resistance of casual footwear: implications for falls in older adults. Gerontology 47(3): 145-149.

Wunderlich, G.E., Cavanagh, P.R. (2001). Gender differences in adult foot shape- implications for shoe design. Med. Sci. Sports Exerc. 33: 605-611.

CHAPTER 21

Relationship Between Arrangement of Functional Joint Rotation Centers and Body Dimensions

Kei Aoki, Makiko Kouchi, Masaaki Mochimaru

Digital Human Research Center
National Institute of Advanced Industrial Science and Technology
2-3-26, Aomi, Koto-ku, Tokyo, JAPAN

ABSTRACT

Functional dimensions, such as the thumb tip reach length are not represented accurately by commercial digital manikins, although they are generated from body dimensions. To solve this technical issue, we proposed the functional center of rotation (FCR) of joints instead of anatomical location of joints. The FCR is determined from motion data of joint rotations. It was difficult to generate FCRs for boundary manikins, because boundary manikins are generated from only body dimensions based on large-scale anthropometric database. In this study, both of FCRs of whole body joints and body dimensions were measured for 20 participants. Information of individual variation of FCR locations was compressed by principal component analysis (PCA). The obtained principal components were represented by body dimensions using multiple regression functions. Subsequently, boundary manikins were generated from body dimensions based on Japanese anthropometric database, and their FCR locations were estimated by the multiple regression functions.

Keywords: Functional Joint Rotation Center, Digital Manikin, Body Dimension, Principal Component Analysis

INTRODUCTION

A problem of existing digital manikin systems is inaccurate functional body dimensions (Kouchi et al., 2004). This is partly because digital manikins assume constant segment length while segment lengths of the human change with motion because of complex joint structure of the shoulder. On the other hand, a digital manikin with constant segment lengths and simple joint structure such as a hinge or a ball joint is much easier to use, and digital manikin models must be able to represent body forms that statistically represent a population in terms of body dimensions. A desirable digital manikin should have simple joint structures, constant segment lengths, and accurate functional measurements. Our purpose is to develop a method to create members of boundary family that satisfy the following requirements: (1) to have functional centers of rotation (FCR) of joints that cause minimal errors in functional body dimensions, (2) to have constant segment lengths, and (3) to be created from body dimensions. For this purpose we conducted the following four studies: (1) Develop a method to calculate the joint rotation centers of a digital manikin with constant segment lengths and simple joint structure from motion capture (mocap) data of a specific subject, (2) Create a database of body dimensions and the locations of calculated rotation centers based on mocap data, (3) Develop formulae to estimate the positions of joint rotation centers from body dimensions, (4) Calculate the rotation centers using these formulae from a set of body dimensions for members of boundary family that statistically represents a population. In our previous study (Aoki et al. 2008), we reported the results of study (1), the development of a method to estimate the functional center of rotation of a joint (FCR). In this paper, we report the results of studies (2), (3) and (4).

MEASUREMENTS

SUBJECTS AND BODY DIMENSIONS

Subjects were 20 healthy Japanese adult males (age: 27±6 [yrs.]). A written informed consent was obtained from each subject. Table 1 and Figure 1 show the measured body dimensions. These 23 body dimensions are used for creating different body forms in a conventional digital manikin system. Chest dimensions are measured in detail because they are good predictors of the shoulder FCR (Aoki et al., 2007).

Table 1: Measurement of body dimensions

	Body dimensions	Avg.	Max.	Min.	S.D.
1	Height [mm]	1724.7	1823.0	1616.0	57.5
2	Height of waist circumference [mm]	1041.1	1112.0	945.0	46.5
3	Iliocristale height [mm]	962.4	1025.0	880.0	40.0
4	Crotch height [mm]	776.1	832.0	704.0	34.4
5	Gluteal furrow height [mm]	749.1	802.0	677.0	34.7
6	Sitting height [mm]	931.7	985.0	889.0	29.1
7	Total head height [mm]	238.2	258.0	233.0	9.0
8	Head length [mm]	191.7	206.0	172.0	7.2
9	Bideltoid breadth [mm]	466.8	562.0	428.0	31.9
10	Chest breadth at mesosternale [mm]	312.1	410.0	259.0	37.3
11	Chest breadth at xiphiale [mm]	295.9	415.0	259.0	36.1
12	Hip breadth [mm]	334.3	382.0	295.0	23.0
13	Chest circumference at mesosternale [mm]	923.8	1287.0	817.0	106.8
14	Chest circumference at xiphiale [mm]	844.9	1202.0	731.0	109.3
15	Minimum abdominal circumference [mm]	776.3	1236.0	646.0	136.2
16	Chest depth [mm]	202.0	354.0	161.0	42.7
17	Wall-acromion distance, arms hanging freely [mm]	106.4	523.0	62.0	99.1
18	Upper arm length [mm]	314.0	344.0	282.0	15.1
19	Elbow to middle fingertip length [mm]	443.7	482.0	422.0	16.7
20	Knee height, sitting [mm]	519.5	560.0	480.0	22.3
21	Foot breadth [mm]	99.7	106.0	92.0	3.6
22	Sphyrion height [mm]	69.2	78.0	59.0	5.0
23	Weight [kg]	67.0	123.6	49.0	17.1

$n=20$

ESTIMATION OF FCR

Seventy-eight markers were attached to the subject (Figure 2). A specific three-dimensional (3D) motion named "positioning motion" was captured with a Vicon MX system (Vicon, UK) using a sampling rate of 200Hz. Using the mocap data, FCRs and the constant lengths of the segments were estimated kinematically (Aoki et al., 2008). Using the same marker set, the displacement of the FCRs in a specific posture can be calculated. In the present study we used the erect standing posture as the standard posture.

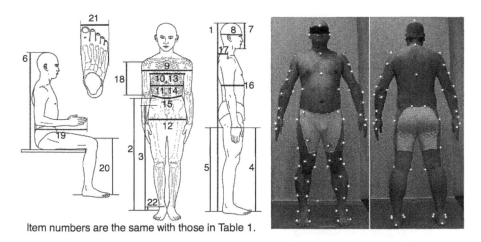

Item numbers are the same with those in Table 1.

Figure 1. Definitions of the body dimensions   Figure 2. Marker set (78 points)

FCR RECONSTRUCTION METHOD FROM BODY DIMENSIONS

BODY ORIENTED COORDINATE SYSTEM

We obtained the locations of 16 FCRs in the standard posture from the mocap data. To define the lengths of most distal segments (head, hands, and feet), we defined 5 more locations. Figure 3 shows the 22 locations. Broken lines indicate the left side of the body. The joint positions are head-neck, neck-chest, chest-abdomen, abdomen-pelvis, right and left shoulders, right and left elbows, right and left wrists, right and left hips, right and left knees, and right and left ankles. The end points of the most distal segments are the middle of the forehead (head), the midpoint of the second metacarpal head and the fifth metacarpal head (hand), and the midpoint of the first metatarsal head and the fifth metatarsal head (toe). The intersection point of a perpendicular line from the abdomen-pelvis joint and the line between the right and left hips is also included in the target positions (pelvis center in the following text). Thus, the dependant variables to be estimated are the 3D locations of the 22 points. These 22 points were described using a common body oriented coordinate system (Figure 4). The origin is the midpoint of the right and left anterior iliac spines (ASIS) projected on the standing surface. The Z-axis is perpendicular to the standing surface. The Y-axis is the line connecting the right and left ASIS projected on the standing surface (right side is positive). The X-axis is decided by the outer product of the Y and Z axes (backward is positive).

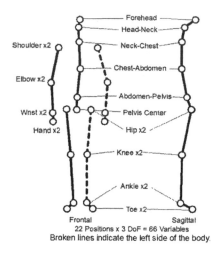

Figure 3. Location of FCRs

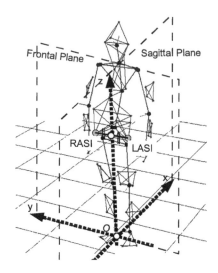

Figure 4. Definition of body oriented coordinate system

STATISTICAL ANALYSIS

The 66 independent variables (x, y, z coordinates of the 22 points) are not independent with each other. Therefore, we tried to compress the information carried by the 66 variables using principal component analysis (PCA). When a PCA starts from variance-covariance matrix, the original locations of the 22 points can be reconstructed from principal component (PC) scores. If a fewer number of PCs compared to 66 explain most of the variance, we only need to estimate fewer number of variables to reconstruct locations of the 22 points. Linear multiple regression functions to estimate the PC scores from 23 body dimensions (Table 1) were calculated.

PRINCIPAL COMPONENT ANALYSIS

There were 12 PCs with more than average contribution rate. The first six PCs with contribution rate larger than 5% were selected for estimating the arrangement of FCRs. The cumulative contribution rate of the first six PCs was 74.2 %. When all the PC scores are zero, the arrangement of FCRs of the whole body is the average arrangement of 20 subjects, as shown in Figure 5. In order to interpret the first six PCs, virtual FCR arrangements with the scores of ±2 standard deviation for only one PC and zeros for other PCs were calculated.

MULTIPLE REGRESSION ANALYSIS

If a set of these six PC scores is given, the arrangement of FCRs can be reconstructed. These PC scores are related to body size, proportions and postures (Fig. 5). Table 2 shows the results of multiple regression analysis. Body dimensions with coefficients significantly different from 0 are shown. PCs can be interpreted based on the results of multiple regression analysis (Table 2) and virtual FCR arrangements (Fig. 5) as follows: PC1: whole body size; PC2: contrasts tall and lean, or short and broad; PC3: the proportions of the upper body especially the shoulder breadth and upper limb. This PC also may be related to the length asymmetry of legs; PC4: backward inclination of the torso; PC5: antero-posterior relationships between torso FCRs; PC6: the proportion between trunk length and leg length. Multiple correlation coefficients are not very high for some PCs, especially PC5, probably because they are related to the posture that cannot be explained by body dimensions.

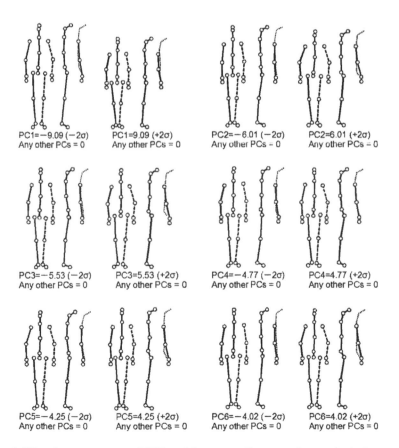

Figure 5. Virtual arrangements of FCRs with very small or very large principal component scores for only one principal component

Table 2: Results of multiple regression analysis

	Adj. R²	Body dimensions	+/-	|p|
PC1	0.94	Height	-	***
		Weight	-	***
PC2	0.79	Upper arm length	-	***
		Minimum abdominal circumference	+	***
		Hip breadth	-	**
		Height of waist circumference	+	*
PC3	0.78	Total head height	+	***
		Chest breadth at xiphiale	-	***
		Chest circumference at xiphiale	+	***
		Wall-acromion distance, arms hanging freely	+	***
		Upper arm length	-	***
		Elbow to middle fingertip length	+	***
		Sphyrion height	-	***
		Bideltoid breadth	-	**
PC4	0.75	Total head height	-	***
		Sitting height	-	***
		Upper arm length	+	***
		Sphyrion height	+	***
		Foot breadth	-	**
PC5	0.53	Total head height	-	**
		Sitting height	+	**
		Weight	-	*
		Chest circumference at mesosternale	+	*
		Wall-acromion distance, arms hanging freely	+	*
PC6	0.81	Height	-	***
		Sitting height	+	***
		Foot breadth	+	***
		Chest breadth at xiphiale	-	*
		Minimum abdominal circumference	+	*
		Knee height, sitting	+	*

***:$|p|<0.001$, **:$|p|<0.01$, *:$|p|<0.05$, Adj. R^2: adjusted R^2

DISCUSSION

ACCURACY OF THE RECONSTRUCTION

The accuracy of the reconstruction of FCRs was evaluated using the data from a subject (subject A: 1.728 m, 66.6 kg), whose data were not used for calculating estimation functions (see Figure 6). In Figure 6, black lines indicate the reconstructed arrangement, and gray lines indicate the measured arrangement. Results of subject A shows that the reconstructed whole body proportion was close to the actual data (Figure 6, frontal view). However, there were large differences in the posture of the upper arm and torso (Figure 6, side view). These differences may be caused by low reproducibility of the posture from body dimensions as shown by low multiple correlation coefficients for PC5 (Table 2). Figure 7 shows the difference between the reconstructed and measured segment lengths of FCRs of subject A. The solid line in Figure 7 indicates the error between measured and reconstructed segment lengths, and the broken line indicates the error relative to the measured segment lengths. The average difference was 12.79 ± 10.56 [mm] and the maximum difference was 39.83 mm at abdomen. Furthermore, the maximum rate of the error was 32.1 % at pelvis. These results show that the accuracy of the present method for estimating the arrangement of the FCRs is good except for the lengths of lower body segments and posture. Especially, the large error of the segment lengths at abdomen and pelvis seems related to the inaccurate estimation of the posture of waist.

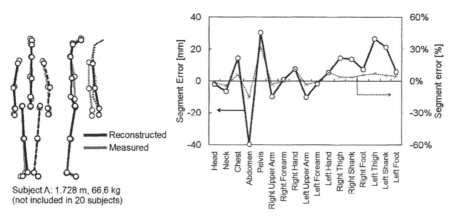

Subject A: 1.728 m, 66,6 kg
(not included in 20 subjects)

Figure 6. Comparison with reconstructed and actual arrangement of FCRs

Figure 7. Difference between the reconstructed and the measured segment lengths for Subject A

ARRANGEMENT OF FCRs IN BOUNDARY FAMILY

By using the multiple regression equations, the arrangement of FCRs can be reconstructed from body dimensions. The 23 body dimensions for 9 members of a boundary family for young adult male Japanese were calculated using factor analysis (Kouchi et al., 2004). Using these body dimensions and multiple regression functions, the 3D coordinates of the 22 points were estimated for the members of the boundary family. Figure 8 shows the arrangement of FCRs of the boundary

194

family which is located in the two-dimensional distribution map. In (Kouchi et al., 2004), the horizontal axis indicates the height, and the vertical axis indicates the thickness of the body. These interpretations are similar to those of PC1 and PC2 in this study, respectively. Therefore, it seems that boundary family from FCRs can be exchanged for boundary family from body dimensions.

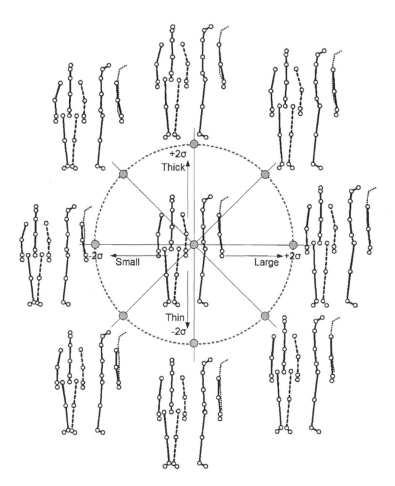

Figure 8. Arrangement of FCRs of Japanese boundary family

COMPARISON WITH GEBOD

GEBOD (Generator of body data) (Cheng et al., 1996) has been extensively used to estimate joint locations from body dimensions. GEBOD needs 32 body dimensions in order to estimate other body dimensions and inertial properties. From these geometric properties, each body segment is modeled as an ellipsoid. The joint locations are decided as intersection points between 2 ellipsoidal segments. Thus, GEBOD can create a body form and rigid link model simultaneously. However, the joint locations of GEBOD do not coincide with anatomy, posture and actual movement of joint. On the other hand, FCRs in the present study reflect the actual kinematic body information. Therefore, our method is more suitable to simulate functional dimensions accurately, and GEBOD is more useful to solve kinetic problems such as impact dummy modeling.

CONCLUSIONS

In this study, we clarified a relationship between an arrangement of FCRs and body dimensions. As a result, the first principal component is related to the height and weight, and the second is related to the depth and breadth of the torso. By using this relationship, the arrangement of FCRs could be estimated for not only an actual human, but also 9 members of a boundary family whose body dimensions are determined statistically. A digital manikin with this arrangement of FCRs is expected to simulate functional dimensions more accurately. In the future, we will examine the relationship between the arrangement of FCRs and 3D body shape data or landmarks, which contain more information than body dimensions.

ACKNOWLEDGEMENTS

This study was a part of the research project carried out by Digital Human Technology Consortium Japan.

REFERENCES

Kouchi, M., Mochimaru, M. and Higuchi, M. (2004). A validation method for digital human anthropometry: towards the standardization of validation and verification, SAE 2004 Transaction Journal of Aerospace, pp.254-259

Aoki, K., Kawachi, K, Kouchi, M. and Mochimaru, M. (2008). Functional Joint Rotation Centers for Whole Body Digital Manikin, Proceedings of SAE DHM, pp.2008-01-1859

Aoki, K., Kawachi, K, Kouchi, M. and Mochimaru, M. (2007). DhaibaShoulder: A Scalable Shoulder Model for Accurate Reach Envelope Using the Orbital Surface of the Functional Joint Center, SAE 2006 Transactions Journal of Passenger Cars: Electronic and Electrical Systems, pp.1081 -1086

Cheng, H., Obergefell, L. and Rizer, A. (1996). The development of the GEBOD program, Biomedical Engineering Conference, 1996., Proceedings of the 1996 Fifteenth Southern, pp.251-254.

CHAPTER 22

Modeling the Impact of Space Suit Components and Anthropometry on the Center of Mass of a Seated Crewmember

Christopher Blackledge[1], Sarah Margerum[2],
Mike Ferrer[1], Richard Morency[3], Sudhakar Rajulu[3]

[1]MEI Technologies
Anthropometry and Biomechanics Facility
NASA Johnson Space Center
Houston, TX 77058, USA

[2]Lockheed Martin
Anthropometry and Biomechanics Facility
NASA Johnson Space Center
Houston, TX 77058, USA

[3]National Aeronautics and Space Administration
Anthropometry and Biomechanics Facility
NASA Johnson Space Center
Houston, TX 77058, USA

ABSTRACT

The Crew Impact Attenuation System (CIAS) is the energy-absorbing strut concept that dampens Orion Crew Exploration Vehicle (CEV) landing loads to levels sustainable by the crew. Significant center of mass (COM) variations across suited crew configurations would amplify the inertial effects of the pallet and potentially create unacceptable crew loading during launch and landing. The objective of this study was to obtain data needed for dynamic simulation models by quantifying the effects of posture, suit components, and the expected range of anthropometry on the COM of a seated individual.

Several elements are required for calculation of a suited human COM in a seated position: anthropometry, body segment mass, suit component mass, suit component location relative to the body, and joint angles defining the seated posture. Three-dimensional (3-D) human body models, suit mass data, and vector calculus were utilized to compute the COM positions for 12 boundary manikins in two different seated postures.

The analysis focused on two objectives: (1) quantify how much the whole-body COM varied from the smallest to largest subject and (2) quantify the effects of the suit components on the overall COM in each seat configuration. The location of the anterior-posterior COM varied across all boundary manikins by about 7 cm, and the vertical COM varied by approximately 9 to 10 cm. The mediolateral COM varied by 1.2 cm from the midline sagittal plane for both seat configurations. The suit components caused an anterior shift of the total COM by approximately 2 cm and a shift to the right along the mediolateral axis of 0.4 cm for both seat configurations. When the seat configuration was in the standard posture the suited vertical COM shifted inferiorly by as much as 1 cm, whereas in the CEV posture the vertical COM had no appreciable change. These general differences were due to the high proportion of suit mass located in the boots and lower legs and their respective distance from the body COM, as well as to the prevalence of suit components on the right side of the body.

Keywords: Center of mass, COM, space suit, anthropometry, boundary manikins

INTRODUCTION

This paper describes a method by which three-dimensional (3-D) coordinates for the center of mass (COM) of a seated human were calculated in unsuited and suited conditions. Designers for the Crew Exploration Vehicle (CEV) requested suited human mass properties to perform their analysis of the Crew Impact Attenuation System (CIAS). The CIAS pallet is the current method by which the entire seated crew would be safely supported during launch and landing operations. This COM data was meant to serve as an input for dynamic modeling of the Orion seat pallet stroking mechanism. The Human Systems Integration Requirements (HSIR) document (NASA, CxP 70024, 2009) and the Man-Systems Integration Standards NASA-STD 3000 (MSIS) (NASA, 1995) the precursor to HSIR, define the requirements for human space flight. However, mass properties information about suited crewmembers does not exist in either document for the current suit architecture or posture required for the CIAS dynamic modeling. As part of a previous project in the Anthropometry and Biomechanics Facility (ABF), a group of three-dimensional boundary manikins were developed that represent the critical anthropometric dimension extremes found in the HSIR database (Young, Margerum, Barr, Ferrer, Rajulu, 2008 [A], [B]). The critical anthropometric dimensions for 6 of the male manikins and 6 of the female manikins were used in this analysis for calculation of whole body COM. The suit information used to calculate the mass properties was taken from data provided by the Extravehicular Activity Project Office (EVA) on the most recent Cx Launch, Entry, and Abort (LEA) configuration suit (NASA, CxE-EM-2009-0001, 2009).

METHODOLOGY

The general approach used in this paper to calculate the human mass properties was to use PolyWorks® software to segment scans of boundary manikins and determine the center of volume (COV) of body segments, apply a density function to determine the COM, and then use a custom MATLAB® script to rotate the COM positions into the seated posture. Suit components were treated as point masses and their positions along the body segments were based on that segment's length and COM. The composite COM position was then calculated in 3-D space.

ASSUMPTIONS

Several simplifying assumptions are required for the calculation of mass properties for human data. Without these assumptions the mass properties would have to be empirically measured using methods exceeding the resource limitations of the current project. The following assumptions are applicable to the data reported. 1) The human body is modeled as a rigid object composed of linked segments; soft tissue deformation, spinal curvature, and movement of internal masses are not accounted for. 2) The density of the human body is assumed to be a homogeneous 1000 kg/m^3. This value was taken from similar human mass studies (Chandler et al., 1975; Young et al., 1983) and is applicable within the range of suited pressures defined in the HSIR (NASA, CxP 70024, 2009). 3) It is also assumed that the body segment planes are close to joint centers of rotation.

AXIAL SYSTEM

The axial coordinate system used for all data presented in this paper has its origin at the whole-body COM and is related to the typical body planes of symmetry. The positive x-direction is described as extending in the anterior direction out of the chest in the sagittal plane and perpendicular to the frontal plane. The y-direction extends laterally out of the left side of the body in the frontal plane and perpendicular to the sagittal plane. The positive z-direction is described as extending superiorly out of the top of the head in the frontal plane and perpendicular to the transverse plane. The axial directions can be seen in Figure 1.

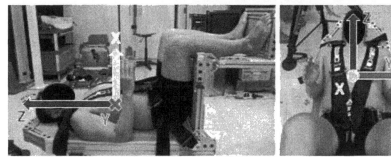

Figure 1. Example of seated axial system

CEV SEATED POSTURE

The assumed seated posture for the current CEV configuration was a variation on the standard seated posture described in the MSIS (Figure 2), herein called the MSIS posture. The axial system was anatomically based, and COM locations were referenced to the seat pan and seat back, and thus were independent of seat positioning (i.e. recumbent versus upright). Because the seat configuration affects the leg posture, the joint angles used to describe the leg seated posture were taken from a previous ABF evaluation of the Orion CEV mockup. A list of the joint angles and associated anatomical landmarks used for the CEV seat configuration are provided in Table 1.

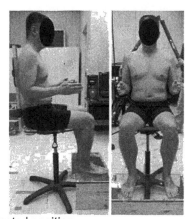

Figure 2. MSIS standard seated position

Table 1: Calculated CEV Seated Configuration Joint Angles

Joint Angle	Associated Points	Angle Value (degrees)
Shoulder	Elbow, Acromion, Side of Torso	0
Elbow	Acromion, Elbow, Wrist	90
Hip	Torso, Hip, Knee	86
Knee	Hip, Knee, Ankle	75

BODY SEGMENTS

The method used to divide the body into rotatable segments followed the body segmentation used in the MSIS and HSIR. The body was divided along joint centers of rotation or anatomical planes into 17 segments. The segmentation divisions are shown in Figure 3. The body scan segmentation was done using the PolyWorks® software to construct bisecting planes at the anatomical reference points. Examples of these planes along with the associated T-pose posture in PolyWorks® are also shown in Figure 3.

Figure 3. Body segments and PolyWorks® T-Pose and segment planes

DETERMINATION OF SEGMENTAL COM

After the body scan segmentation was complete, the open segment ends were closed to create solid segment volumes. PolyWorks® was used to output the COV coordinates, relative to a body coordinate system centered at the navel, for each of these volumes. As a constant density of the body segments was assumed, the segment COV location coincided with the segment COM location. The resulting segmental COM locations and anatomical body landmarks were exported from PolyWorks® for incorporation in a MATLAB® program.

SUITED COMPONENTS

The suited data used for this paper was based on the most current Cx LEA suit configuration information available. Because the suit architecture was still in an early stage of development during the writing of this paper, an assumption was made that the suit masses were applicable for the entire size range of manikins used, essentially creating a "one-size fits all" suit. No sizing rings or reduction of materials were accounted for in the application of suited components.

For the COM calculation, all suit components were considered point masses and were applied to either the associated body segment COM; translated along the long axis of the body segment some distance from the COM based on anthropometry; or were positioned in reference to a set of anthropometric points and measures. A diagram illustrating the COM line-of-action translation is shown in Figure 4 below. The percentages for translation distance were determined from conversations with EVA personnel familiar with the suit components and suit fitting. Position vectors in the body centered coordinate system were determined to mathematically apply the suit to the boundary manikins.

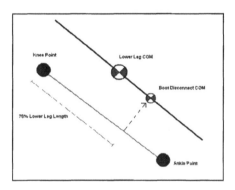

Figure 4. Suit segment translation diagram

Suit components that are mostly uniform in composition and are evenly distributed over the body had their total mass divided by a percentage per body segment, and that percentage of component mass was added to the mass of the corresponding body segment. This resulted in no direct change to the body segment's composite COM position.

Rigid suit components such as bearings and disconnects had their point masses positioned in-line with the applicable body segment COM as a function of that segment's anthropometric length, as shown in Figure 4.

The flexible material of the suit located between segment disconnects and bearings, referred to as soft goods, were treated differently from the soft components distributed evenly over the body. Certain soft good components differed in thickness and were segmented differently because of the locations of disconnect points and bearings. For these soft good components, a mass per suit segment was provided and positioned in line with the body segment COM, at the midway point between an anatomical landmark and a bearing or disconnect.

Miscellaneous rigid suit components, except for the helmet, were placed using a methodology similar to the one used for bearings and disconnects. The difference with these components is that they were not placed along the body segment COM. Most of these components were positioned along the body surfaces. Vector calculus and available anatomical reference points were used to determine a displacement position vector for each of these suit components on a case-by-case basis. The helmet and its associated components were assumed to be evenly distributed on the head and were attributed directly to the head's COM.

SEGMENTAL ROTATION

Once the body was segmented and suit component parts were positioned, all the COM points were rotated from the standing T-pose posture into the MSIS seated and CEV seated posture. The rotations were accomplished by multiplying the position vectors by a series of rotation matrices hierarchically down the body at anatomical joint rotation center points.

WHOLE BODY COM

The calculation for the whole-body suited COM in the MSIS and CEV seated postures is based on the general equation for composite centroid calculation in three dimensions as seen in Equation 1 below.

$$\overrightarrow{COM}_{Whole\_body} = \frac{\sum(mass_i * \overrightarrow{COM_i})}{\sum mass_i}$$

Equation 1. 3D composite centroid equation

This equation was populated with all the body segment COMs and suit segment COMs along with their position vectors to output a position vector describing the composite COM in relation to the anatomical origin established in PolyWorks®.

Once the whole-body COM position vector was determined (Figure 5), it was set as the global origin for the coordinate system. The resulting segmental COM position vectors and body landmark vectors could then be broken down into axial components for each selected coordinate frame of reference. The COM locations were referenced to an artificial seat pan and seat back to provide a frame of reference for modeling purposes. The unsuited total COM, suited total COM, and body landmarks were then exported as the final results. The unsuited COM results obtained from the method described here were checked for validity against the regression equations provided in the MSIS (NASA, 1995, p. 3-65) and were analogous to the COM displacement value range provided by the MSIS equations.

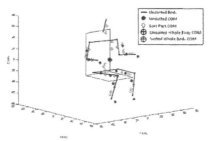

Figure 5. Representative COM output from MATLAB® of a suited individual seated in the MSIS posture

RESULTS

MSIS WHOLE BODY COM

The range of the whole-body COM of manikins using the MSIS standardized sitting posture (Figure 2) in both the unsuited and suited configurations is provided in Table 2. Some data is provided with respect to selected seat hardware locations, referenced as such in order to reconstruct position and placement using the CAD model of the CIAS.

Table 2: COM range for 12 boundary manikins of a MSIS Suited Individual

	Unsuited Minimum (cm)	Unsuited Maximum (cm)	Unsuited Range (cm)	Suited Minimum (cm)	Suited Maximum (cm)	Suited Range (cm)
Seat Back	18.0	24.9	6.9	19.8	27.0	7.2
Seat Pan	24.0	32.8	8.8	23.0	32.8	9.8
From Hip Midline[1]	-0.3	0.5	0.8	-0.5	0.9	1.4
From Shoulder Midline[1]	-0.1	1.1	1.2	0.3	1.4	1.2
From Unsuited COM as Midline[1]	0.0	0.0	0.0	0.3	0.5	0.2

1- A negative value means the BODY COM is to the left of the midline sagittal plane, a positive value means the BODY COM is to the right of the midline sagittal plane.

CEV WHOLE BODY COM

The range of the whole-body COM using the CEV sitting posture (Table 1) in both the unsuited and suited configurations is provided in Table 3. Identical to the MSIS posture in Table 2, some data is provided with respect to selected seat hardware in order to assist in the reconstruction of the COM position and placement within the CAD models.

Table 3: COM range for 12 boundary manikins of a CEV Suited Individual

	Unsuited Minimum (cm)	Unsuited Maximum (cm)	Unsuited Range (cm)	Suited Minimum (cm)	Suited Maximum (cm)	Suited Range (cm)
Seat Back	17.6	24.4	6.8	19.2	26.3	7.1
Seat Pan	24.6	33.8	9.2	23.8	34.0	10.2
From Hip Midline[1]	-0.8	0.5	1.2	-0.5	0.9	1.4
From Shoulder Midline[1]	-0.1	1.1	1.2	0.3	1.4	1.2
From Unsuited COM as Midline[1]	0.0	0.0	0.0	0.3	0.5	0.2

1- A negative value means the BODY COM is to the left of the midline sagittal plane, a positive value means the BODY COM is to the right of the midline sagittal plane.

DISCUSSION

In general, for the MSIS posture, the suited configuration shifted the COM forward by approximately 2 cm relative to the seat back when compared to the unsuited configuration. This general change was due to the mass located in the boots and lower legs of the suit and its corresponding distance from the body COM. The suited configuration also shifted the COM down towards the seat pan by about 1 cm. Again, this was due to the mass located in the boots and lower legs of the suit and their corresponding distance from the body COM. The manikins with longer lower leg lengths showed the greatest amount of change in this regard. Finally, addition of the suit shifted the COM to the right of the midline sagittal plane by approximately 0.4 cm. This shifting of the COM corresponded to the extra components located on the right side of the body, yielding an asymmetry in the suited COM.

The CEV posture, in comparison, shifted the suited COM forward by an average of about 1.9 cm relative to the seat back. This general change was again due to the mass located in the boots and lower legs of the suit and its corresponding distance from the body COM. However, this change did not match the change associated with the MSIS posture because the knees had been drawn closer to the chest by the acute hip angle effects. Unlike the MSIS configuration, the CEV manikins did not have any appreciable change in the vertical COM caused by the suit. This was attributable to the pulling of the legs toward the chest, thus shifting the relative weight vertically. Similar to the MSIS posture, the CEV posture had the shift of the body COM toward the right of the midline sagittal plane because of the extra suit components on the right side.

Other interesting points to note in the data are that the COM locations varied across the subjects for both the unsuited and suited conditions in each seat posture (Tables 2 and 3): the anterior- posterior COM varied by approximately 7 cm, the vertical COM varied by approximately 9 to 10 cm, and the right-left COM varied by approximately 1.2 cm around the midline sagittal plane over the range of subjects. The scale of variation, especially in the anterior-posterior direction, was not anticipated in the initial hypothesis.

CONCLUSION

Because of the variation observed in the results, it is highly recommended that during incorporation of individual crewmember mass and COM data, care is exercised in assessing the impact of overall crew mass and COM locations for a crew complement of 2, 3, or 4 during dynamic modeling of the CIAS. Care must be taken with regard to proper setup and validation of the assorted crew combinations due to the variation in individual sitting position within the CEV seats, the individual variation of the COM placement relative to the seat, the effects of the various suit components, and the overall group variation in body anthropometry in the CEV. The benefits of this study are twofold: first, the methods used to predict overall COM for unsuited and suited individuals in a unique posture underwent a proof of concept, and second, the effect of the suit on the COM of a seated individual was estimated. The results can be refined further as more definitive suit mass components and their associated center-of-mass locations are developed in the prototype phase of the suit development process. This study is just the preliminary step in assessing the impact of the suited crewmembers on the larger vehicle as a whole.

REFERENCES

Chandler, R.F., et al. (1975). *Investigation of Inertial Properties of the Human Body*. Aerospace Medical Research Laboratory, Aerospace Medical Division, Air Force Systems Command, AD-A016485. Wright-Patterson Air Force Base, Ohio.

National Aeronautics and Space Administration (NASA). (2009). *Human-Systems Integration Requirements* (HSIR), CxP 70024, Revision C. Houston, Texas.

National Aeronautics and Space Administration (NASA). (1995). *Man-Systems Integration Standards* (MSIS) NASA-STD-3000, Volume II, Revision B. Houston, Texas.

National Aeronautics and Space Administration (NASA). (2009). *EVA Suit Segment/Seat Interface Analysis, ODAC-4 Initialization Data*, CxE-EM-2009-0001. Houston, Texas.

Young, J.W., et al. (1983). *Anthropometrics and Mass Distribution Characteristics of the Adult Female*. FAA Civil Aeromedical Institute, Federal Aviation Administration, AD-A143096. Oklahoma City, Oklahoma.

Young K., Margerum S., Barr A., Ferrer M., Rajulu S. (2008). *Generation of Boundary Manikins Anthropometry*. International Conference on Environmental Systems. San Francisco, California. [A]

Young K., Margerum S., Barr A., Ferrer M., Rajulu S. (2008). *Derivation of Boundary Manikins: A Principal Component Analysis*. Digital Human Modeling for Design and Engineering Conference. Pittsburgh, Pennsylvania. [B]

CHAPTER **23**

4D Anthropometry: Measurement and Modeling of Whole Body Surface Deformation for Sports Garment Design

Masaaki Mochimaru[1], Sang-Il Park[2]

[1]Digital Human Research Center
National Institute of Advanced Industrial Science and Technology
2-3-26 Aomi, Koto-ku, Tokyo 135-0062 Japan

[2]College of Electronics & Information Engineering
Sejong University
98 Gunja-Dong, Guwangjin-Gu
Seoul 143-747 Republic of Korea

ABSTRACT

3D body scanners are widely used for apparel design. Only static body shapes are used, although wear should be fit in motion. In this study, whole body surface deformation in motion was measured by a 4D measurement method based on dense static shape scanning and sparse dynamic motion capturing. Skin deformation was calculated by the principal strain of surface triangles and visualized by the original software. Visualized results were utilized for sports garment design considering skin movements.

Keywords: Digital human, anthropometry, apparel design, sports biomechanics

INTRODUCTION

3D body scanners are widely used for sizing surveys. Body dimensions are derived from 3D body shape, moreover 3D shape information can be analyzed using homologous modeling and statistics (Mochimaru and Kouchi, 2009). In apparel industries, those 3D body shape models are utilized for designing wear. Although wear should be fit in motion, only static data are used. Motion effects are concerned as allowances by expert experiences. Motion effects are more important for sports garment design. A sports equipment company, Mizuno developed sports garments considering skin deformation (http://design-viz.com/case_study/case_01.html, in Japanese). In this case, the skin deformation was synthesized by computer graphic technologies, not measured actual human deformation in motion. It suggests that apparel industries have technologies and know-how for designing sports garments based on the local skin deformation data. Unfortunately, they have only synthesized data not actual data. In this study, we measured whole body surface deformation in sports motion by a 4D measurement method based on dense static shape scanning and sparse dynamic motion capturing. Skin deformation was calculated by the principal strain of surface triangles and visualized by the original software.

BACKGROUND

4D measurement technologies, measurement methods of time-sequential shape deformation have been developed in two approaches. One is speed-up of a static body scanning method. The other is a combination method of dense static shape scanning and sparse dynamic motion capturing. Representative methods of the first approach, multiple light patterns are projected onto body surface very quickly, and a pixel location is calculated by triangulation of a projector and a camera. These methods are used for modeling of facial expression (Boehnen and Flynn, 2005), dynamic deformation modeling of foot (Coudert et. al, 2006, Schmeltzpfenning et. al, 2009, Kimura et. al, 2009). Using a high-speed projector and a high-speed camera, body deformation can be measured in over 200 Hz. The whole shape of a body part was measured by multiple projector-camera sets. Dense surface points can be measured by these methods without markers, whereas shape representation methods based on anatomical correspondences are not mature. Measurement methods of the second approach aim to obtain dense, dynamic and anatomically homologous models of body deformation. Dense data is obtained by 3D scanning. Dynamic and anatomically homologous data is obtained by motion capturing. Those methods are used in computer graphics, such as facial expression. The facial measurement method was expanded to the whole body measurement (Park and Hodgins, 2006). In this study, we used this method for measurement of whole body surface deformation in sports motion. With this method, time-sequential body shapes can be represented by a common homologous model.

4D MEASUREMENT METHOD

Details of the measurement method should be referred to the original paper (Park and Hodgkin's, 2006). In this section, a summary of the method related to deformation modeling is mentioned. Static body shape of the reference pose is measured by a 3D scanner (FIGURE 1(a)). A total number of data points of the body surface is around one million. Then small mocap (motion capture) markers are located on the body surface like a mesh (FIGURE 1(b)). The number of mocap markers is around 400. At that time, it is not required to place the mocap markers on anatomical landmarks. The reference pose is measured by a motion capture system. Then, the mesh connection information of mocap markers is defined manually. Moreover the near-rigid segment is defined as the mocap marker group according to the body segment (FIGURE 1(c)). In the next process, dense data points measured by the 3D scanner are represented by sparse mocap markers. The near-rigid segments of mocap markers are moved to fit into the dense shape data automatically. Through this process, small difference of two reference poses between 3D scanning and motion capturing are corrected. After pose correction, dense data points are represented by the neighboring local coordinate system of mocap markers. One dense data point is represented by over three neighboring local coordinate systems of mocap markers. In motion sequence, the mocap makers are moved from the reference pose, then the dense point coordinate is moved according to the local coordinate systems of mocap markers based on initial representation of the reference pose. It is the base estimation of the deformation. Actually, mocap markers on the skin do not move rigidly. Thus, the remaining error caused by the skin deformation is modeled by a quadratic transformation matrix and the remaining error calculated from the model was added to the base estimation (FIGURE 1(d)).

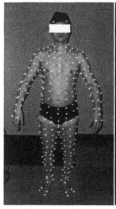
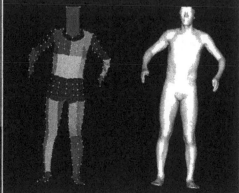

(a) Static body shape in the reference pose

(b) mocap markers placement

(c) near-rigid segmentation by mocap marker grouping

(d) Body shape deformation based on mocap markers

FIGURE 1. Mocap markers and segmentation

SKIN DEFORMATION MODELING

Time-sequential homologous shape models are obtained by the 4D measurement method. Skin deformation can be calculated from corresponding triangles. Two corresponding triangles $\mathbf{P}_1$ and $\mathbf{P}_2$ are converted into the same 2D coordinate system as $\mathbf{Q}_1$ and $\mathbf{Q}_2$. The triangle $\mathbf{Q}_2$ is registered to minimize the summation of Euclidean distance between corresponding vertices of $\mathbf{Q}_1$ and $\mathbf{Q}_2$. The triangle $\mathbf{Q}_2 = \{\mathbf{Q}_{2u}\}$, $\{\mathbf{Q}_{2v}\}$ is relocated into $\{\mathbf{Q}'_{2u}\}$, $\{\mathbf{Q}'_{2v}\}$. Then, two corresponding triangles are described as follows:

$$\{\mathbf{Q}_{1i}\} = (Q_{1iu}, Q_{1iv}), \ \{\mathbf{Q}_{1j}\} = (Q_{1ju}, Q_{1jv}), \ \{\mathbf{Q}_{1k}\} = (Q_{1ku}, Q_{1kv})$$
$$\{\mathbf{Q}'_{2i}\} = (Q'_{2iu}, Q'_{2iv}), \ \{\mathbf{Q}'_{2j}\} = (Q'_{2ju}, Q'_{2jv}), \ \{\mathbf{Q}'_{2k}\} = (Q'_{2ku}, Q'_{2kv})$$

The strain tensor between two triangles can be calculated by the following formula:

$$\varepsilon = \begin{Bmatrix} \varepsilon_u \\ \varepsilon_v \\ \gamma_{uv} \end{Bmatrix} = [\mathbf{B}]\{\mathbf{d}\}$$

$$[\mathbf{B}] = \frac{1}{2A} \begin{bmatrix} Q_{1jv} - Q_{1kv} & 0 & Q_{1kv} - Q_{1jv} & 0 & Q_{1iv} - Q_{1jv} & 0 \\ 0 & Q_{1ku} - Q_{1ju} & 0 & Q_{1iu} - Q_{1ku} & 0 & Q_{1ju} - Q_{1iu} \\ Q_{1ku} - Q_{1ju} & Q_{1jv} - Q_{1kv} & Q_{1iu} - Q_{1iv} & Q_{1kv} - Q_{1iv} & Q_{1ju} - Q_{1iu} & Q_{1iv} - Q_{1jv} \end{bmatrix}$$

$$\{\mathbf{d}\} = \begin{Bmatrix} u_i \\ v_i \\ u_j \\ v_j \\ u_k \\ v_k \end{Bmatrix} = \begin{Bmatrix} Q'_{2iu} - Q_{1iu} \\ Q'_{2iv} - Q_{1iv} \\ Q'_{2ju} - Q_{1ju} \\ Q'_{2jv} - Q_{1jv} \\ Q'_{2ku} - Q_{1ku} \\ Q'_{2kv} - Q_{1kv} \end{Bmatrix}$$

where A is the area of the triangle $\{\mathbf{Q}_{1i}\}$ and calculated by the following formula:

$$A = \frac{1}{2} \begin{vmatrix} 1 & Q_{1iu} & Q_{1iv} \\ 1 & Q_{1ju} & Q_{1jv} \\ 1 & Q_{1ku} & Q_{1kv} \end{vmatrix}$$

Then the principal strain and the direction of the principal strain can be calculated by the following formula:

$$\left.\begin{array}{c}\varepsilon 1\\ \varepsilon 2\end{array}\right\} = \frac{\varepsilon_u + \varepsilon_v}{2} \pm \sqrt{\left(\frac{\varepsilon_u - \varepsilon_v}{2}\right)^2 + \gamma_{uv}^2}$$

$$a_n = \frac{1}{2}\tan^{-1}\frac{2\gamma_{uv}}{\varepsilon_u - \varepsilon_v} \pm \frac{\pi}{2}$$

The principal strain and its direction were calculated for all corresponding triangles and visualized as FIGURE 2 by the original software. The color intensity indicates the magnitude of the principal strain and the small line indicates the direction of the principal strain. The body shape can be visualized with the principal strain information from any point of view and any point in time. In visualization, the small line of the direction of the principal strain is located a little bit far from the polygon surface to avoid occlusion of the line from the surface polygons.

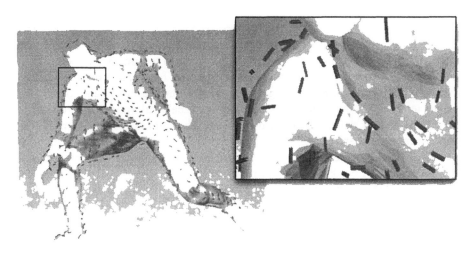

FIGURE 2. Visualization of skin deformation

EXPERIMENTS

Three expert male athletes (1 badminton, 1 soft tennis and 1 tennis) participated in the experiment. A subject wore a short swim pants as shown in FIGURE 1(b). Then the body shape of the reference pose was measured by a 3D scanner (Hamamatsu BL-Scan). After 3D scanning, a measurer placed around 400 mocap markers (4[mm] in the diameter) onto the subject's body surface as shown in FIGURE 1(b). The following 8 motions were measured by a motion capture system (Vicon Motion Systems Vicon-MX, 200Hz, 12 cameras). Each motion was started with the reference pose.

> Motion-1: Forehand cut/stroke (cut: badminton, stroke: tennis, soft tennis)
> Motion-2: Backhand cut/stroke (cut: badminton, stroke: tennis, soft tennis)
> Motion-3: Forehand lob
> Motion-4: Backhand lob
> Motion-5: Smash
> Motion-6: Jumping smash
> Motion-7: Left-side jumping smash (only for badminton player)
> Motion-8: Right side jumping smash (only for badminton player)

Three complete trials were measured for above 8 motions. In total, it took 2 hours for one subject.

RESULTS AND DISCUSSION

Data for one subject contained 3D body shape data (around one million data points x 1 posture) and motion data (around 400 mocap markers x around 1000 frames x 8 motions x 3 trials). The original 3D body shape data had too much data points, thus the number of polygons were reduced to shorten the data processing time. With Geomagic Studio software (Geomagic Inc.), around one millions data points were reduced to around 34000 data points. An operator labeled all mocap markers at the reference pose and defined near-rigid segment. This process required around 40 hours. Then the mocap markers of the reference pose were modified to fit into the body shape data of the reference pose obtained by the 3D scanner. This process was fully automatic and needed around 30 minutes. Through the process around 34000 data points of the body surface were represented by mocap markers' reference frames. This information of representation was saved as a template file for the subject. Then, time-sequential mocap marker locations were tracked and labeled automatically considering marker missing. It required 30 minutes. After this process, complete motion data (around 400 mocap makers x around 1000 frames) for one trial were obtained. Using the template file, dense surface shape data in motion were calculated from the motion data. As a result, shape deformation data (around 34000 data points x around 1000 frames) were obtained. It took around 30 minutes for computing one trial data.

The principal strains and their directions were calculated and visualized for all triangles of the shape deformation data. The strain was calculated between each shape data at the moment (Q_2) and the shape data of the reference pose (Q_1). In visualization, the direction of the principal strains showed sparsely for easy understanding. Visualized directions were calculated from mocap data polygons.

Representative results are shown in FIGURE 3 and 4.

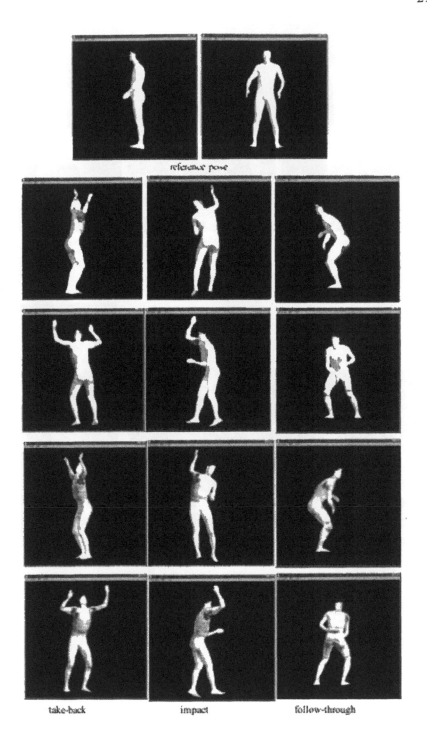

FIGURE 3. Results of skin deformation (Smash motion, badminton)

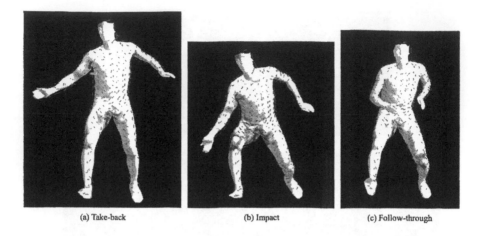

(a) Take-back   (b) Impact   (c) Follow-through

FIGURE 4. Results of skin deformation (Forehand cut motion, badminton)

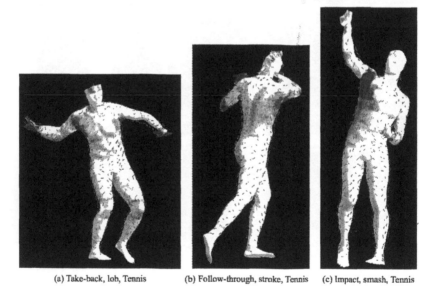

(a) Take-back, lob, Tennis   (b) Follow-through, stroke, Tennis   (c) Impact, smash, Tennis

FIGURE 5. Results of skin deformation (Tennis)

A large skin deformation can be observed in the armpit, the breast (pectoralis major muscle) and the posterior part of the thigh (hamstrings). Several body parts had a large deformation only in the specific sport. In badminton motion, the large deformation was observed in the lower extremity. Especially, the crotch and the anterior part of the knee (patella) had a large skin deformation. In tennis motion, whereas, large deformation was observed in the upper extremity. The directions of the principal strains were observed as the direction to twist the upper body. In soft tennis, larger deformation was observed in the shoulder part.

Calculated skin deformation was caused by two principal factors. One is skin lengthening by the joint flexion. The deformation is observed due to posture change. It is considered that the skin deformation at the shoulder part, the crotch part and the knee part was caused by this factor. The other factor is skin deformation due to muscle protrusion. In the standing posture of knee flexion, the quadriceps femoris activates and the skin deformation of this part is observed. It is difficult to separate these two factors from calculated skin deformation by this method. It is considered that the effect of the first factor can be larger than the second factor. Because the principal strain of the armpit area is over 3.0, that is 5 times larger the principal strain of the anterior part of the thigh due to quadriceps femoris activation.

Results were utilized for designing a new sports garment by a collaboration company. High performance uni-direction extensible materials were used for the area of large skin deformation considering the direction of the principal strains of the area. Materials were selected based on the magnitude of the principal strains. Moreover, seam lines were designed carefully considering the direction of the principal strains, because the seam lines were not extensible along the seam.

In future works, we plan to validate performance of the new sports garment. Furthermore, a common homologous body model will be used for all individuals. In this study, an individual dense body shape was generated from individual 3D scan data. It was commonly used in the same subject, however different homologous models were used for different individual subjects. Thus it was impossible to compare the skin deformation for different individuals. Based on common homologous modeling, statistical analysis of skin deformation will be available.

CONCLUSION

Whole body surface deformation in sports motion was measured by the method using dense static shape scanning and sparse dynamic motion capturing. Skin deformation was calculated by the principal strain of surface triangles and visualized by the original software. Results were used for sports garment design. Traditional anthropometry is limited to dimensions. With 3D anthropometry, body dimensions and static body shape information are available. We propose the next generation technologies, 4D anthropometry through this study. Not only static shape but also deformation in motion are available.

ACKNOWLEDGEMENT

A part of this study was supported by Yonex Company, Ltd.

REFERENCES

Boehnen, C. and Flynn, P. (2005). Accuracy of 3D Scanning Technologies in a Face Scanning Scenario. The 5th International Conference on 3-D Digital Imaging and Modeling, Ottawa.

Coudert , T., Vacher, P.   and Van der Zande, M. (2006), A method to obtain 3D foot shape deformation during the gait cycle. 9th Symposium on 3D Analysis of Human Movement, Valenciennes, France.

Kimura, M., Mochimaru, M. and Kanade, T. (2009), Measurement of 3D Foot Shape Deformation while Running - Simultaneous Measurements by Multiple Projector-Camera Systems -. IEEE International Workshop on Projector Camera Systems PROCAMS2009, Miami Beach, FL.

Mochimaru, M. and Kouchi, M. (2009), Statistics of 3-D body shapes using PCA or MDS anr their applications, 17th World Congress on Ergonomics (IEA 2009), Beijing, China.

Park , S. I. and Hodgins, J. K. (2006). "Capturing and Animating Skin Deformation in Human Motion." acm Transactions on Graphics (ACM SIGGRAPH 2006) 25(3): 881-889.

Schmeltzpfenning, T., Plank, C., Krauss, I., Aswendt, P. and Grau, S. (2009). Dynamic foot scanning: A new approach for measurement of the human foot shape while walking. ISB 9th Footwear Biomechanics Symposium, Stellenbosch, South Africa.

CHAPTER 24

Simulation of the Body Shape after Weight Change for Health-Care Services

Makiko Kouchi, Masaaki Mochimaru

Digital Human Research Center
National Institute of Advanced Industrial Science and Technology
Tokyo 135-0064, Japan

ABSTRACT

Increase of people with overweight and obesity is a serious health problem in many countries. An effective solution to this problem is for people to change their lifestyle. Strong fear appeal and high-efficacy message often produce such positive actions. Seeing one's own current body shape and body shape after gaining or losing weight may motivate people to start and continue actions that may improve health conditions. Targeting middle-aged males, we have developed a system to simulate the body shape after gains or loss of weight. The simulation is based on the information of inter-individual body shape variation described by principal components. We estimated seven circumferences after weight change, and these values as well as changed weight and height were used to estimate principal component scores. The body shape reconstructed from estimated principal component scores is adjusted for the segment lengths of the body that would not change after weight change. Finally seven circumferences are adjusted for the estimated target values. The developed system will be usable in fitness clubs for advising people for health actions.

Keywords: Body shape, Body scan, Intra-individual shape change, Health action

INTRODUCTION

Increase of people with overweight and obesity is a serious health problem in many countries. An effective solution to this problem is for people to change their lifestyle. Strong fear appeal and high-efficacy message often produce such positive actions (Witte and Allen, 2000). Seeing one's own current body shape and body shape after gaining or losing weight may motivate people to start and continue actions that may improve their health conditions. Therefore we have developed a system to simulate the body shape of a person after he/she gains or loses weight. We set middle-aged males as the target population because the rate of overweight in Japan is highest in males aged 30-59.

Inter-individual variations in the human body shape can be represented by a small number of principal components (Allen et al., 2003; Xi et al., 2007). We utilize principal components that represent inter-individual shape variation to simulate the body shape after the weight change.

SIMULATION OF THE BODY SHAPE AFTER WEIGHT CHANGE

The body shape after weight change is simulated according to the following procedure: (1) Obtain body shape models by fitting high resolution template meshes, and analyze these models by principal component analysis (PCA), (2) Calculate multiple regression equations to estimate principal component (PC) scores from nine selected body dimensions (weight, height, and seven circumferences), (3) Calculate simple linear regression equations to estimate changes in the seven circumferences after weight change, (4) Estimate PC scores of the changed body shape from the nine body dimensions (changed weight, height, and seven circumferences after weight change), (5) Reconstruct the body shape model from the estimated PC scores, (6) Adjust the lengths of the internal skeleton of the reconstructed body shape model to those of the original body shape model. (7) Adjust the seven circumferences of the modified reconstructed body shape model to the estimated target values.

We used scan data of 209 males aged 19-59. Each subject wore shorts and swimming cap and was scanned by a bodyline scanner (Hamamatsu photonics K. K.) without any marker stickers on landmark locations. Locations of 64 landmarks and body dimensions were automatically calculated using Bodyline Manager (Hamamatsu photonics K. K.) Weight was measured using a scale. A body shape model was created for each subject by fitting high-resolution template meshes to his scan (Mochimaru, 2008) (Figure 1). A digital man model of Dhaiba (Digital Human Aided Basic Assessment system) (Mochimaru 2006) was used as the template. Dhaiba model has internal skeleton, functional rotation centers of 17 joints, and skinning parameters for smooth surface deformation according to joint flexion.

Subjects were divided into three age groups (age<40, 40≤age<50, age≥50), and all analyses were conducted for three age groups separately. Coordinates of vertices of body shape models were analyzed by PCA. In all analyses 12 or 13 PCs with contribution larger than 1% explained over 90% of the total variance.

Shape change after the cessation of height growth is mainly due to the increase of soft tissues. Therefore we selected seven circumferences as well as body height and weight for estimating PC scores after the weight change. Left side measurements were used for bilateral measurements. We calculated multiple regression equations to estimate PC scores from the selected nine body dimensions.

Simple linear regression equations were calculated to estimate changes in the seven circumferences from the change in the weight. For each circumference, the value after the weight change is calculated by adding the estimated change and the original value. Body height was set constant.

Principal component scores after the weight change were calculated using the multiple regression equations. However, the scores were set to 0 for PCs with very small contributions (31st and over) or PCs whose scores could not be estimated by multiple regressions ($R^2<0.1$ or coefficients of all measurements were not significant at the 5% level). The body shape model after the weight change was reconstructed using these estimated principal component scores.

Principal component scores after the weight change were calculated using the multiple regression equations. However, the scores were set to 0 for PCs with very small contributions (31st and over) or PCs whose scores could not be estimated by multiple regressions ($R^2<0.1$ or coefficients of all measurements were not significant at the 5% level). The body shape model after the weight change was reconstructed using these estimated principal component scores.

In order to keep the integrity of skeletal structure, lengths of the internal skeleton of the body shape model after weight change were adjusted to the values of the original body shape model. When the values of seven circumferences obtained from thus adjusted body shape model differ from the target values, the body shape model was adjusted automatically by local deformation to minimize the differences.

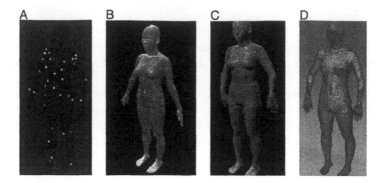

Figure 1. Creating an individual body shape model. A: original scan with landmarks, B: template meshes with landmarks, C: adjusting the posture and segment lengths, D: created individual model.

A HUMAN BODY SHAPE SIMULATION SYSTEM

A human body shape simulation system was developed using the above-mentioned method. This system has functions of (1) visual presentation of a scanned body shape (current body shape), (2) calculation and visual presentation of the estimated body shape after weight change, and (3) comparing two body shapes and their body dimensions. Figure 2 shows a comparison between a scanned body shape (169 cm, 69.5 kg, BMI=24.3) and a simulated body shape after weight increase of 15.5 kg (169 cm, 85.0 kg, BMI=29.8).

DISCUSSION

Simulated body shapes after the weight change reconstructed from PCs are sometimes unnaturally different from the original body shapes. This is due to the fact that inter-individual shape variation is different from intra-individual shape variation. To solve this problem, lengths of the internal skeleton of the body shape model reconstructed from PC scores were adjusted to the values of the original shape model. When longitudinally collected body shape database becomes available, actual intra-individual shape variation can be used for the simulation. Simulated results in the present study need to be validated using such data.

We used automatically calculated landmark locations and body dimensions in the present system. Scan-derived landmark locations or body dimensions are not necessarily as accurate as those measured by expert anthropometrists (Kouchi and Mochimaru, 2008). We assume that systems such as developed in the present study will be used in fitness clubs to support people to start and continue taking a health action. Since such systems need to be low-cost and operator-free as well as accurate, we used automatic calculation in expense of the accuracy. Future development of scanner systems may solve this problem.

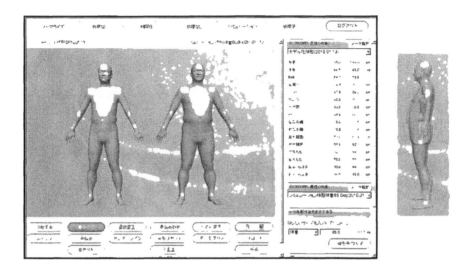

Figure 2. Comparison of scanned body shape (blue figure) and simulated body shape after weight increase of 15.5 kg (orange figure).

CONCLUSIONS

A system that simulates and presents the body shape after weight change was developed. The target of the present system was middle-aged male population, but a similar system for females was also developed. The present system will be usable in fitness clubs for supporting people to start and to continue taking a health action. It is also expected that a large-scale body shape database will be accumulated through this type of health-related services.

ACKNOWLEDGEMENT

A part of this study was supported by Nihon Unisys, Ltd.

REFERENCES

Allen, B., Curless, B., Popovi´c, Z. (2003) The space of human body shapes: reconstruction and parameterization from range scans. ACM SIGGRAPH 2003, 27-31.

Kouchi, M., Mochimaru, M. (2008) Evaluation of accuracy in traditional and 3D anthropometry. SAE Technical Report 2008-1-1882.

Mochimaru, M. (2006) Dhaiba: functional human models to represent variation of shape, motion and subjective assessment. SAE Technical Paper 2006-01-2345.

Mochimaru, M. (2008) Digital human modeling for product design. Design Engineering (JSDE journal), 43, 2-8. (in Japanese)

Mochimaru, M., Kouchi, M. (2009) Statistics of 3-D body shapes using PCA or MDS and their applications. IEA2009.

Witte, K., Allen, M. (2000) A meta-analysis of fear appeals: implications for effective public health campaigns. Health Education & Behavior, 27: 591-615.

Xi, P., Won-Sook, L., Shu, C. (2007) Analysis of Segmented Human Body Scans. Proceedings of Graphic Interface 2007 Conference.

CHAPTER 25

EARS: Toward Fast Analysis of 3D Human Scans

Xuetao Yin[1], Brian D. Corner[2], Anshuman Razdan[1]

[1]Xuetao.Yin@asu.edu; Razdan@asu.edu
Imaging and 3D Data Exploitation & Analysis Lab,
Arizona State University
7171 E. Sonaran Arroyo Mall, Mesa, AZ 85212, USA

[2]Brian.Corner@us.army.mil
US Army Natick Soldier RDEC
1 Kansas St, Natick, MA 01760, USA

ABSTRACT

We present the Enhanced Anthropometric Rating System (EARS), an integrated collection of tunable semi-automatic procedures to compute, visualize, and evaluate the geometric information of a 3D human body scan. To the best of our knowledge, EARS is the first complete system dedicated to fast evaluation and analysis of the quality of a human scan data. EARS is able to detect and remedy scan flaws, perform fast anatomically guided segmentation, analyze the posture of the scanned subject, evaluate the quality of the triangle surface, and provide real-time feedback on the quality of a human body scan mesh. We have tested EARS on a set of 100 female and 100 male subjects randomly drawn from the CEASER database. The average run time on a model is less than 30 seconds. EARS is robust on a test data set of unusual poses. EARS presents intuitive GUI and can also be run in command mode. It is compatible with the Cyberware Inc. CyScan software and will be employed by the US Army during the upcoming large scan anthropometric survey.

Keywords: 3D Human Body Scan, Mesh Segmentation, Posture Analysis

INTRODUCTION

The last decade has seen an increasing use of 3D human scans in Anthropometric research and studies. 3D scans fully preserve the body shape measurements, and are the data-of-choice of large archives for analyzing human physical dimensions and statistics, such as the well known CEASER (Allen, 2003) project.

Recently, a similar project is underway at the U.S. Marines and U.S. Army. The Ergonomics team at the Natick Soldier RDEC is conducting a survey of anthropometrics for the soldier of the 21st century. The survey will help understand the impact of gender, ethnic and racial diversity of today's armed forces. It also aims at creating a long term database for future applications, such as clothing and equipment design and load bearing simulation.

Quality is the primary concern in building a database of digital scans. During data collection, scans of poor quality can be easily produced with various types of flaws (as shown in figure 1.1). These scans might be admitted into the database, especially when there is a limited time for the human operator to inspect the scan and make decision, in our case, 30 seconds. Besides, a human operator might give subjective and inconsistent score.

In an attempt to assist the human operator, we developed a software system, called Enhanced Anthropometric Rating System (EARS), a novel quality control tool that will detect the error and flaws of an incoming scan in a consistent way and provide *go/no go* suggestion in near real time.

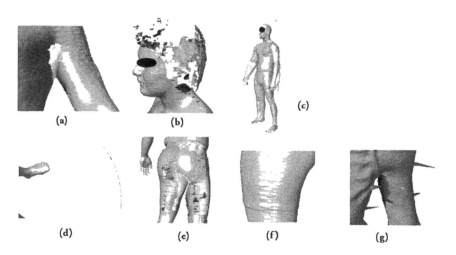

FIGURE 1.1 Sample data flaws: (a), (b), and (e) show variety of missing areas and voids; (c) shows stitching problem when registering scan patches; (d) shows outliers and (f) and (g) show abnormal rough surface caused by calibration issues.

Since the last release (Yin, 2009), we have introduced new void filling algorithm (Section 3) and posture analysis (Section 5), boosted the performance of anatomically guided segmentation, and developed new segmentation algorithm based on Geometry information (Section 4).

The procedures in EARS are organized as a decision tree with 5 major levels as shown in figure 1.2. At the end of each stage, EARS answers a series of inquiries with binary responses based on a set of thresholds predetermined from a learning dataset; and decides whether the scan data is acceptable for continue processing. Each of the following chapters is dedicated to one stage. We discuss the key actions and processes performed and how EARS makes decisions at each stage.

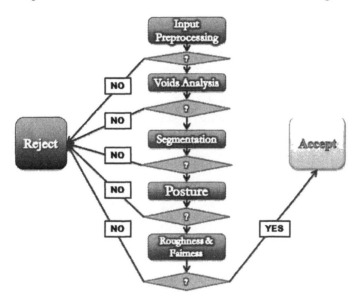

FIGURE 1.2 EARS's decision tree (question marks denote the questions asked at each stage).

STAGE 1: INPUT PREPROCESSING

EARS starts by **preprocessing** the input. EARS loads the PLY file generated by the scanner and performs a series of procedures. EARS stores the mesh using CGAL's halfedge data structure (CGAL), a representation that can obtain the adjacency information in near constant time. During loading, EARS fixes the topological errors such as intruding facets, conflicting face normals, and outliers. EARS searches for the largest connected component and establish it as the main body. Elements outside the main body are discarded as background noise.

The questions that are asked are: (i) Are the numbers of elements (vertices, edges, and faces) within reasonable range? (ii) Are the numbers of deleted elements reasonable? (iii). Does the major connected component contain a significant majority of the total facets? These questions determine whether the scan was properly registered by the scanner software. If the answer of any of these questions is negative, the mesh is rejected and is not processed further.

STAGE 2: VOID ANALYSIS

Voids are internal loops formed by connected boundary edges. They are caused generally by camera occlusion during scanning such as the crotch and armpit areas.

EARS collects information of the voids by detecting and filling them. EARS applies a novel greedy advancing front algorithm with awareness of back-facing triangles to fill the voids. Front-facing means that the facet has a consistent normal orientation with its vicinity.

The algorithm proceeds by adding one triangle to the void boundary at a time; at each iteration, the algorithm chooses to add the facet that has the minimal maximal inner angle among all front-facing candidates. This guarantees that the filled patch is composed of triangles that are as equilateral as possible. The algorithm will only produce back-facing facets when it runs out of front-facing candidates. EARS's filling algorithm runs faster than a dynamic programming filling algorithm (Liepa, 2003) and has reasonably good quality (Figure 3.1).

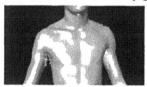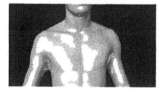

FIGURE 3.1 (Left) the original model with voids; (Middle) a naïve filling where a center point is created and all boundary vertices is connected to that point; (Right) EARS's filling algorithm.

The affected area of voids is computed as the total area of filled facets. Its ratio to the whole body area is also measured. The quality questions at this stage are: (i). Are the area and its ratio below predetermined thresholds? (ii). Do our introduction of back-facing triangles into the mesh is acceptable? The input is rejected if any of these questions were answered negative by EARS.

STAGE 3: SEGMENTATION

EARS provides feedback in terms of body parts, such as "excessive void found in upper inside of the left arm". To achieve this, EARS performs mesh segmentation to identify the human body parts. Mesh segmentation has received a lot of attention in the past century with many generic algorithms proposed. EARS's segmentation procedure focuses on the specific type of input and takes advantage of the human body shape. The segmentation is composed of four steps: PCA, anatomically guided segmentation, head segmentation, and tier-2 segmentation.

First, Principle Component Analysis (PCA) is performed to align the models in a consistent coordinate system.

Second, EARS performs fast anatomically guided segmentation. This procedure makes *cuts* based on topology changes. The algorithm intersects the mesh with a series of planes perpendicular to the z axis spanning from the top to the bottom; each of these planes forms one or many strips of connected faces when they intersect with the main body. The algorithm cuts the mesh into two parts whenever the number of strips changes. The anatomically guided segmentation divides the model into torso, arms, and legs. The anatomically guided segmentation has been the bottleneck of performance of EARS as it used to take 15-20 seconds to finish. The costliest part of the algorithm is that each plane has to search a large set of facets for intersections. EARS has introduced an optimization by identifying the *wave front*, the strip generated by the current cutting plane (figure 4.1). The optimized algorithm keeps track of the wave front; to form the next one, it only needs to search for the facets that are immediately below the current wave front. It is a localized search and stops when all the facets in consideration are below the current cutting plane. The optimization has boosted the speed dramatically as it reduces the running time to 2-3 seconds, tested on a Lenovo T61P Laptop with Intel Core 2 Duo 2.0 GHz, NVidida Quadro FX 579m, 2GB RAM, and 32b Windows Vista OS.

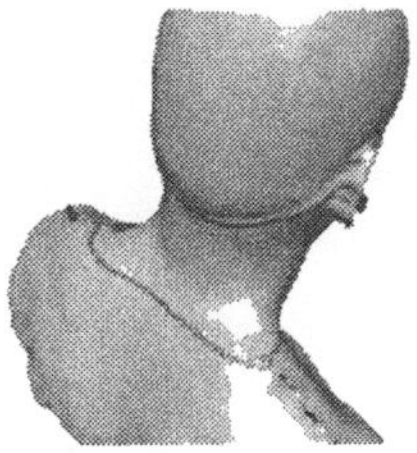

FIGURE 4.1 Visualization of a wave front. The green strip of triangle it the wave front.

Third, EARS divides the head and the torso based solely on geometry information. EARS introduces a scheme using B-splines curve fitting. A B-spline is a spline function that has the minimal support with respect to a given degree, smoothness, and domain partition. It has become the standard curve and surface descriptionin the field of CAD and Computer Graphics (Farin 2003).

EARS collects the left-most and right-most points during anatomically guided segmentation and forms two profiles by connecting them. These profiles are then fitted by B-splines. The B-splines are able to capture the shape of the profiles while reducing scan noise. From there, we have developed schemes to find the division planes of the head at two points, the shoulder point and the chin point. Anthropologists provided the definitions of these points.

The **shoulder point** is the most prominent inflection point on the profile B-spline. Note that there can be many inflection points due to noise and shape features as shown in Figure 4.2. Aside from the shoulder point, there are other inflection points at either end of the curve. Based on our experiments, we have formulated a procedure to locate the shoulder point consistently. We choose 20 control points for the B-spline and run three consecutive Lowess smoothing (Cleveland, 1979) to the control polygons as it will attenuate the prominence of the false inflection points.

The shoulder point is the inflection point whose curvature changes from positive to negative after the longest section of positive curvature. Figure 4-3 shows sample results.

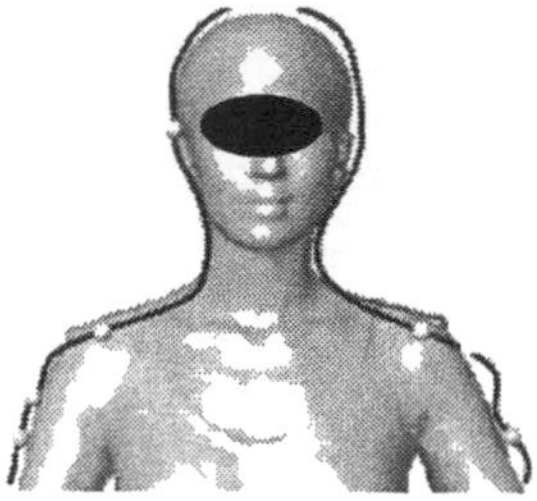

FIGURE 4.2 B-splines and Inflection points. The Green and Red curves are the original profiles; the blue curve is fitted B-spline; the golden balls represent the Inflection points.

FIGURE 4.3 Sample results of the shoulder cut scheme.

The **chin point** is determined by fitting B-splines to the frontal profiles. Based on our experiment, we again choose 20 control points. On the fitted B-spline of the frontal profile, we first identify the longest retreating span. The point of the most significant derivative in that span is marked as the chin point. This scheme is illustrated in figure 4.4.

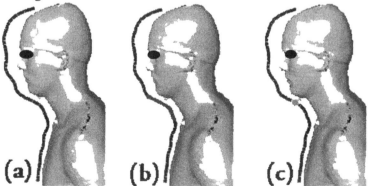

FIGURE 4.4 Illustration of Chin Cut. (a) the B-spline (blue). (b) the longest retreating span in red; (c) Chin point is marked by yellow circle on the curve.

Our experiments show robust alignment of the mathematic models and the actual Anthropometric land marks, however, problem occurs when the candidate's chin occludes part of his/her neck. We employ a slanted cut plane by lowering the front of the plane by 2 centimeters, and elevating the rear by 2 centimeters heuristically (Figure 4.5). More results in figure 4.6.

FIGURE 4.5 Slanted cutting plane (left) original scan, (middle) horizontal cut plane, (right) slanted cut plane.

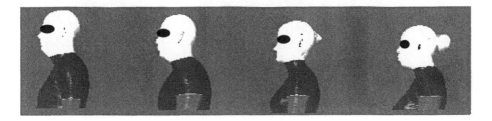

FIGURE 4.6 Chin cut of the same models as in Figure 4.3.

Now we have separated the torso and the extremities (tier-1) into body parts, the last step is to segment further into body patches (tier-2) as shown in Figure 4.7.

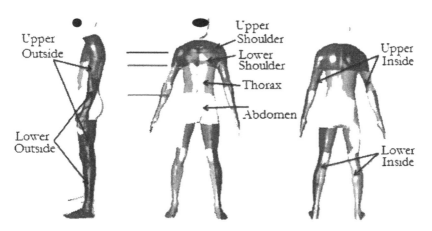

FIGURE 4.7 Tier-2 segmentation body patches shaded in different colors.

Torso is further segmented into 8 sub-parts, 4 on each side. These are referred to as (left/right) upper shoulder, lower shoulder, thorax, and abdomen from top to bottom. **Each extremity** is further divided into 4 sub-parts: upper inside (darker gray), lower inside (lighter gray), upper outside (lighter colors), and the lower outside (darker colors). The division planes are constructed based on PCA axes of local body part.

The quality questions asked at this stage are: (i). Is the number of body parts equal to six? (ii). Does each body patch have reasonable amount of elements? Meshes with negative answers are rejected.

The segmentation scheme proposed by EARS is robust as tested on a set of abnormal poses (figure 4.8).

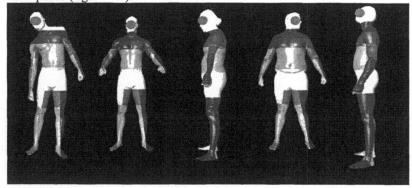

FIGURE 4.8 Results on abnormal poses.

STAGE 4: POSTURE ANALYSIS

The pose of the person being scanned affects the quality of the scan and may result in incorrect measurements. The key to posture analysis is to compute the angles between the major axis of each body part and the z-axis. First, each body part's major axis is extracted using localized PCA; then the spatial angles between it and the z-axis is computed and reported. EARS evaluates two angles: the front/back and left/right.

EARS presents a novel visualization of the postures as *skeleton* of the scanned object. In this visualization, we represent the centroids and ends of each body part with gold balls connected with blue stick along their local major axis. The surface is rendered translucently using OpenGL stencil buffer and alpha matting. Note that the head angle is computed using the major axis of frontal profile rather than the complete head surface as shown in figure 5.1.

For each angle that EARS computes, there is an acceptable range. If the measurements fall outside of the range, the mesh is rejected.

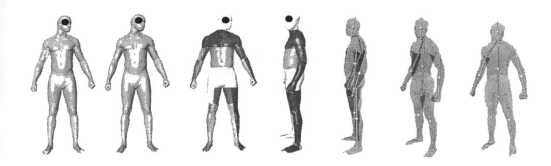

FIGURE 5.1 Visualization of the complete process of segmentation and posture visualization. (subject looking towards his left rather straight ahead; arm bended rather than straight); From left to right, the original mesh, the mesh with voids filled, segmentation front view, segmentation side view, three different views of posture abstraction.

STAGE 5: ROUGHNESS AND FAIRNESS

EARS evaluates the surface roughness (Yin, 2009) (Kushnapally, 2007) using Biquadratic curvature (Razdan 2005). The regions with high roughness are highlighted in red in figure 6.1.

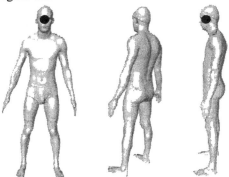

FIGURE 6.1 Rough regions highlighted in red.

The fairness is computed as statistic average and the variance of the triangle shape indicator, edge length and triangle area. These measurements represent the distribution of scan points.

The quality questions at this stage are: (i). Are the surface of side thigh or arm smooth? This is because these are the regions where abnormal rough patches appear most commonly. (ii). Are the statistics of the scan points reasonable? EARS rejects scans that do not conform to a set of predetermined thresholds.

CONCLUSION

EARS introduces fast void filling, novel human 3D mesh segmentation and posture visualization techniques. Based on our test, EARS is able to perform quality control in new real time (within 30 seconds for each scan) and demonstrates robustness in case of bad poses. All EARS procedure can be tuned by providing new thresholds and parameters. Our next step would be to fully explore the potential of automatic landmark searching in EARS.

ACKNOWLEDGEMENT

We would like to acknowledge the U.S. Marine Corps Project Manager-Infantry Combat Equipment (USMC PM-ICE) for funding the work under grant SSP-TCN-08007, and the U.S. Army Natick Soldier Research, Development, and Engineering Center for providing the data and anthropological guidance.

REFERENCES

Allen, B. and Curless, B. and Popovi, Z (2003), "The space of human body shapes: reconstruction and parameterization from range scans." *ACM Transactions on Graphics,* 22(3), 587-594.

CGAL, *CGAL-The Computtational Geometry Algorithms Library.* http://www.cgal.org/

Cleveland, W. (1979), "Robust Locally Weighted Regression and Smoothing Scatter-Plots", *AJASA, 74 freeway corridor surveillance information and control system.* Research Report No. 488-8, Texas Transportation Institute, College Station, Texas.

Farin, G. (2003). *Curves and Surfaces for CAGD: A Practical Guide, 5th Edition,* New York, Morgan Kaufmann Publishers.

Kushnapally, R and Razdan, A. and Bridges, N. (2007). "Roughness as Shape Measure". *Computer-Aided Design & Applications, Proceedings of CAD Conference.* 4, 295-310.

Razdan, A. and Bae, M.-S. (2005). "Curvature Estimation Scheme for Triangle Meshes Using Biquadratic Bezier Patches". *Computer Aided Designs,* 37(14), 1481-1491.

Yin, X. and Razdan, A. and Corner, B.D. (2009) "EARS: A System for Geometric and Anthropometric Evaluation of Human Body Scans". *Computer-Aided Design and Applications. Proceedings of CAD Conference.* 6(4), 431-445.

Chapter 26

Comparison of Anthropometry Obtained from a First Production Millimeter Wave Three-Dimensional Whole Body Scanner to Standard Direct Body Measurements

Brian D. Corner[1], Peng Li[1], Megan Coyne[1],
Steven Paquette[1], Douglas L. McMakin[2]

[1]US Army Natick Soldier Research
Development & Engineering Center
Natick, MA 01760, USA

[2]Pacific Northwest National Laboratory
Richland, WA 99354, USA

ABSTRACT

This study compared nineteen (19) human body measurements obtained from a new three-dimensional (3d) whole body scanner to the same measurements obtained by a qualified anthropometrist using standard direct body measurement techniques. The prototype scanner was based on millimeter wave technology (MMW) developed by Pacific Northwest National Laboratory and their commercial partner

Intellifit. An important benefit of MMW scanning system over light-based scanners is the ability of low-power millimeter waves to pass through most clothing fabric and hair. Over 12 years of scanning experience the authors have observed that requiring subjects to change into body conforming scan wear such as spandex shorts is a major impediment to acceptance of 3d whole body scanning technology by the general public. The MMW scanner allows subjects to wear more comfortable clothing such as sweatpants or running shorts and a t-shirt. However, the advantage of MMW scanning is only realized if body measurements extracted from the surface data are comparable to standard direct anthropometric measurements. This study was undertaken to determine body measurement performance of the MMW system. Direct measurements served as the standard against which the scanner-derived measurements were compared.

Keywords: Millimeter wave, 3d human body scan, human body measurement

INTRODUCTION

To capture the surface of a human body a typical 3d scanner system has one or more emitters that direct low powered energy to the body and one or more sensors to capture the energy as it reflects from the body's surface. For example, the Cyberware WB4 scanner utilizes four digitizing heads that emit low power visible and infrared laser light to illuminate a horizontal stripe on the object. Cameras in the system contain charged-couple devices that detect the laser light as it passes over the body and through triangulation computes surface points. A whole body point cloud of 3d XYZ coordinates is created as the four digitizing heads sweep down over the object. Other systems may use white light stripes or patterned light, but all return a 3d point cloud for visualization and body measurement.

MILLIMETER WAVE SCANNING TECHNOLOGY

The MMW system emits low power, non-ionizing, electromagnetic waves at 9–18 gigahertz (GHz) that reflects off the target and is detected with a sensitive antenna array. It is, in essence, a kind of RADAR system. The wavelength of the signals emitted by the scanner is in the same range as that used for mobile telephones; the scanning millimeter wave power, however, is about only 1/350th of that emitted by a typical mobile telephone. These emitted electromagnetic waves are considered harmless to humans at these power levels (IEEE Std C95.1™-2005). The scanner signal passes through most clothing fabric but does not penetrate the water in the skin. The return wave is captured by an antenna array and the signal is processed digitally using cylindrical holographic image reconstruction techniques to produce a 3d image whole body image (Sheen et al., 2001, 2000, 1997, and McMakin et al., 2007, 1998). A point cloud from this 3d image is generated from which measurements are extracted.

The MMW scanner was housed in a cylindrical booth (kiosk) seen in Figure 1. The sensor array rotated twice around the subject and collected approximately 200,000 three-dimensional coordinates from the body. Scan time was 20 seconds. A scanner that can detect the body beneath clothing raises privacy concerns; the MMW 3d point cloud data are not of a high enough resolution to be able to recognize a subject from their scan (Fig. 2). As with other three-dimensional scanning systems, software was written to interrogate the resulting point cloud and extract body measurements automatically. The MMW software computed 29 measurements from the captured scan data.

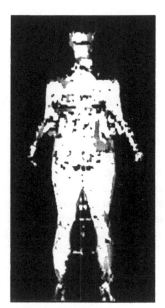

Fig 1. Intellifit scanner kiosk showing movement of the sensor array and subject position

Fig 2. Point cloud from MMW scanner.

BODY MEASUREMENT AND SCANNING METHODS

Twenty-seven (27) Army and civilian test participants volunteered to participate in the study. Twenty-five (25) were male and two (2) were female. Sexes were combined for analysis due to the small number of females.

Of the 29 MMW-based measurements automatically computed from scan geometry, 19 were comparable to standard direct measurement definitions (Gordon et al., 1989) and were selected for this study. Those not selected were special purpose measurements created for other applications by Intellifit. Direct body measurements were obtained by a trained anthropometrist (BDC). Practice sessions were held until repeat measurements were within US Army Anthropometric Survey guidelines (Gordon et al., 1989).

For MMW scanning, test participants wore whatever clothing they had on when they arrived. Civilians typically arrived in their work clothes while Soldiers wore their Advance Combat Uniform (ACU) or athletic shorts and t-shirts (physical training (PT) attire). Participants were asked to remove their shoes and anything metallic. They then entered the MMW booth and assumed the standard MMW scan position- legs about shoulder width apart, arms pulled away from the side of the body laterally, fingers close with hands flexed inward lightly touching the legs (Fig 1). The pose helped to maximize clean signal return. Participants were also scanned in standard surface scan attire- spandex shorts for men, and shorts and spandex exercise bra for women. The second scan was taken to standardize clothing between Cyberware WB4 and MMW scanners for comparison of surface geometry that will be the subject of another paper.

BODY MEASUREMENT RESULTS

Descriptive statistics for standard direct (direct) anthropometry and MMW-based measurements are given in Table 1. Measurements are given in millimeters (mm). Measurements include heights to the level where circumferences were taken. Note that there were differences in body posture between direct and scanner-based measurements that impacted height measurements.

Mean absolute differences (MAD) were computed between direct and scanner-based measurements (direct-MMW) to compare body measurement results. Descriptive statistics for the MAD results are given in Table 2. For the most part, there were considerable differences between direct and MMW-based body dimensions. Neck base circumference difference was among the smallest and an interesting result given that it is one of the more difficult dimensions to measure directly. The greatest divergence related to waist and thigh measurements. Stature values were relatively similar despite the differences in pose. MMW stature was effected by stray points in the scans that added to the height measurement. When considering pose, it is notable that the height where head circumference was taken was very close in the two data sets; although head circumference itself was not consistent. However, the fact that head circumference may be taken at all is notable. Covered or uncovered, hair on the head does not permit accurate cranial circumference measurement with most surface scanners.

Table 1. Descriptive statistics for direct and MMW-based measurements (mm).

Measurement	Direct Measurements				MMW-based Measurements			
	Avg	sd	Min	Max	Avg	sd	Min	Max
Arm Len	592.9	36.2	523	665	581.2	37.8	513	660
Bicep Circ	332.8	24.2	288	382	266.3	25.6	222	315
Calf Circ Ht	338.4	28.8	264	387	401.1	39.6	325	477
Calf Circ	396.6	26.8	351	470	463.1	4.4	443	465
Chest Circ Ht	1298.3	59.8	1151	1416	1329.2	63.1	1199	1453
Chest Circ	1024.6	81.8	872	1211	1034.3	71.4	869	1190
Head Circ Ht	1716.2	76.6	1526	1868	1714.4	73.1	1534	1859
Head Circ	576.3	17.5	550	608	658.1	16.6	627	687
Inseam	798.6	51.0	706	903	845.2	51.9	721	945
Knee Mid Ht	496.7	37.9	406	552	538.3	40.4	463	606
Neck Base Circ	423.3	28.8	327	477	412.5	19.9	347	447
Seat Circ Ht	1023.6	61.8	909	1142	929.1	57.7	792	1041
Seat Circ	1078.6	57.6	945	1180	1015.6	53.6	923	1108
Shoulder Circ	1206.7	75.4	1015	1376	1257.4	64.4	1157	1388
Stature	1788.8	74.8	1611	1946	1819.2	74.2	1641	1966
Thigh Max Circ Ht	621.7	48.0	505	713	809.4	52.4	701	925
Thigh Max Circ	908.6	56.8	775	1034	653.9	35.8	589	730
Waist Om Circ Ht	889.1	145.1	311	1089	1122.5	77.8	985	1280
Waist Om Circ	798.3	52.6	688	898	920.3	77.5	781	1072

DISCUSSION AND CONCLUSIONS

Body measurements from the MMW scanner departed from direct measurements in most cases. Differences, however, were close enough to suggest the MMW scanner has promise as a measurement tool. It is clear that the point cloud captured by the MMW system was too sparse to match direct body measurement results (Fig. 3). Improvements may be made by increasing the power to enhance the signal and by refining the signal processing algorithm. An additional emitter and antenna pair will also improve body coverage by reducing the vertical space between signals put it in line with other whole body surface scanning systems.

Table 2. Direct minus scanner-based body measurement mean absolute difference, MAD.				
Measurement (mm)	Avg	sd	Min	Max
Arm Len	21.0	14.5	3.8	55.3
Bicep Circ	66.5	28.4	9.9	127.4
Calf Circ Ht	62.7	37.7	9.2	152.5
Calf Circ	66.9	27.3	1.6	106.8
Chest Ht	31.9	17.9	2.9	67.1
Chest Circ	22.1	12.5	0.8	47.7
Head Circ Ht	13.4	8.8	2.2	33.3
Head Circ	81.7	9.0	59.4	105.7
Inseam	49.0	20.0	15.4	113.0
Knee Mid Ht	42.9	23.9	7.8	77.7
Neck Base Circ	17.1	10.3	2.6	39.4
Seat Circ Ht	97.8	64.0	1.0	258.1
Seat Circ	81.1	44.0	0.1	180.5
Shoulder Circ	57.7	44.7	1.4	155.1
Stature	30.4	12.9	6.8	53.0
Thigh Max Circ Ht	187.7	63.0	58.0	306.7
Thigh Max Circ	254.6	50.7	114.7	352.7
Waist Om Circ Ht	232.6	149.7	32.1	786.3
Waist Om Circ	123.0	70.5	14.0	281.3

The ability to obtain accurate anthropometry through clothing and hair on the head makes MMW scanner technology worth pursuing. Obtaining accurate body measurements is a crucial step in providing appropriately sized clothing and equipment items. Unlike denim jeans, for example, where an individual has a good chance of finding a manufacturer who makes a product that fits better than the others, groups that distribute uniforms or other items with one or two manufacturers have a limited choice which makes getting a good fit less easy. In this case, improved fit may come when many accurate body measurements are available for size selection. Groups such the military and first responders would benefit if their constituents were more likely to get scanned and measured. Not having to change into tight fitting skimpy scan wear is a major hurdle that is overcome by MMW technology. Thus, we are working to improve MMW technology and demonstrate its utility to the logistic community.

239

ACKNOWLEDGEMENTS

We thank Intellifit and Pacific Northwest National Laboratory for the MMW scanner loan to the US Army Natick Soldier Center. We also thank the volunteers whose participation made this study possible.

REFERENCES

Gordon CC, Bradtmiller B, Churchill T, Clauser CE, McConville JT, Tebbetts I, and Walker R (1989) "1988 Anthropometric Survey of U.S. Army Personnel: Methods and Summary Statistics". Technical Report, Natick/TR-89/044, U.S. Army Natick RDEC, Natick, MA.

IEEE International Committee on Electromagnetic Safety (SCC39), *IEEE Standard for Safety Levels with Respect to Human Exposure to Radio Frequency Electromagnetic Fields, 3kHz to 300 GHz,* IEEE Std C95.1™-2005, Institute of Electrical and Electronics Engineers, Inc., New York.

McMakin DL, DM Sheen, TE Hall, and RH Severtsen (1998) Cylindrical holographic radar camera. *The International Symposium on Enabling Technologies for Law Enforcement and Security*, Proceedings of the SPIE, Vol. 3575.

McMakin DL, DM Sheen, TE Hall, MO Kennedy, and HP Foote (2007) Biometric identification using holographic radar imaging techniques. *Proc. SPIE,* Vol. 6538.

Sheen DM, DL McMakin, and TE. Hall (1997) Cylindrical millimeter-wave imaging technique for concealed weapon detection. *Proceedings of the SPIE - 26th AIPR Workshop: Exploiting new image sources and sensors*, Vol. 3240, pp. 242-250.

Sheen DM, DL McMakin, and TE Hall (2000) Combined illumination cylindrical millimeter-wave imaging technique for concealed weapon detection, *Proceedings of the SPIE - Aerosense 2000: Passive Millimeter-wave Imaging Technology IV*, Vol. 4032.

Sheen DM, DL McMakin, and TE Hall (2001) Three-dimensional millimeter-wave imaging for concealed weapon detection, *IEEE Transactions on Microwave Theory and Techniques*, Vol. 49, pp. 1581-92.

Chapter 27

Creating an Average Shoulder Model from Three-dimensional Surface Scans

Peng Li, Brian Corner, Steven Paquette

US Army Natick Soldier Research
Development & Engineering Center
Natick, MA 01760, USA

ABSTRACT

This paper describes a practical method for generating an average shoulder model from an existing database of three-dimensional (3D) whole body scans. The resulting model was then applied to design and evaluate a series of protective shoulder plates. The proposed method involves segmenting body scans using well defined landmarks and then re-sampling each shoulder surface from the chosen sample to create surface correspondence across the sample. From the established correspondence, an average surface model is created.

Keywords: 3D Scan, Human Surface Modeling, Surface Averaging

INTRODUCTION

With the advance of 3D scanning technologies and rapidly growing 3D human body scan databases in recent years, new research interest has developed to analyzing and quantifying these data to generate meaningful digital human models at the population level. Translating a sample of individual 3D scans into statistically and anthropometrically relevant models (physical and digital) has became an active area [Azouz 2002, Allen 2004].). While possessing an accurate representation of the full range of body size and shape of the user population is crucial for effective design and evaluation of products intended for use by the vast majority of individuals, it is common practice to originate a given design near the center of body size/shape distribution and then grade or scale the item toward the small and large extremes. Other approaches to using 3D body scans in product design involve selecting specific individual scans from the larger database that met some predefined set of statistically derived size/shape criteria. Examples of defining average shape using 3D body scans include the use of spatial analysis [Chen 1989], computation of a distance field [Marquez 2005] and principle components analysis applied to scan mesh vertices [Azouz 2002, Allen 2004].

In this paper we attempt to derive an anthropometrically based model of the shoulder region based on landmark correspondence among the chosen sample. This approach establishes surface correspondence by incorporating landmark alignment and fitting a surface to a shape matrix. The surface correspondence algorithm utilizes a few key landmarks as reference points and re-samples individual shoulder surfaces uniformly. A shape matrix obtained from the sample of raw scan data facilitates the creation of a new surface from the averaged 3D coordinates of all individuals.

METHODS

To segment a shoulder surface we identified several landmarks (bust point and scye point) on each scan automatically using a custom software program [NatickMeasure 2005]. We then sampled the shoulder segment with a B-spline curve approximation. A new surface grid was obtained through a unified data generation process. The surface grid established anatomical correspondence between shoulder surfaces. Finally, from the averaged surface grid we generated an average shoulder surface.

THE LANDMARK FINDING AND BODY SEGMENTATION

We selected 265 male scans as our working sample. These scans were captured with a Human Solutions Vitus Smart scanner [www.human-solutions.com]. The raw data from Vitus Smart consists of eight camera images that have significant overlap between adjacent cameras. We first merged these images to reduce data and converted the data into a standard PLY format. Then we performed a body segmentation procedure [Nurre et al. 2000] to isolate the torso (Figure 1). Whole body segmentation identified a number of landmarks automatically: crotch point, left & right scye point, and neck (cervicale) point.

Additional segmentation was performed to define the shoulder surface from the torso. The torso mesh was trimmed with a vertical plane through bustpoint and a horizontal cut plane at scye level as shown in Figure 2.

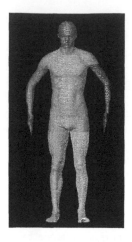

FIGURE 1. A segmented whole body scan

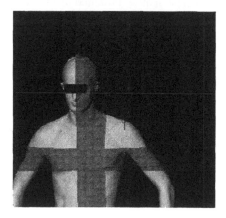

FIGURE 2. Clipping the shoulder with two planes.

FILLING SHOULDER SURFACE VOIDS AND RE-SAMPLING

The shoulder surface in the 3D scans captured with the Vitus Smart system contained missing (void) areas because the surface was parallel to the horizontal plane of the scanner (Figure 3a). The voids are distributed along the shoulder line from the lateral neck point to acromion. Therefore it was best to fit data perpendicular to the shoulder line for a better support to curve fitting. We applied a B-spline curve approximation to a series of vertical sections of the shoulder surface. During the B-spline curve fitting process shoulders were filled as shown in Figure 3b. A third order B-spline function naturally bridged gaps and followed the original shape. With new data points generated from the above curve fitting process we applied a B-spline surface approximation to create a smooth shoulder surface (similar to Li & Jones 1997). From the B-spline surface representation we can generate a uniform data grid as described below.

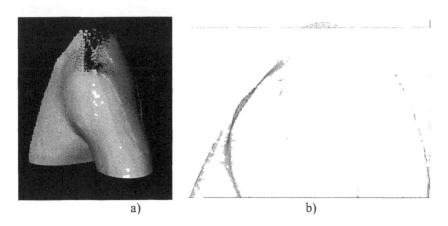

a)   b)

FIGURE 3. a) Voids over the shoulder, b) B-spline fitted cross-sections

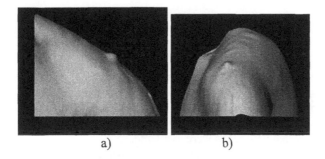

a)   b)

FIGURE 4. Shoulder voids filled with a B-spline surface (the lump at acromion point is a physical hemispherical landmark), a) front view, b) side view,

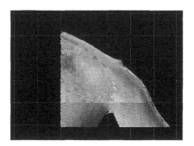

FIGURE 5. B-spline surface superimposed on the original data

CREATING AN AVERAGE SHAPE

The final shoulder surface points were arranged in matrix form making it easy to establish point-to-point correspondence between scans. In the surface grid matrix, each row represents a vertical cross section and contains the same number of points. In each column, new surface points were aligned in equal angular spaces as illustrated in Figure 6. In order to obtain such a shape matrix from the B-spline surface of the shoulder, we applied a second sampling process in a B-spline surface generator. We supplied two parameters, number of rows and number of columns to the generator. Using the number of rows, the generator created equal spaced vertical slices. Using the number of columns, the generator created equal angular spaced points along each slice.

FIGURE 6. Angular sampling illustrated

The result was a rectangle wireframe represented by a MxN shape matrix of 3D coordinates, where M was the number of cross-sections and N the number of samples from each cross-section. For our application we choose M and N equal to 60. With this structure, we were able to process any number of scans and update the average shape when new scans were available.

The average shape is the centroid of all surface grid points from the sample population. The calculation of the centroid is trivial. One may also apply PCA to these surface grids. The average shape from PCA will be the same, but PCA offers the advantage of constructing surfaces along any of the principal axes.

CONSTRUCTION OF THE AVERAGE SURFACE

Construction of the surface from the averaged grid matrix is relatively straight forward. If the density of vertices in the new average is high enough, a mesh generation procedure (Fuchs et al. 1977) was called to construct the average 3D surface. Otherwise, a B-spline surface interpolation (Piegl & Tiller 1997) was used to increase the vertex number. Figure 7 shows final averaged surface.

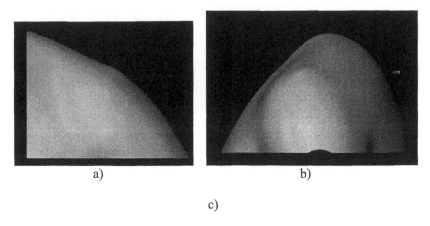

a)   b)

c)

FIGURE 7. Final averaged surface from 265 scans, a) front view, b) side view

With the above generated model we have designed a number of shoulder protective plates prototypes using a commercially available CAD package SolidWorks. We first imported the model into SolidWorks, and then created a base geometry of the plate conforming to the model shape. The final plates resulted from adding thickness and trimming boundaries to the base geometry. Two of prototypes are shown in Figure 8.

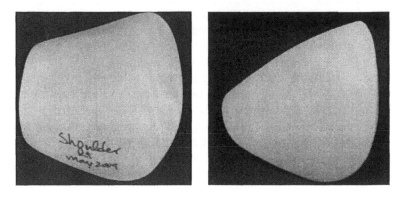

FIGURE 8. Shoulder plates designed on the averaged surface model

CONCLUSIONS

The success of a shape averaging program relies on a consistent surface correspondence. Various methods have been developed to meet this need, but from our experience there is no universal method suitable for every shape. Especially in the human body modeling area where landmarks play an important role for shape comparison, a surface correspondence method needs to take landmarks and shape factors into consideration. Utilization of anthropometric landmarks to control region selection and alignment resulted in more realistic shape averaging. The surface sampling method selected for this application was largely determined by shoulder shape. In the future we will explore other possible ways to achieve a general human body surface correspondence for the purpose of model development.

REFERENCES

Allen, B., Curless, B., Popović, Z. (2004), "Exploring the space of human body shapes: data-driven synthesis under anthropometric control", SAE Digital Human Modeling, 2004

Azouz, Zouhour B., Rioux, M., Lepage, R. (2002), "3D Description of the Human Body Shape Using Karhunen-Loève Expansion", in International Journal of Information Technology, Vol. 8, No. 2, Sept. 2002.

Chen, Shenchang, Parent, R. (1989), "Shape Averaging and it's Applications to Industrial Design", IEEE Computer Graphics and Applications, 1989, Vol.9, No.1,pp. 47-54

Fuchs, H., Kedem, Z., and Uselton, S. (1977), Optimal Surface Reconstruction from Planar Contours, Graphics and Image Processing, Vol. 20, No. 10, pp. 693-702, October 1977.

HumanSolutions, http://www.human-solutions.com/apparel/technology_scanning_vxxl_en.php

Li, Peng & Jones, P.R.M. (1997), "Automatic editing and curve-fitting of 3-D raw scanned data of the human body', International Conference on Recent Advances in 3-D Digital Imaging and Modeling, Ottawa, Canada, May 1997, pp 296-301

Marquez et al. (2005), "Shape-Based Averaging for Craniofacial Anthropometry", Proceedings of the 6th Mexican International Conf. on Computer Science, pp 314-319, 2005 ISBN~ISSN: 1550-4069, 0-7695-2454-0

NatickMeasure: A 3-D Visualization and Anthropometric Measurement Tool, Techincal Report, Natick/TR-05/021L, US Army NSRDEC, Natick, MA, 2005

Nurre, J.H., Connor, J., Lewark, E.A. and Collier, J.S. (2000), "On segmenting the three dimensional scan data of a human body", IEEE Transactions on Medical Imaging, Vol. 19 No.8, pp. 787-97

Piegl, Les, & Tiller, W. (1997), "The NURBS Book", 2nd Edition, Springer.

CHAPTER 28

Skeleton Eextraction From 3D Dynamic Meshes

Weihe Wu[1,2], Aimin Hao[1], Zhao Yongtao[1]

[1]State Key Laboratory of Virtual Reality Technology and System
Beihang University
Beijing 100083,China

[2]School of Art &Design
Beijing University of Technology
Beijing 100029, China

ABSTRACT

This paper proposes a novel method for automatic extraction of human body skeletons from 3D dynamic shapes by taking advantage of geometry properties of static shapes, knowledge of anthropometry (priori knowledge of human body anatomy), and kinematics characteristics of dynamic shapes. The method consists of three steps. In Step I, the human body is divided into five parts including four limbs and the torso based on the Reeb graph created by a Morse function, which is defined on the surface mesh of a reference pose as geodesic distance. Then, limbs and torso are sub-segmented using statistical anthropometric proportions. In Step II, seed points that exhibit the best rigid transformation are identified in the middle parts of the sub-segments by the mesh edge-length deviation induced by its transformation through time. In Step III, we use these seed points to determine joint locations of the skeleton. An empirical experiment demonstrates that the proposed method is able to extract skeletons from 3D dynamic meshes automatically. The accuracy of the proposed method appears to be superior to that of the existing methods.

Keywords: Skeleton extraction, Dynamic shape, Anthropometry, Geodesic distance

INTRODUCTION

Shape skeleton extraction and segmentation are the basis for a variety of applications including character animation, motion analysis, and computer-aided therapy. Quite a few approaches have been developed for extracting skeleton from a static shape (Yu et al., 2008; Oliveira et al., 2003; Wang et al., 2003; Zhong et al., 2006; Aujay et al., 2007), which includes only one pose without temporal information. It remains a challenge to extract kinematic skeleton from a 3D dynamic shape.

Several methods have been developed for extracting skeleton. A 3D dynamic shape, such as an animated mesh sequence comprising of N frames or a set of examples which reveal local rigid transformations of human body, is mostly modeled by constant connectivity surface meshes with varying geometry which display different poses. Most existing methods choose one pose as the reference pose, and then calculated the deformations or matrices of vertices or edges or triangles over the surface mesh from the reference pose to all other poses. These deformations or matrices reveal local rigid transformations. Clustering techniques are then used to segment the meshes. Each joint position is determined by the information about segments that are adjacent to the joint. For example, (Aguiar et al., 2008) select a subset of vertices as seed points, which distribute evenly over the mesh and their number is in the range of 0.3-1% of the total vertex count of the model. (Kirk et al., 2005) get seed points from motion capture markers. For both methods, all mutual Euclidean distances between seed points and their standard deviations are calculated, based on which spectral clustering is used to segment the mesh into approximately rigid parts. Joint positions between interconnecting parts are estimated by minimizing the variance in joint-to-vertex distance for all seed vertices of the adjacent parts at all frames. (Schaefer et al., 2007) estimate rigid transformations of the bones using a face clustering approach based on rigid error function. (James and Twigg, 2005) extract rigid bones using mean shift on the rotation sequences of the mesh triangles over time.

However, most of these existing methods are computationally expensive, because they mainly rely on human body motion characteristics in terms of local rigid transformations and clustering techniques to segment a mesh into parts affected by the same rigid transformation. Meanwhile, due to lack of priori knowledge, many methods involve all vertices or triangles on the mesh for clustering and all edges between vertices on the mesh need to be constructed, even for vertices respectively located at the head and the feet that are clearly independent in motions. In addition, although seed points instead of all vertices can reduce computation, seed points evenly chosen from the mesh do not exhibit the optimal rigid transformation. Recently, (Tierny et al., 2008) present a combined dynamic and static approach that identifies the articulations of an object through time. They determine the contours whose average edge-length deviation is locally maximal as motion boundaries, but that is not suitable for some human body surface meshes such as in SCAPE's data set, since there are no clearly maximums of contours average edge-length deviation.

This paper presents a novel method for automatic extraction of human body skeletons from 3D dynamic shapes by taking advantage of the geometry property of static shape, knowledge of anthropometry, and kinematics characteristics of dynamic shapes. In this method, we use the anthropometry information including the proportions between bones of human skeleton to segment surface meshes in place of a clustering method. The human body is divided into five parts including limbs and torso based on the Reeb graph created by the Morse function, which is defined on the surface mesh at a reference pose as geodesic distance. After limbs and torso are sub-segmented using statistical anthropometric proportions, seed points that exhibit the best rigid transformation are identified in the middle parts of the sub-segments Finally, we use these seed points to determine joint locations of the skeleton.

MESH SEGMENTATION

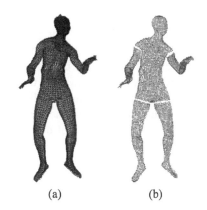

(a)   (b)

Figure 1. Feature points (1(a), in red) of the DANCE shape and its five segments (1(b)).

A 3D dynamic shape is represented by a constant connectivity closed surface mesh $M = (V = vertices, T = triangulation)$, whose vertices' positions vary through time $p_t(v_i) = (x_i, y_i, z_i)_t$, $\forall v_i \in V$. Let $P_t = \{p_t(v_i)\}$ be the coordinate sets at time step t, and P_0 stands for the reference pose. For the extraction of feature points and segmentation mesh, we only consider the reference pose P_0, and employ a method similar to (Yu et al., 2008). First, we extract five feature points, which are located at the extremity of prominent components, shown in red in Fig. 1(a). In order to avoid more than one feature point being extracted at one prominent component, we add a constraint condition for extracting the fifth feature point. The geodesic distance between vertex x and y is expressed as $g(x,y)$. Our algorithm is described as follows.

(1) Initialization

The set of feature points: $F = \{\phi\}$

(2) Calculating the first feature point f_1

Select an arbitrary point v' , $\forall v' \in V$

$f_1 = \arg\max_{v_i}(g(v',v_i))$, $v_i \in V$

$F = F \bigcup \{f_1\}$

(3) Calculating the second, third, forth feature points f_2, f_3, f_4

For $i = 2$ to 4

$$f_i = \arg\max_{v_i}(\sum_{j=1}^{i-1} g(f_j,v_i)) , v_i \in V , \forall f_j \in F$$

$$F = F \bigcup \{f_i\}$$

End for

(4) Calculating the fifth feature point f_5

$g_{\min} = \alpha g(f_1,f_2)$

$f_5 = \arg\max_{v_i}(\sum_{j=1}^{4} g(f_j,v_i))$, $v_i \in V$ and $g(v_i,f_j) > g_{\min}$, $\forall f_j \in F$

$F = F \bigcup \{f_5\}$

Here $\alpha = 0.28$, and the results of our experiment confirm that if we add the constraint condition at step 4, we can extract the only feature point at one protrude component. When five feature points are extracted, the feature point located at the top head can be recognized according to the symmetry of the human body, and the feature points located at the hands or feet can be identified through the fact that the geodesic distance from the top head to one hand is less than the geodesic distance from the top head to one leg. Then, the Morse function based on the geodesic distance of human body shape is calculated by taking the feature point on the top head as the source point, and the Reeb graph of the shape can be constructed automatically according to Morse theory. Finally, the shape is divided into five parts including limbs and torso based on the evolvement of Morse function isolines, and there is a unique feature point for each part, as shown in Fig. 1(b).

SEED POINT SELECTION

(Winter, 1990) provides an average set of segment lengths expressed as a percentage of body height which can serve as a good approximation in the absence of better data, but it is difficult for us to directly use this anthropometry data to determine joints, because a 3D human body shape is modeled by surface meshes and the length between two joints is measured with Euclidean distance. We instead try to measure the length of each segment between two joints using the geodesic distance along the surface mesh, and calculate the proportion of joint's geodesic distance to the geodesic distance L between two source points at the hands. Our results show that the above proportion is relatively stable in human body with different shape and posture. So we further sub-segment the five parts according to the stable proportion and estimate the positions of joints based on that.

Let J_{ij} be the j joint at part i, $i > 0$, J_{i0} be the feature point at part i, and we assign the geodesic distance g_{ij} and the relative geodesic distance rg_{ij} for J_{ij} in Formula 1.

$$g_{ij} = g(v, f_i), \ rg_{ij} = g(J_{ij})/L, v \in C_{J_{ij}} \qquad (1)$$

Where f_i is the feature point at part i (we don't take the feature point at the head as a source point for all joints to avoid accumulated error), and L is the geodesic distance between two feature points at the hands , and $C_{J_{ij}}$ is the geodesic isoline whose centroid is the Euclidean distance-closest point among all centroids of isolines from J_{ij}. Therefore, the joint J_{ij} has the same geodesic distance value with the vertex v in the isoline $C_{J_{ij}}$, and the isoline $C_{J_{ij}}$ is called the separating isoline whose centroid is used as an approximation of the joint's location during joint determination.

In practice, we use the average relative geodesic distances derived from twenty models as an approximation of rg_{ij} and L calculated in the previous phase to calculate g_{ij} for each joint J_{ij}, and then the separating isolines are located by g_{ij}. Furthermore, we can sub-segment each part by these separating isolines, and each sub-segment corresponds to a scope of the geodesic distance. For example, the sub-segment between J_{ij} and J_{ij+1} corresponds to $[g_{ij}, g_{ij+1})$.

Usually, we think that vertices located at the middle part between two joints can reveal the high rigid motion of the skeleton, and this is why these vertices only have one weight in animation driven by skeleton (SSD). Therefore, we choose seed points from the middle of each sub-segment. We determine the candidate scope for the sub-segment between J_{ij} and J_{ij+1}, $[(g_{ij} + g_{ij+1})/2 - s/2, (g_{ij} + g_{ij+1})/2 + s/2]$. Where s is the length of the candidate scope measured in geodesic distance, and is decided by the required number of seed points and the deviation of the anthropometry data, $s = (g_{ij+1} - g_{ij})/2$ (which we found to be satisfactory in our experiments).

Due to rigid transformation belonging to the isometric maps, we further choose seed points that exhibit the high rigid transformation from the candidate scope by the edge-length deviation induced by its transformation through time. We use the following formula (Tierny et al., 2008) to calculate the edge-length deviation for each vertex in the candidate scope, where F is the number of frames, $N(v_i)$ is the set of vertices sharing an edge with v_i and d is the Euclidean distance:

$$l(v_i) = \frac{1}{F} \sum_{t=1}^{F} \frac{1}{|N(v_i)|} \sum_{v_j \in N(v_i)} \left| d(p_t(v_i), p_t(v_j)) - d(p_0(v_i), p_0(v_j)) \right| \quad \dots \dots (2)$$

The vertices whose edge-length deviations are smaller are chosen as seed points.

JOINT DETERMINATION

A joint connecting two bones has the property that the local transform for one part relative to the other can be represented with a rotation transform base on the joint. Using this property, we calculate the local transformed positions of seed points by two ways that move the least with respect to the two ways, which leads to the quadratic minimization problem. For each time step t we first compute the rigid transformation matrix $Z^t = \{R^t, T^t\}$ for each sub-segment that transforms the positions of the seed points from the reference time step to the t time step (Horn, 1987), where R^t and T^t are the rotation and translation matrix respectively. Thus, $P_t(v_i)$ can be approximated with:

$$p_t'(v_i) = Z^t p_0(v_i) = R^t p_0(v_i) + T^t \qquad (3)$$

Let sub-segment A and B be connected by joint O, and the sets of seed points for them are V_A and V_B. At time step t, their transformation matrices are $Z_A^t = \{R_A^t, T_A^t\}$, $Z_B^t = \{R_B^t, T_B^t\}$ respectively.

On the first way, for each time step t the local transform for sub-segment A relative to B can be expressed with $(Z_B^t)^{-1}$:

$$p_t^0(v_i) = (Z_B^t)^{-1} p_t(v_i) = (R_B^t)^{-1}\big(p_t(v_i) - T_B^t\big), \quad v_i \in V_A \qquad (4)$$

On the second way, the same transform can be expressed as a rotation by $R_A^t(R_B^t)^{-1}$ based on joint O.

$$p_t^0(v_i) = R_A^t(R_B^t)^{-1}\big(p_0(v_i) - p_0(O)\big) + p_0(O), \quad v_i \in V_A \qquad (5)$$

Therefore, we solve for $p_0(O)$ by minimizing:

$$f(O) = \sum_t \sum_{i \in V_A} \left| R_A^t(R_B^t)^{-1}(p_0(v_i) - p_0(O)) + p_0(O) - (R_B^t)^{-1}(p_t(v_i) - T_B^t) \right|^2 + \alpha \left| p_0(O) - b_{AB} \right|^2 \quad (6)$$

where b_{AB} is the centroid of the separating isoline which is used as the boundary for sub-segment A and B. In some cases, if the relative motion between adjacent sub-segments acts as hinges, this will induce no unique minimizers or the minimizer which is far away from the actual position. Thus, we add the second item to minimize the distance to the centroid of the boundary. In our experiment, without the second item ($\alpha = 0$) we also get the unique minimizer for two dynamic shapes.

After obtaining the optimal joint location at the reference time step, we can easily find the joint positions at all other time steps by $p_t(O) = Z_A^t p_0(O)$. Finally, we apply the above procedure to all adjacent body parts to extract the skeleton.

EXPERIMENTS AND RESULTS

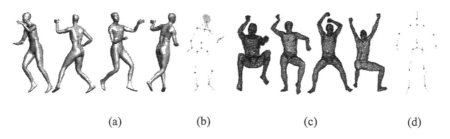

(a)   (b)   (c)   (d)

Figure 2. 3D dynamic surface meshes (2(a) for DANCE, 2(c) for SCAPE) and their extracted skeletons (2(b) for DANCE, 2(d) for SCAPE).

To demonstrate the effect of our algorithm, we applied it to SCAPE scanned human body shapes with 70 frames and DANCE with 201 frames. Fig. 2 shows their 3D dynamic surface meshes in a few time steps and their extracted skeleton. The Red dots indicate the joints positions and the green lines correspond to the skeletons.

SEGMENT LENGTH PROPORTION IN GEODESIC DISTANCE

Table 1 Distribution of the relative geodesic distance of joints. It is measured with the geodesic distance L between two feature points at the hands.

	Head	Neck	Waist	hip	Wrist	Elbow	Shoulder	Ankle	Knee	Thigh
Average	0.178	0.228	0.443	0.541	0.105	0.250	0.388	0.156	0.383	0.595
Standard deviation	0.0086	0.0078	0.0193	0.0284	0.0083	0.0083	0.0154	0.0120	0.0236	0.0278

We calculate the relative geodesic distance of the joints with 20 human body models of various shapes and postures. Tab. 1 shows that the relative geodesic distance for joints are stable for different models, and the deviations for knee, leg, waist, hip, shoulder are more obvious, but still they are negligible for joint determination, since the relative geodesic distance is used just for determining the length and center of the candidate scope of seed points, not for estimating directly the location of joints.

COMPUTATION TIME

Table 2 Running time for the various phases of our algorithm.

Sequence	Frames	Vertices	triangles	F-Point	Segment	S-Point	Joint
SCAPE	70	12.5K	25K	3.15s	10.3s	15s	0.12s
DANCE	201	7.06K	14.1K	0.37s	2.26s	37s	0.15s

Given an animated mesh sequence with vertices(Vertices), triangles (Triangles), and frames (Frames), the running time for extracting feature points (F-Point), segmenting the mesh into limbs and torso (Segment), selecting seed points(S-Point), and estimating the positions of the joints (Joint) are shown.

The column F-Point lists the time needed to extract five feature points from the reference pose and distinguish them. Column S-Point lists the time needed to read files from the hard drive, calculate the edge-length, and choose seed points. Reading files is time costly and if we exclude it, this column will be 0.25s for the SCAPE model and 0.27s for the DANCE model, respectively. The total time needed to extract the skeleton using our method is almost the same as (Tierny et al., 2008). All run times were measured on a PC with 3.4GHz CPU and 2G memory.

ACCURACY OF THE EXTRACTED SKELETON

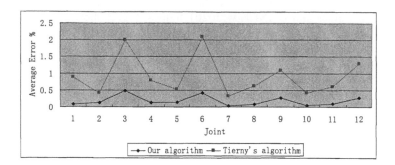

Figure 3. The average Euclidean distance error between the joint positions transformed by parts A's and part B's matrix in percent of the maximal side length of the overall bounding box(L_{BB}) for the DANCE model by our algorithm and (Tierny et al., 2008) respectively.

Owing to the lack of the true joint position for the DANCE model, we use the average Euclidean distance error between the joint positions transformed by parts A's and part B's matrix for all joints at limbs.

$$f(O) = \frac{100}{L_{BB} * F} \sum_t \left| R_A^t O + T_A^t - R_B^t O - T_B^t \right| \qquad (7)$$

Where O is the position of the joint estimated for the DANCE model by our algorithm and by (Tierny et al., 2008) respectively. The plot in Fig. 3 illustrates the accuracy of our method, our average error is a couple of times less than (Tierny et al., 2008).

For the SCAPE model, the average Euclidean distance error between the true and estimated joint positions in percent of the maximal side length of the overall bounding box is 1.1%, and the smaller is about 0.5% for knee, elbow, and the bigger is near 2% for shoulders and thighs.

NUMBER OF SEED POINTS

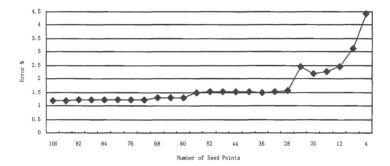

Figure 4. The average Euclidean distance error between the true and estimated joints positions for SCAPE model with a variable number of seed points. The error is measured as a percentage of the maximal side length of the overall bounding box.

Fig. 4 shows a plot of the average Euclidean distance error between input and estimated joints positions for the SCAPE model as the number of seed points decreases. As we decrease the number of seed points from 100 to 28, the variation of errors is negligible. Therefore, increasing the number of seed points is not always desirable and 30 seed points are all right. In practice, especially for joints at wrist, elbow, ankle and knee, 8 seed points is enough. While for joints at shoulder and thigh, because one of the parts adjacent to the joint belongs to torso where seed points usually are affected by more than two bones and the rigid transformation does not exhibit clearly, therefore, a small number of seed points may suffer significant errors and instability of results.

CONCLUSIONS

We present a fully-automatic method to extract a kinematic skeleton from 3D dynamic surface meshes. This method takes advantage of the following properties of the human body shape: (1) Geometry property of a static shape in that the human body can be divided into five parts including limbs and torso based on the Morse function defined as the geodesic distance. (2) Knowledge of anthropometry in that the relative geodesic distance for joints are stable for different models. (3) Kinematic characteristics of a dynamic shape in that the motions of the human body can be defined in terms of local rigid transformations. Our experiments demonstrated the accuracy of its results and the rapidity of this method as a combination of many characteristics of shape model. Still, there are some limitations in our approach. For example, if the model is significantly exaggerated so that its segment length proportions are far from the average one, it will not be able to find the correct joint positions. For future work, we would like to use the skeletons extracted from dynamic shapes to analyze human body motions in medical field.

ACKNOWLEDGMENTS

This work is supported by Beijing Municipal Commission of Education of Science and Technology Program of No.KM200910005020; Special Foundation of the '211 Project' Subject Construction for Young Researcher in Beijing University of Technology. We would like to acknowledge the help of MIT CSAIL Graphics Lab and James Davis, who provided us with the human model data.

REFERENCES

Yu, Y., Mao, T., Xia, S., Wang, Z. (2008). A Pose-Independent Method of Animating Scanned Human Bodies, CGI 2008, Istanbul,Turkey.

Oliveira, J.F., Zhang, D., Spanlang, B., Bernard, F. B. (2003). Animating Scanned Human Models. WSCG 2003.

Wang, C. L., Chang, T. K., Yuan, M. F.(2003). From laser-scanned data to feature human model: A system based on fuzzy logical concept, Computer-Aided Design, 35(3): 241-253.

Zhong, Y., Xu, B. (2006). Automatic segmenting and measurement on scanned human body, International Journal of Clothing Science and Technology, Vol 18 P. 19 – 30

Aujay, G., H'etroy, F., Lazarus, F. and Depraz, C. (2007). Harmonic skeleton for realistic character animation. In Symposium on Computer Animation, pages 151–160.

Aguiar, E., Theobalt,C., Thrun, S., Seidel,H.-P. (2008). Automatic Conversion of Mesh Animations into Skeleton-based Animations. EUROGRAPHICS 2008 ,Volume 27, Number 2

Kirk, A. G., O'Brien, J. F., Forsyth, D. A.(2005). Skeletal parameter estimation from optical motion capture data. In CVPR 2005 (June 2005), pp. 782–788.

Schaefer, S., Yuksel, C. (2007). Example-based skeleton extraction.In SGP '07 (Aire-la-Ville, Switzerland), pp. 153–162.

James, D. L., and Twigg, C. D. (2005). Skinning mesh animations. ACM Trans. Graph. 24, 3, 399–407.

Tierny, J.; Vandeborre, J.-P.; Daoudi, M. (2008). Fast and precise kinematic skeleton extraction of 3D dynamic meshes. ICPR 2008: 1-4

Winter, D.A. (May 1990). Biomechanics and Motor Control of Human Movement. [M]. 2nd Edition, John Wiley & Sons Canada, Ltd.

Horn, B. (1987). Closed-form solution of absolute orientation using unit quaternions. Journal of the Optical Society of America 4(4), 629–642.

CHAPTER 29

Hand Posture Estimation for the Disposition Design of Controls on an Automobile Steering Wheel

Natsuki Miyata, Masaaki Mochimaru

Digital Human Research Center
National Institute of Advanced Industrial Science and Technology
Tokyo 135-0064, JAPAN

ABSTRACT

This paper proposes a method using a limited number of measured postures to estimate the hand posture necessary to push a button on an automobile steering wheel by the thumb tip. The estimation is divided into two processes. First, the approximate posture to push a button is calculated by interpolation using measured reference postures. Then, the posture is adjusted so that the thumb tip reaches a given goal as nearly as possible using optimization method. The accuracy of the proposed method is validated by comparing the estimated postures with measured ones that are not used as reference postures.

Keywords: posture prediction, geostatistical interpolation, reachability

INTRODUCTION

More and more controls have recently been added on automobile steering wheels for safer and more comfortable driving by reducing the repositioning of drivers' hands and the subsequent distraction. Some guidelines exist for designing the disposition of push-buttons on a steering wheel. These guidelines are based on observations of the hand postures of a small number of subjects (Takeuchi, 2000). Because the observation results depend on the subjects' hand size, it is necessary to observe the hand postures of subjects having the hand size of the targeted consumers. If the subject's posture can be estimated in not-observed condition, an arbitrary button position can be tested in product design considering the observed subjects' hand size at the least. In addition, a smaller number of measurement conditions would help lower the experimental cost.

Therefore, the purpose of this research is to develop a method to estimate an individual's hand posture to push a button placed at an arbitrary position on a steering wheel by using a small number of measured postures.

Though several methods have been proposed for hand posture estimation (Koura, 2003) (Miller, 2003) (Lee, 2005) (Miyata, 2006) (Yang, 2006) (Endo, 2008), it is difficult to directly apply them because of the complexity of the wrist configuration for operating switches on a steering wheel. Except for the free operation with the fingertip, under the assumption that the opposite hand has a stable grasp of the steering wheel, the driver's wrist configuration changes to compensate for the thumb tip. The button position must satisfy the limitations of the fingers and palm on the steering wheel as well as the upper limb of the seated driver. Because the configuration of the wrist strongly affects the reachability of the thumb tip, a reasonable determination of the wrist configuration is an important issue.

We take a two-step approach for this posture estimation problem: interpolation of the measured postures and adjustment of the interpolated posture to achieve the given target position by optimization. By interpolating the measured postures, as in (Rose, 2001) and (Mukai, 2005), a reasonably precise posture is acquired as an initial estimate. As the thumb tip reach is not guaranteed by interpolation, optimization is introduced, which can be used to check whether the thumb tip can actually reach the goal position.

After showing the detail of each step in the next section, our method is validated by comparing the estimated postures with the measured results.

POSTURE ESTIMATION METHOD

INITIAL POSTURE ESTIMATE BY GEOSTATISTICAL INTERPOLATION OF REFERENCE POSTURES

Because postures are always collected explicitly with the push-button position, we employ the geostatistical interpolation method proposed in (Mukai, 2005). This interpolation method calculates a weighted average of reference postures. When given a target position, the kernel function value (weight coefficient vector) is calculated by solving the kriging system, which is defined by a variogram that expresses the dissimilarity of the postures according to the change of control parameters (i.e., button position). In this paper, we employed the simplest variogram that assumes linearity. More complex but precise and effective variograms can be estimated, as proposed in (Mukai, 2005).

How to calculate end point and orientation

As stated in (Takeuchi, 2000), a different area on the thumb tip touches a button surface and the thumb tip orientation differs according to the button position. The end point and orientation are determined by interpolation using the same kernel so that the optimization process can calculate the "reach."
The following explanation describes how to find the end point and orientation for each reference posture. Once the reference posture is reconstructed with an individual hand model that does not deform at fingertip by touch, we can estimate the interference of the thumb tip with the plane with its height the same as that of the upside pushed-down button (see Figure 1(a)). We define the deepest point in the normal direction of the button top plane as an end point. As an end orientation, the thumb tip orientation with respect to the button plane is directly employed for interpolation to reduce error accumulation effect.

OPTIMIZATION TO ADJUST INITIAL POSTURE

In the optimization process, we calculate the required modification of the interpolated postures so that the thumb tip can reach, if possible, the given goal position within the joint angle limit. The posture is generated by minimizing the following performance index function P.

$$P = w_{gp}P_{\text{reach\_goal\_pos}} + w_{go}P_{\text{reach\_goal\_ori}} + w_{sp}P_{\text{similar\_pos}} + w_{kr}P_{\text{keep\_within\_rom}}$$

$P_{\text{reach\_goal\_pos}}$ is the index to achieve the given goal position calculated in the previous subsection. Similarly, $P_{\text{reach\_goal\_ori}}$ is the index to achieve the goal orientation of the thumb tip. $P_{\text{similar\_pos}}$ is the index to keep similarity of the generated posture from the initial guess by interpolation. $P_{\text{keep\_within\_rom}}$ is a penalty term to keep the condition that the posture of all joints satisfies the range of motion; w_{gp}, w_{go}, w_{sp} and w_{ky} are weight coefficients.

Directly optimized variables are the thumb joint angles and the angle around the virtual joint whose axis is defined by the metacarpo phalangeal (MP) joints of the index and middle finger, as shown in Figure 1(b). To express realistic upper limb behavior that allows the hand movement, the wrist posture is determined as follows (see Figure 1(b)). A virtual joint at the MP joints of the index and middle finger is fixed with respect to the coordinate frame of the steering wheel, and the rotation at the virtual joint determines the absolute palm position and orientation. The "new" elbow position is then derived so that it remains at a constant distance on the line connecting the new wrist and the original elbow position at the beginning of the optimization. The elbow coordinate frame is determined by minimum rotation to superpose the line from the elbow to the wrist, and the wrist posture is consequently derived.

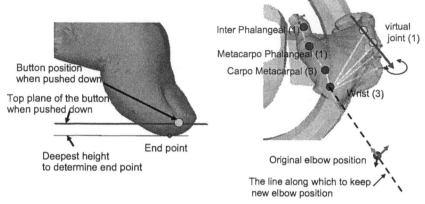

(a) End point determination   (b) Joints to be optimized and how to determine the wrist and the elbow posture

Figure 1 Variables and definition in posture prediction

ESTIMATED AND MEASURED POSTURES

ACTUAL OPERATION MEASUREMENT

To collect postures for reference and evaluation, the hand postures of one subject were measured using the motion capture system as shown in Figure 2. The coordinate frame of the steering wheel was set as shown in the left of Figure 3. Push buttons with a 2 mm stroke were set every 10 mm in the X and Y directions. Three depths in the Z direction were tested and the subject pushed all the reachable buttons at each depth.

As is well known, the steering habits of drivers vary greatly. Therefore, we regulated a basic steering style for the button pushing examination. From the viewpoint of safe driving operation, the subject was asked to hold a steering wheel with fingers as in the center and right of Figure 2 so that the driver can "steer" as he/she chooses while pushing a button.

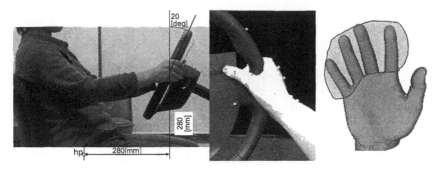

Figure 2 Operation measurement by motion capture system

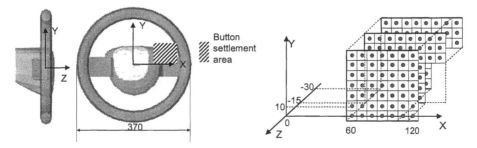

Figure 3 Coordinate frame of the steering wheel and button positions on each z-depth with respect to the steering wheel's coordinate frame

MEASURED POSTURES AND SELECTION OF REFERENCE POSTURES

Figure 4 shows the button positions that the subject's thumb tip could reach. The squares with lines running through them (on the right side) are the button positions that were not accessible because of the steering wheel form. The number of measured postures was 81.

Because our method uses interpolation, the reference postures are selected to cover a wide range of button positions. In this case, the buttons are spread in all x, y, and z directions with respect to the steering wheel coordinate frame. Considering the simplicity of the experimental planning, we decided to use eight reference postures, such as the vertices of the rectangular parallelepiped.

To show the effect of the difference of the posture selection on the estimation result, we examined three sets of eight reference postures (see Figure 4). In posture set 1, each two of the selected button positions were on the same Y axis. Posture set 2 represented a slight change in the posture selection from posture set 1. Posture set 3 was one of the examples of the relatively small area covered by the reference postures and required extrapolation. All the postures were used to validate the estimation accuracy.

262

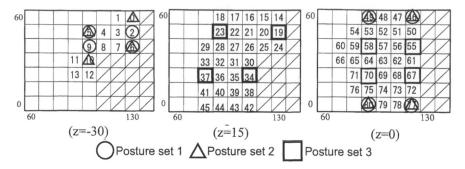

(z=-30)   (z̃=15)   (z=0)

○ Posture set 1   △ Posture set 2   ☐ Posture set 3

Figure 4 The measured posture IDs and Three posture sets

POSTURE ESIMATION SIMULATION

Using one of the reference posture sets, the postures were estimated for 9×7×3 button positions so that the distribution adequately covered the measured area. In this subsection, we show the results when posture set 1 is used.

Figure 5(a) shows an example of the estimated posture. As shown in Figure 5(b), even though the end point of the interpolated posture does not reach the goal position completely, that of the optimized posture does reach the goal. The distance from the interpolated posture's thumb end point to the goal was an average of 5 mm and a maximum of 11 mm.

Figures 6 and 7 show the reachability and the margin of the wrist joint angle to the joint angle limit, respectively. The deeper in the Z direction the goal positions are, the less area the thumb tip reaches. Also, as the goal positions extend further in the upper and left areas, the wrist joint margin becomes smaller, which agrees with the result in (Takeuchi, 2000).

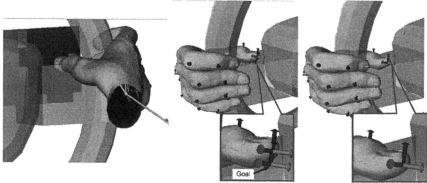

(a)Estimated hand posture (b) Interpolated (left) and Estimated (right) thumb tip position

Figure 5 Example of the estimated posture

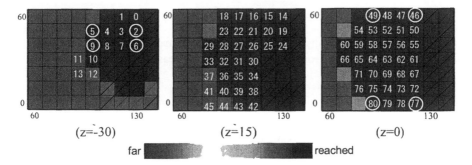

Figure 6 Reachability index of the simulated postures with the measured posture ID when posture set 1 is used

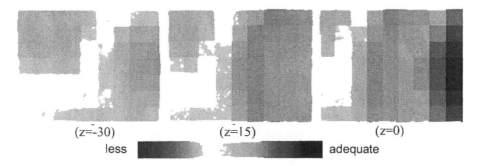

Figure 7 Margin of the wrist joint angle to the joint angle limit when posture set 1 is used

VALIDATION OF THE ESTIMATED RESULT COMPARED WITH THE MEASURED

Using Figure 6, we can compare the reachability of the estimated result with the measured one when posture set 1 is used. The superimposed white numbers (the same numbers as those shown in Figure 4) are the measured posture IDs that correspond to the measured reachability. The numbers with white circles correspond to the reference postures in posture set 1. Again, slashes in the figure show the button positions inaccessible because of the steering wheel form.

The number of goal positions that were reachable in the experiment but estimated to be "not reachable" can be obtained by counting the red square with the white number. It was 7 out of 81, the number of the postures that "did reach" the goal. In total, 93% of the area was correctly estimated to be reachable.

The average and maximum values of the absolute joint angle error are summarized in Table 1. When the goal positions were at the same height as the reference postures ($z = -\square30$ or 0), the average posture errors were approximately 2 degrees. At the height that no reference postures were included ($Z=\square-15$), the average posture errors were 5 to 6 degrees. In total, the joint angle error of the estimated posture was approximately 3 degrees.

Considering the general posture reproducibility, these results show the appropriateness of the result of our method that estimates the postures to push a button on a steering wheel from a limited number of measured data.

DISCUSSION

Effect of the reference posture selection

Table 2 summarizes the absolute joint angle error for different posture sets. A comparison between the result of posture set 1 and 2 suggests that a 10 mm change of the corresponding button positions of the reference postures rarely affects the results. The result of posture set 3 shows a worse joint angle error compared with that of posture set 1. This is because the corresponding button positions of the reference postures covered a smaller area and thus requires extrapolation when given the buttons on the plane whose z is equal to □30 mm.

Refinement of the interpolated posture

A large error of the estimated posture tends to occur when the interpolated posture as an initial guess already has a relatively large error compared with other goal positions. For example, the maximum joint error of the estimated posture occurred in posture ID 45, where the interpolated posture also had the maximum joint error.

Two countermeasures can be devised to avoid this tendency. One is to increase the number of reference postures. A smaller estimation error can be expected, though a smaller number of reference postures is preferable in terms of measurement cost. The other is to estimate and use a variogram that fits the measured data, as mentioned in the previous section.

Extended application by the measurement condition increase for the reference posture

As the proposed method utilizes the geometric distribution of the reference posture's measurement condition, it can be used if a button is set with a gradient, which often appears on the central part of today's steering wheels. Similarly, it can also deal with the handling poses that change in relative position to a steering wheel in a circle (see Figure 8). Also, the target control to be operated is not limited to a push type of button.

Data collection and comparison between subjects for the future extension to a representative hand model

Now that the posture to push a button at an arbitrary position can be estimated with certain accuracy, the comparison between subjects becomes easier. For example, our method can compare the postures at the goal positions that are normalized by the subjects' hand attributes such as hand length, even if the postures at the point have not been observed directly. We believe that such detailed analysis with our system can lead to the development of automatic generation of the posture to push a button on a steering wheel by representative hand models of the target population.

CONCLUSION

This paper proposes a method to estimate a certain individual's hand posture to push a button on a steering wheel by the thumb tip. The method uses a limited number of measured reference postures. The posture to push a button was estimated in two steps. First the measured posture was interpolated using a kriging system, which is a geostatistical interpolation method, to obtain the approximate posture at a given goal position. The end point and orientation of the thumb tip were also determined by the interpolation. Then, the posture was modified to enable the thumb tip end point and orientation to reach the goal position and orientation. Compared with the measured results, the proposed method was valid to at least 93% accuracy in reachability and approximately a 3-degree difference in joint angle estimation on average.

The authors are now extending the method for estimating the postures of representative hand models of the target population as well as collecting validation data.

Table 1: Absolute joint angle errors when posture set 1 is used

Average (Maximum) [deg.]

Z (plane height) [mm]	Interpolated		Estimated	
-30	2	(11)	3	(15)
-15	5	(37)	6	(42)
0	3	(22)	4	(17)

Table 2: Absolute joint angle error comparison between three posture sets

266

Average (Maximum) [deg.]

Z (plane height) [mm]	Posture set 1		Posture set 2		Posture set 3	
-30	3	(15)	2	(19)	4	(15)
-15	6	(42)	6	(41)	5	(47)
0	4	(17)	4	(17)	4	(18)

Figure 8 An example of different relative position to a steering wheel in a circle on edge

ACKNOWLEDGMENTS

This work was supported by Nissan Motor Co., Ltd.

REFERENCES

Endo, Y., Kanai, S., Miyata, N., Kouchi, M., Mochimaru, M., Konno, J., Ogasawara, M., and Shimokawa, M. (2008), "Optimization-Based Grasp Posture Generation Method of Digital Hand for Virtual Ergonomics Assessment." *Proc. SAE DHM 2008*, 2008–01–1902.

Koura, G.E., and Singh, K. (2003), "Handrix: Animating the Human Hand." *Proc. Eurographics/SIGGRAPH Symp. Computer Animation*, 110–119.

Lee, S.W., and Zhang, X. (2005), "Development and evaluation of an optimization-based model for power-grip posture prediction." *Journal of Biomechanics* 38, 1591–1597.

Miller, A.T., Knoop, S., Allen, P.K., Christensen, H.I. (2003), "Automatic Grasp Planning Using Shape Primitives." *Proc. IEEE Int. Conf. Robotics and Automation*, 1824–1829.

Miyata, N., Kouchi, M., and Mochimaru, M. (2006), "Posture Estimation for Design Alternative Screening by DhaibaHand - Cell Phone Operation." *Proc. SAE DHM 2006*, 2006–01–2327.

Mukai, T., and Kuriyama, S. (2005), "Geostatistical Motion Interpolation", *ACM*

Transactions on Graphics, 24(3), 1062–10

Rose III, C.F., Sloan, P.P.J., Cohen, M.F. (2001), "Artist-Directed Inverse-Kinematics Using Radial Basis Function Interpolation." *Computer Graphics Forum 2001*, 20(3), 239–250.

Takeuchi, S., Nishikawa, M., Suzuki, T., Shinzato, T., and Miyata, M. (2000), "Ergonomic Considerations in Steering Wheel Controls." *Proc. SAE DHM 2000*, 2000–01–0169.

Yang, J., Pena Pitarch, E., Kim, J., and Abdel-Malek, K. (2006), "Posture Prediction and Force/Torque Analysis for Human Hands." *Proc. SAE DHM 2006*, 2006–01–2326.

CHAPTER 30

Potential Improvements to the Occupant Accommodation Design Process in Vehicles Using Digital Human Modelling

Steve Summerskill[a] Russell Marshall[a], Keith Case[b]

Departments of: [a]Design & Technology, [b]Mechanical and Manufacturing Engineering, Loughborough University, Loughborough, LE11 3TU, UK

ABSTRACT

The process of designing products has changed drastically over the past 20 years, with an increasing reliance on virtual processes such as Computer Aided Design (CAD) software. These tools have been embraced by automotive manufacturers, who benefit from the ability to accurately model the thousands of parts that interact to produce an automobile. The virtual automotive design process has reduced the need for physical mock-ups that were traditionally produced to support user testing. It is now common practice amongst automotive manufacturers to use Digital Human Modelling systems to replace the early user testing that was traditionally performed. The design of occupant accommodation in automobiles is supported by the widespread adoption of Society of Automotive Engineers standards for seat and steering wheel adjustability etc. These standards are applied worldwide, and yet the data that drives the standards is based upon the US population, with some dimensions being 'estimated' in the 1960's, potentially leading to poor accommodation of international populations. The following paper discusses the use of DHM systems to design vehicle interiors for international populations, whilst allowing compatibility with the many aspects of car design that are affected by the SAE standards.

Keywords: DHM, occupant accommodation, multivariate, anthropometry

INTRODUCTION

The process of designing cars has changed drastically over the past 20 years, with an increasing reliance on virtual processes such as Computer Aided Design (CAD) software. CAD software allows designers and engineers to produce virtual parts, and combine these into assemblies that can represent a complete product such as an automobile. These powerful tools have been embraced by automotive manufacturers, who benefit from the ability to accurately model thousands of parts and how they interact. The virtual automotive design process has reduced the need for physical mock-ups that were traditionally produced to support the engineering design process. A side effect of this is that the user testing opportunities that were afforded by the production of these mock-ups have also been reduced. The ability to perform user testing early in the design process has many potential benefits in terms of the identification of issues of fit, control layout and safety. An added benefit of early user testing is that the issues that are identified can be solved before the majority of the vehicle has been designed. This avoids the costly redesign of parts that are associated with fixing errors that are identified through later user testing.

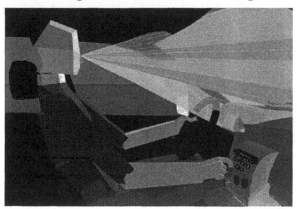

FIGURE 1. The SAMMIE DHM system being used to analyze a vehicle interior

It is now common practice amongst automotive manufacturers to use Digital Human Modelling (DHM) systems (see Figure 1) to replace the early user testing that was traditionally performed with real people. DHM systems provide the designer or engineer with CAD based virtual people that can be changed in size, and postured to replicate human activity. DHM systems can therefore be used to assess issues such as seat adjustability ranges, reach to controls, and vision of displays.

The following paper has been produced by members of the Design Ergonomics Research Group (DERG) based in the Dept. of Design and Technology at Loughborough University in the UK. The DERG has a long history of developing and applying DHM systems to the design of vehicles. Through interaction with a number of automotive manufacturers, and attendance at conferences that focus on the use of DHM, it has been noted that there is the potential for the capabilities of DHM to be exaggerated. For example, DHM systems are used to simulate the task of vehicle ingress and egress. The task of entering a vehicle is dependent upon a number of variables such as body size and shape, muscle strength, joint flexibility and individual behaviour. The data on the variability of joint motion, force application ability and behaviour used by current DHM systems is not sufficiently detailed to allow an accurate prediction of how a person will get into or out of a car, especially when one considers the simulation of the abilities of elderly people. This highlights that it is important for DHM users to be aware of the limitations of the data that drives them.

DIGITAL HUMAN MODELLING

DHM systems generally allow the evaluation of fit, posture, reach and vision of workstation elements, such as reach to control switches and vision of displays. The size variability of different national populations can be represented. Typically, by using a range of DHMs to explore the limits of a single, or multiple populations, conclusions can be drawn about the ability of the vehicle being investigated to accommodate that population. DHMs are able to interact with CAD models of workstation designs by having the ability to change posture, enabling the assessment of control reach-ability, and visibility. In this way workstations can be assessed for fit before they ever take physical form, reducing the need to build expensive prototypes and enabling a rapid iterative problem solving design process. However, DHMs are not replacements for physical mock-ups and user trials with real people. Their benefit is in establishing an accessible and accommodating design early in the development process whilst changes can be made easily and cheaply. At this stage alternatives can easily be explored and the issues fully understood. At an appropriate point, when the design is reasonably mature, user trails should be conducted to elicit the rich data provided by real people including feedback on comfort, and other cognitive and emotional issues, the analysis of which are currently lacking from DHM systems. DHM systems are currently used to analyse a wide range of design problems. However, caution should be employed with DHM system use. The validity of any DHM analysis relies upon the quality of the data that drives the system. This requires the user of a DHM system to understand the limitations of the data that drives it, such as the age of the anthropometric data used, the understanding of user behaviour, the limitations of the joint constraint system etc. There is a danger that engineers use DHM systems taking the results at face value. The use of DHM systems is likely to increase as there is pressure to reduce the use of expensive and time consuming user testing. It is therefore the responsibility of Human Factors specialists to design thorough DHM analysis protocols and to promote these as good practice for occupant accommodation.

MULTIVARIATE ACCOMMODATION

Anthropometric data is the key source for DHM and without it all DHMs would be a purely notional representation of the human form that would essentially look like a human but with no actual basis in reality. Typically anthropometric data are collected for a number of variables in body size such as knee height, stature and sitting height, among hundreds of others. These data are generally collect from thousands of members of a particular population (e.g. US, UK, German, Japanese). The statistical treatment of the data allows the user to understand how variable any one dimension can be, e.g. the range of stature found in any population. However, the typical presentation of individual measures in tables of data removes some crucial information about the population, i.e. the variability in body proportion. It is not possible to understand how the ratio of stature to sitting height may vary using standard anthropometric data because the ability to model any one person has been removed. Every one of us has variability in the percentile of each body dimension. A 5th percentile female (based upon stature) is unlikely to have 5th percentile values for other measures such as knee height, sitting height etc. As such, when designing from 5th percentile female to 95th percentile male a different 5% will be designed out for every dimension considered. For example, Herman Miller the provider of office furniture and services performed a chair design exercise using 5th to 95th percentile values. They found that when using only four variables: popliteal height (seat height), buttock to popliteal length (seat depth), elbow height, and lumbar height, the design only accommodated 68% of the population even though the starting intention was to design for 95% (Stumpf et al. 2001). The issue of multivariate accommodation is rarely addressed in the use of DHM. Generally the percentile of all of the dimensions of DHMs will be the same as the stature percentile. This has particular impact when designing products that have a number of variables that require adjustment (multivariate), such as an office chair or a car seating position. A car seat is generally adjustable forwards and backwards to allow correct reach to the pedals, and also allows adjustability of the seat height, and back rest angle. The user must be able to reach to the steering wheel, which may also have some form of fore/aft adjustability. When using anthropometric data to design the adjustability ranges that are built into the seat and steering wheel, it is useful to understand the prevalence of people that would be considered 'worst case scenarios' for such design activity. An example of this would be a tall driver, with long legs and long body, but relatively short arms. The long legs take the user further away from the steering wheel and the roof line forces the user to recline the seat more than usual to allow sufficient head clearance, again taking the user further from the steering wheel. The relatively short arms of the 'worst case scenario' would then generally find it difficult to reach the steering wheel, forcing a slumped posture that is likely to cause lower back problems. If the DHM system user is not easily able to model these 'worst case scenarios' then multivariate accommodation becomes extremely difficult to do with any degree of confidence. Assessments can be performed but their representativeness of the population is limited. A recent development that addresses one of the concerns with multivariate accommodation is the A-CADRE family of DHMs (Bittner 2000). A-CADRE is a statistically derived family of 17 DHMs that have been designed to represent both the breadth of the population but also more accurately represent the extremes of the population. In particular they represent 'interesting' body proportionality such as people with relatively long legs and tall bodies and short arms, or short legs, tall bodies and long arms. The validity of the A-CADRE data set was tested by designing a workstation (a helicopter cockpit) using the dataset, and then testing the accommodation of a randomly generated sample of 400 users. 99% of the 400 users were found to be accommodated by the design produced using the ACADRE set. This is in stark contrast to the findings of Stumpf et al (2001) which showed that 30% of a population are excluded from the comfortable use of a chair designed using univariate 5th%ile to 95%ile values.

SAE STANDARDS ON VEHICLE DESIGN

The approach for the design for occupant accommodation in vehicles is defined by a number of Society of Automotive Engineers (SAE) standards. The standards provide guidance for the definition of seat adjustability, steering wheel adjustability, vision of displays and reach to control panels, amongst others. The contents of the various standards provide template data that can be used in the CAD systems and DHM systems that are used to design vehicles.

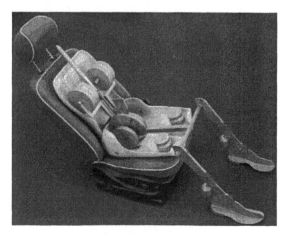

FIGURE 2. The SAE H-point manikin

The key tool defined by the SAE standards is the H-Point Manikin (SAE standard J826, 1995). The SAE Hip point Manikin (see Figure 2) is a mechanical device that can be used to simulate the size of drivers' buttock knee length and knee height for a range of the population of the USA (10th female to 95th male). The dimensional referencing system used in the SAE standards (SAE J110, 2005) relies upon the hip locations, or H-Points, that are derived from the use of the SAE manikin.

The H-point manikin and the design process that it supports are used by the majority of automotive manufacturers, as it is inherently linked to processes such as crash testing, and yet the data that was used to define the adjustability ranges of the SAE manikin are not well defined.

For example, the source of the 95th%ile value for the adjustability of the lower leg of the H-point manikin is described as follows;

> *"Values for the 95th percentile leg lengths were developed on the basis of best judgment of available data by the Design Devices Subcommittee of the SAE Human Factors Engineering committee at the July 1968 and March 1969 meetings"* SAE Standard J826

The quoted dimension for the 95th%ile lower leg length is 459.1mm. This is 24.9 millimetres longer that 95th%ile lower leg dimension found in the anthropometric data source ADULTDATA (1998) for the US population. Other issues exist with the H-point manikin. For example, the straight legged posture that is adopted by the H-point manikin does not replicate the actual posture used by drivers. The leg posture of larger drivers tends to include some rotation of the upper thigh, with the heel located between the accelerator and brake pedals to allow the foot to pivot when changing between accelerator and brake use. DHM systems have the potential to simulate more accurate postures for a range of driver sizes than a design process that relies upon the use of SAE H-Point manikin alone.

This has been illustrated in Figure 3 using the System for Aiding Man Machine Interaction Evaluation (SAMMIE) DHM system (Summerskill et al, 2009). The figure shows two human models that have been generated to represent the 95th%ile US male, and 10%ile US female, driving a car that represents the adjustability ranges of a current production vehicle. The human models were postured using data on preferred driving postures collected from a sample of 56 people with a stature %ile range of 1st-99th (Porter and Gyi, 1998). The postures are therefore exhibited by real people. The white human model shown in Figure 3 is a standard 2d template based upon the size data provide by a number of SAE standards. The ellipses in front of the face of the SAE manikin represent the range of eye positions derived from the SAE data and presented as a design tool. The DHM human models and the SAE manikin represent the same range of the US population accommodation according to the source data, and yet it is clear to see that eye positions of the two human models are outside the zones defined by the SAE eye position data. The more up to date anthropometric data and simulation of posture used in the DHM system produced very different results to those derived from the SAE data. The differences that are highlighted for key design variables such as eye point are exacerbated when one considers the wider range of driver sizes found for international populations. The SAE H-point manikin was updated in 2005, but still uses the same anthropometric data and manikin posture as those shown in Figure 2. This issue has been identified by researchers in the area of occupant accommodation. For example, Parkinson et al (2006) discussed the univariate nature of the SAE templates, and the potential for DHM systems to provide more accurate simulations of people at the limits of the percentile range.

FIGURE 3: A comparison between the SAE eye location data and the eye points found for a 95th%ile US male and a 10th%ile US female DHMs

The more accurate replication of the human form that is possible using DHM systems can complement the valuable referencing system that is used in many car design processes. The wide variety of uses for the data derived from the SAE manikin, from vehicle design, to crash testing, makes it unlikely that the anthropometry of the manikin will be changed. The following section discusses an approach to occupant accommodation that has been defined to provide more accurate driver modelling.

THE APPROACH USED BY THE DERG FOR OCCUPANT ACCOMODATION IN VEHICLES

The DERG process used to design vehicle occupant accommodation has been applied to a number of vehicle design projects performed with automotive manufacturers. The method combines the use of a number of datasets. The dataset used to define the size of human models is ADULTDATA (1998). This contains anthropometric data for a number of international populations that can be used to create DHMs. As discussed, only univariate DHMs can be produced from standard anthropometric data. The issue of multivariate accommodation is addressed using the A-CADRE human model dataset, with certain A-CADRE models being used to examine certain situations. The dataset used to define the posture of the human model were produced by the DERG and disseminated in a paper by Porter and Gyi (1998).

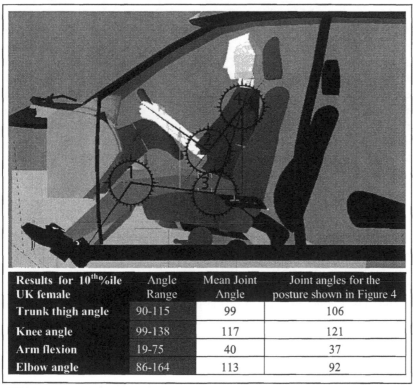

Results for 10th%ile UK female	Angle Range	Mean Joint Angle	Joint angles for the posture shown in Figure 4
Trunk thigh angle	90-115	99	106
Knee angle	99-138	117	121
Arm flexion	19-75	40	37
Elbow angle	86-164	113	92

FIGURE 4. A posture analysis being performed using the Porter and Gyi joint data

The study provided data on the preferred driving posture for a large range of user sizes (1st-99.9%ile UK). These data were gathered using a rig that allowed free adjustment of the pedal and steering wheel positions. These data were used to generate mean joint angle values for the ankle, knee, hip, upper arm and elbow, as well as the range found for those values across the 56 participants. Combining these data with the anthropometry from each of the 56 sample members provides a method for the testing of occupant accommodation that augments the statistically derived size range of A-CADRE. In addition, the ability to represent the real body proportionality and preferred driving posture of an individual removes the assumptions that are made that relate to how the posture of a driver is affected by their size. For example, the SAE standard that explains the use of the SAE manikin does not account for a posture that is affected by limited head room (Parkinson, 2006). Figure 4 shows an image from a DHM analysis being performed to determine the minimum sized UK female that can be accommodated by a particular vehicle interior. This vehicle interior had been designed using SAE data. The analysis identified the 10th%ile UK female as the minimum sized driver. It was also determined that smaller human models could only be accommodated by changing the adjustability ranges of the steering wheel and the seat, or by moving the pedals rearwards in the car. The testing was performed by iteratively increasing the size of the DHM until the pedals could be effectively reached (clutch fully depressed) from the foremost seat position, the adopted posture was within the joint angle ranges defined by Porter and Gyi and that there was sufficient steering wheel clearance. Once this is established options for the improvement of the minimum accommodated percentile were supplied to the manufacturer. These included increasing the range of the fore-aft adjustability of the steering wheel, moving the pedals more rearward in the car, and changing the angle of the central value for steering wheel rake adjust. These recommendations would increase the proportion of the population that would be able to effectively drive the car. This example of DHM use reflects the reactive mode of ergonomic intervention in the design process. However, a proactive mode of ergonomic design does occur.

The DERG have been involved in design processes for vehicles that start with the definition of seat and steering wheel adjustability ranges in DHM systems, around which vehicles are then designed. This proactive process involves the use of the various data sources discussed above, being used in collaboration with SAE standard design limits and referencing systems. By designing using multivariate human models as described by Porter and Bittner, combined with standard univariate DHMs it is anticipated that the final results will be an interior design that accommodates a larger proportion of the population, with a smaller proportion of users needing to adapt using compromised postures. As with any study performed by the DERG, the final SAMMIE designs were tested using a full size mock-up with a sample of 30 users that exhibited the stature variation of 5th%ile UK female to 99th%ile UK male. The designed package was found to be suitable for all users in terms of posture comfort, and reach and vision of controls. The DERG process for occupant accommodation has been shown to improve occupant accommodation through user testing of specific design examples. However, more research is required that allows a deeper understanding of the prevalence and effects of multivariate accommodation issues such as body proportionality. As one might expect, the increased adjustability ranges for seat and steering wheel that are likely outcomes from the DERG process have cost implications for the manufacturer. It would therefore be extremely useful to be able to define the proportion of the population that will be designed out due to multivariate accommodation as justification for the added expense involved in increasing adjustability ranges. For this to be possible it is first necessary to understand the prevalence of the body proportionality that is exhibited by models such as A-CADRE. There is the possibility that new data sources, gathered using 3D body scanners, such as CAESAR (2010), could be processed to provide an understanding of these issues.

CONCLUSIONS

This paper has highlighted the benefits of improved occupant accommodation through the use of DHM systems and anthropometric data for international populations, and multivariate human models. However, the constraints imposed by the SAE process make it likely that vehicle design for the generally smaller populations of emerging economies such as China will exclude a larger proportion of the population. For example, the minimum design limit defined by the SAE manikin (10th%ile US female) is equivalent in size to a 35th%ile Chinese female based upon data available in Adultdata (1998). This means that the SAE design process has the potential to exclude more than a third of the Chinese female population from driving a car. It is only by gathering up to date anthropometric data for these emerging economies, and accounting for the international variability in body proportionality that this situation can be improved.

REFERENCES

Adultdata. (1998). The handbook of adult anthropometry and strength measurements – data for design safety. eds. L. Peebles and B. Norris. Department of Trade and Industry.

Bittner, A.C. (2000). A-CADRE: Advanced family of manikins for workstation design. *Proceedings of the IEA 2OOO/HFES 2000 Congress,* San Diego. (4): 774-777.

CAESAR. (2010). Civilian American and European Surface Anthropometry Resource. http://store.sae.org/caesar/ [Accessed 28/02/2010].

Parkinson, M., and Reed, M., (2006). "Optimizing vehicle occupant packaging". *SAE Transactions: Journal of Passenger Cars–Mechanical Systems,* 115.

Porter, J.M., Gyi, D.E., (1998). Exploring the optimum posture for driving comfort. International Journal of Vehicle Design 19 (3), 255–266.

Society of Automotive Engineers, (1995). Motor Vehicle Dimensions. SAE J1100. SAE Standard.

Society of Automotive Engineers, (1995). Manikins for use in defining vehicle seating accommodation. SAE J826. SAE Standard, Vehicle Occupant Restraint Systems and Components.

Stumpf, B. Chadwick, D. and Dowell, D. 2001. The Anthropometrics of Fit: Ergonomic Criteria for the Design of a New Work Chair. Herman Miller White paper. http://www.hermanmiller.co.uk/our-business/white-papers/ (accessed 21/02/2010).

Summerskill, S.J., Marshall, R., Case, K., Gyi, D.E., Sims, R.E. and Davis, P. (2009). Validation of the HADRIAN System using an ATM evaluation case study. Lecture Notes in Computer Science: Proceedings of the Second International Conference, ICDHM 2009, Held as Part of HCI International 2009, San Diego, CA, USA, July 19-24. 727-736. Springer: Berlin.

Chapter 31

Determining Weights of Joint Displacement Function in Direct Optimization-Based Posture Prediction-A Pilot Study

Qiuling Zou[1], Qinghong Zhang[2], Jingzhou (James) Yang[1]

[1]Human-Centric Design Research (HCDR) Lab
Department of Mechanical Engineering
Texas Tech University
Lubbock, TX 79409
*james.yang@ttu.edu

[2]Department of Mathematics and Computer Science
Northern Michigan University
Marquette, MI 49855

ABSTRACT

Human posture prediction can often be formulated as a non-linear multi-objective optimization (MOO) problem. The joint displacement function is considered as a benchmark of human performance measures. The hypothesis is that human performance measures govern how human move as they do. Therefore, when joint displacement is used as the objective function, posture prediction is a MOO problem. The weighted sum method is commonly used to find a Pareto solution of this MOO problem. Within the joint displacement function, the relative value of the weights represents the relative importance of the joint. Usually, weights are determined by trial and error approaches. This paper proposes a systematic approach to determine the weights for the joint displacement function in optimization-based posture prediction where a realistic posture is given. It can be formulated as a two-level optimization problem. The design variables are joint angles and weights. The cost function is the summation of the differences between joint angles and a realistic posture. Constraints include (1) normalized weights within limits; (2) an inner optimization problem to solve for joint angles where joint displacement is the objective function, and constraints include that the end-effector reaches the target point and joint angles are within their limits. Additional constraints such as weight limits and weight linear equality constraints obtained through observations are also implemented in the formulation to test the method. A 21 degree of freedom (DOF) upper human model and three target points in-vehicle are used to illustrate the procedure of the method.

Keywords: MOO, weights, weighted sum, posture prediction.

INTRODUCTION

Digital human modeling and simulation has proven to revolutionize the way new vehicles are designed, built, operated, and maintained. The expected benefits entail the reduction of development time, the reduction of development costs and increasing quality. Posture prediction is a main component in digital human modeling. Posture prediction aims to predict a single static posture for a specific scenario. Significant research has focused on posture prediction models. Three types of approaches are generally utilized. In the empirical-statistical approach, data are collected either from thousands of experiments with human subjects or from simulations with three-dimensional computer-aided human-modeling software (Porter et al., 1990; Das and Sengupta, 1995). The data are then analyzed statistically in order to form predictive posture models. These models are implemented in the simulation software along with various methods for selecting the most probable posture given a specific scenario (Beck and Chaffin, 1992; Das and Behara, 1998; Faraway et al., 1999). In the direct inverse kinematics approach, the position of a limb is modeled mathematically with the goal of formulating a set of equations that is used to solve for joint variables (Jung et al., 1992; Jung et al., 1995; Kee et al., 1994; Jung and Choe, 1996; Wang and Verriest, 1998; Wang, 1999; Tolani et al., 2000). Griffin (2001) reviewed the validation of biodynamic models. Wang et al. (2005) demonstrated the validation of the model-based motion in the REALMAN project. In the direct optimization-based approach, posture prediction is considered to be an optimization problem in which humans choose a posture to minimize certain objective functions. In the biomechanics literature, significant efforts have focused on static lifting posture prediction using three different behavioral criteria or objective functions. The first criterion assumes that subjects choose a posture which requires the minimum overall effort (Byun, 1991).

The second criterion assumes that subjects minimize local effort or fatigue (Bean et al., 1998; Park, 1973). The third criterion assumes that subjects choose the posture with the greatest stability (Kerk, 1992). Dysart and Woldstad (1996) compared these three models. In our previous research, a multi-objective optimization (MOO)-based kinematic posture prediction model was proposed (Yang et al., 2004; Marler et al., 2007; Yang et al., 2007; Yang et al., in press).

Different human performance measures have been proposed for direct optimization-based posture prediction. Jung et al (1994) proposed a joint displacement function that is the weighted sum of the differences between current joint angles and angles that constitute a predetermined neutral position. A perceived discomfort about the joint movement from a regression model was developed by Kee and Jung (1996) and Choe (1996). Visual displacement, delta potential energy, and musculoskeletal discomfort were proposed (Yang et al., 2004; Marler et al., 2006; Yang et al., in press). Within direct optimization-based methods, different human performance measures were considered as the cost functions and it is a MOO problem. How to combine different human performance measures is a key issue for the direct optimization-based approach. Several methods were used to find some Pareto solutions of the multi-objective optimization problem, such as the weighted sum method, the global criterion method and the min-max method. Among them, the weighted sum method is commonly used. Weights are used to indicate the relative importance of the cost functions. The value of each weight is only significant relative to the other weights and relative to the value of its corresponding objective function. Several approaches have been used to determine the weights in the weighted sum method. Consistency Ratio method was proposed by Saaty (1977; 1991) to determine the values in Analytic Hierarchy Process. It is required to construct a hierarchy matrix to express the relative values of a set of attributes, perform pair-wise comparisons to obtain the weights of importance for all the factors, and finally calculate a Consistency Ratio to measure how consistent the judgments have been relative to large samples of purely random judgments. Zhang et al. (1998) proposed a method to estimate weights based on the minimization of the time-averaged mean square error for the difference between predicted angles and a set of empirical observations. This set of weights forms a pseudo-inverse of Jacobian and represents the participation of each degree of freedom in the model on the instantaneous effort that is performed. Zhang et al. (2001) introduced an improved weighting method with multi-bounds formulation and convex programming for multi-criteria structural optimization, and an analytical approach in which the lower and upper bounds were analytically calculated in a closed form and the effects of scaling factors upon the bounds of the ineffective weights were revealed. A trial and error method is a common method that has been used in human modeling and simulation and other areas (Yang et al., 2004; Messac and Mattson 2002). Kim and Weck (2004; 2005) and Khan and Ardil (2009) proposed an adaptive weighted-sum method for bi-objective optimization and multi-objective optimization. It focuses on unexplored regions by changing the weights adaptively rather than by using a priori weight selection. Zhang and Gao (2006) integrated adaptive weightings in a min-max method for optimization. Dong (2008) adopted orthogonal interactive genetic algorithm to calculate the weights of different factors which affect posture choosing.

There is lack of research effort to systematically determine the weights for the weighted sum method in literature. This paper is a pilot study to investigate a systematic mathematical model to determine the weights for the joint displacement function in direct optimization-based posture prediction. If this method is effective for the joint displacement function, then we can extend it for any multiple objective functions.

This paper is organized as follows: Section 2 briefly introduces a DH-based (Denavit and Hartenberg, 1955) digital human model. Section 3 describes the problem definition. Section 4 gives detailed formulation as a two-level optimization problem. Section 5 illustrates examples to demonstrate the efficiency and effectiveness of the proposed method.

HUMAN MODEL

Human body can be modeled as a kinematic system, which is a series of links connected by revolute joints that represent musculoskeletal joints such as spine, shoulder, arm, and wrist. The rotation of each joint in the human body is described as a generalized coordinate q_i.

In this study, we only focus on the human upper body and a 21-DOF model is used and shown in Fig. 1, where the spine has 12 DOFs and the right arm has 9 DOFs. Joint angles are defined as $\mathbf{q} = \begin{bmatrix} q_1 & \cdots & q_{21} \end{bmatrix}^T$. According to the DH method, the position vector of a point of interest on the end-effector of a human articulated model (e.g., a point on the thumb with respect to the torso coordinate system) can be written in terms of joint variables as

$$\mathbf{x} = \mathbf{x}(\mathbf{q}), \tag{1}$$

where $\mathbf{q} \in \mathbf{R}^{21}$ is the vector of 21-generalized coordinates, and $\mathbf{x}(\mathbf{q})$ can be obtained from the multiplication of the homogeneous transformation matrices defined by the DH method as

$$^{0}\mathbf{T}_{n} = {}^{0}\mathbf{T}_{1}\,{}^{1}\mathbf{T}_{2} \ldots {}^{n-1}\mathbf{T}_{n} = \begin{bmatrix} {}^{0}\mathbf{R}_{n}(\mathbf{q}) & \mathbf{x}(\mathbf{q}) \\ 0 & 1 \end{bmatrix}, \tag{2}$$

where $^{i}\mathbf{R}_{j}$ is the rotation matrix relating coordinate frames i and j. The vector function $\mathbf{x}(\mathbf{q})$ characterizes the set of all points touched by the end-effector. The transformation matrix can be defined as

$$^{i-1}\mathbf{T}_{i} = \begin{bmatrix} \cos\theta_i & -\cos\alpha_i \sin\theta_i & \sin\alpha_i \sin\theta_i & a_i \cos\theta_i \\ \sin\theta_i & \cos\alpha_i \cos\theta_i & -\sin\alpha_i \cos\theta_i & a_i \sin\theta_i \\ 0 & \sin\alpha_i & \cos\alpha_i & d_i \\ \hline 0 & 0 & 0 & 1 \end{bmatrix} \tag{3}$$

where $q_i = \theta_i$ since all joints are revolute joints θ_i, depicted in Fig. 1, is the joint angle from x_{i-1} to the x_i axis, d_i is the distance from the origin of the $(i-1)^{th}$ coordinate frame to the intersection of the z_{i-1} axis with the x_i, a_i is the offset distance from the intersection of the z_{i-1} axis with the x_i axis, and α_i is the offset angle from the z_{i-1} axis to the z_i axis.

In Fig. 1, q_1 through q_{12} represent the spine, q_{13} through q_{17} represent the shoulder and clavicle, q_{18} through q_{21} represent the right arm.

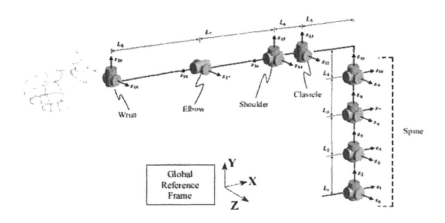

Fig. 1 A 21 DOF human upper body model

PROBLEM DEFINITION

The problem is defined as: Given a set of joint angles $\mathbf{q}^*$ (posture), the objective is to determine the weights in the joint displacement cost function in the following direct optimization-based posture prediction formulation (Yang et al., 2004):

Find: Joint angles $\mathbf{q}$

Minimize: $f(\mathbf{q}) = \sum w_i (\dfrac{q_i - q_i^N}{q_i^U - q_i^L})^2$ (4)

Subject to: $d^2 = (\mathbf{x} - \mathbf{x}^{\text{Target}})^2 = 0$

$\qquad\qquad q_i^L \le q_i \le q_i^U$, $i = 1, \cdots, n$

Where $\mathbf{q}^{\mathrm{N}}$ denotes a relatively comfortable position (neutral posture). q_i^L and q_i^U are the lower and upper boundaries of q_i, $\mathbf{w} = \begin{bmatrix} w_1 & \cdots & w_n \end{bmatrix}^T$.

FORMULATION

Based on the above problem definition, we can formulate it as a two-level optimization problem as follows:

Find: $\mathbf{q}$, $\mathbf{w}$

Minimize: $\displaystyle\sum_{i=1}^{n} w_i (\frac{q_i - q_i^*}{q_i^U - q_i^L})^2$ $\hspace{3cm}$ (5)

Subject to: $\displaystyle\sum_{i=1}^{n} w_i = 1$, $w_i \geq 0$

$\hspace{2cm}$ $\mathbf{q}$ is a solution of the following optimization problem

$\hspace{3cm}$ Find: Joint angles $\mathbf{q}$

$\hspace{4cm}$ Minimize: $f(\mathbf{q}) = \displaystyle\sum_{i=1}^{n} w_i (\frac{q_i - q_i^*}{q_i^U - q_i^L})^2$

$\hspace{4cm}$ Subject to: $g(\mathbf{q}) = (\mathbf{x} - \mathbf{x}^{\mathrm{Target}})^2 = 0$

$\hspace{4cm}$ $q_i^L \leq q_i \leq q_i^U$, $i = 1, \cdots, n$.

It is difficult to directly solve Eq. (5). Therefore, we transfer this formulation to the following format by using the Lagrange and KKT theory:

Formulation I:

$\hspace{1cm}$ Minimize: $f(\mathbf{q}) = \displaystyle\sum_{i=1}^{n} w_i (\frac{q_i - q_i^*}{q_i^U - q_i^L})^2$

$\hspace{8cm}$ (6)

Subject to: $\displaystyle\sum_{i=1}^{n} w_i = 1$, $w_i \geq 0$

$\hspace{1.5cm}$ $\displaystyle\sum_{i=1}^{n} w_i \nabla f_i(\mathbf{q}) + \lambda \nabla g(\mathbf{q}) + (\mu_1, \mu_2, \cdots, \mu_n)^T + (v_1, v_2, \cdots, v_n)^T = 0$,

$\hspace{1.5cm}$ $g(\mathbf{q}) = (\mathbf{x} - \mathbf{x}^{\mathrm{Target}})^2 = 0$;

$\hspace{1.5cm}$ $\mu_i(q_i - q_i^L) = 0$ $v_i(q_i^U - q_i) = 0$ $u_i \geq 0, v_i \geq 0$ $q_i^L \leq q_i \leq q_i^U$ $i = 1, 2, \ldots, n$;

$\hspace{1.5cm}$ λ_i, μ_i, and v_i are Lagrange multipliers and $f_i(q) = (\frac{q_i - q_i^*}{q_i^U - q_i^L})^2$.

Formulation II:

The optimization formulation is the same as Eq. (6), but we add additional constraints by observations:

$$w_i^L \leq w_i \leq w_i^U.$$

(7)

Formulation III:

The optimization formulation is the same as Eq. (7), but we add additional constraints by observations:

$$w_1 = w_4 = w_7 = w_{10} = 100w_{21}$$

(8)

$$w_2 = w_5 = w_8 = w_{11} = w_1$$

$$w_3 = w_6 = w_9 = w_{12}$$

$$w_{18} = w_{19} = w_{20} = w_{21}$$

$$w_{14} = w_{15} = w_{16} = w_{21}.$$

The total number of linear equality constraints should be less than the number of DOF.

ILLUSTRATIVE EXAMPLES

We selected three in-vehicle tasks that test both the simple and the complex functionality of the human simulations (Yang et al., in press). Fig. 2 shows the three tasks that were chosen for the experiment. Task 1 requires reaching the point at the top of the A-pillar, a simple reach task. Task 2 requires reaching the radio tuner button, a slightly difficult reach task. Task 3 requires reaching the glove box handle, a difficult reach task. In the motion capture process, a number of reflective markers are attached over bony landmarks on the participant's body, such as the elbow, the clavicle, or the vertebral spinous processes. In this work, redundant markers (more than the minimum required) were used to compensate for occluded markers. The time history of the location of the reflective markers was collected using a Vicon motion capture system with eight cameras at a rate of 200 frames per second. In the plug-in gait protocol, markers are attached to bony landmarks on the subject's body to establish local coordinate systems on various segments of the body. Joint centers and joint profiles can then be obtained using these coordinate systems. Methodologies for calculating joint center locations and link lengths of humans are available and have been somewhat successful (Halvorsen et al., 1999). In this work, due to the complexity of the capturing environment for a seated person inside a car and due to the limited number of cameras available at the time of the experiments (eight), redundant markers were attached to the upper part of the subject's body to estimate joint center locations and to compensate for the missing markers.

To cover a larger driver population, auto designers choose a range of percentiles from 5% female to 95% male. Therefore, in our experiment, we chose four different populations, all Americans (Caucasians): 5th-percentile female, 50th-percentile female, 50th-percentile male, and 95th-percentile male. Three subjects were selected within each percentile, for a total of 12 participants. For males, the average height was 182.9 (SD 8.3) cm and the average mass was 90.7 (SD14.4) kg. For females, the average height was 157.7 (SD 7.2) cm and the average mass was 58 (SD 5.9) kg. They were well distributed between the ages of 26 and 48 years. All participants gave informed consent prior to the study. In this paper, we only choose one subject (50th-percentile male) as an example to illustrate the proposed method. The target points are $\mathbf{x}^{\text{Target1}} = \begin{bmatrix} 28.95 & 50.23 & 45.54 \end{bmatrix}^T$, $\mathbf{x}^{\text{Target2}} = \begin{bmatrix} -50.65 & 11.85 & 61.03 \end{bmatrix}^T$, $\mathbf{x}^{\text{Target3}} = \begin{bmatrix} -78.89 & -7.65 & 64.74 \end{bmatrix}^T$. The postures obtained through this experiment for the three tasks are shown in Fig. 2.

To study which formulation is the best choice for predicting the weights, we need compare the experimental postures $\mathbf{q}^*$ with those $\mathbf{q}$ predicted by using the obtained weights. We use two different approaches to compare these postures: joint error and linear regression. We plot the joint error and posture correlation for all cases in Figs. 4 to 9. Neutral posture (position) is from Yang et al. (2004): $q_i^N = 0$ for $i = 1, 2, ..., 12$, $q_{13}^N = -0.143623$, $q_{14}^N = -0.480856$, $q_{15}^N = 0.314316$, $q_{16}^N = 0.425406$, $q_{17}^N = -1.16848$, $q_{18}^N = -0.495404$, $q_{19}^N = -0.258396$, $q_{20}^N = 0.029583$, $q_{21}^N = 0.307632$. Formulation I and II have similar results, which have the maximum absolute joint error 1.4 radians and $R^2 = 0.52$. While in Formulation III, the maximum absolute joint error is 0.85 radians and $R^2 = 0.70$ in Fig. 3. Therefore, Formulation III is the idea formulation to determine weights.

The joint angles for each of these three postures were used in Eqs. (6), (7), and (8) independently to obtain the corresponding weights.

Using Formulation III, for each target point, one can obtain one set of weights. If we choose the average from all target points, the average weights will be used for posture prediction for all target points. The maximum absolute joint error is around 0.05 radians and $R^2 = 0.9999$ (Figs. 4 and 5). Therefore, the weights for the joint displacement function in the optimization-based posture prediction should be determined from Formulation III. In addition, the final weights should be the average of the weights for all target points.

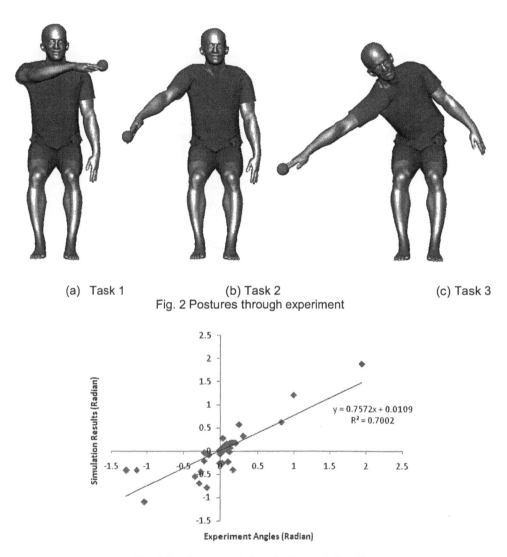

(a) Task 1   (b) Task 2   (c) Task 3

Fig. 2 Postures through experiment

Fig. 3 Posture correlation for Formulation III

The predicted postures based on the average weights obtained from Formulation III are shown in Fig. 6. Visually they are really close to those in Fig. 2.

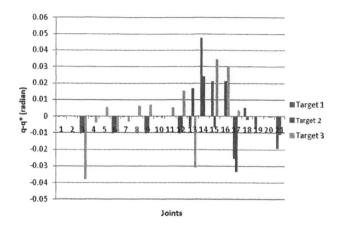

Fig. 4 Posture comparison for Formulation III using average weights from all tasks

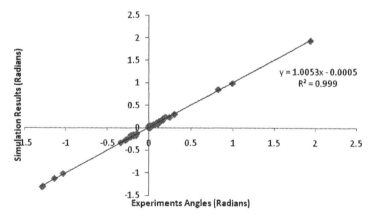

Fig. 5 Posture correlation for Formulation III using average weights from all tasks

(b) Task 1   (b) Task 2   (c) Task 3
Fig. 6 Predicted postures based on the average weights in Formulation III

CONCLUSION

This paper proposed a systematic method to determine the weights of the joint displacement function in direct optimization-based posture prediction approach. The problem was formulated as a two-level optimization problem and then transferred to an optimization problem by using Lagrange theory and KKT theory. Additional constraints had been added by observations. Three cases were studied and examples were used to demonstrate the formulation and results were compared in terms of joint error and linear correlation.

Weight limits and some weight linear constraints are important information to be considered as more constraints in the optimization problem. Formulation III is the idea formulation to determine weights of the joint displacement function. By deploying this formulation, each task generated different sets of weights. The final weights for posture prediction were obtained by averaging all weights for different tasks. In addition, for the final posture prediction, only one neutral posture was selected for all tasks and one set of weights (average weights) was implemented. It was shown that the predicted posture based on the obtained weights correlated the experimental posture very well and the coefficient of correlation R^2 is 0.999.

The proposed method was demonstrated by only one subject and three target points. The future work will include extending the method to larger size of subjects with different percentiles and a larger number of target points in vehicle. We will also extend this method to other human performance measures in posture prediction and motion prediction.

ACKNOWLEDGEMENTS

This research was partly supported by the startup fund of Texas Tech University (TTU), a grant from National Science Foundation (NSF) awarded to TTU, and the Faculty Research Grant from Northern Michigan University.

REFERENCE

Bottaso, C.L., Prilutsky, B.I., Croce, A., Imberti, E., Sartirana, S. (2006), "A numerical procedure for inferring from experimental data the optimization cost functions using a multi-body model of the neuro-musculoskeletal system." *Multibody Syst. Dyn.* 16: 123-154.

Craig, J.J. (2005), *Introduction to Robotics: mechanics and control,* Pearson Education Asia Limited and China Machine Press, Beijing.

Choi, J., Armstrong, T.J. (2005), "3-D Dimensional kinematic model for predicting hand posture during certain gripping tasks." *ASB 29th Annual Meeting,* Cleveland, Ohio, 665.

Das, B., Behara, D.N. (1998), "Three-dimensional workspace for industrial workstations." *Human Factors,* Vol. 40, No.4, pp. 633-646.

Das, I., Dennis, J. (1996), "Normal-Boundary Intersection: an Alternate Method for Generating Pareto Optimal Points in Multicriteria Optimization Problems." *NASA Contract No.* NASI-19480.

Dong, Z., Xu, J., Zou, N., Chai, C. (2008), "Posture Prediction Based on Orthogonal Interactive Genetic Algorithm." *Fourth International Conference on Natural Computation.* vol. 1, pp.336-340.

Dysart, M.J., Woldstad, J.C. (1996), "Posture prediction for static sagittal-plane lifting." *Journal of Biomechanics,* 29(10), 1393-1397.

Faraway, J.J., Zhang, X.D., Chaffin, D.B. (1999), "Rectifying postures reconstructed from joint angles to meet constraints." *Journal of Biomechanics,* Vol. 32, pp. 733-736.

Gill, P., Murray, W., Saunders, A. (2002), "SNOPT: An SQP Algorithm for Large-Scale Constrained Optimization." *SIAM Journal of Optimization,* Vol. 12, No. 4, pp. 979-1006.

Halvorsen, K., Lesser, M., Lindberg, A. (1999), "A new method for estimating the axis of rotation and the center of rotation." *Journal of Biomechanics,* Vol. 32, pp. 1221-1227.

Hetier, M., Wang, X. (2004), "Parametric posture prediction module methodology for development of a predictive driver's position system in real time." *TIP3-CT-2004-506503.*

Jung, E. S., Park, S. (1994), "Prediction of human reach posture using a neural network for ergonomic man models." *Computers & Industrial Engineering*, 27 (1-4), 369-372.

Jung, E.S., Choe, J. (1996), "Human reach posture prediction based on psychophysical discomfort." *International Journal of Industrial Ergonomics*, Vol. 18, pp. 173-179.

Khan, S.U., Ardil, C. (2009), "A Weighted Sum Technique for the Joint Optimization of performance and Power Consumption in Data Centers." International *Journal of Electrical, Computer, and Systems Engineering* 3:1.

Kim, I.Y., Weck, O.L. (2009), "Adaptive Weighted Sum Method for Multiobjective Optimization." *AIAA*, 2004-4322.

Kim, I.Y., Weck, O.L. (2005), "Adaptive weighted-sum method for bi-objective optimization: Pareto front generation." *Struct Multidisc Optim*, 29, 149–158.

Ma, L., Wei, Z., Chablat, D., Bennis, F., Guillaume, F. (2009), "Multi-Objective Optimization Method for posture prediction and analysis with consideration of fatigue effect and its application case." *Computers & Industrial Engineering*, Vol. 57(4), pp. 1235-1246.

Marler T., Farrell K., Kim J., Rahmatalla S., Abdel-Malek K. (2006), "Vision Performance Measures for optimization based posture prediction." *Digital Human Modeling for Design and Engineering conference,* Lyon, France, SAE 2006-01-2334.

Mi, Z., Yang, J., Abdel-Malek, K. (2002), "Real-Time Inverse Kinematics for Humans." *Proceedings of 2002 ASME Design Engineering Technical Conferences,* DETC2002/MECH-34239, Montreal, Canada.

Messac, A., Mattson, C.A. (2002), "Generating Well-Distributed Sets of Pareto Points for Engineering Design Using Physical Programming." *Optimization and Engineering*, 3, 431-450.

Saaty, T.L. (1997), "A Scaling Method for Priorities in Hierarchical Structures." *Journal of Mathematical Psychology*, 15: 57-68.

Saaty, T.L., Vargas, L.G. (1991), "The Logic of Priorities: Applications of the Analytic Hierarchy Process in Business." *Energy, Health, & Transportation.* Pittsburgh, PA: RWS Publications.

Tolani, D., Goswami, A., Badler, N. (2000), "Real-Time Inverse Kinematics Techniques for Anthropomorphic Limbs." *Graphical Models*, 62, 353–388.

Wang, Q., Xiang, Y.J., Kim, H.J., Arora, J., Abdel-Malek, K. (2005), "Alternative formulations for optimization-based digital human motion simulation." SAE Technical Paper 2005-01-2691.

Wang, X. (1999), "Behavior-based inverse kinematics algorithm to predict arm prehension postures for computer-aided ergonomic evaluation." *Journal of Biomechanics*, v32 n 5,pp. 453-460.

Yang, J., Marler, R.T., Kim, H., Arora, J., Abdel-Malek, K. (2004), "Multi-Objective Optimization for Upper Body Posture Prediction." *10th AIAA/ISSMO Multidisciplinary Analysis and Optimization Conference*, Aug. 30-Sept. 1, 2004, Albany, New York, USA.

Zeleny, M. (1973), *Compromise Programming in Multiple Criteria Decision Making*, pp. 262–301 University of South Carolina Press: Columbia, SC.

Zhang, J., Fan, Y., Jia, D., Wu, Y. (2006), "Kinematic simulation of a parallel NC machine tool in the manufacturing process." *Front. Mech. Eng*, China, 2: 173–176, 2006.

Zhang, K.S., Li W.J., Song W.P. (2008), "Bi-level Adaptive Weighted Sum method for Multidisciplinary Multi-objective Optimization." *AIAA*, 2008-908.

Zhang, W.H, Yang, H.C. (2001a), "A study of the weighting method for a certain type of multicriteria optimization problem." *Computers and Structures*, 79, 2741-2749.

Zhang, W.H., Gao, T. (2006), "A min–max method with adaptive weightings for uniformly spaced Pareto optimum points." *Computers and Structures,* 84, 1760–1769.

Zhang, W.H., Domaszewski, M., Fleury C. (2001b), "An improved weighting method with multibounds formulation and convex programming for multicriteria structural optimization." *International Journal for Numerical Methods in Engineering,* 52:889–902.

Zhang, X., Chaffin, D.B. (1996), "Task effects on three-dimensional dynamic postures during seated reaching movements: an analysis method and illustration." *Proceedings of the 1996 40th Annual Meeting of the Human Factors and Ergonomics Society*, Philadelphia, PA, Part 1, Vol. 1, pp. 594-598.

Zhang, X., Kuo, A., Chaffin, D. (1998), "Optimization-based differential kinematic modeling exhibits a velocity-control strategy for dynamic posture determination in seated reaching movements." *Journal of Biomechanics* 31, 1035-1042.

Chapter 32

Toward a New Digital Pregnant Woman Model and Kinematic Posture Prediction

Bradley Howard, Jingzhou (James) Yang, Jared Gragg

Department of Mechanical Engineering
Texas Tech University
Lubbock, TX 79409

ABSTRACT

Significant research efforts have been given to general digital human model and occupational accommodation for vehicles and workstations. However, no one has developed digital pregnant woman models or studied issues associated with pregnant women. This paper presents a new digital pregnant woman model and the study and simulation of seated posture prediction for pregnant women in order to explore the effects of pregnancy size and shape on seated posture and provide insight for future studies. This newly developed digital pregnant woman model has considered not only the anatomy of a general human with joints and links but also the changing size and shape throughout the course of the pregnancy. In simulation, the digital model is required to touch specific points in space from a seated position. Optimal joint angles are then calculated by simultaneously minimizing joint displacement and the distance between the right and left hands and their targets using multi-objective optimizations techniques. This optimization is naturally subject to physical and kinematical constraints. The physical constraints applied to a pregnant woman obviously vary from that of conventional posture prediction due to the unique physical movement limitations that apply only to this demographic.

As such, a self collision avoidance subroutine is implemented to allow these unique constraints to be enforced. In order to truly characterize the effect of the pregnancy, simulations for each of the three trimesters of pregnancy are run. The results of the pregnant woman simulations are compared to the results of regular female simulations to illustrate the extent of the effects of the pregnancy on posture. A 48 degree of freedom (DOF) full body female model is used in this study.

Keywords: Pregnant woman model, human modeling, posture prediction, multi-objective optimization.

INTRODUCTION

Digital Human modeling has quickly gained momentum over the past few years. Various commercial software packages are available, e.g., Jack, Ramsis, Safework, and Santos. However, none of them have pregnant woman models. Digital human modeling has a wide variety of applications to which it applies, whether it is for product design or biomechanical purposes. As the technology advances and computational capability increases, the application spectrum is growing without bounds. The use of more than one human performance measure to predict human posture was laid out by Yang et al. (2004). One off shoot of human modeling is seated posture prediction. Seated posture is important due to the many tasks that humans complete while seated. For instance, the full adjustment range of drivers' seats in cars has recently been studied (Gragg et al. 2009). Also general seated stability in vehicles has been studied recently (Kim et al. 2009). A demographic that has not been considered in these studies is pregnant women. Due to the unique physical constraints of pregnant women, the seated posture of a model could change drastically throughout the course of the pregnancy.

Previous work has been done with pregnant women in the automotive industry. An anthropometrical pregnant woman model was generated based on experiments at the University of Michigan Transportation Institute (Klinich et al. 1999). Out of the same institute, a pregnant woman crash dummy was created to try to simulate car crashes and determine the effect of vehicle restraints (i.e. seat belts and airbags) on the abdomen (Rupp et al. 2001). These studies are purely experimental but provide the information needed to create a sufficiently accurate digital model. The ability to accomplish this type of work in simulation would be invaluable.

This work develops a new digital pregnant woman model and lays a foundation for seated pregnant posture prediction in order to apply it to more complex problems. It is the goal that future avenues of study will be presented by the results of these simulations. This paper is organized as follows: We first generate a skeletal model considering size and shape based on pregnant woman's anatomy. Then, an optimization-based formulation is presented followed by posture prediction examples. Finally, a conclusion is given.

HUMAN MODEL

ANATOMICAL CONSIDERATIONS

Before the actual kinematics and mathematics of the digital model considered in this study are discussed, it is important to be familiar with the anatomy of pregnant women, especially their abdomen because human models depend on their anatomical structure. The anatomy of joints and bones for pregnant women are the same as regular people. Therefore, the skeletal structure should be the same as was presented in previous work (Yang et al., 2005). The only different part for pregnant women is the continuously changing abdomen. Fig. 1 shows the general anatomical composition of a pregnancy (Rupp et. al, 2001). The fetus is contained within the uterus, which at the beginning of the pregnancy has volume of around 5 mL. By the end of the pregnancy, the volume of the uterus can increase to as much as 10 L (Rupp, et al., 2001). It is this drastic size change that leads to the concerns of seated posture in pregnant women.

The model used in this study consists of a pregnant woman at the 3 month, 6 month, and 9 month stages of gestation. Fig. 2 illustrates the general shape of a seated pregnant model at each of the months of interest (Culver and Viano 1990). Culver and Viano estimated the pregnancy shape in the form of ellipses. However, it is important to note that the size of the abdomen is not dependent on the size and stature of the mother, but rather depends solely on the size of the fetus (Klinich et al 1999). Consequently, the abdomen sizes used in these simulations is a spherical estimation based on this past research of the anthropometry of pregnant women. The goal was to create a computer model that generally followed the trends of the outward facing portion of the pregnancy.

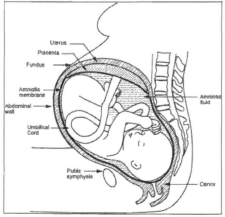

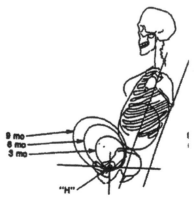

Fig.1 Anatomy of a pregnant abdomen   Fig. 2 Estimated profiles

SKELETAL MODEL

Once the anatomical considerations are understood, the kinematics behind the model can be developed. The human body can be modeled as kinematic chain consisting of revolute joints connected by links that represent the musculoskeletal joints and bones, respectively. A body fixed, local coordinate frame is attached to each link and the simulated motion is produced by rotating each of the joints until the desired orientation is achieved. The amount of rotation in each joint is measured about the local Z axis and is denoted by a generalized coordinate q_i.

These generalized coordinates make up the components of a vector $\mathbf{q}$ $= \begin{bmatrix} q_1 \cdots q_n \end{bmatrix}^T \in R^n$ that represents a specific posture. $x(q) \in R^3$ Is the position vector in Cartesian space that describes the location of an end-effector as a function of generalized coordinates (Yang et al 2004).

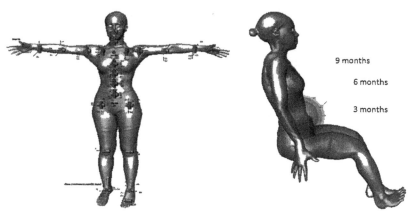

Fig. 3 Skeletal model   Fig. 4 Model considering shape and size

As the generalized coordinates are measured in a local frame, a unique transformation is needed to calculate $x(q)$. This is achieved by using the Denavit-Hartenberg (DH) Method. The DH method is a unique modeling method based on four parameters that are associated with each link in the kinematic chain; representing joint angle, link twist, link offset, and link length. These four parameters, θ_i, α_i, d_i, and a_i are used to create arbitrary homogeneous transformation matrices (T) that will relate any two adjacent coordinate frames through rotation and translation. The general form of this transformation matrix is shown in equation 1.

$$T_i^{i-1} = \begin{bmatrix} \cos\theta_i & -\cos\alpha_i \sin\theta_i & \sin\alpha_i \sin\theta_i & a_i \cos\theta_i \\ \sin\theta_i & \cos\alpha_i \cos\theta_i & -\sin\alpha_i \cos\theta_i & a_i \sin\theta_i \\ 0 & \sin\alpha_i & \cos\alpha_i & d_i \\ 0 & 0 & 0 & 1 \end{bmatrix} \quad (1)$$

A table is setup so that the four parameters are numerically defined between each adjacent frame in the kinematic chain, creating a transformation matrix for each degree of freedom (Denavit and Hartenberg, 1995). By simply multiplying the transformation matrices together, the local coordinates of any frame can be transformed into any other frame of reference including the global frame. Equation (2) illustrates this point.

$$x(q) = \left(\prod_{i=1}^{n} T_i^{i-1} \right) x_n \quad (2)$$

The spatial human body model considered in this study is the full body model shown in Fig. 3. It consists of 48 revolute joints (48 DOF) simulating the musculoskeletal joints: six DOF in the right leg, 12 DOF in the spine, nine DOF in each of the arms, five DOF in the neck and head, and seven DOF in the left leg. Single joints that have three DOF such as the shoulder are simplified down to three separate revolute joints with a link length of zero. This particular model is setup to simulate car driving conditions where the right foot remains on the gas pedal. Therefore the heel of the right foot is considered as the origin of the system. This point is fixed in space and the angle of the right foot is controlled by a constant, β (Gragg et al, 2009). It is for this reason that the left leg contains one more degree of freedom than the right.

For organization during coordinate transformation, the model is broken down into 6 separate kinematic chains: the right leg, spine, right arm, left arm, neck, and left leg. The end-effectors of this particular model are broken into two groups: objective targets containing the right and left hands, and constraint targets containing the right and left knees, hip point, and left foot. The nature of and reasons for this definition is explained more in the formulation section of this paper.

FORMULATION

Before the mathematical formulation of the problem is explained, the conceptual formulation will be discussed. In the previous section, a definition was made between objective and constraint end effectors. The reasoning behind this stems from the nature of the problem being studied. As the goal was to show the effects of the pregnancy size and shape had on posture, it is important to distinguish between goals and constraints. Treating the right and left knees, hip point, and left foot as end-effectors was necessary to define and hold the seated posture. Consequently, the distance between those end-effectors and the target points had to be absolutely defined. Conversely, the distance between the right and left hands

298

and their respective targets was the means to evaluate the effects of the pregnancy. Some target points, depending on the stage of pregnancy, may not be able to be reached. Therefore, hitting the targets with these end-effectors is treated as an objective and isn't necessarily required. The mathematical formulation of the optimization problem consists of design variables, constraints, and objective functions. The design variables are the generalized coordinates, q_i or local rotations measured in degrees, discussed in the human model section of this paper. The vector **q** will then represent the optimized posture. The constraints and objective functions are based directly from the conceptualization previously discussed.

The first constraint is simply that the distance between the constraint end-effectors and their respective targets is close to zero, since the numerical convergence to zero is impossible. The second constraint is a self collision avoidance constraint. This limits the posture such that the body cannot intersect itself. This is accomplished by placing virtual spheres throughout the body and simply limiting the distance between their centers to be greater than or equal to their respective radii. It is through this constraint that the pregnant posture is realized. The sphere placed in the abdomen of the model represents the fetus, and its radius is changed to accommodate for the different stages of gestation. The placement of the spheres is shown in Fig. 5, and the sphere shown in the abdomen is further detailed in Fig. 3. Together these constraints make up the constraint vector $g(q)$ which is a function of q_i. The q_i needed to satisfy $g(q)$ have to fall within the upper and lower bounds q_i^U and q_i^L , representing the extent of motion for each musculoskeletal joint, such that the resulting posture is realistic for humans.

There are three performance measures, or objective functions, that are defined for this system. They are joint displacement, the distance between the objective end-effectors and their targets, and visual displacement.

Fig. 5 General Sphere positions and sizes for self collision avoidance

They are simultaneously minimized using multi-objective optimization (MOO) techniques. Joint displacement refers to the total amount of rotation in each kinematic joint from the neutral position q_i^N. It is based on a weighted sum method, where each rotation is multiplied by a specific weight, then added together. The purpose of the weights is to prioritize the joints used in the simulated motion. Table 1 gives the weights used in this work.

Because it rational to assume that some joint rotations will be in the negative direction, the displacement in each joint is squared in order to avoid numerical problems (Yang et al., 2004,). Equation_ shows the equation that is minimized.

$$JD = \sum_{i=1}^{48} w_i \left(\frac{q_i - q_i^N}{q_i^U - q_i^L} \right)^2 \tag{3}$$

Table 1: Joint weights used in joint displacement objective function

Joint Variable	Joint Weight	Comments
q_1, q_6, q_{44}, q_{46}	300	
q_2, q_{47}	500	
$q_3, q_{20}\text{-}q_{22}, q_{24}\text{-}q_{27}, q_{29}\text{-}q_{31}, q_{33}\text{-}q_{36}, q_{39}, q_{41}, q_{45}$	1	
q_4, q_5, q_{42}, q_{43}	50	For both positive and negative
$q_7, q_{10}, q_{13}, q_{16}, q_{37}, q_{40}$	100	values of $q_i\text{-}q_r$
$q_9, q_{12}, q_{15}, q_{17}$	5	
q_{19}, q_{28}	75	
q_{48}	1400	
$q_8, q_{11}, q_{14}, q_{17}$	1000	if $q_i\text{-}q_r < 0$
	100	if $q_i\text{-}q_r > 0$
q_{23}, q_{32}, q_{38}	50	if $q_i\text{-}q_r > 0$
	1	if $q_i\text{-}q_r < 0$

The distance between the objective end-effectors and their targets, referred to as the distance values DV^{RH} and DV^{LH} for the left and right hands, respectively. The equation below illustrates the mathematical representation of the right hand distance value:

$$DV^{RH} = \left(\prod_{i=1}^{End-Effector} T_i^{i-1} \right) x_{end-effector} - Target^{RH} \tag{4}$$

The visual displacement objective is the last objective function used in this formulation. Proposed by Marler et al. (2006), visual displacement essentially minimizes the angle between a vector that stems from the line of sight of the model, and a vector that connects the model's eyes with the target point. This method makes it possible for the model to attempt to look at a target point even if that point is out of the field of view, much like a real person would. The equation for visual displacement is shown below.

$$VD = 10 \left(1 - \cos \left[\frac{\theta(q)}{2} \right] \right) \tag{5}$$

Using MOO (Yang et al., 2004), each objective function is optimized simultaneously. The equation used to accomplish this is as follows:

$$f_{objective} = \left\{ \begin{bmatrix} W_1 \left(\dfrac{JV}{JV_{max}} \right) + 1 \end{bmatrix}^2 + \begin{bmatrix} W_2 \left(\dfrac{VD}{VD_{max}} \right) + 1 \end{bmatrix}^2 + \begin{bmatrix} W_3 \left(\dfrac{DV^{LH}}{DV_{max}^{LH}} \right) + 1 \end{bmatrix}^2 + \begin{bmatrix} W_4 \left(\dfrac{DV^{RH}}{DV_{max}^{RH}} \right) + 1 \end{bmatrix}^2 \right\}^{1/2} \tag{6}$$

W_1, W_2, W_3, and W_4 are weights used to prioritize the optimization. In this case, W_3 and W_4 were set to 10.0 and the other two were defined to be 1.0. This is because the model reaching the targets (i.e. accomplishing the required task) is much more important than the joint displacement and the visual displacement. Each function is divided by its maximum possible value in order to keep the all of the objectives on the same order, eliminating the possibility of un-prescribed priority for optimization.

Finally the optimization problem can be developed:

Find: $q \in R^{DOF}$

Minimize: $f_{objective}$

Subject to: $g(q); q^L \leq q_i \leq q^U$

The numerical problem is solved using SNOPT, a black box numeric algorithm that solves constrained optimization problems (Gill et al 2002). The objective and constraint functions are as previously defined. The explicit gradients of the objective and constraint functions are also calculated in subroutines and are used by SNOPT to define search directions.

RESULTS

The achieved results are discussed in this section. The visualizations were created using Autodesk Maya Simulation Environment. Other than testing the new digital model, one of the purposes of this study was to investigate the feasibility of future work with this particular demographic. Target points were chosen in order to explore the possibility of unique design and/or health and safety issues that might arise due to the size and shape of a pregnancy.

Two simulation cases were run in this study. In the first case, a target points were chosen such that all of the subjects could reach the point. This was to done to see if the pregnancy had an effect on calculated posture for "reachable targets." In the second case, targets were chosen such that the non-pregnant subject can reach without over extension (i.e. without reaching the extent of reach). The purpose is to show if at any time reachable targets for one subject become unreachable due to the shape of the pregnancy.

The target points for the first case were selected to simulate a person picking something off of the floor while seated. As stated, it was selected such that all of the subjects could reach the point. As can be seen in Fig. 6, the pregnancy definitely had an effect on the predicted posture. The case of 3 months and non-pregnant models used more frontwards bending in the hips in order to reach the point whereas the models of nine and six months bend sideways and use more arm movement in order to reach the point. This provides an insight into health and safety issues that may affect a woman as the pregnancy progresses. If, like in case 1, a woman has to use more sideways bending to accomplish tasks, it may cause back problems. Also seated stability issues arise because of the nature of the resulting posture. A greater moment will be applied about the forward looking axis.

Fig. 6 Posture results from case 1

In the second case, the targets of case 1 were modified slightly, as if the object on the ground was moved forward. The point was chosen such that the non-pregnant model could reach the point without being at the extent of reach in order to show that pregnancy size and shape may cause tasks that are possible at the beginning of the pregnancy become impossible. As can be seen in Fig. 7, the subject of 3 months is able to complete the task, however the subjects of three and nine months are not. This presents and interesting insight into design challenges that may arise for pregnant women. For example, a woman may be able to reach the steering wheel of a car at the beginning of her pregnancy, but towards the end she will not.

Fig. 7 Posture results from case 2

CONCLUSION

A new digital model for pregnant women was developed and tested and is based on past anthropometric research of pregnant women. Because the size of the uterus/abdomen of pregnant women depend on the size of the fetus instead of the size and stature of the mother, it is feasible to use a general estimation for the size and shape of the pregnancy. The use of spheres place in the abdomen appears to be sufficient in the kinematic posture prediction for pregnant women.

The posture prediction portion of this study was created using MOO. Joint displacement is very reliable in posture prediction due to the fact that humans try to accomplish tasks with the least amount of work possible. Also selecting the distance value as an objective rather than a constraint, although not typical in other posture prediction studies, was a necessary part of determining the effects of seated posture prediction, because all tasks may not be possible from a seated position. In the MOO formulation it is imperative the weights placed on these objectives are the largest as it is most important that the model accomplish the task given to it rather than the amount of work needed to do it.

The results obtained illustrate the need for the future study of posture prediction in pregnant women. Unique health and safety and design problems affect this demographic due the unique physical limitations that exist because of the pregnancy. Also stability concerns of seated posture are brought into question because of these limitations. Ongoing research is being conducted to improve the digital model presented in this paper. Other human performance measures are also being considered to improve the posture prediction aspect of this study. The joint weights used in the joint displacement function are being improved upon to more closely shadow that of real human motion.

ACKNOWLEDGMENTS

This research is partly supported by National Science Foundation (Award # 0926549)

REFERENCES

1. Culver, C.C., Viano, D.C., "Anthropometry of Seated Women during Pregnancy," Human Factors, Vol. 32, No. 6, 1990, 625-636.
2. Denavit, J., and Hartenberg, R. S. (1955), "A Kinematic Notation for Lower-Pair Mechanisms Based onMatrices," Journal of Applied Mechanics, 22, 215-221.
3. Gill, P.E.; Murray, W.; Saunders, M.A., "SNOPT: An SQP algorithm for large-scale constrained optimization," SIAM J. Optim., 12 (2002).
4. Gragg, J., Yang, J., Long, J., "Optimization-Based Approach for Determining Driver Seat Adjustment range for Vehicles," International Journal of Vehicle Design, in print, 2010.
5. Kim, J., Yang, J., Abdel-Malek, K., "Multi-objective Optimization Approach for Predicting Seated Posture Considering Balance," International Journal of Vehicle Design, Vol 51, No. 3-4, Aug. 2009, 278-291.
6. Klinich, K.D., Schneider, L.W., Eby,B., Rupp, J. D., Pearlman, M.D., Seated anthropometry during pregnancy. University of Michigan Transportation Institute, NO. UMTRI-99-16, 1999.

7. Marler, T., Farrell, K., Kim, J., Rahmatalla, S., Abdel-Malek, K., "Vision Performance Measures for Optimization-Based Posture Prediction," Digital Human Modeling for Design and Engineering Conference, Jul. 4-6, 2006, Lyon, France.

8. Rupp, J. S. (2001). Design, development, and testing of a new pregnant abdomen for the Hybrid III small female crash test dummy. University of Michigan Transportation Institute.

9. Yang, J., Marler, R.T., Kim, H., Arora, J., and Abdel-Malek,K., "Multi-Objective Optimization for Upper Body Posture Prediction," 10th AIAA/ISSMO, Multidisciplinary Analysis and Optimization Conference, Aug. 30-Sept. 1, Albany, New York, USA.

10. Yang, J., Abdel-Malek, K., and Nebel, K., "Reach Envelope of a 9 Degree of Freedom Model of the Upper Extremity," International Journal of Robotics and Automation, Vol. 20, No.4, 2005, 240-259.

Chapter 33

Toward High Fidelity Headform and Respirator Models

Zhipeng Lei, Jingzhou (James) Yang

Department of Mechanical Engineering
Texas Tech University
Lubbock, TX 79409

ABSTRACT

Respirator fit and comfort are two important performances for designers, manufacturers, users, and standard developers. Simulation-based approach to predict respirator fit and comfort has advantage over traditional experiment method because simulation-based method can not only predict fit and comfort, but also give the insights of whys. For simulation-based approach, the fidelity of the models of respirator and headform is the key to ensure that the predictive models are accurate. In this paper, we will present the systematic procedure and method to generate high fidelity models for respirators and headforms. First, scanned models are obtained by reverse engineering technology. Second, further manipulation is required to remove deficiencies and generate accurate CAD models. Third, complex multi-layer headform and respirator models are developed within the LY-DYNA environment. The deformable tissue consists of dermis layer, fatter tissue layer, muscle layer and bone layer and the whole head is divided into five segments (two segments for maxilla, one segment for frontal, one segment for mandible, and one segment for back head). All five segments have different thicknesses on each layer. The headform model can be customized and used for simulation study.

Keywords: Headform, respirator, high fidelity model, fit and comfort.

INTRODUCTION

The development of respirator is a complex task, because the respirator must provide the requisite degree of protection from chemical and biological agents. One requirement is the sealing ability that respirators should block those agents. Another requirement is that respirators should provide feeling of comfort to its wearer. Respirator comfort and fit is associated with the contact pressure between a respirator and a human face. Simulation method to investigate the interaction mechanism between a respirator and a headform is easier and faster than experimental method. However, due to the complex anatomy of the human face and the fabric structure of the respirator, realistic models for headforms and respirators are critical for simulation approach. This paper presents a systematic procedure to generate high fidelity respirator and headform models.

Challenges of headform models are connected to the geometry modeling with respect to head anatomy, the contacting behavior between different tissues, and the viscoelastic mechanical characteristics of soft biological tissues. Different face models were introduced in the literature. Keeve el al. (1998) proposed a deformable model of facial tissue with unitform viscoelasitic properties, which is described by one nonlinear function, for all soft tissue. Couteau et al. (2000) started from patient data (CT scans and MRI) and generated finite element model of the patient face by "Mesh-Matching" algorithm. Payan et al. (2002) modeled different tissue layers and explicit representation of face muscles by distinct constitutive equations. Luboz et al. (2005) introduced a mesh generating algorithm that can automaticly mesh patient's face for facial tissues modelling. Barbarino et al. (2008) implemented nonlinear constitutive equations into the numerical algorithms for modeling the mechanical behavior of facial tissue. However, none of these models have been considered in the interaction analysis of a respirator and a headform.

A few researchers simulated the interaction between a respirator and a headform. Bitterman (1991) used finite element (FE) method to calculate the pressure between oxygen mask and pilot's face. The model of pilot's face was simplified as a rigid surface without deformation. Piccione et al. (1997) applied a finite element analysis tool, DYNA3d, which is an explicit 3D code and an old version of LS-DYNA, to create deformable human face and respirator models. Yang et al. (2009) used headform and respirator models from 3D scanner and built FEA models as shell elements. However, none of previous work has considered a realistic headform model and respirator model.

The objectives of this paper center on developing high fidelity headform and respirator models based on human face anatomy and respirator structure. These models will be used for investigating predictive models of respirator comfort and fit based on simulation approach.

We first present the anatomy of human head/face and structure of respirator 3M8210. Then, a methodology has been developed to generate the high fidelity models. The methodology includes 3D scanning technology, reverse engineering technology, and finite element analysis software application. Based on the procedure CAD models are generated. The respirator model has inner seal face and outer comfort layers. The headform is divided into frontal, maxilla, mandible, and back head parts. Each part has multilayer including dermis, fatter tissue, muscle, and bone layers. Section 5 presents the FE models. Finally a conclusion is given.

ANATOMY OF HUMAN HEAD

In order to build an accurate deformable headform model, anatomy of human head has to be studied and well understood. Epidermis, dermis and hypodermis are three layers of tissue of facial skin shown in Fig. 1(a) (Keeve et al, 1998). Outer layer is the epidermis, a superficial 0.1 mm thick layer of dead cells. The underlying dermis layer is 0.5–3.5 mm thick and is responsible for the elasticity of the skin. The inner hypodermis layer consists of fatty tissues and is connected to the skull. As shown in Fig. 1(b), many facial muscles are existed between those skin layers and the underlying bone structure. Thus, dermis, fatter tissue, muscle and bone are four main layers we need to model. The biomechanical behavior of the facial soft tissue is based on three substances: actin, elastin and collagen. Their mechanical properties can be found in literature.

In the case of respirator contacting, majority of the intervention areas happen on frontal, maxilla and mandible (Gray, 1918), shown in Fig. 1 (c). Thus, these three areas should be modeled as deformable model, and other areas of head can be simplified as rigid body.

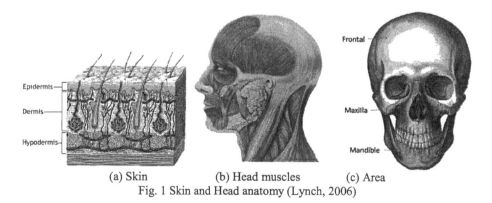

(a) Skin   (b) Head muscles   (c) Area
Fig. 1 Skin and Head anatomy (Lynch, 2006)

RESPIRATOR STRUCTURE

This work adopts respirator 3M8210, which is a N95 filtering facepiece mask. The prototype and structure of respirator 3M8210 can be seen in Fig. 2. It has several layers, including soft inner seal, comfortable inner, filtration and antimicrobial coated scrim layers. Flexible nosepiece and elastic head straps are also featured. The contact behavior is mainly relative to soft seal, comfortable layers, and head straps, while other components of respirator can be neglected.

METHODOLOGY

GENERAL PROCEDURE

Fig. 3 shows a general procedure of developing our high fidelity headform and respirator models. Models go through 3 stages. Firstly, the original models of point cloud are obtained by scanning headform and respirator prototype. Secondly, CAD models are generated from the scanned data; meanwhile, CAD models' quality is increased by removing deficiencies in shape. Thirdly, FEA models are built by element meshing and generation and by applying various parameters. Technical difficulties appear in the second and third steps. Holes, spikes and noises, which commonly arise in 3D scanned data, decrease CAD models' quality, and geometry complexity in surfaces of headform and respirator attributes to the difficulties of meshing their surfaces and generating 3D solid elements. Thus, in next sections, we will describe these two steps in details and introduce treatments for these technical difficulties.

CAD MODELS

3D scanning is the most feasible and accurate way of modeling surfaces of complex shapes such as respirator and headform. A 3D scanner emits some kind of radiation and detects its reflection for probing the surface of an object. So, the data obtained from the 3D scanner are point clouds.

Polygon and NURBS models are two stages between point cloud and CAD models, as shown in Fig. 4. The polygon model is made of a large number of triangles which are generated by connecting adjacent 3 points in point clouds. And then, NURBS surface model is created by applying a reproducible surface over its underlying polygons. CAD model can be obtained by transformation of NURBS model.

Holes, spikes and noises in polygon stage can cause deficiencies in CAD models. Complexity of respirator and headform shapes makes holes, spikes and noises hard to be found in the early stage. So, if deficiencies of CAD models cannot be overcome, it takes a few iterations.

Fig. 4 shows a point cloud model of respirator, which were generated by scanning the shape of respirators 3M8210. Fig. 5 gives its polygon model, in which holes are filled, spikes and noises are removed, and surface is relaxed. This polygon model is a manifold closed surface, having both inner and outer face. Since we are designing to create a multilayer FE respirator model, only inner face is kept for further usage and outer face is deleted. Fig. 6 presents the CAD model of respirator as a single shell that comes from the surface of the inner side of respirator. This shell consists of 47 curved faces, perfectly manifest the shape of respirator 3M8210.

Fig. 7 shows the polygon model of a medium size headform with 232937 triangles. Using the similar process of respirator model yields CAD models. But several additional treatments need to be done for complex shape parts for the headform. Ear areas are the most intricate parts in the headform. Polygon model of headform, which comes from the scanned point cloud, has low quality in ear areas, as shown in Fig. 8. Spikes and holes are carefully removed and the shape is relaxed to the extent that does not diminish its main shape features. Fig. 8 shows the repaired ear area in polygon model. Eyes, nose and mouth parts are also the complex ones as the eye part. Fig. 9 gives the CAD model of headform with 759 curved faces.

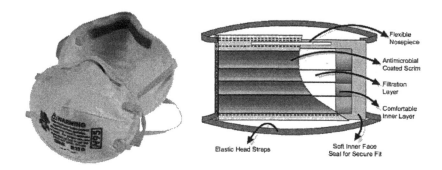

Fig. 2 Prototype and structure of respirator 3M8210

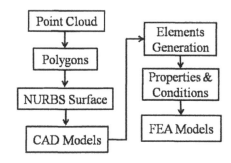

Fig. 3 Flow Chart of modeling procedure

FEA MODELS OF THE RESPIRATOR AND HEADFORM

Our objective in this paper is to create high fidelity models for the interaction between a respirator and headform. FEA models are the most useful way to solve contact problems. So, this paper introduces the FEA models for both respirator and headform. This section discusses the main procedure of building FEA model: (a) surface meshing, and (b) solid element generating.

ELEMENTS GENERATION

Through the study of respirator structure, FEA model of respirator is designed to have inner face sealing layer and outer comfortable layer. In previous section, the CAD model of respirator 3M8210 is a shell surface which has many curved faces. After importing it into LS-DYNA preprocess software, we mesh the shell surface into triangle elements. By offsetting shell surface, 2 layers of tetrahedron solid element are obtained and assigned two different materials. As shown in Fig. 10, the inner layer of respirator has 1 mm thickness and the outer layer of respirator has 2 mm thickness.

As discussed in Section 2 (face anatomy), human headform can be divided into 5 segments with different thickness. In the beginning, the CAD model of headform is trimmed and divided using LS-DYNA preprocessing software. Fig. 11 shows the divided headform model with 5 segments, including frontal, left maxilla, right maxilla, mandible, and back head. Then, LS-DYNA's preprocessor meshes the surfaces of these 5 parts into triangle shell elements, which all have the same size for maintaining consistence. Next, layers are created by offsetting shell elements in each part, except back head that is treated as rigid body. In Fig. 12, the frontal part has 3 mm thick dermis and 1 mm thick bone; left and right maxilla parts have 3 mm thick dermis, 3 mm thick muscle, 3 mm thick fatter tissue, and 1 mm thick bone; each mandible has 3 mm dermis, 1 mm thick fatter tissue, and 1 mm thick bone. Note that some areas in layers are holed based on head anatomy. Finally, headform FEA model is made up of 36782 tetrahedron solid elements for finite element analyzing. The layers are shown in Fig. 13.

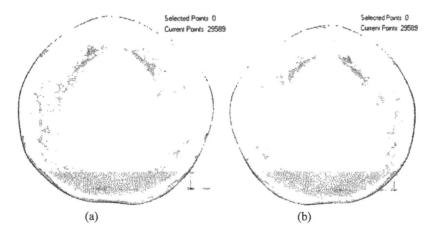

(a)   (b)

Fig. 4 Point cloud model of respirator 3M8210: (a) front view; (b) back view

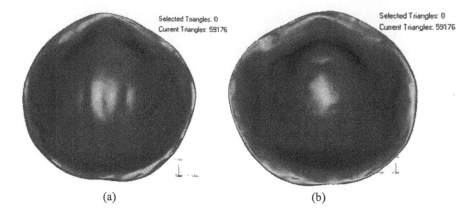

(a)   (b)

Fig. 5 Polygon model of respirator 3M8210: (a) front view; (b) back view

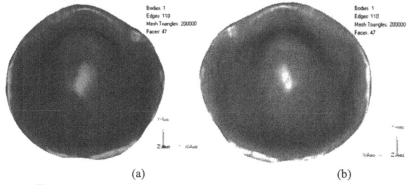

(a)   (b)

Fig. 6 CAD model of respirator 3M8210: (a) front view; (b) back view

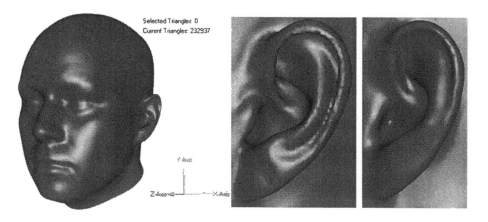

Fig. 7 Polygon model of head   Fig. 8 Polygons model of ear before and after repairing

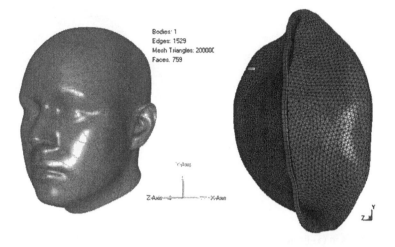

Fig. 9 CAD model of the headform   Fig. 10 FEA model of respirator 3M8210

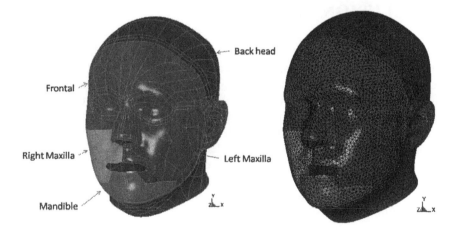

Fig. 11 CAD Model of divided headform   Fig. 12 Surface mesh of headform

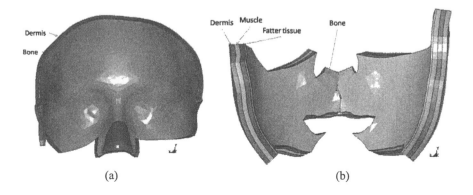

(a)   (b)

312

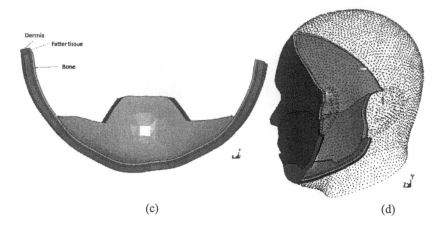

(c)   (d)

Fig. 13 Layers in (a) Frontal part; (b) Maxill part; (c) Mandible part; (d) 4 parts as a whole

CONCLUSION

Human headforms maintain complete features of head morphology. From 3D scanning models of the respirators and headforms, we build their CAD models and FEA models. The deformable tissue contains dermis layer, fatter tissue layer, muscle layer and bone layer and the whole head is divided into five segments (two segments for maxilla, one segment for frontal, one segment for mandible, and one segment for back head). All different layers have been incorporated in the new headform model. Also, the respirator model consists of inner seal layer and outer comfortable layer. Furthermore, simulating conditions can be conveniently defined for variety of cases. Future work will include 1) incorporating straps in the FE model; 2) determining accurate material properties in literature; 3) considering head movement for the simulation.

ACKNOWLEDGEMENTS

This research was partly supported National Institute for Occupational Safety and Health (NIOSH) project (Contract No. 254-2009-M-31878).

DISCLAIMER

The findings and conclusions in this paper are those of the authors and do not necessarily represent the views of the National Institute for Occupational Safety and Health.

REFERENCES

Barbarino, G., Jabareen, M., Trzewik, J., & Mazza, E. (2008). Physically based finite elmenet model of the face. Springer-verlag Berlin Heidelberg , 1-10.

Bitterman, B. H. (1991). Application of Finite Element Modeling and Analysis to the Design of Positive Pressure Oxygen Masks. Air force institue of technology .

Couteau, B., Payan, Y., & Lavallee, S. (2000). The mesh-matching algorithm: an automatic 3D mesh generator for finite element structures. Journal of Biomechanics , 33 (8), 1005-1009.

Gray, H. (1918). Anatomy of the Human Body. Philadelphia: Lea & Febiger.

Keeve, E., Girod, S., Kikinis, R., & Girod, B. (1998). Deformable modeling of facial tissue for craniofacial surgery simulation. Computer aided surgery .

Luboz, V., Chabanas, M., Swider, P., & Payan, Y. (2005). Orbital and Maxillofacial Computer Aided Surgery: Patient-Specific Finite Element Models To Predict Surgical Outcomes. Computer Methods in Biomechanics and Biomedical Engineering , Volume 8 (4), 259-265.

Lynch, P. (2006, 12 23). Facial muscles. Retrieved from wikipedia: http://en.wikipedia.org/wiki/Facial_muscles

Payan, Y., Chabanas, M., Pelorson, X., Vilain, C., Levy, P., Luboz, V., et al. (2002). Biomechanical models to simulate consequeces of maxillofacial surgery. Académie des sciences , 407-417.

Piccione, D., moyer Jr, E. T., & Cohen, K. S. (1997). Modeling the Interface between a Respirator and the Human Face,. Human enginerring and research diretiorate, Maryland: Army research Laborary.

Yang, J., Dai, J., & Zhang, Z. (2009). Simulating the Interaction between a Respirator and a Headform Using LS-DYNA. Computer-aided design and applications , 539-551.

CHAPTER 34

Controls-Based Motion Prediction in the Presence of External Forces

Katha Sheth, Faisal Goussous, Rajankumar Bhatt,
Soura Dasgupta, Karim Abdel-Malek

University of Iowa
Iowa City, IA 52246, USA

ABSTRACT

We consider a new approach to digital human simulation, using Model Predictive Control (MPC). This approach permits a virtual human to react online to unanticipated disturbances that occur in the course of performing a task. In particular, we predict the motion of a virtual human in response to two different types of real world disturbances: impulsive and sustained. This stands in contrast to prior approaches where all such disturbances need to be known *a priori* and the optimal reactions must be computed off line. We validate this approach using a planar 3 degrees of freedom serial chain mechanism to imitate the human upper limb. The response of the virtual human upper limb to various inputs and external disturbances is determined by solving the Equations of Motion (EOM). The control input is determined by the MPC Controller using only the current and the desired states of the system. MPC replaces the closed loop optimization problem with an open loop optimization allowing the ease of implementation of control law. Results presented in this chapter show that the proposed controller can produce physically realistic simulations of a planar upper limb of digital human in presence of impulsive and sustained disturbances.

Keywords: Model Predictive Control (MPC), digital human, upper limb, measured and known external forces.

INTRODUCTION

Digital modeling and simulation techniques have brought about a major change in the product development cycle for all industries including engineering, defense, medicine and entertainment. Considering the increased momentum of technological enhancements in digital human modeling and simulation, the days are not far when the virtual avatars would be able to interact with the material world, extending the horizons of virtual susceptibility. All these efforts are extended following the eternal human grail since existence, of improvising the current methods of development using optimal amount of resources. However, even with the current advent of simulation methods, there are considerable challenges towards predicting and controlling human like motion. Different methods have been tried, combined and improvised to predict and simulate human motion considering the dynamic response in presence of external and internal forces. The most widely used method in the animation and gaming industry is Motion Capture (MoCap) which records-processes human motion and translates it to virtual avatars in real time.

The unique feature of MoCap (Sturman 1999) systems to becoming so popular in entertainment industry is its ability to concatenate and play monotonous motions in real time. However, MoCap system and software is expensive with additional requirement of a well maintained environment to capture human motion using sensors. In order to capture a motion, time intensive experiments must be conducted and large amounts of data need to be collected, processed and stored. The science fiction epic film "avatar" required over a petabyte of memory and each minute occupied 17.28 gigabytes of storage memory (Masters 2009) for creating the motion of the characters and the virtual world. In addition to these large storage, time and cost requirements, such data driven approaches are generally specific to anthropometries and have difficulties with replaying the same motion for varying anthropometries and body types. Additionally, some postures and motions cannot be captured by current MoCap techniques, requiring additional artificial constraints and estimation for motion reconstruction (Sergio Ausejo 2009). As dynamics (external and internal forces) play a major role in human motion, MoCap techniques fail partially in predicting non-static postures (Karim Abdel-Malek 2009). Additional output in terms of actuation torque requirements and spine shear and compression forces are also necessary in order to employ the digital avatar in ergonomics field to answer questions like whether a 50th percentile female can perform a task or not. As a result, MoCap is not used as a sole technique in predicting human motion but, in many cases, combined with physics based optimization by using the kinetic data available from MoCap (Chaffin 1999) or for validating the output of a predictive methodology (Salam Rahmatalla 2008).

One of the successful physics based optimization techniques is the Predictive Dynamics (Yujiang Xiang 2009) approach. Predictive Dynamics approach minimizes the objective function which is a function of the joint angle profiles over the entire simulation time. In complex models as humans, solving a single optimization problem over the entire motion forms a highly non-linear and non-convex minimization problem. To address this problem, the motion is predicted over a small future interval using linearized dynamics model and the process is repeated at regular intervals, incorporating the disturbance as well as the changes in the system states(M. da Silva 2008). Since Predictive Dynamics optimizes the motion over the entire simulation time, it can react to only those external disturbances on the virtual avatar that are known *a priori*. In our controls-based approach, we incorporate the ability of the virtual avatar to predict motion under the effect of external disturbances that can be applied online when the motion is being simulated. This helps the avatar to stabilize or produce counter forces while performing a task, avoiding the need to recompute the motion for the whole task with the disturbances. In particular, our current work in this paper is limited to predicting and controlling the motion of a planar upper limb. We would extend the algorithms developed in this work and imbibe it in predictive dynamics based tasks like stairs climbing (Rajankumar Bhatt 2008), walking (Yujiang Xiang 2009) so as to predict the motion of these tasks in the presence of unknown disturbances on the virtual avatar.

In this approach, a novel model-based MPC Controller (Faisal 2009) depicts the brain of the plant. The plant is 3 degrees of freedom (dof) serial-link structure with 3 revolute joints that mimics the planar upper limb model of the virtual avatar/digital human. The goal of a controller is to provide appropriate control inputs to the plant so as to generate desired behavior of the plant. MPC comprises of a predictor and an optimizer component. The predictor component linearizes the non-linear and complex upper limb plant model about the operating point while the optimizer uses the estimate from the linear model and the information about the desired motion to calculate the future control inputs. Linear MPC has also been used for the dynamic control of flexible-link manipulators for space applications (Silva 2008) where gravity loads do not dominate the structures. A non-linear MPC approach for a flexible joint robot has been further discussed in (Bumin J. Song 1999). However, for the purpose of this paper, we assume that the links in human body are rigid and the joints are pure revolute joints.

METHOD

[1] THE PLANT

Figure 1 shows the system block diagram. The plant is the digital human avatar - Santos^{TM} developed by the Virtual Soldier Research team at the University of Iowa. Given some controlled torque $\tau(t)$ from the controller, the avatar simulates precise movements of the joints. So, the mathematical model of the plant consists of the equations of motion (EOMs) of the virtual avatar. The EOMs of a system describe the relationship between the joint angles, velocities, accelerations and the joint torques applied to the avatar as a function of time. These equations are solved for joint angle positions, velocities, and accelerations using Differential Algebraic Equation (DAE) Solver.

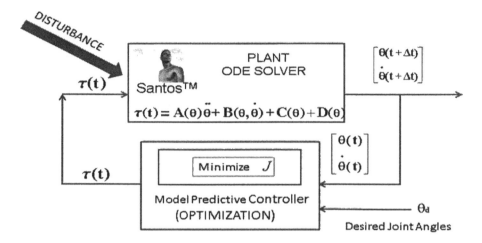

Figure 1. Block diagram representation of the Model Predictive Control.

The EOM are derived using Lagrange-Euler method. For a n dof model, a typical EOM can be written as follows:

$$\tau(t) = A(\theta)\ddot{\theta} + B(\theta,\dot{\theta}) + C(\theta) + D(\theta) \tag{1}$$

where,

$\tau(t) = n \times 1$ vector of joint torques,

$\theta = n \times 1$ vector of joint angles,

$\dot{\theta} = n \times 1$ vector of joint velocities,

$\ddot{\theta} = n \times 1$ vector of joint accelerations,

$A = n \times n$ mass-inertia matrix,

$B = n \times 1$ vector representing the coriolis and centrifugal forces,

$C = n \times 1$ vector of gravity forces, and

$D = n \times 1$ vector representing the external disturbances acting on the system.

The detailed EOMs for the plant are given in one of our earlier papers(Faisal 2009). These EOMs are solved using the open source solver IDA obtained from Suite of Nonlinear and Differential/Algebraic Equation Solver (Hindmarsh 2000)

[2] MODEL PREDICTIVE CONTROL (MPC) BASED CONTROLLER

The controller produces the control input, i.e. the actuation torques, for the plant based on the desired motion and the current states of the upper limb. It consists of a predictor component and an optimizer component.

The Predictive Component

At any current time t with time step size Δt , the predictive component of the controller predicts future values of the plant movements according to the linearized model of the system plant, at every incremental time step Δt for the selected future horizon. A second order Taylor's series expansion, as shown in (2) is used to approximate the plant movements i.e. the joint angles and joint velocities depending upon the torques and joint angle acceleration values.

$$\theta(t+\Delta t) = \theta(t) + \frac{\Delta t}{1!} \times \dot{\theta}(t) + \frac{\Delta t \times \Delta t}{2!} \times \ddot{\theta}(t)$$

$$\dot{\theta}(t+\Delta t) = \dot{\theta}(t) + \frac{\Delta t}{1!} \times \ddot{\theta}(t)$$

(2)

where,

$$\ddot{\theta}(t) = A^{-1}(\theta) \left[\tau(t) - B(\theta, \dot{\theta}) - C(\theta) - D(\theta) \right]$$

The Optimization Component

The optimization component determines the optimal values of the torques, which are the design variables, while minimizing an objective function and maintaining all the constraints. SNOPT software (Philip E. Gill 2002) has been used to solve the optimization problem using a sequential quadratic programming algorithm. The objective function being minimized for the current problem is:

$$J_y = \sum_{y=1}^{k} \left[J_{y-1} + \sum_{i=1}^{nDOF=3} \left[W_1(\theta_{di} - \theta_{yi})^2 + W_2(\dot{\theta}_{di} - \dot{\theta}_{yi})^2 + W_3(\tau_i(t))^2 \right] \right]$$

(3)

where,

k = optimization window size = finite number of time steps Δt

$i = 1, 2...$ndof

W_1, W_2, W_3 = weights of the objective function

θ_d = desired joint angle

$\tilde{\theta}, \dot{\theta}$ = second order approximation at $t = t + \Delta t$ for $y = 1\,to\,k$

$\tau(t)$ = joint actuation torques which are also the design variables.

Every human has natural joint and torque limits which should not be violated while performing any task. These limits are enforced in our formulation in terms of joint angle position and joint actuation torque limits as follows:

Joint angle position limits $\theta_{low,i} \le \theta_i \le \theta_{high,i}$

Joint actuation torque limits $\tau_{low,i} \le \tau_i \le \tau_{high,i}$

Since the objective of the controller is to track the desired motion, the difference between future predicted joint angles θ_{yi} with their velocities $\dot{\theta}_{yi}$ and the desired joint angles θ_{di} and their velocities $\dot{\theta}_{di}$ is minimized. In addition, exceedingly high values of control inputs are avoided by penalizing the design variable $\tau(t)$. For an easy numerical minimization, our interest lies in obtaining a squared valued function, So minimizing a weighted $(\theta_{di} - \theta_{yi})^2$, $(\dot{\theta}_{di} - \dot{\theta}_{yi})^2$, $(\tau_i(t))^2$ at every optimization window, would yield a better future tracking.

Steps for MPC algorithm

1. Input to MPC at current time t : θ_d , $\theta(t)$, $\dot{\theta}(t)$
2. Prediction of: $\theta(t + \Delta t)$, $\dot{\theta}(t + \Delta t)$ for an increment in time step Δt over an optimization window.
3. Optimize the objective function
4. Repeat steps 2-3 until the maximum optimization window size is reached
5. Output $\tau(t)$

[3] THE DISTURBANCE FORMULATION

The main objective of the controller is to produce physically realistic motions for a given task under the effect of disturbances. The disturbances are classified on a broader basis into impulsive and sustained. In this work, the controller response in the presense of sustained disturbance is presented. The sustained disturbance could be known or measured. In the case of known sustained disturbance, the amount and point of application of disturbance is known to the controller and hence, appropriate counter torques are produced without any delay. However, for sustained measured disturbance, the controller can only measure the disturbance after it has been applied. Hence, the response of the controller is delayed by a single window size.

A general formulation for the disturbance acting as shown in EQ. (1) is as follows:

$$D = J^T (-F)$$

where,

F = magnitude and direction of the external disturbance force

$$\mathbf{J} = \begin{bmatrix} \dfrac{dx}{d\theta_1} & \dfrac{dx}{d\theta_2} & \dfrac{dx}{d\theta_3} & \cdots & \dfrac{dx}{\partial\theta_i} \\ \dfrac{dy}{d\theta_1} & \dfrac{dy}{d\theta_2} & \dfrac{dy}{d\theta_3} & \cdots & \dfrac{dy}{d\theta_i} \end{bmatrix} \quad i = 1, 2\ldots\text{ndof}$$

x, y = location of the external disturbance on any of the links

RESULTS

CASE STUDIES

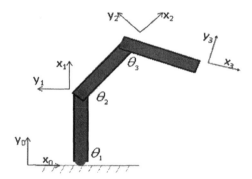

Figure 2. 3-link serial chain model of plant with revolute joint.

A planar 3- link, 3 revolute jointed serial link structure which mimics an upper limb model of a human being as shown in Figure 2. The corresponding anthropometric data is shown in Table 1. This upper limb model is simulated under the effect of no disturbances as well as different sustained disturbance cases: known, measured and unknown. Results present the response of the proposed control algorithm to real world disturbances acting on human upper limb, by producing optimized counter torques so as to reduce the joint angle error square closer to zero. While joint torque based impulsive disturbances were considered in (Faisal 2009), in this work, we consider an external force anywhere on the body segment in any random direction as disturbance.

So starting from any initial joint angles θ_0, the objective of the system is to obtain the desired motion (the desired joint angles θ_d). The respective data can be found in Table 2. For the sustained disturbance cases, an external force $|\mathbf{F}|$ of magnitude 0.25N is applied at the end of link 1 at 0° for time duration from 4-10 seconds of a total 15 seconds simulation time. The initial transient response of the plant has not been shown in Figure 4 to avoid misleading data.

Table 1: Anthropometric data of an upper limb of a digital human

Links	Link 1	Link 2	Link 3
Link lengths (m)	0.258638	0.247374	0.165099
Link masses (kg)	2.8	1.6	0.6

Table 2: Case Study parameters

Time step (sec)	Initial joint angles θ_0	Desired joint angles θ_d	Objective function weights		
			W_1	W_2	W_3
0.05	1.57,0,0	1.57,-0.523, -1.57	3282.06	10	1

[1]Case: No disturbance (comparison of 3 optimization window sizes)

With the increase in the future horizon, the controller produces better response of the plant. As seen in Figure 3(a), the angle error squared reduces with the increase in window size. The maximum torque required to sustain the disturbance is also less as the window size increases as seen in Figure 3(b). The effect of the change in the size of time step on the system response has also been shown in (Faisal 2009).

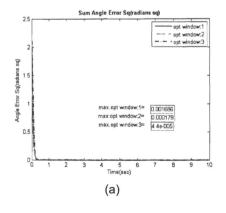
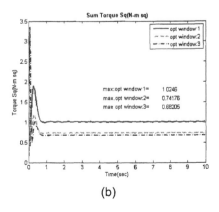

(a)   (b)

Figure 3. Comparison of the output of three optimization windows under no disturbance. (a) angle error square (b) torque square

[2]Case: Comparison of sustained type disturbances (known, measured and unknown) for optimization window size 1

All three types of disturbances: known, measured and unknown are compared to test the robustness of the controller in Figure 4. A known disturbance optimizes quickly producing a high transient joint torque. Measured disturbance shows larger angle error, but requires less amount of joint torque for a greater duration than known case. Unknown case takes more time to optimize than both the above cases requiring a higher joint torque.

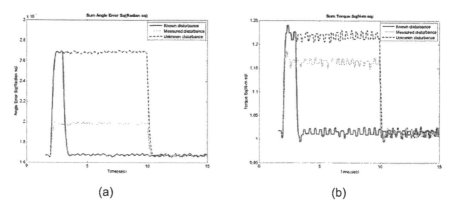

(a)   (b)

Figure 4. Comparison of three types of sustained disturbances (known, measured and unknown) for optimization window size 1. (a) angle error square and (b) torque square

CONCLUSION

We simulate a planar 3-link, 3 degrees of freedom upper limb motion under the effect of known and measured disturbances using linear Model Predictive Controller. Also the effect of increase in optimization window size on the response of the system has been studied. When the response of different types of disturbances are compared, the system intercepts and reacts differently in real time, based upon the physics and bio-mechanical requirements of a human upper limb affected by the dynamics to give a physically realistic motion. A future blend of the Model Predictive Controls based technique with the previously discussed physics based optimization technique: predictive dynamics, in a 3D environment is an ongoing work, which is imagined to create a benchmark in the world of digital human modeling and simulation.

REFERENCES

Bumin J. Song, A. J. K. (1999). "Non-linear Predictive Control with Application to Manipulator with Flexible Forearm." IEEE transaction of Industrial Electronics.

Chaffin, X. Z. a. d. (1999). "A three-dimensional dynamic posture prediction model for simulating in-vehicle seated reaching movements: development and validation." Ergonimics 43(9): 1314-1330.

Faisal (2009). Model Predictive Control for Human Motion Simulation. 2009 SAE International.

Hindmarsh, A. C. (2000). The PVODE and IDA algorithms.

Karim Abdel-Malek, J. A. (2009). Physics-Based Digital Human Modeling:Predictive Dynamics. Handbook of digital human Modelling. V. G. Duffy.

M. da Silva, Y. A. a. J. P. (2008). Simulation of Human Motion Data using Short-Horizon Model Predictive Control.

Masters, T. (2009). " Will Avatar crown James Cameron 'King of the Universe'? BBC News." Will Avatar crown James Cameron 'King of the Universe'? BBC News, from Romina D'Ugo.

Philip E. Gill, W. M., Michael A. Saunders, Arne Drud, Erwin Kalvelagen (2002). "Snopt: An SOP algorithm for large-scale constrained optimization." Society for Industrial and Applied Mathematics 47(1): 99-131.

Rajankumar Bhatt, Y. X., Joo Kim, Anith Mathai, Rajeev Penmatsa, Hyun-Joon Chung, Hyun-Jung Kwon, Amos Patrick, Salam Rahmatalla, Timothy Marler, Steve Beck, Jingzhou Yang, Jasbir Arora, Karim Abdel-Malek, John P. Obusek (2008). Dynamic Optimization of Human Stair-Climbing Motion.

Salam Rahmatalla, Y. X., Rosalind Smith, Jinzheng Li, John Meusch, Rajan Bhatt, Colby Swan, Jasbir Arora, Karim Abdel-Malek (2008). A Validation Protocol for Predictive Human Location. 2008 SAE International.

Sergio Ausejo, X. W. (2009). Motion Capture and Human Motion Reconstruction. Handbook of digital human modeling. V. G. Duffy.

Silva, T. F. a. C. W. d. (2008). "Dynamic Modelling and Model Predictive Control of Flexible-Link Manipulators." International Journal of Robotics and Automation 23(4): 206-3149.

Sturman, D. J. (1999). "A Brief History of Motion Capture for Computer Character Animation." from Yujiang Xiang, H.-J. C., Joo H. Kim, Rajankumar Bhatt, Salam Rahamatalla, Jingzhou Yang, TImothy Marler, Jasbir S. Arora, Karim Abdel-Malek (2009). "Predictive dynamics: an optimization-based novel approach for human motion simulation." Structural and Multidisciplinary Optimization.

Yujiang Xiang, J. S. A., Salam Rahmatalla, karim Abdel-Malek (2009). "Optimizaiton-based dynamic human walking prediction: One step formulation." International Journal for Numerical Methods in Engineering.

CHAPTER 35

The Virtual Profiler: Capturing Profiling Expertise for Behavioral Modeling

Marta S. Weber, William N. Reynolds

Least Squares Software, Inc.
12231 Academy Road, NE #301-192
Albuquerque, NM 87111
bill@leastsquares.com

ABSTRACT

This paper has two aims: to review the utility of behavioral profiling and to suggest a practical means of placing profiling tools in the hands of operational personnel through behavioral modeling methods. The success and value of profiling for behavioral forecasting has been demonstrated in business and intelligence domains over several decades. Standardized, transferable methodologies to produce that value have not been developed. We here propose a methodology to make behavioral profiling techniques available to a range of operational applications

INTRODUCTION

BEHAVIORAL PROFILING
History

The ability to know in advance what key people, collectives and organizations will do is a human striving much older than the social, political and psychological theories and methods we now call upon for those purposes. Divination for predictive purposes gave way to observation and logic in the investigations of ancient strategists, among whom the wisdom of Sun Tzu *(The Art of War)* is a stellar exemplar. The great Greek philosophers contributed to the knowledge base underlying behavioral forecasting two major insights into the human species:

- the durable, repetitive nature of behaviors in stable personalities, and
- the ways in which important elements cluster in personalities across all cultures

The first provides the basis for an educated guess about future behaviors based on past actions and the second allows a keen observer to extrapolate from one personal characteristic to another within an identified cluster.

Fast-forwarding with stops along the way to acknowledge the contributions of such luminaries as Machiavelli – so often cited and maligned, so poorly read and understood – and Freud, of whom much the same could be said, we arrive at the present day with more than a century of learning to underpin our inquiries. The social sciences, notably psychology, social psychology, sociology, anthropology and political science, with contributions from linguistics, economics and human geography, allow us to locate our subjects of interest within their cultural and social frames of reference and to analyze aspects of individual motivations and abilities for the purpose of forecasting future actions. The various formal processes for this research fall within the rubric of profiling.

There are several distinct types of profiling, distinguished by variables including: a) whether or not the subject is known to the profiler, b) whether or not the subject is a witting participant in the process, c) the focus of concern, e.g. political actions, decision-making, criminal modus operandi and signature styles, and d) the various methodologies employed. In this paper we are concerned with remote, i.e. at-a-distance profiling where the subject is neither aware nor involved in the inquiry. We distinguish a profile from a portrait of an individual or group by the key element of assessment aimed at behavioral *forecasting* that is its objective.

Applications

Not surprisingly, the history of formal psychological profiling in the modern era has been richest in the arenas of international relations, conflict – especially war – and diplomacy. Along with psychodynamic theory, political psychology has made the largest single contribution to leadership and key decision-maker profiling approaches. One of the first major profiling successes was scored by WWII Allied Intelligence experts who used a novel method to produce the famous profile of Hitler that accurately predicted his reactions to a major disinformation operation. (Post, 2005) The work of Margaret Herman over several decades produced the seminal insights into leadership traits, elements of which have been successfully extended beyond the narrowly political realm to leadership in many contexts, e.g. non-state actors (NSAs) including CEOs of multinational institutions, global NGOs, multinational corporations and illicit organization (Hermann, 1977, 1980, 2001) Among these are several examples that were not only successes in themselves, but also benchmarks in the refinement of the techniques. For instance, the advance personality profiles of Menachem Begin and Anwar Sadat guided the breakthrough achievements of the Camp David accords. (Post, 2005) Specialized units within the U. S. intelligence community continue to use various forms of profiling for the assessment of important political figures including the known leadership of major terrorist organizations.

In evaluating leaders and key individuals within their larger organizational contexts, organizational theory supplies contextual structures, processes and dynamics. Anthropology provides the necessary cross-cultural lens through which to adjust approaches that are anchored in Western cultural assumptions when looking cross-culturally. The fundamental lens for the development of individual and collective profiles is applied psychology.

Today, behavioral profiling is also increasingly in demand for various business and other organizational applications. In business/organizational applications, HR departments routinely employ a type of direct profiling for recruitment, selection and assignment. Typically, candidates take various psychological tests designed to identify and measure temperament, decision-making methods, communication style and general attitudes toward specified realms, such as work. There are two generally accepted categories of personality profile tests, trait and type, each with a distinct theoretical base and each validated through accepted social science research methodology. This form of direct, subject-engaged profiling does not concern us further here except to note that test questions aimed at a subject's key attributes can be embedded in elicitation and other techniques to remotely produce indications of the characteristics sought by the inquiry.

The remote profiling and behavioral forecasting of competitors is often a key element in competitive intelligence that may be sought by businesses, NGOs and other organizations and institutions operating in a competitive landscape. One of the authors of this paper has developed a detailed methodology for remote behavioral profiling that has set the standard in competitive intelligence applications. (Weber, 2004)

The profiling of leaders, technical experts and key decision-makers produces a multi-axial view of the subject or subjects in which the objective is to forecast specified behaviors. Specialists usually do not offer "prediction" which really means singling out a specific action at a specific time – almost always a near impossibility. Capture of the most likely actions in specified scenarios is the core approach. The possible scenarios are ranked in order of likelihood; the approach is made more robust and self-evaluative by including analysis of competing hypotheses for each component. The mapping of subjects' specific objectives to underlying motivational drivers as well as capabilities, both individual and organizational, offers insights of high value.

Several generalizations that can be made about human behavior inform this methodology:

Assuming personality stability, the single best indicator of future actions is previous actions.

Individuals and groups prefer to remain in their comfort zones of perception and action.

Adjusting for cultural variables, individuals often telegraph their future actions in their verbal statements, both wittingly and unwittingly.

Individuals and groups tend to repeat perceived successes and to avoid perceived failures.

Changes in MO, style, focus or apparent priorities signal important shifts that may ramify beyond the specific instance at hand. They should never be accepted at face value, but analyzed in depth.

Motivation is always at least as important as capabilities. It is not what your subject *can* do that matters most; it's what he *will* do.

Values and beliefs influence perception more than the reverse dynamic.

The need for cross-cultural literacy is crucial to profiling success on a global scale. While we have become individually and, as a team, more cross-culturally knowledgeable, we strongly believe that cultural expertise requires subject matter expertise. A key skill for any researcher is knowing what you don't know and adequately compensating for your knowledge gaps. Our own guideline is that if we don't have the language, we don't have the cultural access we need without additional assistance. In those cases, we turn to the cultural experts.

Over time, discernible patterns have emerged among leaders, followers and other key players we've profiled. Leadership needs in the 21st century increasingly differ from those of the 20th and the rate of change appears to be accelerating. (Weber, 2010) Among the multiple drivers of these changes are advancing technology, providing both opportunity and challenge, and the evolving nature of organizations. Networks have supplanted monolithic organizational structures in many domains, while the character and function of collectives varies by time, place and circumstances. In addition to personality factors, cultural influences play at the macro level and within individual organizations and collectives. The extensive work in political psychology led by the U. S. intelligence community as well as academic centers of study has provided a knowledge base that can be utilized in analysis of not only political leaders, but others as well. Numerous cultural studies aid in the clarification of values, norms and other grounding bases for individual and collective behavior. In all cases, empirical evidence has to anchor the interpretation of traits, attributes and other individual variables, as well as situational determinants.

Ongoing opportunities are rich for cross-fertilization between academic study, practical methods developed by the U. S. intelligence community (IC) and the techniques refined in the private sector for competitive intelligence applications.

FROM PROFILING TO STANDARDIZED MODELING METHODOLOGIES

In sum from the foregoing, there is good empirical evidence that the behavior of groups and individuals can be understood, modeled and incorporated into effective operational plans. Moreover, the ability to forecast the actions and behaviors of key individuals and groups yields intelligence benefits of inestimable value for military, diplomatic and economic objectives. The value of forecasting is recognized, but the IC has not developed standardized, transferable methodologies to produce that value, principally because modeling the behavior of individuals and groups is a notoriously difficult process. While there have been some signal successes in modeling decision processes of small groups, (Reynolds, 2010; Feder, 1987) most quantitative approaches for modeling individual and group behaviors have been notable for their lack of success. (Zacharias, 2008). A further problem is the lack of codified, quantifiable modeling and simulation (M&S) methodologies that has resulted in poor institutional memory; the result is that each generation has had to rediscover or reinvent effective approaches to behavioral modeling

Two central questions arise: Why have operational successes been so difficult to translate into the quantitative realm of M&S? Having identified these failures, how can they be addressed? The first question has been definitively addressed in a recent paper authored by an expert panel assembled by the National Academy of Sciences [NAS]. (Zacharias, 2008). This extensive (329-page) report provides an exhaustive survey and critique of current approaches to Individual, Organizational and Societal (IOS) modeling. To address shortcomings of current approaches, they make a number of recommendations. Those most relevant to our arguments here are these:

- Keep the model as simple as possible for its purpose. (Zacharias, 2008) [1]
- Monolithic, static approaches are inappropriate for IOS modeling[2]

---

[1] An IOS model does not have to be complex. Parsimonious models are preferred." *NAS Report*, op cit., p. 8-28.

[2] "Flexible, adaptable components and semantically interoperable models will potentially do much to avoid this pitfall." *NAS Report*, op cit., p.10-3.

- Use an action validation approach as well as triangulation[3]
- Avoid "kitchen sink" models[4]

Methodology-based modeling tools focus less on fidelity to the system they are modeling and more on improving the quality of reasoning of their users. To illustrate, Landscape-Decision developed by the other author of this paper (Reynolds) models coalition formation in a political system by enumerating and exploring the space of possible coalitions. Traditional M&S approaches, often based on opaque algorithms and assumptions, present their user with a prediction of the "correct" coalition. In the methodological approach, it is more important that the analyst-user of the product be exposed to many novel hypotheses about potential coalitions. Emphasis is not on whether a coalition is "realistic" – that is left for a human expert to judge. In the language of the NAS report, this approach to modeling is called *validation for action*, where the impact on the reasoning of the user becomes the metric, rather than the model's fidelity to reality.

Recently we have conducted research into extending analytic methodologies to capturing expert knowledge to support operational decision-makers. This effort was delivered to a USG customer. Working with experts from four domains of social science – psychology, political science, anthropology and economics – we tried to understand the issue of non-state actors in Iraq. The effort, termed the *Cultural Order of Battle (COOB)*, was framed in terms of a question: If you could tell a commander in Iraq to incorporate one important fact from your area of expertise, what would it be?

The COOB resulted in three simple methods for structuring reasoning about non-state actors. The first focused on *cultural segmentation* (McCallister, 2007) in order to understand the different cultural segments (Arab/non-Arab, Shi'a/Sunni, tribal/non-tribal, etc.) that contribute to actor identity. Conflict is most likely to occur across segments, and initiatives are most likely to succeed within segments.

---

[3] "Failure to appreciate the extent to which IOS models differ from physical models has led to inappropriate expectations regarding VV&A for IOS models." Also, "This report recommends an action validation approach (see Chapter 8) that requires a clear specification of the purpose of the model and validates the usefulness of the answers provided by the model against that purpose. We also recommend triangulation, in which models are reviewed by multiple types of experts, compared with qualitative and theoretical studies as well as quantitative results, and similar models are compared with each other (docking).: *NAS Report*, op cit., p. 10-6.

[4] The kitchen sink tactic [adding variables to a model in a hodgepodge fashion] is based on a misconception about the relation between model features and variables and about the model's ultimate usefulness for providing information about behavior in cases beyond those used for testing." Also, "Models can become unwieldy when weighed down by a proliferation of features and variables." *NAS Report*, op cit., p.10-6.

Although simple, this method demonstrates that even a few actors and attributes can have an extremely rich segmentation structure. The second focused on *Prospect Theory* (Kahneman, 1979) which is based on the empirical observation that people value things relative to some reference, rather than in absolute terms. This has important consequences when pursuing hearts and minds because efforts to curry favor, such as reconstruction, must be measured in terms of the target population's reference point or points. The final approach was based on institutional networks (Khoja, 2008) linking non-state actors into a marketplace of needs and capabilities, where they exchange services, such as security, logistics or trade goods. The actors act as nodes in a network, which can self-assemble, fitting together like toy Lego bricks. There are potentially an enormous number of networks, as a novel calculation with Lego bricks illustrates – six bricks can be assembled in over 600 million ways. (Durhuus, 2008)

THE VIRTUAL PROFILER – CAPTURING AND QUANTIFYING THE CRAFT OF BEHAVIORAL PROFILING

We argue that it is possible to develop decision-support methodologies that systematically capture expert knowledge and tradecraft on behavioral forecasting, organized as small "method modules" that capture the relevant data and process used. The method modules could provide context-specific processes and data for predicting behavior. Each module would be modest in the aspect of behavior or culture that it addresses.

As illustrated in Figure 1, the CONOPS underlying the approach is that, for a given problem, method modules would be selected and combined to provide an approach tailored to the operational requirements. We do not suggest utilizing a unifying computational framework, as such approaches have been found to be ineffective and are explicitly proscribed by the NAS report. Instead we would capture and organize proven processes and data systematically, and use them as the basis of a quantitative approach to modeling specific behavioral prediction problems. Validation would focus on the impact of the methods on the final "Policy to Influence Behavior" oval. Metrics would focus on 1) whether employing the proposed methods provided a positive impact on policy-making, and 2) assessing the cost incurred by the analysis.

The method modules would consist of small encodings of different aspects of the behavioral profiler's craft, comprised of explanatory text, motivating and explanatory examples, data structures for capturing information relevant to the module, and recommendations and examples of using the modules for particular modeling applications. To provide an example of how a module might look, we consider four fundamental aspects of behavioral profiling:

1. How has the subject perceived events in the past?
2. How has the subject responded to events in the past?
3. What are the subject's motivational drivers and how are they expressed in the subject's specific intentions/objectives that are the focus of the inquiry?
4. How are the subject's capabilities aligned with those intentions/objectives?

The central thrust of the research we advocate here is to explore the idea of the method modules to capture behavioral modeling, and our central research question would be this: Does the idea work? In more detail, the research would address the following questions:

- Can we modularize the process of behavioral profiling into distinct elements suitable for data capture and process description?
- Can we develop techniques for developing quantitative modeling approaches based on these modules?
- How do we address the related problems of selecting and combining modules and then implementing algorithms appropriate for a given problem?

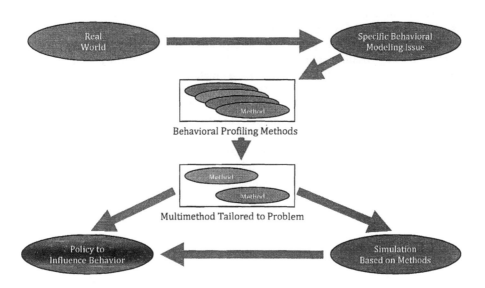

Figure 1: CONOPS for Profiling Methodology. Different profiling techniques would be captured, which could be combined in response to specific questions. The resulting multimethod could be used as a basis for simulation or to inform policy directly.

EXAMPLE NARRATIVE

We provide here a notional narrative of how the suite of tools in the virtual profiler might be applied to a real decision problem. A commander of a composite aviation squadron is tasked with supporting British forces conducting counter-insurgency and counter-narcotics operations in Afghanistan's Helmand province. The mission goals lead to conflicting objectives actions that are likely to disrupt narcotics production are also likely to alienate the local population, exacerbating the insurgency.

Focused air strikes on heroin production facilities in northern Helmand are considered. A particular concern is the reaction of the Afghan central government, and its leader, Hamid Karzai. The first step is to predict the response of Karzai and other key decision groups to such an attack. To determine which profiling methods to use, a master list of personality attributes is consulted. Applying the selected attributes to Karzai's historical behavior leads to the assessment that he is a leader-*mediator* and a *responsive learner*, which suggests that two methods should be used. The first identifies the stakeholders in Helmand province and their relationship to Karzai, and the second enumerates possible responses by Karzai to various contingencies based on the stakeholders' reactions.

Due to Karzai's intelligent and responsive profile, the second method implies an in-depth study of possible contingencies. Each stakeholder is assigned a potential favorable/unfavorable response, and all possible response configurations are enumerated. Morphological analysis (Ritchey, 1998) is used to reduce the possible number of configurations – yielding ten possible outcomes. Using Karzai's profile as a leader-mediator, each scenario is considered under the hypothesis that Karzai will do his best to satisfy each stakeholder. It is observed that in each scenario the civilian farmer stakeholders will be particularly impacted by the deterioration in security, and it is predicted that Karzai will demand increased security for them. Providing an increase in security by moving British bases into towns in the north of Helmand would address this concern. Plans are then made to redeploy ground forces after the air campaign begins; this will enhance Karzai's prestige when it appears that forces are redeploying in response to his demands.

OPERATIONAL PLAN

We argue for the value proposition of a multi-phase study. The initial phase of the study would be focused on the systematic capture of individual profiling techniques. During this phase, we would:

- Develop theories – this will use a case study-based approach. Using a number of historical examples of profiling, we will identify the processes used, segregate the process into modules and develop schema for data capture.
- Apply the virtual profiler modules to a challenge problem – either by using the modules to directly support analysis or by using them as a basis for developing an IOS simulation.
- Validate the generated results through Subject Matter Expert (SME) face-validation, triangulation and validation for action.

Subsequent phases would focus on applying the same approach to the behavioral modeling of groups, communities and cultures. We would research individual behavior first because we feel that individual profiling has, in general, been very successful in non-quantitative spaces. Groups have much more complex contextual issues, such as collective dynamics, institutional relationships, influential organizational structure, and psychological makeup of leaders and cultural segmentation. We envision a process by which this effort would be informed by the initial work, and additional SMEs would be brought in to advise this effort. The exact nature of these experts would be informed by the results of the first phase.

CONCLUSIONS

A number of valid underlying theories and practical methods support profiling aimed at behavioral forecasting. Numerous successes in the work of U. S. governmental agencies and in the private sector can be cited. The value proposition has been established and recognized. The IC has not developed standardized, transferable methodologies to produce that value, principally because modeling the behavior of individuals and groups is a notoriously difficult process. We propose an approach to develop decision-support methodologies that systematically capture expert knowledge and tradecraft on behavioral forecasting, organized as small "method modules" that capture the relevant data and process used. The method modules could provide context-specific processes and data for predicting behavior. Each module would be modest in the aspect of behavior or culture that it addresses. Method modules would be selected and combined to provide an approach tailored to the operational requirements. Validation would focus on the impact of the methods on the final "Policy to Influence Behavior" oval. Metrics would focus on 1) whether employing the proposed methods provided a positive impact on policy-making, and 2) assessing the cost incurred by the analysis.

REFERENCES

Durhuus, Bergfinnur and Eilers, Sren, 2008, "On the Entropy of LEGO,arXiv.org/pdf/math.CO/0504039 on 5/13/08.
(LEGO is a trademark of LEGO Company.)
Feder, Stanley (1987). "Factions and Policon: New Ways to Analyze Politics" in *Inside CIA's Private World*, H. Bradford Westerfield, Ed. (1995), *Declassified Articles from the Agency's Internal Journal*.
Hermann, Margaret, 1977, *A Psychological Examination of Political Leaders*; News York, Free Press
_____, 1980, Explaining Foreign Policy Behavior, *International Studies Quarterly*, 24:7-46
_____, 2001, *Leaders, Groups and Coalitions: Understanding the People and Processes in Foreign Policy-making*. New York, Free Press
Kahneman, Daniel and Amos Tversky, 1979, "Prospect Theory: An Analysis of Decision under Risk,"
Econometrica, XLVII , pp 263-291. Last accessed at http://www.hss.caltech.edu/~camerer/Ec101/ProspectTheory.pdf on 5/14/08.
Khoja, Faiza and Lutafali, Shabnam, 2008, "Micro-financing: an innovative application of social networking," Entrepreneur.com, January-February 2008. Last accessed at http://www.entrepreneur.com/tradejournals/article/175632909.html on 5/14/08.
McCallister, William S., 2007, *COIN and Irregular Warfare in a Tribal Society*, Last accessed at http://www.smallwarsjournal.com/documents/coinandiwinatribalsociety.pdf on 5/14/08.
Post, Jerrold M., 2005, *The Psychological Assessment of Political Leaders: With Profiles of Saddam Hussein and Bill Clinton*, Ann Arbor, University of Michigan Press, 2005, pp. 53-61.
Reynolds, William N. and Moore, David T., 2010, "Advancing the Practice: Multi-methodological Analysis For Intelligence," in preparation.
Ritchey, T., 1998, *General Morphological Analysis: A general method for non-quantified modeling*.
(Adapted from the paper "Fritz Zwicky, Morphologie and Policy Analysis," presented at the 16th EURO Conference on Operational Analysis, Brussels,1998http://www.swemorph.com/m.html, last accessed on 5/13/08.
Waters. T.J., 2010, *Hyperformance: Using competitive intelligence for Better Strategy and Execution*,
Chapter 7: Psychological forensics – CSI Meets P&G *passim* New York, Jossey-Bass
Weber, Marta S., 2004, "Profiling for Leadership Analysis" in *Competitive Intelligence Magazine*, 7(4),
July-August 2004.

_____, 2010. in *2010 Anthology of Competitive Intelligence*, Chapter 7, *passim*, Washington, D.C., SCIP

Zacharias, Greg L., et al. Eds. 2008 (Committee on Organizational Modeling from Individuals to Societies, National Research Council of the National Academies), *Behavioral Modeling and Simulation: From Individuals to Societies* (Prepublication copy, uncorrected proofs), National Academies Press.   last accessed on 4/21/08. Subsequently cited as *NAS Report*.

_____,*NAS Report*, op. cit.

CHAPTER 36

New Developments with Collision Avoidance for Posture Prediction

Ross Johnson, Carl Fruehan, Matt Schikore,
Tim Marler, Karim Abdel-Malek

Virtual Soldier Research (VSR) Program
University of Iowa
USA

ABSTRACT

A substantial advantage of predictive virtual human models is the ability to adapt to changes in a virtual environment automatically, and with respect to posture prediction and analysis, this ability hinges on collision avoidance. Collision avoidance must be robust enough to accommodate various types of geometry, must apply to the avatar (self-avoidance) as well as virtual objects, must not detract from real-time operation, and must be suitable for a variety of real-world scenarios. Thus, while leveraging optimization-based posture prediction and a unique method for collision avoidance with increase computational speed we present new developments in this arena. A new sphere-filling algorithm is presented with increased speed and fidelity for creating surrogate geometry, which is critical for any type of collision avoidance or detection. The collision avoidance algorithm is implemented for self-avoidance. And, the new capabilities are demonstrated on automotive and motorcycle examples for ergonomic analysis. The results not only involve realistic predicted postures and novel forms of human-performance feedback, but also reflect real-time operation.

Keywords: Human Modeling, Optimization, Collision Avoidance, Sphere Filling

INTRODUCTION

A critical advantage to using digital human models is the ability to evaluate new products and processes virtually. However, fully recognizing this advantage requires the virtual human to interact with digital models in a 3-D environment. This interaction can be useful for identifying design issues relating to human factors and ergonomics, and can provide insight into human behavior. Such interaction between virtual humans and products often depends on predictive capabilities and the tendency of changes in the virtual environment to affect predicted responses.

Thus, this paper presents new developments with predictive collision avoidance capabilities, in the context of optimization-based posture prediction. Collision avoidance includes the avoidance of collisions with external objects, as well as the avoidance of collisions between one part of the digital human and another. The collision avoidance described here does not describe just the ability to detect when two objects collide, which is known as collision detection, but entails formulating the simulation in such a way that obstacles are actually considered as input to the problem, and therefore affect the results of a given task.

The optimization-based posture prediction has many advantages and can be adapted to solve a variety of digital human modeling problems (Yang et al., 2006; Marler et al., 2007; Marler et al., 2009). One main advantage of this technique is the relative ease with which one can expand the accuracy and utility of a simulation by adding new mathematical constraints that represent different real-world factors. These constraints can represent anything from location-specific targets (Farrell et al., 2005) to equations of static equilibrium (Liu et al., 2009), but there are practical limits to the number of constraints that can be added while still maintaining acceptable software runtimes.

Consequently, this paper presents new developments that increase both the speed and accuracy of collision avoidance for the optimization-based posture prediction approach. This work builds and expands on posture prediction capabilities for Santos™, a high-fidelity predictive human model (Abdel-Malek et al., 2006; Marler et al., 2008), so the optimization-based method for predicting posture is first summarized. Fundamental to most collision detection or avoidance approaches is the use of surrogate geometry. Consequently, we outline a new sphere-based algorithm for approximating geometry, which can be integrated in the optimization formulation. We then extend the multi-run obstacle avoidance method (Johnson et al., 2009) to function with self-avoidance. These improvements increase the accuracy, speed, and value of human modeling and simulation capabilities, and these improvements are demonstrated with practical examples.

OPTIMIZATION-BASED POSTURE PREDICTION

In this section, an overview of human optimization-based posture prediction is discussed. This includes a brief description of the skeletal model, as well as the

338

final optimization formulation.

Simulating human posture depends largely on how the human skeleton is modeled. One way to view a skeleton is as a kinematic system, or series of links with each pair of links connected by one or more revolute joints. Therefore, a complete human body can be modeled as several kinematic chains, formed by series of links and revolute joints.

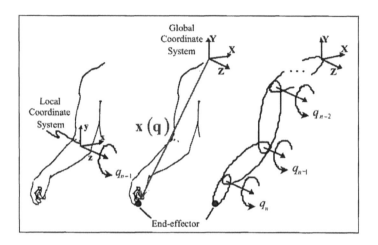

FIGURE 1 A kinematic chain of joints

q_i is a *joint angle* and represents the rotation of a single revolute joint. There is one joint angle for each degree of freedom (DOF). $q = [q_1 \quad \cdots \quad q_n]^T \epsilon \mathbb{R}^n$ is the vector of joint angles in an n-DOF model and represents a specific posture. Each skeletal joint is modeled using one or more kinematic revolute joints. $x(q)\epsilon\mathbb{R}^3$ is the position vector in Cartesian space that describes the location of an end-effector with respect to the global coordinate system. For a given set of joint angles q, $x(q)$ is determined using the Denavit-Hartenberg (DH)-method (Denavit and Hartenberg, 1955). With this work, a 55-DOF model for the human torso, arms, legs, and neck is used. This also includes six global DOFs, three for translation of the hip point and three for rotation about the hip point. The posture of this model is determined by solving the optimization problem formulated as follows.

The design variables for the problem are q_i, measured in units of radians. One constraint, called the *distance* constraint, requires the end-effector to contact a target point. In addition, each joint angle is constrained to lie within predetermined limits. q_i^U represents the upper limit, and q_i^L represents the lower limit. The basic benchmark performance measure, which serves as the objective function in the optimization problem, is joint displacement. This performance measure is proportional to the deviation from a *neutral position*, which is selected as a relatively comfortable posture, and is denoted q_i^N for a particular joint. Because some joints articulate more readily than others, a weight w_i is introduced to stress the relative stiffness of a joint.

The optimum posture for the system is then determined by solving the following problem:

Find: $q \epsilon \mathbb{R}^n$

To minimize: $f_{Join\ tDisplacement}\ (q) = \sum_{i=1}^{n} w_i(q_i - q_i^N)^2$

Subject to: $distance = \left\| x(q)^{end\ -effector} - x^{target\ point} \right\| \leq \varepsilon$ $\qquad$ (1)

$\qquad\qquad q_i^L \leq q_i \leq q_i^U; i = 1,2,\dots,n$

where ε is a small positive number that approximates zero and DOF is the total number of degrees of freedom. (1) is solved using the software SNOPT (Gill et al., 2002), which uses a gradient-based method. Thus, analytical gradients are determined for all objective functions and for all constraints.

SPHERE-FILLING ALGORITHM

The optimization formulation for posture prediction can be extended to include constraints that prevent an avatar from intersecting objects in the environment; however, these constraints must be continuous, differentiable functions. Objects in the 3D environment are internally stored as sets of primitive polygons, but attempting to formulate constraints that use these primitives would prove to be computationally intractable, especially when striving for real-time performance. The proposed approach represents an object in the environment with a set of spheres, such that the union of the sphere volumes approximates the shape of the object. The sphere representation allows for a relatively simple constraint formulation. In addition, it allows for variable fidelity, depending on the requirements of the problem being solved. Sphere filling algorithms are numerous, each with advantages and disadvantages, and many different techniques have been tested in the context of the above-mentioned posture prediction problem.

The current approach for representing surrogate geometry is sphere shelling (Johnson et al., 2009), which generates a large number of spheres that cover the surface of the object. Sphere shelling is fast but generates a large number of spheres, which greatly increases the runtime of posture prediction. Another disadvantage of sphere shelling is that it overestimates the shape of the object, as the spheres are on the surface and are not necessarily contained inside the object's surface. This overestimation prevents the digital human from getting close to object edges and occasionally creates an infeasible problem where conceptually there should be a solution. Thus, an advanced sphere filling algorithm has been implemented that specializes in generating efficient representations of geometry. This adaptive medial-axis approximation is based on three-dimensional Voronoi diagrams (Bradshaw et al., 2004) and generates close representations of objects with as few spheres as possible. The drawback of this medial-axis algorithm is that it takes a relatively long time to generate spheres, even for very simple meshes. Consequently, a two-step hybrid sphere-filling algorithm was created that utilizes sphere inflation and sphere culling, described as follows.

Here, we describe the first stage of the hybrid approach. This method uses a grid of points (voxels) to create the initial spheres, as with sphere-shelling, but then expands spheres inside the object until they touch the edge of the mesh. The positioning of the spheres may not be as optimal as with the medial-axis method, but the slight loss of representation efficiency is compensated for with the speed increase. The following pseudo-code describes this inflation method.

inflate(K: *mesh*, M: *integer*)
    VOXELS: *grid* ← impose M×M×M grid over bounding box of K
    SPHERES: *sphere set* ← ∅

    for each P: *point* **in** VOXELS
        DIST: *float* ← signed distance from P to closest triangle in K
                (Bærentzen & Aanæs, 2002)
        if (DIST < 0)
            // The voxel is inside the mesh
            SPHERES ← SPHERES ∪ { **sphere**(P,DIST) }

    return SPHERES

This sphere inflating algorithm provides a set of spheres, the centers of which are placed on a Cartesian grid. As seen in FIGURE 2**Error! Reference source not found.**, there can be a large number of spheres, many of them redundant. Thus, the second phase of the hybrid approach involves culling redundant spheres.

FIGURE 2 A sphere-filled cube before culling

When culling spheres, the algorithm ensures that the center of each voxel is filled (remains contained within at least one sphere). A greedy algorithm is used to select the final set of spheres. During each iteration the algorithm chooses the sphere that includes the highest number of previously unfilled voxels, and adds it to the final set of spheres. This is repeated until all voxels are filled. Various intermediate results for culling a cube are shown in FIGURE 3.

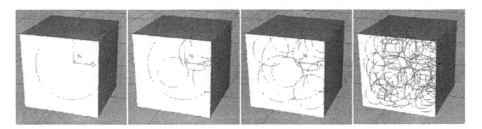

FIGURE 3   The first 1 (a), 5 (b), and 10 (c) spheres selected, and the final sphere set (d)

COLLISION AVOIDANCE ALGORITHM

Conceptually, collision avoidance is modeled by adding an additional constraint to the optimization problem (see equation (1)) for every pair of spheres that should not intersect. The use of spheres as surrogate geometry greatly increases the simplicity of calculating the constraint functions and their gradients. Given an obstacle sphere O with a global position and a body sphere B with joint-relative position, the constraint function preventing them from intersecting is given as:

$$f(q) = position(O) \cdot position(B, q) - \left(radius(O) - radius(B)\right)^2 \geq 0$$

Similarly, given two body spheres (approximating the avatar mesh) B_1 and B_2, with positions given locally, the constraint function preventing their intersection is:

$$f(q) = position(B_1, q) \cdot position(B_2, q) - \left(radius(B_1) - radius(B_2)\right)^2 \geq 0$$

With a basic implementation, one constraint is added to the optimization formulation for every pair of spheres that should not collide. Thus, for m body spheres and n obstacle spheres, $m \times n$ constraints are added for obstacle avoidance. Thus, the optimization running time is at best linear in the number of constraints, or $\Omega(MN)$. In many real-world simulations, it is not uncommon to have thousands of obstacle spheres present in the environment, which can greatly increase the time required to find a solution. Consequently, a multi-run collision avoidance method has been implemented, in which the optimizer is executed in a loop, with each iteration running posture prediction, performing collision detection, and then either adding new necessary constraints to the problem, or returning a satisfactory result (Johnson et al., 2009). This approach requires multiple executions of the optimizer, but it ultimately considers fewer constraints than the basic implementation.

Here, we outline how the multi-run approach is used for self-avoidance, where spheres representing an avatar are restricted from colliding with other spheres in the avatar. This multi-run approach is especially helpful at reducing optimization constraints, because with the basic implementation, the number of constraints required for M body spheres is $O(M^2)$. Thus, implementing the multi-run approach reduces constraints and allows for an increased number of defined body spheres.

One source of difficulty with self-avoidance is that there are certain body-sphere pairs that should never be constrained from colliding. For example, the body spheres that represent the avatar's right forearm need not be constrained from intersecting spheres in the right hand, but they should not collide with spheres that represent the avatar's torso. To handle this, a new grouping approach has been incorporated, whereby the body spheres are grouped based on their physical location and spheres in the same group are not checked for avoidance. In addition, there are specific pairs of body spheres that may reside in different groups but should still not be constrained to avoid one another. The user has the ability to alter the grouping, and this provides significant flexibility in tailoring speed and precision.

A new multi-run approach that incorporates these considerations for self-avoidance is shown in following pseudo-code:

```
avoid( BODY: body spheres ) : posture
    ENABLED: sphere pairs ← ∅
    DISABLED: sphere pairs ← ∅

    for all {X, Y} ∈ P(BODY)
        if ( group(X) ≠ group(Y) ∧ ignore ( X, Y) = false )
            DISABLED ← DISABLED ∪ {X, Y}

    do
        RESULT: posture ← optimizer_solve( ENABLED )
        NEW_COLLISIONS: boolean ← false

        for each  {X: body sphere, Y: body sphere} in DISABLED
            D: float ← |position( X, RESULT )−position( Y, RESULT )|
            if ( D < radius( X ) + radius( Y ) ) then
                NEW_COLLISIONS ← true
                DISABLED ← DISABLED \ {(X, Y)}
                ENABLED ← ENABLED ∪ {(X, Y)}
    while ( NEW_COLLISIONS = true)

    return RESULT
```

RESULTS

The performance benefits of the multi-run approach have already been discussed in a previous paper by Johnson et al., so this section takes the algorithmic advancements discussed thus far and shows their application to biomechanics, design, and analysis. The first two examples display the merits of the new sphere-filling algorithm and independently demonstrates the effects of self avoidance and obstacle avoidance on posture prediction results.

Figure 4 demonstrates a reaching task where the avatar is instructed to touch his seat belt buckle with his left hand. The resulting posture is shown with self-avoidance turned off (Figure 4.b) and with self avoidance enabled (Figure 4.c). The joint displacement objective function was used in this example, with the neutral posture shown in Figure 4.a. The numbers indicate the relative objective function value for the various postures. These numbers show that Figure 4.b is a better posture according to the objective function, but with the self avoidance constraints enabled, the avatar is forced to increase the objective value to avoid collisions.

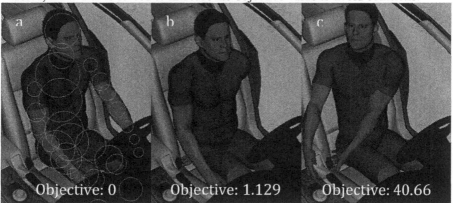

FIGURE 4 Posture prediction results with b. self-avoidance disabled and c. self-avoidance enabled using the neutral posture and body spheres in a.

The second example, shown in Figure 5, demonstates the new sphere filling method and its use with obstacle avoidance. Figure 5.a shows a steering wheel represented filled with 1206 spheres by the sphere-shelling algorithm. Figure 5.b shows the same steering wheel filled with 185 spheres generated by the inflate method. Figure 5.c and d show a posture prediction task where the avatar is reaching to the right of the steering column first without obstacle avoidance and then with obstacle avoidance. The obstacle avoidance in Figure 5.d uses the spheres from the inflate method (Figure 5.b).

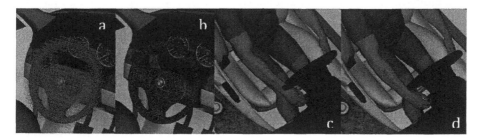

FIGURE 5 Obstacle spheres a. generated by shelling method and b. generated by

inflation method. Posture prediction results c. with obstacle avoidance disabled and d. obstacle avoidance enablde using spheres from b.

The third example, shown in Figure 6, shows a predicted posture on a motorcycle. The avatar Is constrained to touch his right knee with his left hand. Without collision avoidance, the avatar's wrist intersects with the motorcycle gas tank, as shown in Figure 6.a. With collision avoidance and self-avoidance, the avatar avoids the collision with the motorcycle, and he also avoids colliding with himself, as shown in Figure 6.b.

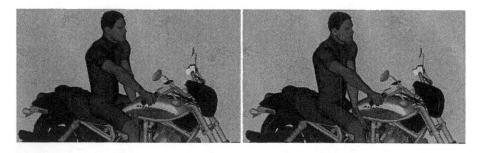

FIGURE 6   Posture prediction with self-avoidance and obstacle avoidance a. disabled and b. enabled.

CONCLUSION

Using optimization-based posture prediction as a foundation, this paper has presented novel advances with collision avoidance, which is a critical advantage of predictive DHM capabilities. Real-time collision avoidance provides one more way in which a user can alter a virtual environment on the fly and see the effects on human performance. A new multi-run approach to obstacle avoidance has been extended to self avoidance. In addition, a new method for developing surrogate geometry has been developed and tested in conjunction with a grouping method for culling surrogate geometry used to represent avatars. The results, which are demonstrated in the context of automotive and motorcycle ergonomic analysis, are quite successful.

The presented method for sphere filling has potential applications that extend far beyond posture prediction. For instance, collision detection is another critical component of virtual modeling and simulation and also requires fast and accurate creation of surrogate geometry. Thus, the proposed sphere-filling method is also used in an algorithm for detecting collisions between the avatar and geometry, when collision avoidance is turned off. This provides users with an indication of geometry that restricts motion and thus area of focus for potential design changes.

With respect to future work, objective validation using motion capture is ongoing, to verify the accuracy of the predicted postures. In addition, it is possible to predict postures, not just of body segments, but of body location and orientation. This capability will be tested with collision avoidance as well. Finally, the ability will be developed to fill geometry automatically as it is loaded. Then, only those spheres within an avatar's immediate reach envelope will be considered for avoidance.

REFERENCES

Abdel-Malek, K., Yang, J., Marler, T., Beck, S., Mathai, A., Zhou, X., Patrick, A., Arora, J. (2006), "Towards a New Generation of Virtual Humans," *International Journal of Human Factors Modelling & Simulation*, 1(1), 2-39.

Bærentzen, J. A., Aanæs H. (2002), "Generating Signed Distance Fields From Triangle Meshes," Technical Report IMM-TR-2002-21, Informatics and Mathematical Modelling, Technical University of Denmark, Lyngby, Denmark.

Meshes Bradshaw, G., O'Sullivan, C. (2004), "Adaptive Medial-Axis Approximation for Sphere-Tree Construction," *ACM Transactions on Graphics*, 23(1), 1-26.

Denavit, J., Hartenberg, R. S. (1955), "A Kinematic Notation for Lower-pair Mechanisms Based on Matrices." *Journal of Applied Mechanics*, 77, 215-221.

Farrell, K., Marler, R. T., Abdel-Malek, K. (2005), "Modeling Dual-Arm Coordination for Posture: An Optimization-Based Approach," *SAE 2005 Transactions Journal of Passenger Cars - Mechanical Systems*, 114-6, 2891, SAE paper number 2005-01-2686.

Gill, P., Murray, W., Saunders, A. (2002), "SNOPT: An SQP Algorithm for Large-Scale Constrained Optimization." *SIAM Journal of Optimization*, 12(4), 97-1006.

Johnson, R., Smith, B. L., Penmatsa, R., Marler, T., Abdel-Malek, K. (2009), "Real-Time Obstacle Avoidance for Posture Prediction," *SAE Digital Human Modeling Conference*, June, Gothenburg, Sweden, Society of Automotive Engineers, Warrendale, PA.

Liu, Q., Marler, T., Yang, J., Kim, H. J., Harrison, C. (2009), "Posture Prediction with External Loads – A Pilot Study," *SAE 2009 World Congress*, April, Detroit, MI, Society of Automotive Engineers, Warrendale, PA.

Yang, J., Marler, T., Beck, S., Abdel-Malek, K., Kim, H. J. (2006), "Real-Time Optimal-Reach Posture Prediction in a New Interactive Virtual Environment." *Journal of Computer Sciience and Technology*, 21(2), 189-198.

Marler, R. T., Arora, J. S., Yang, J., Kim, H. J., and Abdel-Malek, K., (2009), "Use of Multi-objective Optimization for Digital Human Posture Prediction," *Engineering Optimization*. 41(10), 925-943.

Marler, T., Arora, J., Beck, S., Lu, J., Mathai, A., Patrick, A., Swan, C. (2008), "Computational Approaches in DHM," in *Handbook of Digital Human Modeling for Human Factors and Ergonomics*, Vincent G. Duffy, Ed., Taylor and Francis Press, London, England.

Marler, T., Yang, J., Rahmatalla, S., Abdel-Malek, K., Harrison, C. (2007), "Validation Methodology Development for Predicted Posture," *SAE 2007 Transactions Journal of Passenger Cars – Electronic and Electrical Systems*, SAE paper number 2007-01-2467.

ACKNOWLEDGEMENTS

This work has been partially funded by Caterpillar Inc. project: Digital Human Modeling and Simulation for Safety and Serviceability. This support is gratefully acknowledged.

Chapter 37

Optimization-Based Collision Avoidance Using Spheres, Finite Cylinders and Finite Planes

Mahdiar Hariri, Rajankumar Bhatt,
Jasbir Arora, Karim Abdel-Malek

VSR, CCAD
The University of Iowa
Iowa City, IA 52242, USA

ABSTRACT

A digital human must avoid the collision of the body segments with other non-adjacent body segments as well as with the objects in the environment while performing a task. In this research, we develop mathematical models for constraints that can avoid these collisions. The digital human body segments and the obstacles in the environment are modeled using surrogate geometries. The body segments are represented by using one or more spheres rigidly attached to a local reference frame so that these spheres move as the body segments move. The objects in the environment are modeled using one or more of the five primitive geometries: spheres, infinite cylinders, infinite planes, finite cylinders, and finite planes. A generic collision avoidance strategy is developed to avoid spheres with all the five primitive geometries used for representing obstacles.

We use gradient based optimization strategy for predicting the motion of the digital human avatar while performing a task. One of the requirements of a gradient based optimization is use of constraint functions and objective functions with continuous gradients of at least first order. This is equivalent to a requirement for the elements to have smooth surfaces(no edges). But finite cylinders and finite planes do not have smooth edges. Hence, we present a method to smooth out the edges of finite cylinders and planes and consider these modified elements instead, so that the constraint gradients are continuous.

Keywords: Optimization, Collision Avoidance, Sphere, Cylinder, Plane, Finite Cylinder, Finite Plane, Proximity Distance with Continuous Gradient

INTRODUCTION

The computation of distance between two mathematical objects finds many applications in robotics. Most of the effort in robotics in the field of collision avoidance involves the path planning optimization for mobile robots where the path of the robot is normally modified by optimization in order to avoid collision. Using spheres to model obstacles has also been popular in the field of path planning in flying spacecrafts (Singh, 2001, 2002). There have also been several studies which use spheres for modeling objects. For a human avatar, modeled as a robot with multiple branches, the design variables which affect self collision are the joint angles. For obstacle collision avoidance, the global translation/rotation of the avatar is also added to those variables.

(Colbaugh et al., 1989) uses simple geometric primitives to represent the robot arms and its environment for a planar manipulator. The obstacles were represented by circles surrounded by a surface of influence, and the links were modeled by straight lines. A redundancy resolution scheme was proposed to achieve obstacle avoidance. This approach was extended to the 3-D workspace of redundant manipulators in (Shadpey et al. , 1994, 1995) , (Glass et al., 1995).

Using spheres to model links and obstacles for collision avoidance has an important advantage. The advantage is that the optimization constraint that needs to be satisfied to avoid collision between 2 spheres is simply: $d^2 \geq (r_1 + r_2)^2$, where d is the distance between the spheres' centers and r_1 and r_2 are the radii of the spheres. This constraint is simple to calculate and if the motion of the 2 spheres are functions of q_i, then this constraint is also a C^∞ function of q_i (The class of functions with continuous pth-order derivatives is denoted by C^p.).

By using available optimization softwares, collision avoidance can be integrated as a constraint among others. Indeed, one may consider writing these constraints using any available proximity distance algorithm which returns signed proximity distances separating two bodies; see the recent exhaustive books (van den Bergen, 2004) and (Ericson, 2005).

However, while using the gradient-based optimization software, the gradients of the criteria and the constraints need to be continuous with respect to the design variables (generally robot joint angles and joint angle trajectories' parameters). Some of the works in the field of collision avoidance model the surface of objects by many flat surfaces such as polyhedrons. The proximity distance between polyhedrons does not have continuous gradients with respect to the parameters.

Continuity properties of the distance have been merely discussed or even assumed in previous works. For example, in (Lee, 2001), (Choset, 1997), where the obstacle avoidance problem has been addressed in a 2D case, it has been claimed that the distance between convex objects is smooth and thus the gradient is continuous. The latter assertion is not always valid unless one object is strictly convex, the former depends on the continuity properties of both objects' surfaces. It is only in (Rusaw, 2001) that the non differentiability (and non-convexity) of the distance between convex bodies is well addressed and used with non-smooth analysis in the context of sensory-based planning.

A recent paper (Escande, 2007) claims to be the first to treat the problem of ensuring continuous distance's gradients. It draws solution to get rid of the non-differentiability. They build offline strictly convex bounding volume that can be considered as a smooth 'rounding' of the polyhedron convex hull. However, even in this work, the edges between any 2 polyhedrons are not smoothed out. They don't actually need to perform such an operation, as they consider the collision of polyhedrons one by one with each other during the motion.

FORMULATIONS

In this paper, we introduce a new method for obstacle avoidance by combining parts of the surfaces of differentiable primitives (sphere, infinite cylinder, infinite plane) and creating new compound primitives (finite cylinder, finite plane, finite one-sided plane, box) with convex surfaces that have no edges. The distance between the two primitives or compound primitives is always calculated by a single scalar constraint function. Theorem 3 proposed in this paper's appendix states that if two objects have convex surfaces without any edges (C^1 surfaces), then the derivative of minimum distance between them with respect to q_i is a continuous function of q_i. Hence, the proposed method is suitable enough to be used as a constraint in gradient based optimization solver.

Below, we present the formula for collision avoidance constraints. In each case, f is the value of the constraint that should be kept positive to avoid collision. Also, the position vectors defined in the section below are considered to be the functions of the design variables, q_i. If the primitive is associated with a fixed obstacle in space, the position vectors of all points on such primitives are fixed and hence, the gradients of all such points with respect to design variables are zero.

Constraint for sphere-to-sphere collision avoidance:
As shown in figure 1, A and B are the center of the spheres with radii r_1 and r_2 respectively. Let the global position vector of the points A and B be given by $\vec{G}_A$ and $\vec{G}_B$. The constraint function that calculates the distance between the spheres and its gradients with respect to the design variables can be calculated as:

350

$$f = \left(\vec{G}_A - \vec{G}_B \right) . \left(\vec{G}_A - \vec{G}_B \right) - \left(r_1 + r_2 \right)^2 \geq 0$$

$$\frac{\partial f}{\partial q_i} = 2 \left(\frac{\partial \vec{G}_A}{\partial q_i} - \frac{\partial \vec{G}_B}{\partial q_i} \right) . \left(\vec{G}_A - \vec{G}_B \right) \quad \boxed{1}$$

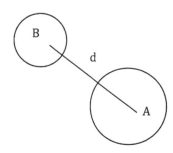

Figure 1. sphere-to-sphere collision

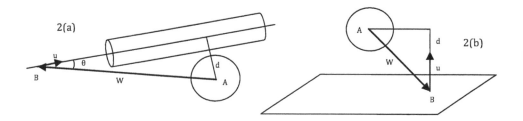

Figure 2. Distance of a sphere from an infinite cylinder and an infinite plane

Constraint for sphere-to-infinite cylinder collision avoidance:
As shown in Figure 2(a), A is the center of the sphere with radius r_1. B is any point on the cylinder's axis with radius r_2. $\vec{G}_A$ is the global position of A. $\vec{G}_B$ is the global position of B. $\vec{u}$ is the unit vector along the cylinder's axis. $\vec{W}$, the vector connecting point A to point B, can thus be written as: $\qquad \vec{W} = \vec{G}_B - \vec{G}_A$

Now, define a vector $\vec{h}$ such that: $\qquad\qquad \vec{h} \equiv \vec{W} \times \vec{u}$

Since $\vec{u}$ is a unit vector: $\qquad\qquad \left| \vec{h} \right|^2 = \vec{h}.\vec{h} = \left| \vec{W} \right|^2 \sin^2 \theta = \left| \vec{d} \right|^2 \qquad \boxed{2}$

Hence, the constraint function and the gradients can thus be calculated as:

$$f = \left| \vec{d} \right|^2 - \left(r_1 + r_2 \right)^2 = \vec{h}.\vec{h} - \left(r_1 + r_2 \right)^2 \geq 0 \qquad\qquad \frac{\partial f}{\partial q_i} = 2 \frac{\partial \vec{h}}{\partial q_i} . \vec{h} \qquad \boxed{3}$$

Constraint for sphere-to-infinite plane collision avoidance:
As shown in Figure 2(b), A is the sphere's center with radius r_1. B is any point on the mid-plane of the plane with thickness $2t_2$. $\vec{G}_A$ is the global position of A. $\vec{G}_B$ is the global position of B. $\vec{W}$ is the vector connecting A to B. $\vec{u}$ is the unit vector prependicular to the plane.

$$\vec{W} = \vec{G}_B - \vec{G}_A \qquad\qquad f = \left(\vec{W}\cdot\vec{u}\right)^2 - \left(r_1 + t_2\right)^2 \geq 0$$

$$\frac{\partial f}{\partial q_i} = 2\left(\vec{W}\cdot\vec{u}\right)\left(\frac{\partial \vec{W}}{\partial q_i}\cdot\vec{u} + \vec{W}\cdot\frac{\partial \vec{u}}{\partial q_i}\right) \qquad\qquad \boxed{4}$$

CONSTRAINTS FOR COMPOUND PRIMITIVES

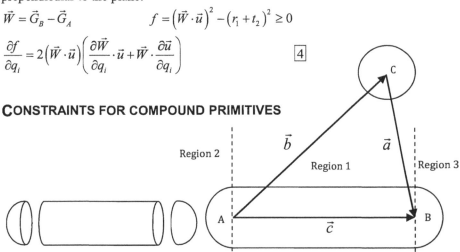

Figure 3. Smoothed finite cylinder with regions defined around it
The primitives presented in the previous sections can also be combined to produce compound primitives like finite cylinders (smoothed) and finite planes (smoothed) with constraints that have C^1 continuity according to theorem 3. The constraints used for such compound primitives are discussed in this section.

Sphere-to-finite cylinder collision:
First, the region of the location of the sphere with respect to the finite cylinder as shown in Figure 3 is determined by evaluating the scalar products of vectors $\vec{a}, \vec{b}$ with vector $\vec{c}$. Based on the region, the value and all the gradients of the collision avoidance constraint are set equal to one of the following constraints:
At Region 1: sphere-to-infinite cylinder   At Regions 2,3 : sphere-to-sphere

Sphere-to-finite plane collision:
First, the region of the location of the sphere with respect to the finite plane as shown in Figure 4 is determined by evaluating the scalar product of vectors $\vec{a}, \vec{b}, \vec{c}, \vec{d}$ with vectors $\vec{u}, \vec{v}$. Based on the region, the value and all the gradients of the collision avoidance constraint are set equal to one of the following constraints:
At Region 1: sphere-to-infinite plane   At Regions 6,7,8,9: sphere-to-sphere
At Regions 2,3,4,5: sphere-to-infinite cylinder

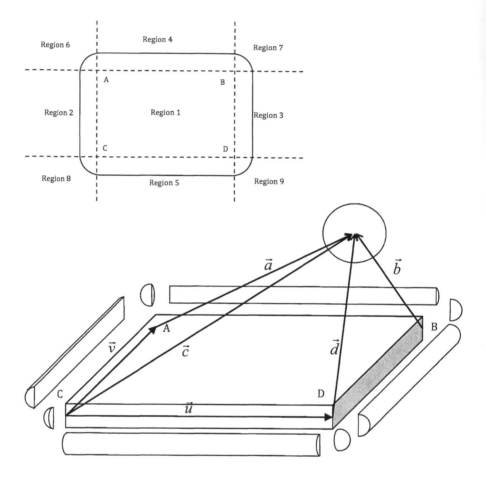

Figure 4. Finite plane (smoothed) to sphere collision avoidance

RESULTS

We apply the proposed obstacle avoidance constraints within the framework of predictive dynamics, a constrained optimization-based approach, with dynamic effort as performance measure subject to different task-based, physics-based and environment based constraints, to predict and simulate digital human motion. This approach to predicting and simulating physics-based motions has been validated (Xiang et al., 2007).

Simulation 1: The results of using the self avoidance modules in the predictive dynamics code is shown in Figure 5. The segments of the human avatar are filled with spheres. The task shown below is a complicated task that requires the avatar to kneel down and touch the right knee with left hand and touch the left midfoot with its right hand. This task cannot be simulated correctly unless self-avoidance constraint is properly modeled and implemented in the predictive dynamics code.

FIGURE 5. The test task with sphere to sphere self collision avoidance

Simulation 2: This task requires the avatar to move its hand from sphere A located at the top of a table to sphere B at the bottom of the table. Figure 6 shows the result of the simulation without using any collision avoidance constraints.

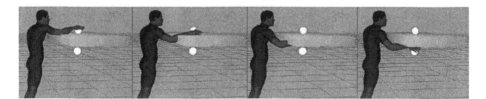

FIGURE 6. Result of simulation without imposing obstacle avoidance constraints

Simulation 3: This task also requires the avatar to move its hand from sphere A located at the top of the table to sphere B located at the bottom of the table. The table is modeled as a finite plane. The right arm is filled with 7 spheres and so, 7 sphere-to-finite plane collision avoidance constraints are imposed during the motion. The cost function minimized during this motion is a combination of dynamics effort and total displacement. Figure 7 shows the result of the simulation using the mentioned collision avoidance constraints.

FIGURE 7. Result of simulation after imposing obstacle avoidance constraints

CONCLUSION

In this work, it is proved that any shape may be used for object definition in optimization-based collision avoidance as long as the shape's surface is covex and has no edges (C^1 surface) so that the gradients of the obstacle avoidance constraint (minimum distance between the surfaces) are continuous.

Finite cylinders and finite planes have edges which violates the C^1 continuity requirements. These edges of the finite cylinder and finite plane can be smoothed out by combining their surfaces with parts of the surfaces of spheres and cylinders as shown in Figures 3 and 4. We call these objects compound primitives. Implementation of other compound objects as shown in Figure 8 is in progress.

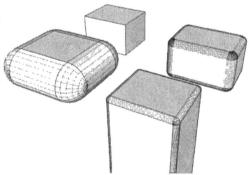

Figure 8. An edged box (unsuitable), a finite plane, a 1-sided finite plane, a box (smoothed)

APPENDIX (THEORETICAL BACKGROUND)

Definition Set 1: P and Q are rigid bodies whose motions are functions of $q_i (i = 1...n)$. Witness points of minimum distance between these two rigid bodies called A, B are defined as the points respectively on P, Q between which the minimum distance of P, Q occurs at any q_i. We denote the global position of points A, B at any q_i by $\overline{A}(q_i)$ and $\overline{B}(q_i)$.

$A'_{q_{i_0}}$ and $B'_{q_{i_0}}$ are defined as the footprint points of A, B on P, Q for $q_i = q_{i_0}$. $A'_{q_{i_0}}$ and $B'_{q_{i_0}}$ are points attached to the surfaces of P, Q which move with P, Q. We denote the global position of $A'_{q_{i_0}}$ and $B'_{q_{i_0}}$ by $\overline{A'_{q_{i_0}}}(q_i)$ and $\overline{B'_{q_{i_0}}}(q_i)$. Since $A'_{q_{i_0}}$ and $B'_{q_{i_0}}$ are the footprints of A, B at $q_i = q_{i_0}$, the following conditions hold:

$$\left\{ \overline{A'_{q_{i_0}}}(q_{i_0}) = \overline{A}(q_{i_0}) \quad , \quad \overline{B'_{q_{i_0}}}(q_{i_0}) = \overline{B}(q_{i_0}) \right\} \qquad \boxed{5}$$

Theorem 1: Using Definition set 1, The minimum distance between any 2 rigid bodies P, Q is a continuous function of q_i

Proof: The footprints of A, B at $q_i = q_{i_0}$ shown by $\overline{A'_{q_{i_0}}}(q_i)$ and $\overline{B'_{q_{i_0}}}(q_i)$ are continuous functions of q_i due to the rules of rigid body motion. The minimum distance between P, Q at $q_i = q_{i_0}$ is called $d_{q_{i_0}}$ and therefore:

$$\forall \Delta q_i : \quad d_{q_{i_0}} = \left\| \overline{AB} \right\| = \left\| \overline{B'_{q_{i_0}}}(q_{i_0}) - \overline{A'_{q_{i_0}}}(q_{i_0}) \right\| \le \left\| \overline{B'_{q_{i_0}+\Delta q_i}}(q_{i_0}) - \overline{A'_{q_{i_0}+\Delta q_i}}(q_{i_0}) \right\|$$

$$\lim_{\Delta q_i \to 0} d_{q_{i_0}+\Delta q_i} = \lim_{\Delta q_i \to 0} \left\| \overline{B'_{q_{i_0}+\Delta q_i}}(q_{i_0}+\Delta q_i) - \overline{A'_{q_{i_0}+\Delta q_i}}(q_{i_0}+\Delta q_i) \right\| \ge d_{q_{i_0}} \qquad \boxed{6}$$

The minimum distance between P, Q at $q_i = q_{i_0} + \Delta q_i$ is equal to $d_{q_{i_0}+\Delta q_i}$ and so:

$$d_{q_{i_0}+\Delta q_i} = \left\| \overline{AB} \right\| = \left\| \overline{B'_{q_{i_0}+\Delta q_i}}(q_{i_0}+\Delta q_i) - \overline{A'_{q_{i_0}+\Delta q_i}}(q_{i_0}+\Delta q_i) \right\|$$

$$d_{q_{i_0}+\Delta q_i} = \left\| \overline{B'_{q_{i_0}+\Delta q_i}}(q_{i_0}+\Delta q_i) - \overline{A'_{q_{i_0}+\Delta q_i}}(q_{i_0}+\Delta q_i) \right\| \le \left\| \overline{B'_{q_{i_0}}}(q_{i_0}+\Delta q_i) - \overline{A'_{q_{i_0}}}(q_{i_0}+\Delta q_i) \right\|$$

$$\lim_{\Delta q_i \to 0} d_{q_{i_0}+\Delta q_i} \le \lim_{\Delta q_i \to 0} \left\| \overline{B'_{q_{i_0}}}(q_{i_0}) - \overline{A'_{q_{i_0}}}(q_{i_0}) \right\| = d_{q_{i_0}} \qquad \boxed{7}$$

And therefore : $\qquad \boxed{6} \, and \, \boxed{7} \Rightarrow \lim_{\Delta q_i \to 0} d_{q_{i_0}+\Delta q_i} = d_{q_{i_0}}$

Theorem 2: Using Definition set 1, if P has a convex surface and Q has a strictly convex surface, then the locations of A,B are continuous function of q_i

Proof: For any q_i :

$$d_{q_i} = \left\| \overline{B'_{q_i}}(q_i) - \overline{A'_{q_i}}(q_i) \right\| \le \left\| \overline{B'_{q_i+\Delta q_i}}(q_i) - \overline{A'_{q_i+\Delta q_i}}(q_i) \right\| \qquad \text{and}$$

$$d_{q_i+\Delta q_i} = \left\| \overline{B'_{q_i+\Delta q_i}}(q_i + \Delta q_i) - \overline{A'_{q_i+\Delta q_i}}(q_i + \Delta q_i) \right\| \le \left\| \overline{B'_{q_i}}(q_i + \Delta q_i) - \overline{A'_{q_i}}(q_i + \Delta q_i) \right\|$$

Since $\left\| \overline{B'_{q_i}}(q_i) - \overline{A'_{q_i}}(q_i) \right\|$ and $\left\| \overline{B'_{q_i+\Delta q_i}}(q_i) - \overline{A'_{q_i+\Delta q_i}}(q_i) \right\|$ are continuous function of q_i according to theorem 1, then there exists some $0 < \alpha < 1$ such that:

$$\left\| \overline{B'_{q_i}}(q_i + \alpha\Delta q_i) - \overline{A'_{q_i}}(q_i + \alpha\Delta q_i) \right\| = \left\| \overline{B'_{q_i+\Delta q_i}}(q_i + \alpha\Delta q_i) - \overline{A'_{q_i+\Delta q_i}}(q_i + \alpha\Delta q_i) \right\| \ge d_{q_i+\alpha\Delta q_i}$$

$$\lim_{\Delta q_i \to 0} d_{q_i+\alpha\Delta q_i} = \lim_{\Delta q_i \to 0} d_{q_i} = \lim_{\Delta q_i \to 0} \left\| \overline{B'_{q_i}}(q_i + \alpha\Delta q_i) - \overline{A'_{q_i}}(q_i + \alpha\Delta q_i) \right\| =$$

$$\lim_{\Delta q_i \to 0} d_{q_i+\Delta q_i} = \lim_{\Delta q_i \to 0} \left\| \overline{B'_{q_i+\Delta q_i}}(q_i + \alpha\Delta q_i) - \overline{A'_{q_i+\Delta q_i}}(q_i + \alpha\Delta q_i) \right\| \qquad \boxed{8}$$

Because P is convex, one can pass a planar surface through $\overline{A'_{q_i}}(q_i + \alpha\Delta q_i)$ and $\overline{A'_{q_i+\Delta q_i}}(q_i + \alpha\Delta q_i)$ such that all points on that surface belong to P.

Also, since Q is strictly convex, one can pass a spherical cap through $\overline{B'_{q_i}}(q_i + \alpha\Delta q_i)$ and $\overline{B'_{q_i+\Delta q_i}}(q_i + \alpha\Delta q_i)$ with the convex side of the cap towards the plane that we have passed through P such that all points on that cap belong to Q.

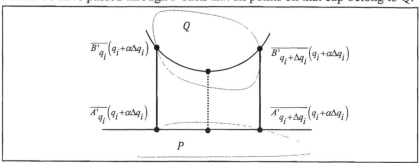

FIGURE 10.   Continuity of witness points positions as a result of surface convexity

But as shown in Figure 10, and if $|\Delta q_i| \ll 1$ the minimum distance between P,Q ($\lim_{\Delta q_i \to 0} d_{q_i+\alpha\Delta q_i}$) is equal to the value calculated in equation $\boxed{8}$, if and only if:

$$\begin{cases} \lim_{\Delta q_i \to 0} \overline{A'_{q_i}}(q_i + \alpha\Delta q_i) = \lim_{\Delta q_i \to 0} \overline{A'_{q_i+\Delta q_i}}(q_i + \alpha\Delta q_i) \\ \lim_{\Delta q_i \to 0} \overline{B'_{q_i}}(q_i + \alpha\Delta q_i) = \lim_{\Delta q_i \to 0} \overline{B'_{q_i+\Delta q_i}}(q_i + \alpha\Delta q_i) \end{cases} \qquad \boxed{9}$$

On the other hand, $\overline{A'_{q_{i_0}}}(q_i)$ and $\overline{B'_{q_{i_0}}}(q_i)$ are continuous functions of q_i due to the rules of rigid body motion as stated in theorem 1 and therefore:

$$\begin{cases} \lim_{\Delta q_i \to 0} \overline{A'_{q_i}}(q_i + \alpha\Delta q_i) = \overline{A'_{q_i}}(q_i) = \overline{A}(q_i) \\ \lim_{\Delta q_i \to 0} \overline{A'_{q_i+\Delta q_i}}(q_i + \alpha\Delta q_i) = \lim_{\Delta q_i \to 0} \overline{A'_{q_i+\Delta q_i}}(q_i + \Delta q_i) = \lim_{\Delta q_i \to 0} \overline{A}(q_i + \Delta q_i) \end{cases} \qquad \boxed{10}$$

$$\begin{cases} \lim_{\Delta q_i \to 0} \overline{B'}_{q_i}(q_i + \alpha \Delta q_i) = \overline{B'}_{q_i}(q_i) = \overline{B}(q_i) \\ \lim_{\Delta q_i \to 0} \overline{B'}_{q_i + \Delta q_i}(q_i + \alpha \Delta q_i) = \lim_{\Delta q_i \to 0} \overline{B'}_{q_i + \Delta q_i}(q_i + \Delta q_i) = \lim_{\Delta q_i \to 0} \overline{B}(q_i + \Delta q_i) \end{cases} \quad \boxed{11}$$

Therefore:

$$\boxed{9} \, and \, \boxed{10} \Rightarrow \quad \lim_{\Delta q_i \to 0} \overline{A}(q_i + \Delta q_i) = \overline{A}(q_i) \left. \right\}$$
$$\boxed{9} \, and \, \boxed{11} \Rightarrow \quad \lim_{\Delta q_i \to 0} \overline{B}(q_i + \Delta q_i) = \overline{B}(q_i) \left. \right\} \Rightarrow \quad \overline{A}(q_i) \, and \, \overline{B}(q_i) \, are \, continuous$$

For an alternative proof for theorem 2, you can see (Escande, 2007).

Theorem 3: If P has a C^1 convex surface and Q has a C^1 strictly convex surface, then the minimum distance between P, Q is a C^1 function of q_i

Proof: According to theorem 2, the witness points A,B move on continuous paths on P,Q during the motion. These paths are C^1, because they are located on P,Q. Therefore $\overline{A}(q_i)$ and $\overline{B}(q_i)$ and therefore the minimum distance between P,Q are also C^1 functions of q_i.

REFERENCES

Choset, H., Mirtich, B., and Burdick, J. (1997), "Sensor based planning for a planar rod robot: Incremental construction of the planar Rod-HGVG" *IEEE International Conference on Robotics and Automation, vol. 4.*

Colbaugh, R., Seraji, H., and Glass, K. (1989), "Obstacle avoidance of redundant robots using configuration control", *Int. J. Robot. Res. 6, 721–744.*

Ericson, C.(2005), "Real-time collision detection", *The Morgan Kaufmann Series in Interactive 3D Technology, D. H. Eberly, Ed. Morgan Kaufmann Publishers.*

Escande, A. , Miossec, S. , and Kheddar, A. (2007), "Continuous gradient proximity distance for humanoids free-collision optimized-postures," *IEEE-RAS 7th International Conference on Humanoid Robots.*

Glass, K., Colbaugh, R., Lim, D., and Seraji, H. (1995), "Real-time collision avoidance for redundant manipulators", *IEEE Trans. Rob. Autom. 11:(10)..*

Lee, J. Y. and Choset, H. (2001), "Sensor-based construction of a retract-like structure for a planar rod robot," *IEEE Transactions on Robotics and Automation, vol. 17, no. 4, pp. 435–449.*

Rusaw, S. (2001), "Sensor-based motion planning in SE(2) and SE(3) via nonsmooth analysis," *Oxford University Computing Laboratory, Tech. Rep,.*

Shadpey, F., Tessier, C., Patel, R.V., Langlois, B., and Robins, A. (1995), "A trajectory planning and object avoidance system for kinematically redundant manipulators: An experimental evaluation" , *AAS/AIAA American Astrodynamics Conference, August 1995, Halifax, NS, Canada.*

Shadpey, F., Tessier, C., Patel, R.V., and Robins, A. (1994), "A trajectory planning and obstacle avoidance system for kinematically redundant manipulators", *CASI Conference on Astronautics, Ottawa, ON, November 1994.*

358

Singh, G., Hadaegh, F.Y. (2001), "Collision Avoidance Guidance for Formation Flying Applications", *AIAA Guidance, Navigation, and Control Conference, Montreal, Quebec, Canada*

Singh, G. (2002), "Collision-avoidance assured path-planning for Starlight interferometer", *International Symposium Formation Flying Mission and Technologies Toulouse, France*

van den Bergen, G. (2004), "Collision detection in interactive 3D environments", *ser. The Morgan Kaufmann Series in Interactive 3D Technology, D. H. Eberly, Ed. Morgan Kaufmann Publishers.*

Xiang, Y., Chung, H., Mathai, A., Rahmatalla, S. F., Kim, J. H., Marler, T., Beck, S., Yang, J., Abdel-Malek, K., Arora, J., and Obusek, J. (2007). "Optimization-based dynamic human walking prediction", *Digital Human Modeling Conference, Seattle, WA, USA.*

Chapter 38

A Study of the Dynamic Model in the Analysis of Whole Body Vibration in Manufacturing Environments

Abbas Mohammadi

mohammadi@perdana.um.edu.my
Manufacturing Department
Faculty of Engineering University Malaya
Kuala Lumpur, Malaysia

ABSTRACT

Many processes and machines generate unwanted vibration. The standards and directions concerning whole-body vibration are designed to control and reduce vibration to a lower level where most workers can perform job tasks without discomfort. As a result of increasing humans face a multiplicity of vibrations at the workplace. It can prevent these effects on human body into two divisions as following: reduction the comfort of the occupants, or Lead to serious physical injuries. The main purpose of this study is to develop a dynamic model to predict the effects of whole body vibration. Because of the difficulty of the model it is necessary to have capable and accurate analytical or numerical procedures to study the dynamic behaviors. The use of Newton's law and Lagrange's equations could be able to for formulating and modeling in human body. On the other hand, comfort criteria regarding to standards and directions for vibration are mentioned in relation to vibration. Indeed, this subject can be effective in the manufacturing processes and their relative activities for upgrading the safety and comfort.

360

Keywords: Whole body vibration, Manufacturing environment, Dynamic model, ISO 2631

INTRODUCTION

Vibration is usually classified based on mediums, forms and sources. It can be caused by a human or a machine. This study focuses on vibration caused by mechanical movement and is directed to a human body through a surface. One of the main sources of vibration is the floor on which machines are working. Forced vibration is the response of a structure to a repetitive forcing function that causes the structure to vibrate at the frequency of the excitation. In forced vibration, there is a relationship between the amplitude of the forcing function and the corresponding vibration level. The relationship is dictated by the properties of the structure. This vibration is coming from the ground to the human body. This chain of mediums can be very complex and is difficult to calculate when evaluating the amount of vibration affecting the human body. In the frequency range of 2 to 8 Hz in which people are most sensitive to vibration, the threshold level corresponds approximately to 0.5 % acceleration due to gravity. Figur1 shows the exposure limit to vibration according to acceleration and frequency. Generally, human response to vibration is taken as the yardstick to limit the amplitude and frequency of a vibrating floor. The present study is mainly aimed at design of a floor against vibration perceived by humans. To design a floor structure, only the source of vibration near or on the floor need be considered.

In a manufacturing environment the vibration is generated continuously and is generally more annoying. In such cases industrial floor systems with a natural frequency less than 8 Hz should be avoided.

Dynamic parameters of machines and floor of workplace have an intrinsic influence on vibration transmission from machines to the floor and finally on the comfort of the operators who work there. Normally, the most important factor in a dynamic analysis is the internal forces in the load bearing members.

This paper is concerned with modeling the vertical vibration of the human body in a standing position such as machine operators. It is then necessary to find out the mechanical properties of the floor and human body as vibratory mediums.

The condition of installation and the type of foundation are the main parameters that can affect the stiffness and actually damping of the floor. In addition, the study of the human body response to vibration involves some different fields like biomechanics, building and mechanical engineering. Figure 2 shows the scope of this study.

The potentially dangerous effects of human exposure to whole body vibration (WBV) have been known for a long time. The quantification of the risk and the accurate prediction of the long-term disorders, however, have been more elusive. The difficulty in risk assessment stems from: 1) a lack of accurate quantification of human response to acceleration input and 2) the lack of a means to correlate immediate human response with long-term disorders.

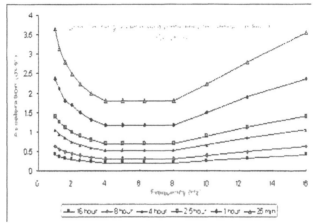

Figure1. Whole body vibration tolerance

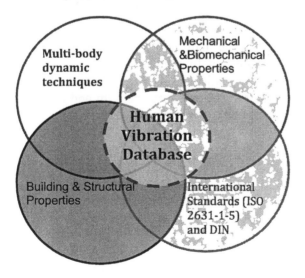

Figure 2. The fields that concern to develop human vibration model in manufacturing environments

Acceptable values of human exposure to continuous vibration are dependent on the time of day and the activity taking place in the occupied space (e.g. workshop, office, residence or a vibration-critical area). Guidance on preferred values for continuous and impulsive vibration acceleration is set out in Table 1.

Table 1. Preferred and maximum weighted rms values for continuous and impulsive vibration acceleration (m/s2)1-80 Hz [26]

Location	Assessment period	Preferred values		Maximum values	
		z-axis	x- and y-axes	z-axis	x- and y-axes
Continuous vibration					
Critical areas[2]	Day- or night-time	0.0050	0.0036	0.010	0.0072
Residences	Daytime	0.010	0.0071	0.020	0.014
	Night-time	0.007	0.005	0.014	0.010
Workshops	Day/night	0.04	0.029	0.080	0.058

CONSTRUCTION OF VIBRATION MODEL OF HUMAN BODY

As previously mentioned, the main aim of this model will be closer to the reality and can be integrated with a Floor–Machine model to evaluate the rate of vibration transmissibility into human body from floor. Figure3 shows the location of machine, man and medium and how they influenced on each other and also define the coordinate axes.

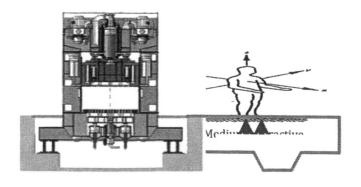

Figure 3, Problem definition

This will help to explain biodynamic response functions simulated in this study. The human body model consists of 8 rigid bodies in the proposal model in this study representing the right leg, left leg, lower torso (pelvis), center torso (abdominal parts), upper torso, head and neck, right arm, and left arm. Efficient analysis procedures have been developed for generating the governing equations of motion of such systems. The design procedure of human model and its main steps is shown in figure 4. The derivation of EOMs of the lumped-parameter models as listed from the forth step of procedure is very straightforward. From the fourth step of procedure in figure 4, the system EOMs can be expressed in matrix form as follows:

Step 1 Segmentation of human body

The standing human body system include 8 main segments as following:
1)Right leg 2)Left leg 3) lower torso 4) Middle torso 5) Upper torso 6) Head 7) Right arm 8) Left hand

Step2 taken mass and stiffness values of the corresponding segments

According to working conditions and using the experimental biodynamic properties can obtain.

In this study these properties are taken from the previous experimental works [10, 11 and 12].

Step 3 lumping the segments at discrete points and connecting the mass by springs and dampers

1. The human body segments are rigid links.
2. Human body can be modeled as a series of rigid bodies.
3. Segment has fixed mass located at a point mass at its location.
4. Joints are hinge or ball and socket.
5. Length of each segment remains constant during movement.
6. According to contact points and force direction, arms are neglected to the model.
7. The segments would only swing back and forth as well as move up and down.
8. Feet and shanks lumped together.
9. Feet would never slip on the floor and there is adequate frictional force at each point of contacts.

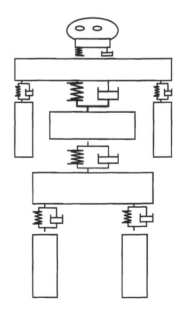

Step 4. Formulation of motion equations and analysis according to the kind of dynamic strategy:
1. Forward dynamic   2. Inverse dynamic   3. Kinematics   4.Trimming

Method: Forward Dynamic

$[M]\{\ddot{z}\}+[C]\{\dot{Z}\}+[K]\{Z\}=F_{GRF}$

Where;

[C]=Damping Matrix
[M]=Mass Matrix
[K] =Rigidity coefficients Matrix

$$m_1\ddot{z}_1+c_1\dot{z}_1-c_1\dot{z}+k_1z_1-k_1z_3=F_G$$

$$m_2\ddot{z}_2+c_2\dot{z}_2-c_2\dot{z}_3+k_2z_2-k_2z_3=F_G$$

$$m_3\ddot{z}+(c_1+c_2+c_3)\dot{z}-c_1\dot{z}_1-c_2\dot{z}_2-c_3\dot{z}_4+$$
$$(k_1+k_2+k_3)z_3-k_1z_1-k_{2,}z_2-k_3z_4=0$$

$$m_4\ddot{z}_4+(c_3+c_4)\dot{z}_4-c_3\dot{z}_3-c_4\dot{z}_5+(k_3+k_4)z_4$$
$$-k_3z_3-k_4z_5=0$$

$$m_5\ddot{z}_5+(c_4+c_5+c_6+c_7)\dot{z}_5-c_3\dot{z}_3-c_4\dot{z}_4-c_5\dot{z}_6-$$
$$c_6\dot{z}_7-c_7\dot{z}_8+(k_4+k_5+k_6+k_7)z_5-k_5z_6$$
$$-k_6z_7-k_7z_8=0$$

$$m_6\ddot{z}_6+c_5\dot{z}_6-c_5\dot{z}_5+k_5z_6-k_5z_5=0$$

$$m_7\ddot{z}_7+c_6\dot{z}_7-c_6\dot{z}_5+k_6z_7-k_6z_5=0$$

$$m_8\ddot{z}_8+c_7\dot{z}_8-c_7\dot{z}_5+k_7z_8-k_7z_5=0$$

Figure 4 Modeling procedure

Where [M], [C] and [K] are 8*8 mass, damping and stiffness matrices $\ddot{z}, \dot{z}$ and z are acceleration, velocity and displacement.

$$[M]\{\ddot{z}\}+[C]\{\dot{Z}\}+[K]\{Z\}=F_{GRF} \qquad (3)$$

According to the equation of motion, all their matrices and vector can be further represented as follows:

$$\begin{vmatrix} m_1 & 0 & 0 & 0 & 0 & 0 & 0 & 0 \\ 0 & m_2 & 0 & 0 & 0 & 0 & 0 & 0 \\ 0 & 0 & m_3 & 0 & 0 & 0 & 0 & 0 \\ 0 & 0 & 0 & m_4 & 0 & 0 & 0 & 0 \\ 0 & 0 & 0 & 0 & m_5 & 0 & 0 & 0 \\ 0 & 0 & 0 & 0 & 0 & m_6 & 0 & 0 \\ 0 & 0 & 0 & 0 & 0 & 0 & m_7 & 0 \\ 0 & 0 & 0 & 0 & 0 & 0 & 0 & m_8 \end{vmatrix} \begin{vmatrix} \ddot{z}_1 \\ \ddot{z}_2 \\ \ddot{z}_3 \\ \ddot{z}_4 \\ \ddot{z}_5 \\ \ddot{z}_6 \\ \ddot{z}_7 \\ \ddot{z}_8 \end{vmatrix} +$$

$$\begin{vmatrix} c_1 & 0 & -c_1 & 0 & 0 & 0 & 0 & 0 \\ 0 & c_2 & -c_2 & 0 & 0 & 0 & 0 & 0 \\ -c_1 & -c_2 & c_1+c_2+c_3 & -c_3 & 0 & 0 & 0 & 0 \\ 0 & 0 & -c_3 & c_3+c_4 & -c_4 & 0 & 0 & 0 \\ 0 & 0 & 0 & -c_4 & c_4+c_5+c_6+c_7 & 0 & -c_6 & c_7 \\ 0 & 0 & 0 & 0 & -c_5 & c_5 & 0 & 0 \\ 0 & 0 & 0 & 0 & -C_6 & 0 & C_6 & 0 \\ 0 & 0 & 0 & 0 & -C_7 & 0 & 0 & C_7 \end{vmatrix}$$

$$\begin{vmatrix} \dot{z}_1 \\ \dot{z}_2 \\ \dot{z}_3 \\ \dot{z}_4 \\ \dot{z}_5 \\ \dot{z}_6 \\ \dot{z}_7 \\ \dot{z}_8 \end{vmatrix} + \begin{vmatrix} k_1 & 0 & -k_2 & 0 & 0 & 0 & 0 & 0 \\ 0 & k_2 & -k_2 & 0 & 0 & 0 & 0 & 0 \\ -k_1 & -k_2 & k_1+k_2+k_3 & 0 & 0 & 0 & 0 & 0 \\ 0 & 0 & -k_3 & k_3+k_4 & -k_4 & 0 & 0 & 0 \\ 0 & 0 & 0 & -k_4 & k_4+k_5+k_6+k_7 & -k_5 & -k_6 & -k_7 \\ 0 & 0 & 0 & 0 & -k_5 & k_5 & 0 & 0 \\ 0 & 0 & 0 & 0 & 0 & 0 & k_6 & 0 \\ 0 & 0 & 0 & 0 & -k_7 & 0 & 0 & k_7 \end{vmatrix} \begin{vmatrix} z_1 \\ z_2 \\ z_3 \\ z_4 \\ z_5 \\ z_6 \\ z_7 \\ z_8 \end{vmatrix} = \begin{vmatrix} F_G \\ F_G \\ 0 \\ 0 \\ 0 \\ 0 \\ 0 \\ 0 \end{vmatrix}$$

The graphical representation with using the equation 12 can solve a dynamic system.

$$M\ddot{\xi}_r + C_r\dot{\xi} + K_r\xi = F_r \qquad (4)$$

For one DOF, the above-mentioned equation can be solved as following:

The experienced floor vibration is often assumed to be a result of the first natural frequency. The natural frequency depends on the stiffness and the mass.

$$v = \frac{\pi}{2}\sqrt{\frac{EI_z}{ml^4}}\sqrt{1+\left\{2\left(\frac{l}{w}\right)^2 - \left(\frac{l}{w}\right)^4\right\}\frac{I_y}{I_z}} \qquad (5)$$

$$FRP = \frac{X_p}{F_q} = \sum_{i=1}^{n}\frac{A_{pq}\mid_i}{\lambda_i^2 - \omega^2} \qquad (6)$$

FRP= frequency response functions

Each result of this equation can solved asunder, identically with the equation of constrained vibration ale of the system with a degree of freedom (DOF) can be developed as followed:

$$Z = Z_0 Cos\sqrt{\frac{k}{m}}*t + \frac{1}{\sqrt{\frac{k}{m}}}\left(\omega_0 \frac{\frac{F_0\omega}{m}}{\frac{k}{m}-\omega^2}\right)Sin\sqrt{\frac{k}{m}}*t + \frac{F_0}{m\left(\frac{k}{m}-\omega^2\right)}Sin\omega t \quad (7)$$

For a harmonic force such as $F(t) = F_0 \sin \omega t$ according to parts of body equation:

$$k_{r1} = k_1 \qquad\qquad k_{r6} = k_5$$
$$k_{r2} = k_2 \qquad\qquad k_{r1} = k_6$$
$$k_{r3} = k_1 + k_2 + k_3 \qquad\qquad k_{r8} = k_7$$
$$k_{r4} = k_3 + k_4$$

Equation 7 is the expresston of system movments. For a supposed case, it is used as following:
Total mass= 80 kg

According to standing position of Havanas's antropotical model the persent of body wight obtains:

$\omega = 2\pi...50$ rad/sec

Fo = 30 N

Figure 5 represent the displacement, velocity and acceleration for pelvis that obtain from expression 3.

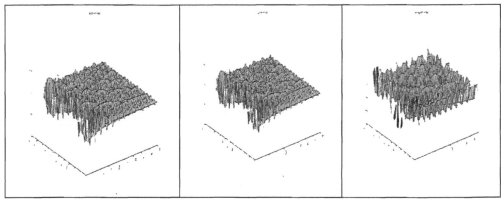

Figure 5, variety of displacement, velocity and acceleration for pelvis

CONCLUSION

There are some effective contrasts in manufacturing environment compare to the other vibratory sources such as vehicles when design a vibrating assessing for these environments should see. For example, interaction among Man-Medium-Machine, production system and management and the kind and level of maintenance can be effective in amount of exposure and actually health risk.

Indeed, it has been our general experience in industrial and academic research that there are so many activity and job when workers are exposed to different unsafe levels of vibration so their effects may be appearing in long time. And unfortunately, the number of researches that have done yet is no enough. In this study, developed a dynamic model consist of 8 segments (legs, arms, neck and head, upper torso, middle torso and lower torso) to predicting the effects of vibration on human body. As a sample, figure 3 represented the displacement, velocity and acceleration of pelvis cause of a harmonic force.

REFERENCES

Bhawani Pathak, whole body vibration, Physical Hazards Group, Canadian Centre for Occupational Health and Safety, Hamilton Ontario.

A. Burdorf and C.T.J. Hulshof, Modelling the effects of exposure to whole-body vibration on low-back pain and its long-term consequences for sickness absence and associated work disability, Journal of Sound and Vibration 298 (2006) 480-491.

Griffin, M.J., 2004. Minimum health and safety requirements for workers exposed to hand-transmitted vibration and whole-body vibration in the European Union: a review. Occupational and Environmental Medicine 61 (5), 387–397.

Raimondas Grubliauskas, Gintas Stankus, Vaidotas Vaišis, and Vytautas Nainys, Estimation of vibration in KLAIPĖDA POWDERY MANURE discharge terminal, Journal of environmental engineering and landscape management, 2006 Vol. XIV, No.2, 95_100.

Ernest P. Hanavan,"A Mathematical Model of the Human Body", Aerospace Medical Research Lab., technical report, No. AMRL-64-102, 1964.

HSE, 2003. Improving health and safety in construction: Phase 2—depth and breadth, Vol. 4. Hand arm vibration syndrome—underlying causes and risk control in the construction industry. HSE (Health and Safety Executive) Research Report 114, HSE Books, ISBN 0-7176-2219-3.

International Organization for Standardization, "ISO 2631, Mechanical Vibration and shock-Evaluation of human exposure to whole body vibration, 1987.

Fang Li, "Constrained Multi-Body Dynamics Method to study Musculoskeletal Disorder Duo to Human Vibration", University of Cincinnati, PhD dissertation, 2007.

Gergana Nikolova, Liliya Stefanova and Yuli Toshev,"3D Model of the human body genterraed within Pro-Engineer environment", problems of Engineering Cybernetics and Robotics, Bulgarian Academy of Science, 2oo5, Sofia.

Chapter 39

A Data-Based Approach for Predicting Variation Range of Hand and Foot Maximum Force on a Control: Application to Hand Brake

Xuguang Wang[1], Caroline Barelle[1], Romain Pannetier[1,2],
Julien Numa[1], Thomas Chapuis[2]

[1]INRETS – LBMC
Bron, France

[2]Renault, Ergonomie
Guyancourt, France

ABSTRACT

The measurement of maximum foot and hand maximum force is important not only for specifying force limit of industrial workers but also for evaluating a hand or foot control which requires high demand of force. Even for the controls requiring low force demand, maximum static strength can be used as an objective indicator for defining discomfort evaluation criteria. Although a large amount of data of muscle strength has been collected in the past, few can be directly used for today's automotive control design. Thus, according to today's car driving conditions, a specific experiment was therefore set up to collect maximum force that a driver can exert on three frequently used automotive controls: clutch pedal, gear lever and hand brake. Another reason that motivated this experiment was the need to collect corresponding body postures so as to predict both maximum static strength and corresponding posture using a digital human model. This paper aims at presenting the data based approach developed for this purpose including data collecting protocol and data processing as well as methods for exploiting the collected data. The approach is illustrated using the hand brake as example.

Keywords: Static strength, Muscle isometric force, Hand and foot, Automotive control, Hand Brake, Discomfort, Digital human models

INTRODUCTION

Digital human models (DHMs) are more and more used in early phrase of product design for ensuring a better consideration of human factors (Chaffin, 2005). However, most of existing DHMs are based on geometric and kinematic human models, and are not able to predict dynamic effects of a task. Predicting maximum foot and hand maximum force is a highly desired functionality not only for specifying force limit of industrial workers but also for evaluating a hand or foot control which requires high demand of force. Even for the controls requiring low force demand, maximum static strength can be used as an objective indicator for defining discomfort evaluation criteria.

A large amount of data of muscle strength has been collected in the past (see for instance, Kumar, 2004). It has been shown that many factors affect the maximum voluntary strength. Apart from personal characteristics such as gender and physical conditions, body posture, force direction, environment (e.g., support, type, handle type) and motivation are among the most influent factors (Kroemer, 1970; Chaffin, 1975 ; Daams, 1994 ; Mital et al., 1998 ; Kumar, 2004). In addition, the assessment methods, such as, test duration, instruction, rest period between trials, can also strongly affect muscle strength, making difficult to compare data from different investigators. This is probably the main reason why few data can be directly used for today's automotive control design. Thus, according to current car driving conditions, a specific experiment was therefore set up at INRETS (French National Institute for Transport and Safety Research) to collect maximum force that a driver can exert on three frequently used automotive controls: clutch pedal, gear lever and handbrake in order to provide more reliable functional data for automotive hand and foot control design.

This paper aims at presenting the data based approach developed for this purpose including data collecting protocol and data processing as well as methods for exploiting the collected data. The approach is illustrated using the hand brake as example.

371

METHODS

PRINCIPLE

The proposed approach is similar to that we have already used for in-vehicle reach motion simulation (Wang et al, 2006) and reach discomfort evaluation (Wang et al, 2007 and 2008). As a data-based approach, the first step is to collect task related data. For this, a sample of subjects representative of drivers has to be defined. A large range of control configurations should be tested. Once the data are collected, the next step is to structure the data according to subject characteristics (e.g, gender, stature, weight, etc), task description (e.g. control position, force direction if relevant), environment (e.g, seat height). If different postural/motion control strategies are observed, data should be classified accordingly. The third step is to define appropriate interpolation methods so as to predict new posture/motion as well as force strength for a new configuration. One of interesting simulation modes proposed in the ergonomic simulation tool RPx, jointly developed by INRETS and Renault, is so-called 'virtual experiment' (Wang et la, 2006; Monnier et al, 2008). In this mode, all subjects who participated in the physical experiment 'come back' to assess the new configuration virtually. In the present work, virtual experiment is used to estimate the range of variation of maximum force of the subject sample.

DATA COLLECTING

Thirty-one voluntary male and female subjects participated in the experiment and were paid for it. They were aged from 20 to 44 and had neither musculoskeletal abnormalities nor any history of trauma. All of them had driving experience of more than one year. They were divided into three groups according to stature and gender of French driver population : 10 short females (<1625 mm), 10 average height males (1705 - 1810 mm) and 11 tall males (>1810 mm). The experimental protocol was approved by INRETS ethical committee. The subjects gave their informed consent before taking part in the study. Each subject was aware of the experimental procedure.

Six, five and eleven locations were defined respectively for clutch pedal, gear level and hand brake, covering a wide range of vehicles. They were defined in a car-fixed coordinate system centered at the nominal seat H-point (Figure 1) located with help of SAE J826 H-point machine. For instance, Figure 1 shows the 5 hand brake positions tested in the vertical plane at y=300 mm. For the gear level, 6 force directions (up and down, left and right, forward and backward) were studied separately. Prior to experimentation, each subject was asked to adjust the seat position longitudinally in a standard driving configuration with respect to the end position of a standard clutch pedal and to the steering wheel.

In addition to the measurement of maximum static force, four intermediate forces (very low, low, moderate, high) were also recorded to define force perception law for some control locations. For the hand brake, two handle types were studied as well as force exertions with and without movement elan. In total, each subject performed 125 trials. To reduce the possible effects of fatigue, trial order was randomized whenever possible. The total duration of the experiment did not exceed 4.5 hours including subject preparation (welcoming subject, cloth change, anthropometric measurement, attachment of markers, etc).

Based on the recommendation for measuring muscle strength by Gallagher et al (2004), the subjects were instructed to exert their maximum force (or an intermediate force level) as fast as possible once a red light turning on and to maintain the maximum force until to the light turning off. The force exertion duration was 5 seconds for each trial. Non verbal encouragement was given during the force exertion. At least two trials were repeated for maximum force measurement. If more than 10% difference between the two repetitions, third measurement was performed. The trial with maximum force level was retained. At least two minutes of rest were imposed between two maximum force trials. For intermediate force levels, at least 40 seconds were proposed. Subjects were asked to keep normal driving posture during force exerting. When the right hand exerting force at the gear level and hand brake, the left hand was on the steering wheel and the two feet on the floor. For the left foot pedal operation, two hands were on the steering wheel and the right foot on the floor. The mean value between 1.5 to 4.5 seconds was calculated and considered as static force.

Three force sensors (two with 3-axes and one with 6-axes) were used for measurement foot and hand force applied on clutch pedal, gear level and hand brake. In order to know all external contact force for estimating joint loads by inverse dynamics, one 6-axes force sensor was put on the steering wheel column for measuring the force transmitted by the hand(s) and a 6-axes force plate on the floor. 44 passive reflective markers were attached to the body and measured using the motion capture system Vicon MX-40 with 10 digital cameras. All measurements were synchronized.

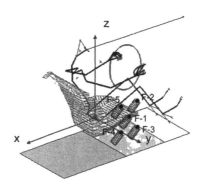

FIGURE 1. Illustration of the 5 locations of handbrake control defined in the plane y=300 mm (plane 1). The same positions in x and z were also tested in the plane

y=400 mm (plane 3). The position 1 was also tested in the plane y=350 (plane 2), forming therefore the middle point of the experimental design.

Using the same procedure as for our in-vehicle reach study (Wang et al, 2008), whole body postures were reconstructed by minimizing model-based and captured markers trajectories. With an inverse dynamic approach, joint forces and moments were also estimated. Figure 2 illustrated the experimental set-up and corresponding reconstructed posture and joint loads.

FIGURE 2. Experimental set-up and reconstructed posture with external contact forces showed as well as joint loads (force and moment) for a handbrake trial.

INTERPOLATION METHOD

For a new control position that is not tested during the experiment, one has to predict both force and corresponding posture from existing data. The interpolation method is based on the principle of case-based simulation and implemented in the motion simulation tool RPx (see Wang et al, 2006; Monnier et al, 2008). Regarding posture prediction, a reference posture that is the most similar has to be selected and then to be adapted to new geometric constraints imposed by a simulation scenario.

Concerning the force prediction, we propose to use inverse distance weighted (IDW) interpolation method for estimating the force F_I at a new control position (x_I, y_I, z_I) :

$$F_I = \frac{\sum_i F_i / D_i^2}{\sum_i 1 / D_i^2}, \ D_i^2 = (x_I - x_i)^2 + (y_I - y_i)^2 + (z_I - z_i)^2 \quad (1)$$

where F_i and (x_i, y_i, z_i) are the measured force at the position i. Here squared distance is used. Inverse distance weighted methods are based on the assumption that the interpolating surface should be influenced most by the nearby points and less by the more distant points. The interpolating surface is a weighted average of the scatter points and the weight assigned to each scatter point diminishes as the distance from the interpolation point to the scatter point increases. Interestingly, if the point to be interpolated exists, the IDW method gives the same value as existing one due to the fact that the weight associates to this point is infinitely high.

PRELIMINARY DATA ANALYSIS

A preliminary analysis was carried out to examine the effects of subject group, control location on maximum force. For the hand brake, the collected data form a full factorial design with 3 subject groups and 11 configurations (locations). A two-way ANOVA (Table 1) shows that hand maximum effort on a handbrake strongly depended on subject group and handbrake configuration. In average, maximum effort for the short female was almost less than half of that of two male groups, confirming large discrepancy in muscle capacity between males and females. The highest maximum effort was observed at the hand brake position 3 (Figure 1), which was the furthest point from the shoulder. The lowest effort was at the position 5, closest to the shoulder.

Table 1. ANOVA of the effects of subject group and hand brake configuration on hand maximum effort

Source	Sum of sq	Dof	Sq. mean	Ration F	Proba.
Group (G)	3.46039E6	2	1.7302E6	210.73	0.0000
Configuration (C)	1.19833E6	10	119833.	14.59	0.0000
G*C	129998.	20	6499.9	0.79	0.7235
Residu	2.43853E6	297	8210.54		
Total	7.29708E6	329			

APPLICATION EXAMPLE

A short matlab program was written to exploit the collected data for predicting maximum effort using the IWD interpolation method. Using the concept of virtual experiment, each subject virtually comes back to test a new configuration. His/her own data are used to interpolate for this new configuration. Then the distribution of the test sample can be obtained. An example for the handbrake is shown in Figure 4.

```
Choose the control (Pedal, GearLever, HandBrake): 'HandBrake'
*****************************************************************
*      Distribution of maximum effort       *
*****************************************************************
Enter the end position of handbrake in X (-220 to -40 mm): -100
Enter the end position of handbrake in Y (300 to 400 mm) : 350
Enter the end position of handbrake in Z (-50 to 200 mm) : 0
Average according to subject groupe (N):
```

```
Short female: 184.9
Average height male: 370.4
Tall male: 445.4

Standard deviation according to subject group (N):
Short female: 55.4
Average height male: 110.9
Tall male: 72.3
```

FIGURE 4. An example of predicting the distribution of maximum effort for the handbrake located at (-100, 350, 0).

CONCLUDING REMARKS

A pragmatic data-based approach has been proposed for predicting both body posture and hand/foot maximum force. Using the concept of virtual experiment, the variability of the sample of participants of the real physical experiment can be simulated. In this paper, the proposed approach is illustrated for predicting the variation range of hand maximum force applied on the hand brake.

As a data-based method, its main limitation is that it depends strongly on the range of collected data. It can hardly be applied outside of the range covered by experimental conditions. Though the proposed approach is useful for hand/foot control design, it is not helpful for understanding hand/foot force production and its interaction with body posture. Further investigation with help of a biomechanical model is needed. Such a research is still on going at INRETS especially for understanding force direction control and force/posture interaction, with the long-term objective being to define discomfort criterion for hand/foot control design.

ACKNOWLEDGEMENT

We would like to acknowledge the technical assistance of Richard Roussillon as well as Jules Trasbot from Renault who initiated this study.

REFERENCES

Chaffin, D.B. (2005), "Improving digital human modeling for proactive ergonomics in design." *Ergonomics*, 48(5), 478-491

Daams, B.J. (1994), *Human Force exertion in user product interaction. Backgrounds for design.* Physical Ergonomics Series. Delftse Universitaire Pers, Delft.

Gallagher, S., Moore J.S., Stobbe, T.J., (2004), "Isometric, isoinertial and psychophysical strength testing: devices and protocols." In *Muscle Strength*, edited by Kumar S., 2004, CRC Press

Kroemer, K.. (1970), "Human strength : terminology, measurement, and interpretation of data." *Human factors*, 12, 3, 297-313.

Kumar, S. (2004), *Muscle strength*. CRC Press. 558p.

Mital, A., Kumar, S. (1998), "Human muscle strength definitions, measurement and usage: Part I-Guidelines for the practioner." *International Journal of Industrial Ergonomics*, 22, 101-121.

Mital, A., Kumar, S. (1998), "Human muscle strength definitions, measurement and usage : Part II – The scientific basis (knowledge base) for the guide." *International Journal of Industrial Ergonomics*, 22, 123-144.

Monnier, G., Wang, X., Trasbot, J. (2008), *RPx : A motion simulation tool for car interior design*. Publisher: Taylor & Francis Group. Editor: Vincent G. Duffy

Wang, X., Chevalot, N., Trasbot, J. (2008), "Prediction of in-vehicle reach surfaces and discomfort by digital human models". SAE International conference and exposition of Digital Human Modeling for Design and Engineering, June 17-19, 2008, Sheraton Station Square, Pittsburgh, Pennsylvania, USA. SAE paper N° 2008-01-1869

Wang, X., Chateauroux, E., Chevalot, N., (2007), "A data-based modeling approach of reach capacity and discomfort for digital human models." HCI2007, First International conference on Digital Human Modeling. Beijing, 22-27 july, 2007, *Lecture Notes in Computer Science*, Vol. 4561, Duffy, Vincent D. (Ed.), 2007, XXIII,.p. 215-223

Wang, X., Chevalot, N., Monnier, G., Trasbot, J., (2006), "From motion capture to motion simulation: an in-vehicle reach motion database for car design." *SAE 2006 Transactions Journal of Passenger Car – Electronic and Electronic Systems*, SAE Paper 2006-01-2362. pp.1124-1130.

CHAPTER 40

Human Simulation Under Anosognosia and Neglect in Stroke Patients

Neus Ticó Falguera [1], Esteban Peña-Pitarch[2]

[1]Xarxa Assistencial de Manresa, Althaia

[2]Escola Politècnica Superior d'Enginyeria de Manresa (UPC)
Av.Bases de Manresa 61-73
(08240) Manresa
esteban.pena@upc.edu

ABSTRACT

A number of patients developing anosognosia and neglect after stroke can benefit from specific physician supervised rehabilitation programs. Assessment of these patients provides data on initial, intermediate and final affectation of such deficits in the upper limbs. We have developed a virtual environment of parametric human arm and hand with 29 degrees of freedom (DOF) to simulate these three phases.

Simulation of functional recovery of stroke patients affected by anosognosia and neglect is crucial for physicians to aid their assessment. The objective of the present paper is to describe a newly devised environment to simulate the initial, intermediate and final affectation of upper limb deficit in stroke patients with anosognosia and neglect. After a time lapse from the initial assessment and on rehabilitation therapy, we perform a new assessment. With the parameters from the two assessments we are then able to simulate the functional recovery of patients within the same time period.

Keywords: stroke, neglect, anosognosia, rehabilitation, virtual physiology simulation.

INTRODUCTION

Stroke or cerebrovascular accident is a brain infarction caused by ischemia or hemorrhage. It is the second cause of death and the first cause of severe disability among adults in developed countries, cause great burden on and resource consumption of healthcare services. It has been estimated that a third of those suffering and surviving stroke will die within the ensuing year of the event, another third will remain dependent on others for care, and another third will fare to good recovery (Garrison, 1993). Treatment improving functional outcomes can significantly reduce severity of sequelae as well as the financial burden of this illness on the individual, the family and society overall. Rehabilitation therapy aimed to reduce post-stroke disability and handicap is recognized as a corner stone of multidisciplinary stroke care (Kalra, 2007).

The mechanisms underlying recovery are may, even if mostly not yet understood, include (i.) resolution of oedema with volume reduction of non-functioning brain area, (ii.) reduction of impairment, and (iii.) acquisition of adaptive techniques to overcome impairment. Emergency medical attention and early active rehabilitation therapy are paramount toward good functional recovery (Catalonia official Clinical Practice Guidelines for stroke GPC, 2007).

Although about half of overall recovery from disability occurs over the course of the first month, it can continue up to 6 months post stroke. After this time frame the occasional patient will show significant improvement, this being the exception and rarely expected. This is well established for activities of daily living (ADL) and applies to all other features, such as aphasia, arm function, neglect and walking (Garrison, 1993).

Many studies have found single impairments to be associated with more severe disability at 6 months, including hemianopia, severe sensory loss, complete motor loss, loss of trunk balance, loss of consciousness, urinary incontinence, neglect and confusion (Wade, 1993). Hemiplegia is a common effect of stroke, which may have significant impact on upper limb function. Neglect is a common behavioural disorder in stroke patients (Plummer, 1993). Stone et al. (1993) defined anosognosia as denial or lack of awareness of a hemiparesis. Patients with neglect, compared with those without neglect, have more motor impairment, sensory dysfunction, visual extinction, basic (non-lateralized) attention deficit, and anosognosia (Tham, 2000).

Unilateral neglect (UN) is characterized by an inability to report on or respond to other individuals or objects presented at the contralateral side in space to the brain lesion. If such failure to respond can be accounted for neither by sensory nor by motor deficit, it is not considered to be neglect (Vahlberg, 2008). UN can be described in terms of the modality in which the behaviour is elicited (sensory, motor, or representational) or by the distribution of the abnormal behaviour (personal or spatial) (Tham, 2000 and Plummer, 2003).

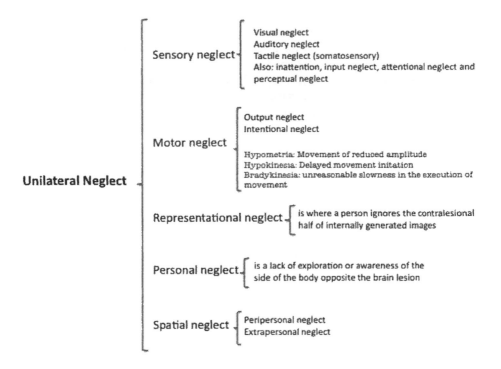

FIGURE 1.1 Classification of unilateral neglect by Plummer et al.

Patients may have one type of neglect or a combination of neglect behaviours. Because UN has a wide variety of clinical presentations, there is not a single test that can reach a comprehensive diagnosis of neglect-like behaviour. Some authors have recommended using a test battery which includes measures of assessment for all types of neglect.

Anosognosia is the lack of awareness of a specific deficit in sensory, perceptual, motor, affective, or cognitive functioning due to brain injury (Babinski, 1914). Previous studies have shown broad variation in the incidence of anosognosia for hemiparesis ranging from 17% to 28% (Stone, 1993 and Appelros, 2006). Cutting found anosognosia in 58% of patients with right hemisphere lesions compared with 14% of patients with left hemisphere lesions. Bisiach and colleagues described a frequency of 33% for moderate or severe anosognosia in patients with right hemisphere lesions. Anosognosia is most commonly reported in association with left hemiplegia and left hemispatial neglect. Anosognosia is also more frequent after right than left hemisphere brain lesions, and may be compounded by disability deriving from the neglect status itself.

MATERIALS AND METHODS

We included four patients who had developed acute stroke within the previous

20 days, with left or right sided hemiparesis or hemiplegia with initial anosognosia and neglect. Inclusion criteria for these patients were: (i.) over eighteen years of age, (ii.) collaborative patients, good cognitive status, and (iii.) functional motor deficit associated with anosognosia and neglect. Patients who were not alert, uncooperative, or had severe aphasia were excluded only if their communication abilities precluded completing even a simple interview. The patients gave informed consent for their participation in the study, which was carried out in accordance with the ethical standards laid down in the 1964 Declaration of Helsinki. The demographic data of patients included were collected.

We used the Oxfordshire Community Stroke Project (OCSP) for clinical classification of the stroke:

- Total anterior Circulation Infarction (TACI)
- Partial Anterior Circulation Infarction(PACI)
- Lacunar Infarction (LACI)

Stroke lesions were documented by magnetic resonance imaging or computed tomography and the severity of stroke was assessed using the National Institutes of Health Stroke Scale (NIHSS).

The Ashworth's scale was used to assess the degree of spasticity, which was scored as 0 for no spasticity to 5 for severe spasticity with joint stiffness.

The motor control for the upper limbs was assessed by the Brünnstrom scale, which has six levels of motor control: 1 for no movement to 6 for normal movement.

The Fugl-Meyer scale was used as a system for assessing motor function, balance, some sensory details, and joint dysfunction in hemiplegic patients. Five levels of motor impairment were identified according to Fugl-Meyer. Severe scores were <50; marked scores were 50 to 84; moderate scores were 85 to 95; slight scores were 96 to 99; and normal motor function score reached 100.

The presences of neglect was assessed with the Bells test (Gauthier 1989).

Anosognosia for hemiplegia was assessed using the anosognosia scale suggested by Bisiach:

- Grade 0 (no anosognosia): the disorder is spontaneously reported or mentioned by the patient following a general question about their complaints;
- Grade 1: the disorder is reported only following a specific question about the strength of the patient's limbs;
- Grade 2: the disorder is acknowledged only after demonstrations through routine techniques of neurological examination;
- Grade 3: no acknowledgement of the disorder can be obtained.

However, the degree of independence and assistance needed during activities of the patients' daily living was measured with ADL scales: Barthel's Index (BI) (Mahoney 1965) and the Functional Independence Measure (FIM). The FIM is an ordinal scale composed of 18 items with 7 levels ranging from 1 (total dependence) to 7 (total independence) and can be broken down into a 13-item motor subscale and a 5-item cognitive subscale. The scoring ranges for are 13 to 91 for the motor and 5 to 35 for cognitive subscale. Total score is 126.

Disability was evaluated with the modified Rankin scale. This is a 6 grade scale, ranging form 0 (independence) to 5 (severe disability).

We also measured the deficits of angles, lengths and range of motion for arm and hand affected.

All of these tests were administered 20 days prior to, at admission and also at two months of stroke. After two months into the rehabilitation program under physician supervision, we took a second measure of the above deficits. These two measures were implemented in a virtual environment with 29 DOF.

RESULTS

Four patients were studied. One man with left hemiplegia, two women with left hemiplegia and another woman with right hemiplegia. Mean age was 67.75 years. Clinical classification by the OCSP is described in Table 1.1. Demographic data and clinical variables of these patients are given in Tables 1.1 and 1.2. Table 1.2 presents profiles of the neglect test and anosognosia at the first assessment and two months later. Table 1.3 presents the profiles of the functional capacities with BI and FIM at the first and second assessments. The table shows that cognitive and motor FIM scores of the second evaluation were higher than those of the first evaluation. BI also showed an improvement in the second assessment. Table 1.3 also describes patient disability evaluated with the Rankin scale. At admission, disability was severe in all of the patients. After rehabilitation, the disability was still severe.

Table 1.1 Demographic data and clinical variables of patients

Patient	Sex	Age	Aetiology	Side of lesion	OCSP	MRI
1	Man	78	Ischemic	Left	PACI	TBG*
2	Woman	38	Ischemic	Left	PACI	TPC*
3	Woman	82	Ischemic	Right	PACI	Protuberance
4	Woman	73	Ischemic	Left	PACI	FPC

TBG= Temporary basal ganglia
TPC= Temporal parietal cortex
FPC= front-parietal cortex

Table 1.2 Clinical variables at the first and second assessment for each patient

Patient	NIHSS	Ashworth	Fugl-Meyer	Brunnstrom	Anosognosia	Neglect
1	17/13	1/2	0/8	1/2	3/2	yes/yes
2	13/2	1/1	8/65	1/6	3/0	yes/yes
3	11/7	1/2	0/21	1/4	1/0	yes/yes
4	21/13	1/1	0/8	1/2	3/2	yes/yes

Table 1.3 Profiles of the functional capacities with BI and FIM at the first and second assessment and disability evaluated with the Rankin scale.

Patient	IB	FIM	Rankin
1	10/24	30/39	5/4
2	31/94	57/122	5/2
3	11/45	58/75	5/4
4	0/24	30/39	5/4

FIGURE 1.2 Left hand of Patient #1 at first and second evaluations.

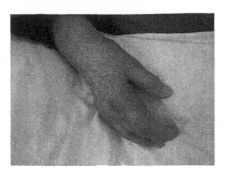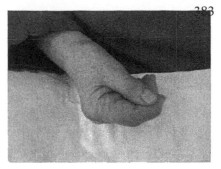

FIGURE 1.3 Hands of Patient #3. Left at first evaluation and right at second evaluation two months into rehabilitation.

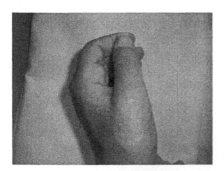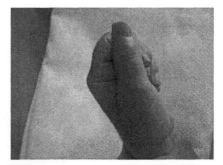

FIGURE 1.4 Hands of Patient #4. Left at first evaluation and right at second evaluation two months into rehabilitation.

It is note that Patient #2 in not shown here because the limb was a normally functional hand.

HUMAN SIMULATION UNDER ANOSOGNOSIA AND NEGLECT

Once the patients were assessed with the scales mentioned above, simulation was implemented in a virtual environment. With two virtual human models, one woman and one man, we implemented the range of motion from the functional capabilities for the positions between the first evaluation and second evaluation.

To simulate the range of motion of these positions, we used a virtual hand and arm with 29 DOF, 25 DOF were used to simulate the hand, 2 DOF for the wrist and 2 DOF for the arm. Shoulder movement was not considered in this simulation.

The virtual hand an arm to simulate the patients' were implemented using the model presented by Peña-Pitarch and colleagues in 2005.

Once the patients were evaluated first 20 days before admission, all the angles of joints for the hand and wrist were introduced in a database, and two months after stroke was read an introduced in the database with the patient data. Both angles depicted in Table 1.4, are the maximum angles reached by the patients. Movements of adduction/abduction (Ad/Ab) for the metacarpophalangeal joint (MCP), Proximal Interphalangeal joint (PIP) and Distal Interphalangeal joint (DIP) in flexion/extension (F/E) are also shown in Table 1.4.

Table 1.4 Maximum angles at first and second evaluation two months later.

Index angles	Patient 1	Patient 2	Patient 3	Patient 4
MCP (Ab/Ad)	13/13	13/42	13/30	13/30
MCP (E/F)	30/30	30/80	30/60	60/80
PIP (E/F)	30/30	30/100	30/70	60/100
DIP (E/F)	10/10	10/90	10/45	10/70

Other fingers depict similar evolutions and are not showed herewith in order not to crowd the manuscript with a similar tables.

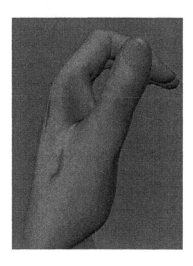

FIGURE 1.4 Simulation of an intermediate position for patients #2, #3 and #4.

DISCUSSION

This is the first study of our group testing how this approach to simulate the movements proposed may work in the clinical setting.

Tables 1.1 to 1.3 show the assessment degrees per each patient. Patient #1 at both assessments show severe stroke; the physicians did not appreciate any variation, simulation of this patient is done by setting the hand and arm in a neutral position where the angles for each joint are 0 or 30 degrees for MCP, PIP and 10 degrees for DIP (Peña-Pitarch et al. 2005).

Patient #2 was a young patient, the stroke was slight and within two months recovery for this patient was complete; simulation was similar to a normally functioning hand.

Patients #3 and #4 had similar evolutions within two months and simulation was based on reading the minimum and maximum angles reached in the first assessment and the minimum and maximum angles two months later. Bearing these parameters we devised a value interval that can be implemented in our virtual environment. Table 1.4 shows the maximum angles at the two assessments. The first assessment for Patient #3 showed, as seen on the left on Figure 1.3, the hand in a neutral position and on the right two months later. The new angles are shown in Table 1.4. Patient #4 showed in the left at the first assessment. The new angles are showed in Table 1.4.

To simulate any one of these patients in a virtual environment, we introduced the initial and final angles for the first assessment and the initial and final angles for the second assessment. These four parameters allowed us to simulate the evolution of recovery for patients and apply the same to other patients with similar assessments. This is the first time that an anticipation of what can occur one month into therapy.

CONCLUSIONS

Virtual human simulation of the arm and hand in patients affected by anosognosia and neglect after stroke supplies an new and objective tool for physicians allowing a simulated evolution of deficits in some patients. Anosognosia and neglect deficit associated with hemiplegia impair functional recovery. Relevance of this simulation for patients affected by these deficits is, upon the first physician interview, to inform on planned evolution as to the time functional recovery will take. The simulation is also relevant because the model arm and hand are implemented with parametric lengths allowing to extrapolate to other patients affected with the same deficits.

ACKNOWLEDGEMENTS

This work was partially supported by the project DPI2007-63665 and project Fundación Mapfre. The authors thank Mr. Joseph A. Graells for language editing.

REFERENCES

Appelros, P., Karlssonb, G.M. and Hennerdalc, S. (2006), "Anosognosia versus unilateral neglect. Coexistence and their relations to age, stroke severity,

386

lesion site and cognition." *European Journal of Neurology*, 14, 54-59.

Ashworth, Bohonnon, Smith (1987), "Interrater reliability of a modified Ashworth scale of muscle spasticity." *Phys. Ther.*, 67, 206-207.

Babinski M.J. (1914). "Contribution a l' etude des troubles mentaux dans l'hemiplegie organique cerebrale (anosognosie)." *Rev. Neurol.*, 27, 845-848.

Bisiah E., Allar G., Perani D., et al. (1986), "Unawareness of disease following lesions of the right hemisphere : anosognosia for hemiplegia and anosognosia for hemianopia." *Neuropsychologia*, 24, 471-82.

Brunnstrom S. (1966). "Motor testing procedus in hemiplegia based on recovery stages." *J, Am.Phys. ther Assoc.*, 46, 357.

Catalonia official Clinical Practice Guidelines for stroke GPC (AATRM) (2004). *Guía practica clínica (GPC) sobre ictus.*(actualizada 2007).

Cutting J. (1978) "Stydy of anosognosia." *J. Neurol Neurosurg Psychiatry*, 41,548-555.

Garrison S.J., Rolak L.A. (1993), "Rehabilitation of the stroke patient." *In Delisa JA, Gans BM, Currie DM, Gerber LH, Leonard JA, MCPhee MC, Pease WS. Rehabilitation medicine. Principles and practice. Philadelphia: Lippincott*, 801-824.

Gauthier L., Dehaut F., Joanette Y. (1989), "The bells test." *Int. J. Clin. Neuropsychol* , 11, 49-54.

Fugl-Meyer A.R., Jääsköl L., Leyman I., Olsson S., and Steglind S. (1975), "The post-stroke hemiplegic patient. I. A. Method for evaluation of physical performance." *Scand J. Rehab. Med.*, 7, 13-31.

Kalra L, Langhorne P. (2007), "Facilitating recovery: Evidence for organized stroke care." *J. Rehabil. Med.*, 39, 97-102.

Mahoney F.I., Barthel D., (1965). "Functional evaluation: the Barthel Index." *Maryland State Medical Journal,* 56-61.

Orfei, M.D., Robinson, R.G., Prigatano, G.P., Starkstein, S., Rüsch, N., Bria, P., Caltagirone, C. and Spalletta, C. (2007), "Anosognosia for hemiplegia after stroke is a multifaceted phenomenon: a systematic review of the literature." *Brain*, 130, 3075- 3090.

Peña-Pitarch, E., Yang, J. and Abdel-Malek, K. (2005), "SANTOS™ Hand: A 25 Degree-Of-Freedom Model." *SAE International, Iowa City, IA, June 14-16, 2005-01-2727 DHM.*

Plummer P., Morris M.E., and Dunai J. (2003), "Assessment of unilateral neglect." *Phys Ther.*, 83, 732–740.

Stone S.P., Halligan P.W., and Greenwood R.J. (1993), "The incidence of neglect o phenomena and related disorders in patients with an acute right or left hemisphere stroke." *Age Aging*, 22, 46-52.

Tham K., Borell L., and Gustawsson A. (2000). "The discovery of disability: A phenomenological study of unilateral neglect." *Am J Occup Ther.*,54, 398-406.

Vahlberg B., and Hellström K. (2008). "Treatment and assessment of neglect after stroke-from a physiotherapy perspective: A systematic review." *Advances in Physiotherapy*, 10, 178-187.

Wade D. (1993). "Stroke." *In Greenwood R, Barnes M, Macmillan T, Ward C. Neurological rehabilitation. New York : Churchill Livingstone.* 451-458.

CHAPTER 41

Flashpoint Organizational Profiling: Structured, Theory-Driven Morphological Analysis of Insurgent Networks

William N. Reynolds[1], Marta Weber[1],
James Holden-Rhodes[1], Leo Felix[2]

[1]Least Squares Software, Inc.
12231 Academy Road, NE #301-192
Albuquerque, NM 87111
bill@leastsquares.com

[2]Arctan Group, LLC
2200 Wilson Blvd Suite 102-150
Arlington, VA 22201

ABSTRACT

We argue that *social science theories of organization structure and function* can be operationalized through the methodology of *Morphological Analysis (MA)*. MA provides a structured and efficient approach for describing complex systems that exhaustively explores hypothesis space for a given set of *units of analysis* (UOA). We will develop means to map concepts from social science theories such as *Contingency Theory* and *Open Systems Theory* to morphological UOAs and demonstrate how operationally useful analytic descriptions of human networks conducting asymmetric operations can be constructed. We outline a set of experiments to test the hypothesis that analysis supported by the framework is superior to unsupported analysis. We specify a software tool implementing the MA-operationalized social science theory.

Keywords: Social science theories of organization structure and function, Morphological Analysis, units of analysis, Contingency Theory, Open Systems Theory

INTRODUCTION

Uncertainty regarding an adversary's capabilities and intentions is a problem as old as conflict itself (Kelley, 2009). Asymmetric conflict involves not only asymmetry in capabilities, but also asymmetries in information. Insurgents have good information about the dispositions of regular forces, who themselves are hobbled by limited information about the human terrain.

In typical applications, the solution is to find more data points. The problem with statistical and pattern-based approaches is that it takes a large number of observations before confidence can be established (in statistical terms, $p<.05$). In the real world, this is often unrealistic -- how many IED attacks does it take before we can begin using regression models? It is a truism that, given infinite time and resources, it is possible to build expansive opposition profiles – practical methodologies must make do without such luxuries.

The alternative to data-driven modeling is to build causal models of adversary networks and use these models to develop response strategies. Since effective asymmetric adversaries are adaptive and will proactively seek out our vulnerabilities, empirical data-based approaches are ineffective and predictive modeling is the only effective counter. The problem with predictive modeling is that in the absence of sufficient data, there is a tremendous uncertainty in models. We acknowledge this by saying there are a large number of possible models that could be constructed that could be constructed consistent with available evidence. These large "possibility spaces" are the defining characteristics of complex systems and a principal confounding factor in intelligence analysis. An effective predictive modeling effort needs to be aware of this problem, and conduct breadth-first surveys of the possibility space in order to effectively counter new threats and mitigate surprise.

For example, a regression-based approach would not consider the possibility of WMD smuggling across the US' Southwest border (SWB), since there are no known cases of WMDs ever crossing the border. Nevertheless, breadth-first reasoning reveals the SWB is one of the best areas to bring contraband into the US. We can deduce this by considering the particular threat - a WMD attack in the US. We examine the requirements to execute that threat - a logistical need to bring materials into the US. After reviewing what the most effective routes to bring in those materials would be, size and porosity suggest the SWB as a prime candidate.

What we have done is break the problem down into *units of analysis* (UOA), in this case, *threat type, threat requirements* and *adversary means*. We then construct a model using these units. The WMD attack via the SWB is only one example. Our methodology is generalizable across all types of threats. A more thorough use of the approach first considers *all* contingencies and then falsifies those that are inconsistent using available evidence, gradually narrowing the possibility space and suggesting the most likely courses of action available to an adversary.

MORPHOLOGICAL ANALYSIS

The approach just outlined - to break a problem down into components and then systematically enumerate all possibilities - is an example of a formal technique known as *Morphological Analysis (MA)*. MA was developed by physicist Fritz Zwicky during the 1950s at CalTech and is an extremely effective technique for generating analytic hypotheses and mitigating surprise. It has come into increasing use as computer technology has facilitated the enumeration of large possibility spaces and researchers have increasingly recognized its utility for understanding complex adaptive systems. MA also provides systematic techniques for "pruning" possibility spaces using available evidence – typically in the form of subject matter expert (SME) knowledge; but structured data can be used for well-understood problems.

A problem with MA is that it remains something of an art. Identifying appropriate units of analysis is a tricky business. If one uses too few units, important explanations will be missed, if too many are used, the possibility space quickly explodes to an unmanageable size. It is possible that as the problem becomes better understood that machine reasoning techniques can help manage large possibility spaces, but that again necessitates a fine-grained understanding of the problem, which should only be attempted after research has generated confidence levels sufficient to justify the investment of analytic resources.

A MORPHOLOGICAL FRAMEWORK FOR ANALYSIS OF ASYMMETRIC HUMAN NETWORKS

We describe an MA-based analytic framework implementing organizational theory that supports analysis of human networks that carry out asymmetric activities. We describe tests of the hypothesis that analysis conducted using this framework is superior to unsupported analysis

Our analytic framework is comprised of several components:
- A set of appropriate units of analysis to describe the capabilities of asymmetric organizations. These units are the minimal set necessary to effectively construct analysis for developing indicators and warnings and

effective counters to asymmetric threats. Once this framework is developed, a key follow-on research question is to determine the typical structure and size of possibility spaces associated with asymmetric networks given available evidence.

- A set of experiments using the framework on known asymmetric networks. This involves identifying open-source data sets, designing and conducting experimental tests, and depicting likely outcome scenarios. We hypothesize that analysis conducted using the MA-based realization is superior to analysis without such support. Testing this hypothesis requires metrics of analytic utility, such as those described by office of Analytical Standards and Integrity (AIS) in the US Office of Director of National Intelligence (ICD 203 2007) . Experimental tests would attempt to falsify the hypothesis by comparisons of analytic products generated with and without the framework.

- Methodologies for employing the resultant framework will need to be developed. This will likely employ a *breadth-depth* approach where initial coarse-grained (breadth) models are used to identify regions of possibility space where more analytic resources should be invested for developing more in-depth models.

- MA can be supported by software tools to simplify the process. Examples include the *Casper* tool developed by Ritchey *et. al.* and the *Morphine* tool developed by Least Squares Sofware, Inc..

Units of Analysis: The Structure and Function of Organizations

We have drawn our UOAs from established theories of organizational structure and function. These theories include:

- **Typologies of organizations**: involving attempts to classify organizations according to a variety of key characteristics, such as who benefits from their operations, or how they obtain compliance from their members. Works by Peter Blau (Blau 1970), Amitai Etzioni (Etzioni 1964), Tom Burns (Burns 1968), and G. M. Stalker (Stalker 1968) are among the most prominent of these studies.

- **Organizations as social systems**: an approach particularly identified with Talcott Parsons' structural-functionalist theory of action (Parsons 1937) and with Philip Selznick and Robert Merton's more focused work on organizations (Merton 1949, 1957, 1986). Organizations consist of social systems in interaction with other social systems (therefore "open systems") whose values and goals are oriented to those of the wider society. According to Parsons, key requirements for organizational maintenance (which is seen to be the overriding goal of any organization) are those which apply to all social systems; namely adaptation, goal attainment, integration, and pattern (or value) maintenance.

Especially relevant is Thomas, Kiser and Casebeer's (TKC) systematic analysis of non-state actors (NSAs). Their groundbreaking 2005 study, *Warlords Rising: Confronting Violent Non-State Actors*, has a particular emphasis on terrorist groups. TKC employ open-systems theory, which fundamentally, "views all organizations as systems interacting with their environments." The system is conceptualized as an "organized cohesive complex of elements standing in interaction," with that interaction referring to two patterns of behavior: a) the relationship between the organization and its environment, and b) the relationships between the elements within the organization. This theory is highly adaptable and broadly applicable, which allows analysts to mold typologies to meet the goals of the specific analysis.

- **Organizations as empirically contingent structures**: an approach particularly associated with the Aston Programme (Pugh 1998), based on research developed in the UK. It applies insights derived from psychology, together with statistical techniques such as scaling and factor analysis, to relate measures of organizational performance to different dimensions of organizational structure. This perspective also views organizations as open systems, drawing upon the work of Bertalanffy (Bertalanffy 1968).

- **Organizations as structures of action**: a group of approaches which focus on the circumstances determining the actions of individuals in organizations. An early contribution was made by Herbert A. Simon in his work on *satisficing* (Simon 1947). Later work, for example by David Silverman (Silverman 1970), is influenced by phenomenological sociology. Several units of analysis within this approach can be applied to our problem.

- **Organizational Architecture Theory:** Conventionally organizational architecture (Nadler, Gerstein, and Shaw 1992) consists of the formal organization (organizational structure), informal organization (organizational culture), operational processes, strategy and human resources. Key organizational attributes in this theory include *simplicity, flexibility, reliability, economy* and *acceptability.*

Taking our UOAs from established theory helps to focus our methodology. Rather than searching aimlessly, the theories point us to what characteristics we should be flagging and looking for. Our UOAs then enable us to deduce information about the adversary networks from the type of attack carried out, allowing our methodology to function very effectively with little data.

NOTIONAL EXAMPLE: RKG-3 ATTACKS IN IRAQ

We show how we can use MA to operationalize organizational theory in the context of organizational networks conducting asymmetric warfare. Consider the adaption of these organizational networks to use new weaponry. As the US has improved its capability in Iraq to detect IEDs and introduced new armored vehicles, such as the MRAP and the Stryker, there has been an increase in attacks against vehicles using the Russian-made RKG-3 anti-tank grenade. The RKG-3 is a tactical/strategic hand launched weapon. The decision to use this type of weapon is typically made at the strategic level by the insurgent organization. Choice of weapons is determined by situation, terrain and availability.

In considering the evidence we have about the RKG-3, employing the RKG-3 is non-trivial, it weighs about 2.5 lbs, can be thrown about 15m and has a lethal blast radius of 20m. Grenadiers have historically been those soldiers who were enlisted for their height and strong throwing arm. The RKG-3 jihadi is a modern day grenadier. He must be well trained, steady in his movement to target and good of eye. Videos of RKG-3 attacks indicate good planning, training and high morale. This implies a well-developed recruiting and training network. Further, the RKG-3 is a complex weapon that cannot be indigenously produced. This implies a logistics system to support the attacks.

Organizational Contingency Theory holds that organizations adapt to new challenges in their environment by differentiating new subsystems to provide new capabilities to meet these challenges. For UOA it indicates *subsystems* within an organization that perform necessary functions. It predicts that the more dynamic and varied an environment, the larger number of subsystems that will develop in the organization. A key prediction of Contingency Theory is that as the complexity of the organization increases, the organization will become more difficult to manage due to problems of *coordination* and *goal substitution*, in which the subsystems pursue their local objectives at the expense of the organization's overall goals.

In the example RKG-3 insurgent network, we analyze how the insurgent organization will adapt from employing roadside IEDs to employing the RKG-3. Contingency theory directs us to consider leadership issues in the adapted organization.

In this illustration, we break down the network that employs RKG-3s in terms of three subsystems: *personnel/recruitment, leadership* and *logistics*. From these UOAs we deduce significant amounts of information about the adversary. The resulting MA is in the figure. For each UOA, we assign 3 possible states, then enumerate all possible configurations - there are 3 x 3 x 3 = 27 possible arrangements - then eliminate inconsistent configurations. For example, state 3 is eliminated because the *logistics* state *new supply network* will require increased control from leadership, and is inconsistent with the *leadership* state *unchanged*. At the conclusion of this analysis, there are 19 possible configurations remaining. Each constitutes a valid hypothesis about the insurgent network that should be examined in more detail. This involves comparison to available evidence and triangulation against other analytic models (including MA models based on other theories). At the end of this process, we are left with a set of explanatory hypotheses. These hypotheses can be used as a basis for developing collection plans, indicator and warning frameworks and constructing courses of action to counter adversary actions. The approach is exhaustive within the context of the theory. This is essential for mitigating cognitive biases in human reasoning.

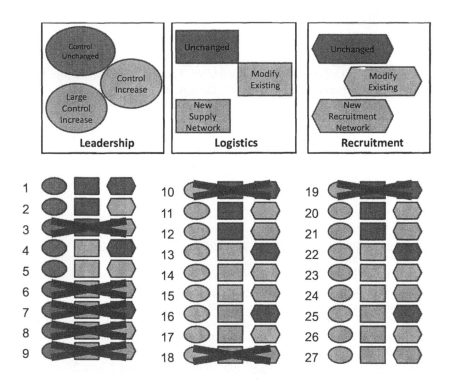

FIGURE 1 Enumeration of Organizational Structures Supporting RKG Attacks.

There are a number of advantages to this approach – it is based on existing theory, leveraging validated knowledge of organizational behavior. It provides an explicit operationalization of theory that can be applied by non-experts. The reasoning is transparent and explicit – the approach is clear about what the UOAs are and what assumptions are made. Evidence is clearly associated with hypotheses. It can be taught. It provides an explicit "audit trail" for analysis. It is flexible and can accommodate a broad variety of circumstances and unstructured evidence.

The approach is not without shortcomings, many of which can be addressed by a software tool. A general problem for MA is that the selection of UOA can be ambiguous and non-obvious. Automated support can mitigate this by identifying problem classes and theories that fit well together and identify UOA, or classes of UOA, appropriate for the application of a theory to a given problem class. Another problem with MA is that real problems will often use many UOA, which leads to explosive growth in the possibility space. This is not actually a shortcoming of the method, but reflects the intrinsic difficulties in reasoning about complex systems - the explosion in possibility space is a signature of complexity. The viability of the approach indicates the need for software approaches for managing large possibility spaces.

REFERENCES

von Bertalanffy, Ludwig, 1968, *General System Theory: Foundations, Development, Applications* New York: George Braziller

Blau, Peter, 1970, *A Formal Theory of Differentiation in Organizations*, New York, John Wiley & Sons

Brewer, John & Hunter, Albert 2006, *Foundations of Multimethod Research: Synthesizing Styles*. Thousand Oaks, CA: Sage Publications.

Burns, Tom and Stalker, G. M., Burns, Tom, 1968, *Managing Innovation*, London, Tavistock Publications

Edmunds, Bruce, 1991, *Syntactic Measures of Complexity*. Doctoral thesis, University of Manchester, Manchester, UK. Available online at http://bruce.edmonds.name/thesis.

http://www.dailymotion.com/video/x7crpa_compilation-rkg3-in-iraq_news last accessed 2 June 2, 2009.

Etzioni, Amitai, 1964, *MODERN ORGANIZATIONS* (Englewood Cliffs, N.J.: Prentice-Hall,

ICD 2007 http://fas.org/irp/dni/icd/icd-203.pdf

Kahneman, Daniel, Slovic, Paul and Tversky, Amos, Eds., 1982, *Judgment Under Uncertainty: Heuristics and Bias*, Cambridge University Press.

Kelley, Patrick A., *Imperial Secrets: Remapping the Mind of Empire,* NDIC Press, 2008. http://www.ndic.edu/press/12053.htm last accessed 2 June 2009.

Lawrence, Paul R and Lorsch, Jay William 1967. *Organization and Environment; Managing Differentiation and Integration*. Boston: Division of Research, Graduate School of Business Administration, Harvard University.

Least Squares Software, 2009 "Breadth-Depth Triangulation for Validation and Verification of Modeling and Simulation of Complex Systems", proposal in response to ONR BAA #09-001. Selected for funding March 2009.

Merton, Robert K., 1949, *Social theory and social structure.* New York: Free Press, Revised, 1957, 1986.

Moore, David T. & Reynolds, William N. , 2006, "So Many Ways to Lie: Complexity of Denial and Deception". *Defense Intelligence Journal*, 15(2), 95-116.

Nadler, David, Gerstein, Marc C. and Shaw, Robert B., 1992 *Organizational Architecture*, Jossey-Bass.

Parsons, Talcott, 1937, *The Structure of Social Action.* New York: McGraw-Hill,

1939, *Action, Situation and Normative Pattern* , New York: McGraw-Hill.

Pugh, Derek S. *ed.* 1998, *The Aston Programme,* Vols. I-III. Singapore and Sydney: Ashgate/Dartmouth. ...

Reynolds, William N., 2009 "Breadth Depth as an Analytic Multimethodology" unpublished.

Ritchey, T. 2006, "Problem Structuring Using Computer-Aided Morphological

Analysis", Journal of the Operational Research Society 57, 792–801.

Selznick, Philip, 1984, *Leadership in Administraiton; A Sociological Interpretation,* Berkeley, Ca, Unmiversity of California Press.

Silverman, David, 1970, *The Theory of Organizations: A Sociological Framework,* London: Heinemann Educational.

Simon, Herbert A., 1947, *Administrative Behavior: A Study of Decision-Making Processes in Administrative Organizations*, MIT Press Cambride, MA, 4th ed. 1997.

Thomas, Troy S., Kiser, Stephen D. and Casebeer, William D., 2005, *Warlords Rising: Confronting Violent Non-State Actors*, Lexington Books.

Zwicky, F. 1966, Discovery, Invention, Research Through the Morphological Approach. New York: Macmillan.

CHAPTER 42

Human Shape Modeling in Various Poses

Zhiqing Cheng[1], Jeanne Smith[1], *Julia Parakkat[2]*,
Huaining Cheng[2], Kathleen Robinette[2]

[1]Infoscitex Corporation
4027 Colonel Glenn Highway
Dayton, OH 45431, USA

[2]711th Human Performance Wing
Air Force Research Laboratory
2800 Q Street
Wright-Patterson AFB, OH 45433, USA

ABSTRACT

Human body shape modeling in various poses was studied in this paper. The investigations on pose modeling were briefly reviewed. A strategy for pose modeling was developed. The issues involved in pose data collection and processing were discussed. Principal component analysis was performed to characterize segment surface deformation. Principal components were used to represent and reconstruct the surface deformation and to predict the body shape in new poses.

Keywords: Human body shape, human body pose, 3-D shape modeling, principal component analysis

INTRODUCTION

Dynamic human shape modeling has drawn great attention recently as it has vast potential for various applications, such as human identification and human intention prediction, virtual reality creation, workstation design and human-machine interfacing, virtual fitting and design of clothing, and virtual figures and animation. Dynamic modeling describes or captures the body shape changes while the human body is moving, changing poses, or performing some actions. Since a human motion or action consists of a sequence of poses, pose modeling, which models the body shape in various poses, becomes an essential part in dynamic shape modeling. The major issues involved in pose modeling include pose definition and identification, skeleton model derivation and surface segmentation, surface deformation modeling, and pose deformation mapping.

Pose definition The human body can assume various poses. Since it is impossible to collect the data or create template models for all possible poses, it is necessary to define a set of standard, typical poses. A convention for pose definition is yet to be established. One approach is to use joint angle changes as the measures to characterize human pose changes and gross motion. This means that poses can be defined by joint angles. Given a set of scan data, imagery, or photos, the corresponding pose can be determined or identified. Mittal et al (2003) studied human body pose estimation using silhouette shape analysis. Cohen and Li (2003) proposed an approach for inferring the body posture using a 3-D visual-hull constructed from a set of silhouettes.

Skeleton model A common method for measuring and defining joint angles is using a skeleton model where each segment is represented by a rigid linkage and an appropriate joint is placed between the two corresponding linkages. Allen et al (2002) constructed a kinematic skeleton model to identify the pose of a scan data set using markers captured during range scanning. Anguelov et al (2004) developed an algorithm that automatically recovers from 3-D range data a decomposition of the object into approximately rigid parts, the location of the parts in the different poses, and the articulated object skeleton linking the parts.

Surface deformation modeling Two main approaches for modeling body deformations are anatomical modeling and example-based modeling. The anatomical modeling is based on an accurate representation of the major bones, muscles, and other interior structures of the body (Aubel and Thalmann 2001). The finite element method is the primary modeling technique used for anatomical modeling. In the example-based approach, a model of some body part in several different poses with the same underlying mesh structure can be generated by an artist. These poses are correlated to various degrees of freedom, such as joint angles. Lewis et al (2000) and Sloan et al (20091) developed similar techniques for applying example-based approaches to meshes. Instead of using artist-generated models, recent work on the example-based modeling uses range-scan data. Allen et al (2002 & 2003) presented an example-based method for calculating skeleton-driven body deformations. Their example data consists of range scans of a human body in a variety of poses. Using markers captured during range scanning, a kinematic skeleton is constructed first to identify the pose of each scan. Then a mutually consistent parameterization of all the scans is constructed using a posable subdivision surface template. Anguelov et al (2005) developed a method that incorporates both articulated and non-rigid deformations. A pose deformation model was constructed from training scan data that derives the non-rigid surface deformation as a function of the pose of the articulated skeleton. A separate model of shape variation was derived from the training data also. The two models were combined to produce a 3-D surface model with realistic muscle deformation for different people in different poses.

Pose deformation mapping For pose modeling, it is impossible to acquire the pose deformation for each person at each pose. Instead, pose deformation can be transferred from one person to another for a given pose. This can be referred to as pose deformation mapping. Anguelov et al (2005) addressed this issue by integrating a pose model with a shape model reconstructed from eigen space. As such, they were able to generate a mesh for any body shape in any pose.

In this paper, based on an analysis of the human body structure, a strategy for pose modeling was developed. The emphasis was placed on pose deformation modeling. A procedure was devised to separate the rigid component from the surface deformation. Based on a set of pose data of a subject scanned in a number of different poses, principal component analysis (PCA) was used to characterize the surface deformation of each segment in all the poses observed. The obtained principal components (PCs) were used to approximate and reconstruct the surface deformation of each segment.

PROBLEM ANALYSIS AND APPROACH FORMATION

In order to develop feasible and effective pose modeling methods, a general analysis of the problem is in order.

- The human body is an articulated structure. That is, the human body can be treated as a system of segments linked by joints.
- The human pose changes as the joints rotate. Therefore, a pose can be defined in terms of respective joint angles.
- The body shape varies in different poses. The variations are caused by two factors: the articulated motion of each segment and the surface deformation of each segment.
- It can be reasonably assumed that the surface deformation of a segment depends only on the rotations of the joint(s) adjacent to the segment. While the surface deformation of certain body regions may still be effected by the rotations of joints that are not directly connected, this assumption is valid for most regions of the human body and thus is often used.

Based on the above analyses, a framework for pose modeling was formulated, as shown in Fig. 1. The core part of pose modeling is to establish a mapping matrix that can be used to predict and construct the body shape model of a particular person at a particular pose. Therefore, in the true meaning of pose modeling, the mapping matrix need to represent the shape changes not only due to body variations of different human and pose deformations at different poses independently, but also resulting from the cross correlations between identity and pose. In reality, it is not feasible to determine the relationship between the pose deformation and the body shape variation using PCA in the same way as used for shape variation analysis, since it is too costly to collect pose data for a large number of subjects. Alternatively, it is possible to collect pose data for several typical subjects (e.g., male, female, tall, short, big, and small) who are selected to represent the entire population. For a particular subject, the mapping matrix for his/her pose deformation can be determined by subject classification based on certain criteria such as nearest neighborhood.

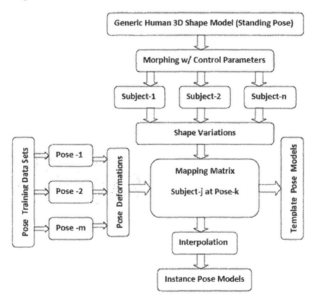

Figure 1. The framework of a pose modeling technology.

DATA COLLECTION AND PROCESSING

In order to create a morphable (for different subjects), deformable (for different poses), 3-D model of human body shape, it is necessary to collect many samples of human shape and pose. Properly sampling the entire range of human body shapes and poses is important to creating a robust model. The CAESAR (Civilian American and European Surface Anthropometry Resource) database provides human shape data for thousands of subjects in three poses. It can be used to train a static shape model and to represent human shape variation. The data sets that are required to establish a pose mapping matrix and to train pose models are not publicly available. Therefore, a research study collecting pose data for different subjects in different poses using a 3-D whole body laser scanner and a motion capture system is being undertaken in the Human Signatures Laboratory of the Air Force Research Laboratory in the United States. The data used in this paper, however, was collected from one subject in 70 poses (Anguelov et al 2005).

In order to use the pose data for pose modeling, processing is usually required. It includes three major tasks: hole-filling, point-to-point registration, and automatic surface segmentation.

Hole-filling Polygonal meshes that are derived from laser scanners frequently have missing data, regions where the laser neither reached nor produced adequate reflectance. This problem occurs more often when a subject is not in the standard pose (the standing pose used in the CAESAR database). Interpolating data into these regions often goes by the name of hole-filling. Several methods have been developed for hole-filling, such as the volumetric method (Davis et al 2002).

Point-to-point registration Polygonal mesh surfaces of the same object, but taken during different scans are not naturally in correspondence. In order to form complete models it is necessary to find this correspondence, i.e., which point on surface A corresponds to which point on surface B. Non-rigid registration is required to bring 3-D meshes of people in different poses into alignment. While many academic papers have been published which describe fully automated methods (Chui and Rangarajan 2000; Rusinkiewicz and Levoy 2001), the complexity of the problem often leads to optimization prone to local minima. Thus most of these methods tend to lack sufficient robustness for unattended real world application. Fortunately, establishing correspondence of a few control points by hand is usually sufficient to insure convergence. Labeling more points insures better convergence.

Automatic surface segmentation Given a deformable surface with multiple poses brought into correspondence, it is possible to segment the surface into disjoint regions. Each of these regions approximates a rigid articulated segment of the human body (Anguelov et al 2004 & 2005). The easiest way to achieve segmentation is to observe that polygons in the same segment tend to move together, that is, their rotation and translation are the same for a given pair of poses. By performing a K-means clustering over all polygons in all poses, and enforcing continuity of segments, the best segmentation is obtainable.

POSE DEFORMATION MODELING

The template model associated with the pose dataset consists of 16 segments, each of which has the pre-defined surface division (Anguelov et al 2005). Identifying the surface for each segment in different poses and establishing point-to-point correspondence for each surface in all observed poses is essential to the pose modeling. The method developed for pose deformation modeling in this paper consists of multiple steps, which are described below.

Coordinate Transformation The body shape variations caused by pose changing and motion can be decomposed into rigid and non-rigid deformation. Rigid deformation is associated with the orientation and position of segments. Non-rigid deformation is related to the changes in shape of soft tissues associated with segments in motion, which, however, excludes local deformation caused by muscle action alone. In the global (body) coordinate system, a segment surface has the articulated motion and surface deformation. However, in the local (segment) coordinate system, a segment surface has deformation only. Therefore, by transforming the global coordinate system to the local system, the effect of the articulated motion on each segment could be eliminated.

Surface Deformation Characterization Suppose the surface deformations of each segment are collected in all poses. Then PCA can be used to find the principal components of the surface deformation for each segment. Figure 2 illustrate the eigen value percentage ratio in each component (total 70) of all segments (total 16). It is shown that for all segments, the variance (eigen value ratio) of principal components increases sequentially, and significant principal components are those from the order of 60 to 70. As PCA exploits the underlying characteristics of a data set, the surface deformation of a segment in all observed poses can be characterized by these principal components. The surface deformation in a particular pose can be decomposed or projected in the space that is formed by the PCs. Each decomposition/projection coefficient represents the contribution or effect from the corresponding PC.

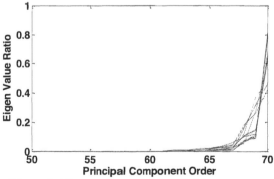

Figure 2. Eigen value ratio for all 16 segments.

Surface Deformation Reconstruction The decomposition/projection coefficients can be used to reconstruct surface deformation. There are two types of reconstruction:

(a) Full reconstruction, which uses all the PCs or eigenvectors; and (b) Partial reconstruction, which uses a number of significant PCs. Figure 3 illustrates the reconstructed shape for 2 different poses. In each row of Fig. 3, the first is the original shape, the second is the shape from full reconstruction, and the third and fourth are the shapes from partial reconstruction with 20 and 10 largest PCs, respectively. Figure 4 displays the sum of square errors of surface vertices for full and partial reconstruction. It is shown that the full reconstruction can completely reconstruct the original surface deformation in all poses, which means it is a perfect reconstruction, and partial reconstruction can provide a reasonable approximation of the original shape. While full reconstruction provides complete reconstruction of the original deformation, it is not necessary in many cases. On the other hand, the accuracy of partial reconstruction can be controlled by selecting a proper number of significant PCs. As partial reconstruction provides a reasonable simplification or approximation to the original deformation, it is often used in practice.

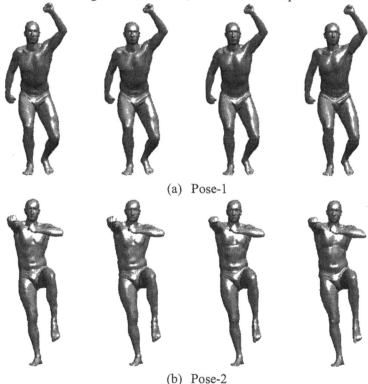

(a) Pose-1

(b) Pose-2

Figure 3. Shape reconstruction using principal components (First column: original shape; second column: full reconstruction; Third column: partial reconstruction with 20 largest PCs; Fourth column partial reconstruction with 10 largest PCs).

Surface Deformation Representation As the surface deformation of a segment is assumed to depend only on the rotation of the joint(s) connected, the relationship between the surface deformation and joint rotations has to be known. Joint rotations can be conveniently represented by their twist coordinates. The surface deformation can be compactly represented by its decomposition/projection coefficients. Ideally, the surface deformation can be expressed as a function of joint rotations. The relation between surface deformation and joint rotations can be linear or nonlinear. An appropriate function needs to be identified. The same function can be applied to all poses. Then, the measurement of surface deformation and joint rotations in all poses can be used to estimate the parameters of the function.

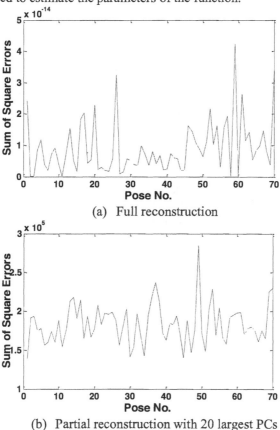

(a) Full reconstruction

(b) Partial reconstruction with 20 largest PCs

404

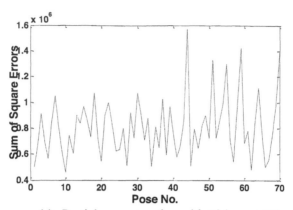

(c) Partial reconstruction with 10 largest PCs

Figure 4. Sum of square errors of shape reconstruction

Surface Deformation Prediction It is not feasible to measure the surface deformation of each segment for all possible poses, because the human body has a large number of degrees of freedom and can take virtually an infinite number of different poses. As a matter of fact, only a limited number of poses can be investigated in tests, but it is often required to predict surface deformation for new poses that have not been observed. Three methods can be used to predict surface deformation.

- Method-1: using principal components. Given the joint twist angles for a segment to define a particular pose, projection coefficients can be estimated. Using the full or a partial set of principal components, the surface deformation is reconstructed.
- Method-2: taking the nearest neighbor pose. Given the joint twist angles, find the nearest neighbor to the prescribed pose and take its surface deformation as an approximation. The neighborhood is measured in terms of the Euclidean distance between the joint twist angles for the two poses.
- Method-3: interpolating between two nearest neighbors. Given the joint twist angles, find two nearest neighbors to the prescribed pose. The pose deformation is determined by interpolating between the deformations of these two neighbor poses.

Figure 5 illustrates the predicted shape for 8 different poses using method-2.

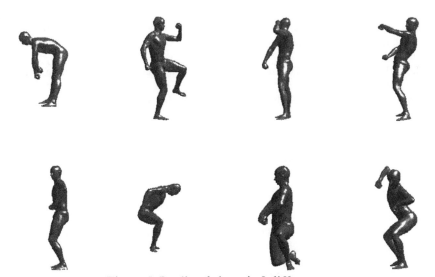

Figure 5. Predicted shape in 8 different poses.

CONCLUSIONS

The body shape variations caused by pose changing and motion can be decomposed into rigid and non-rigid deformation. In the global (body) coordinate system, a segment surface has the articulated motion and surface deformation. In the local (segment) coordinate system, a segment surface has deformation only. Therefore, by transforming the global coordinate system to the local system, the effect of the articulated motion on each segment can be eliminated, and the sole surface deformation can be determined. PCA can be used to characterize the surface deformations of a segment in different poses. The obtained PCs can be used to represent and reconstruct the surface deformation and to predict the body shape in new poses.

ACKNOWLEDGEMENT

The authors wish to thank Dr. James Davis from University of California, Santa Cruz for the help and the pose dataset which was acquired originally by the Stanford BioMotion Lab and processed into a usable form primarily by Dr. Drago Anguelov.

REFERENCES

Allen, B., Curless, B., Popovic, Z. (2002). "Articulated Body Deformation from Range Scan Data", in: ACM SIGGRAPH 2002, 21-26, San Antonio, TX, USA.

Allen, B., Curless, B., Popvic, Z.: The space of human body shapes: reconstruction and parameterization from range scans. In: ACM SIGGRAPH 2003, 27-31 July 2003, San Diego, CA, USA (2003).

Aubel, A., Thalmann, D (2001). "Interactive modeling of the human musculature", in: Proc. of Computer Animation.

Anguelov, D., Srinivasan, P., Koller, D., Thrun, S., Rodgers, J., Davis, J. (2005). "SCAPE: Shape Completion and Animation of People", in ACM Transactions on Graphics (SIGGRAPH) 24(3).

Anguelov, D., Srinivasan, P., Koller, D., Thrun, S., Pang, H. C., and Davis, J. (2004). "The Correlated Correspondence Algorithm for Unsupervised Registration of Nonrigid Surfaces", in: Neural Information Processing Systems (NIPS).

Cohen, I., Li, H.X. (2003). "Inference of Human Postures by Classification of 3D Human Body Shape", in: Proceedings of the IEEE International Workshop on Analysis and Modeling of Faces and Gestures.

Chui, H. and Rangarajan, A. (2000). "A new point matching algorithm for non-rigid registration", in: Proceedings of the Conference on Computer Vision and Pattern Recognition (CVPR)

Davis, J., Marschner, S., Garr, M., and Levoy, M. (2002). "Filling Holes in Complex Surfaces Using Volumetric Diffusion", in: Symposium on 3-D Data Processing, Visualization, and Transmission.

Lewis, J.P., Cordner, M., Fong, N. (2000). "Pose space deformations: A unified approach to shape interpolation and skeleton-driven deformation", in: Proceedings of ACM SIGGRAPH 2000, pp. 165–172.

Mittal, A., Zhao, L., Davis, L. S (2003). "Human Body Pose Estimation Using Silhouette Shape Analysis", in: Proceedings of the IEEE Conference on Advanced Video and Signal Based Surveillance.

Rusinkiewicz, S. and Levoy, M. (2001). "Efficient variants of the ICP algorithm", in: Proc. 3DIM, Quebec City, Canada, 2001. IEEEComputer Society.

Sloan, P.P., Rose, C., Cohen, M. F. (2001). "Shape by example", in Proceedings of 2001 Symposium on Interactive 3D Graphics.

CHAPTER 43

Experimental Research on the Confirmation of Human Body Surface Reachable Grade

Dong Dayong[1, 2], Wang Lijing[1] Yuan Xiugan[1]

[1]School of Aeronautical Science and Engineering
Beihang University, Beijing, China

[2] Institute of Aerospace Science and Technology
SJTU, Shanghai, China

ABSTRACT

A system to measure and evaluate functional reachable area on the human body that meets the special requirements of soldiers and military equipment. The human body was divided into 32 parts based on the equipment and systems carried by a Chinese soldier. Three grades - Comfort Reach, MAX-Reach, and Unreachable – were used to evaluate the reachability of each part of the body. Subjects were evaluated in three positions: standing, kneeling and prone. Data was verified through the maximum membership principal, and the result is satisfactory.

Keywords: Human body surface; reach envelope; enveloping surface; Fuzzy evaluation; Work efficiency

INTRODUCTION

In the designing of human-machine systems or related products, upper body reachable area is a key element. With good accessibility, systems or products can provide the operator a safe, efficient and comfortable work environment (Yuan Xiugan, 2002). This is especially important in workspaces which demand a high degree of safety, like airline cockpits, automobiles, and nuclear power plant control rooms. In such situations, an upper body reachablity analysis is indispensable (Ding Yulan, 2004).

The primary method of performing an upper body reachability analysis is to collect sufficient data about the human body, measure the functional reach envelope, and then determine if all the equipments is within this area (Wang Lijing, 2002).

REACHABILITY ANALYSIS METHOD

The most commonly used method for reachability analysis is to use anthropometric data to determine the upper body effective reach envelope. The effective reach envelope can be determined by using the work space area required by the Chinese government in files GB/T13547-1992, GJB2873-1997 and GJB/Z131-2002.

Additional data can be found in NASA STD 3000, which provides measurements of a sitting person's functional reach envelope measured in 15 degree increments along the body planes. (Wesley E.Woodson, et al 1991, NASA-STD-3000, 1995)

With the development of the computer technology, 3D human models have become available in software like JACK, SAFEWORK, and RAMSIS. But due to the complex nature of the models, the software all use the shoulder as the center, and treat the upper arm, forearm and hand as a rigid linkage in order to calculate the reachable area of the upper body. The human body model based on human motion tracking, developed by Beijing University of Aeronautics and Astronautics, makes it possible to track limb movement and map the surface of the functional reach envelope.

 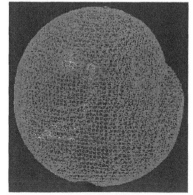 

(a)JACK (b) SAFEWORK (c) BUAA human body model

FIGURE .1 Upper body functional reach envelope

With the upper body functional reach envelope, we can perform reachability analysis simply by determining if the operational controls are located on this surface. At present, this method has been wildly used in the aerospace field and the auto industry. The result of reachability analysis has become a key actor in the Human Engineering Analysis of cockpit layout.

REACHABILITY ANALYSIS OF HUMAN BODY

The Reachability Analyses above are based on defined work space, and the operator will be in a defined posture (sitting or standing position). However, sometimes operators need to access equipment that is carried on their person. For example, a soldier may need to remove magazines from their bandoleer to load their weapon, or reach for the power button of the computer on their back. In such situations, these Reachability Analysis methods become less practical, especially in situations involving the back and leg areas. In addition, because soldiers do not work in a fixed position, sometimes they are laying prone on the ground or sitting, the functional reach area is difficult to determine.

Therefore, I decided to run a study of the reachable area with unfixed positions in order to provide a reference for a Reachability Analysis.

EXPERIMENTS DESIGN

First, the human body surface is divided into 32 parts (see Figure 2) based on how the U.S. military measures the efficiency of portable systems with some adjustments to compensate for differences of Chinese soldiers' portable systems(Everett Harmanc, Peter Frykman, et al. 1999, Michael Lafiandra, Suzanne Lynch, et al, 2003).

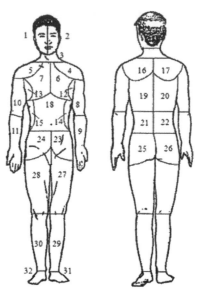

FIGURE 2 human body surface divisions

Table 1 Name of body surface area divided

Number	Body parts	Number	Body parts
1	Right side of head	17	Right shoulder
2	Left side of head	18	Abdomen
3	Neck	19	Left back
4	Left front of shoulder	20	Right back
5	right front of shoulder	21	Left lower back
6	Left chest	22	Right lower back
7	Right chest	23	Left waist
8	Left upper arm	24	Right waist
9	Left arm	25	Left hip
10	Right upper arm	26	Right hip
11	Right arm	27	Left leg
12	Left armpit	28	Right leg
13	Right armpit	29	Left calf
14	Left rib	30	Right calf
15	Right rib	31	Left ankle
16	Left shoulder	32	Right ankle

Subjects were tested under three positions: Standing; Kneeing; and Prone. Standing is when human body is in a natural standing state with unrestricted movement of the upper body; Kneeing is when the subject rests on their right knee. Prone is when subject lays on his stomach. There were 12 male subjects in this test. All their basic information can be seen in table 2. Considering there is little difference between two hands, all the subjects are right-handed.

Table2 Basic info of subjects

Number	Age	height(cm)	Weight (kg)
1	25	167	67
2	24	166	54.2
3	26	174.5	62.3
4	25	173	70.9
5	25	175.5	63.3
6	24	173	65.4
7	25	169	56.3
8	24	169	64.6
9	25	174.5	59.5
10	25	174	64.4
11	25	172	64.3
12	23	167	75.8

TEST PROCESS

During the test, subjects were asked to reach for different body surfaces in each of the three positions and rate the reachability as one of three levels: "easily reachable"; "barely reachable" or "unreachable". "Easily reachable" refers to the parts that can be easily reached; "barely reachable" refers to the parts that can be touched, but not comfortably; "unreachable" refers to the parts that are impossible to reach.

To avoid Exhaustion, all subjects were allowed to rest during the test. Because kneeing posture and prone positions are more physically taxing, subjects were allowed more to rest more frequently during these tests.

DATA PROCESSING

After recording the test data, we were able to calculate the reachability level of each body surface. We chose the evaluation of fuzzy evaluation principle of vector analysis, following the maximum membership degree principle (Hu Yonghong et al, 2000). Of which the main processing steps are:

Based on the test results, assuming the result vector as follows:

$$S = (s_1, s_2, \ldots, s_n) \tag{1}$$

Run Validation, methods as follows:

Define equation (1)

$$\beta = \max_{1 \leq i \leq n}\{s_i\} / \sum_{i=1}^{n} s_i \tag{2}$$

$$\gamma = \sec_{1 \leq i \leq n}\{s_i\} / \sum_{i=1}^{n} s_i \tag{3}$$

In which $\sec_{1 \leq i \leq n}\{s_i\}$ is behalf of the second largest component of S.

Definition:

$$\beta' = \frac{n\beta - 1}{n - 1} \tag{4}$$

$$\gamma' = 2\gamma \tag{5}$$

$$\alpha = \frac{\beta'}{\gamma'} = \frac{n\beta - 1}{2\gamma(n-1)} \tag{6}$$

Determine the validity of the principle of maximum membership degree based on the value of α. As follows :(Table 3)

Table 3 Principle for judging the validity by maximum degree of membership criteria

Value of α	validity of maximum degree of membership criteria
$+\infty$	full validity
$[1, +\infty)$	High validity
$[0.5, 1)$	Common validity
$(0, 0.5)$	Low validity
0	No validity

Run validity test based on the Vector of the number of people who give certain feedback.
Define:

$$R = (r_1, r_2, r_3) \tag{7}$$

In which: R— Vector for certain part's reachablity level; r_1—number of subject who found it "easily reachable"; r_2— number of subject who found it "barely reachable"; r_3— number of subject who found it "unreachable"。

Get the value of α for each surface part. Since all the α values are over 0.5, thus the validity test is successful.

Define: $r_j = \max_{1 \le i \le 3}\{r_i\}$, j equals the reachablity level of certain part.

CONCLUSION

Based on the actual needs of soldier's equipment system, the reachablity level of different parts of human body surfaces was tested with three positions (standing, kneeing and prone). Table 4 shows the reachablity level of every part. In which 2 refers to "easily reachable", 1 refers to it "barely reachable", and 0 refers to "unreachable". '–'indicates the part was not tested.

Table 4 Human body surface reachable grade

Body parts	Reachablity			Body parts	Reachablity		
	Standing	Kneeing	Prone		Standing	Kneeing	Prone
Right side of head	2	2	2	Right shoulder	0	0	0

Left side of head	2	1	1	Abdomen	2	2	0
Neck	2	2	2	Left back	0	0	0
Left front of shoulder	2	1	0	Right back	0	0	0
Right front of shoulder	1	1	1	Left lower back	1	1	0
Left chest	2	2	0	Right lower back	1	1	1
Right chest	1	1	1	Left waist	1	1	0
Left upper arm	1	2	0	Right waist	2	2	2
Left arm	2	2	0	Left hip	1	0	0
Right upper arm	—	—	—	Right hip	2	1	2
Right arm	—	—	—	Left leg	0	2	0
Left armpit	2	2	0	Right leg	0	2	1
Right armpit	1	1	1	Left calf	0	1	0
Left rib	2	2	0	Right calf	0	2	0
Right rib	1	1	1	Left ankle	0	1	0
Left shoulder	1	1	0	Right ankle	0	2	0

(1) Because the symmetry of human body, the reachablity level of left-handed should be the same.

(2) Table 4 shows that some body parts such as chest, abdomen, hip of the working arm side, as well as the arm of the opposite side have a high level of reachability. Potable equipments should be located more in these areas.

(3) If using computer to define body surface reachability, one should consider the relationship between the division of the body's surface and the put different colors on different parts of the body.

REFERENCES

Yuan Xiugan, Zhuang Damin. (2002), *Human Engineering*. Beijing, BUAA Press, 27-43(in Chinese)

Ding Yulan. (2004), *Human Engineering*. Beijing: Beijing University of Technology Press.15-47

Wang Lijing. (2002), *The Study of Computer-aided Ergonomic Analysis and Assessment System for Cockpit of Military Aircraft*, BUAA, pp9-15(in Chinese)

Wesley E.Woodson, Barry Tillman, Peggy Tillman. (1991), *Human Factors Design Handbook*. New York: McGraw-Hill Publishing Company. 547-52

NASA-STD-3000. (1995), *Man-Systems Integration Standards*.NASA-STD-3000, Volume I.3-31, 3-50.

Everett Harmanc, Peter Frykman, et al.(1999), *Physiological, biomechanical, and maximal performance comparisons of female soldiers carrying loads using prototype U.S. Marine Corps Modular Lightweight Load-Carrying Equipment (MOLLE) with Interceptor body armor and U.S. Army All-Purpose*

Lightweight Individual Carrying Equipment (ALICE) with PASGT body armor.

Evertt Harmanc, Peter Frykman, et al. (1999), *Physiological, Biomechanical, and maximal Performance Comparisons of Soldiers Carrying Loads Using U.S. Marine Corps Modular Lightweight Load-Carrying Equipment (MOLLE), and US Army Modular Load System (MLS) Prototypes.*

Michael Lafiandra, Suzanne Lynch, et al. (2003), *A Comparison of Two Commercial off the Shelf Backpacks to the Modular Lightweight Load Carrying Equipment (MOLLE) In Biomechanics, Metabolic Cost and Performance.*

Hu Yonghong, He Sihui. (2000), *Methods of Synthesize evaluation.* Beijing, Science Press. 167-207(in Chinese)

Chapter 44

Human Jumping Motion Analysis and Simulation- A Literature Review

Burak Ozsoy, Jingzhou (James) Yang, Rhonda Boros

Texas Tech University
Lubbock, TX 79409, USA

ABSTRACT

Jumping is a common activity in different sports and is used in trainings for many others. Because jumping activity is involved in impact between lower extremities with the environment, it is prone to injuries. In order to get an insight of parameters associated with performance improvement and injury prevention, significant researches have been devoted to jumping study for decades. These studies cover different types of jumping in different starting positions and conditions, such as: countermovement vertical jumping, squat vertical jumping, and long jumping with free and restricted arm movements. In terms of study methods, two major approaches have been used: Data-driven motion analysis (experimental method) and simulation-based simulation. Each method has its pros and cons. In general, data-driven method can give the answer; however, it cannot predict the motion. Whenever the environment parameters change, a new experiment needs to run. On the other hand, simulation-based method can predict the cause and effect easily. But the validation work is complicated. This paper attempts to have a comprehensive literature survey to summarize different approaches for jumping motion and their pros and cons.

Keywords: Human Jumping, simulation, motion drive analysis.

INTRODUCTION

Jumping is a fundamental human movement and a vital skill, especially in sports, that requires a coordination of lower and upper extremities. This movement consists of three phases: propulsive phase, flying phase and finally landing phase. Because of the contact of the foot with the ground at the end of propulsive phase, it is prone to injuries. In order to prevent the injuries and also find out how to increase the performance, significant research has been done to this area. However, there is lack of a literature survey to summarize different methods and their pros and cons.

PROBLEM DEFINITION

There are different types of jumping such as: countermovement vertical jumping (Fig.1a, Fig. 1b), squat vertical jumping (Fig. 1c, Fig. 1d), and long jumping (Fig. 1e) with free and restricted arm movements. It is mainly divided into three phases: propulsive, flying and finally landing phase. In propulsive phase body is in contact with the ground and at the end of this phase, body loses the interaction with the ground and moves to flying phase. During this phase body moves in the air and at the end of this phase it establishes the contact with the ground again and this final phase is called landing.

In countermovement jumping, the body first moves downward as a preparation step before starting to push off and then moves upward where in squat jumping there is no downward movement. Also there is no horizontal movement at the vertical jump. The performance in the jumping can be explained as the maximum vertical displacement in vertical jumping and maximum horizontal distance of the center of mass of the body.

It is a fact that the countermovement increases the performance and can be explained as producing and using more work at the joints in countermovement jump (Hara et al. (2008), Harman et al. (1990), Bobbert et al. (1996), Bobbert et al. (2005), Khalid et al. (1989)).

Contribution of the upper extremities with arm swing to the performance cannot be overlooked during the jumping. Three theories, force transmission, joint torque/work augmentation, pull/impart energy are questioned (Cheng et al. (2008), Lees et al. (2004)) for the augmentation of arm swing to the performance of the. During the arm swing, shoulder and the elbow rotation will cause increased vertical velocity of center of mass at the take off which cause higher performance.

The purpose of this study is to summarize methods and their pros and cons in literature and future trends although the studies about the muscle behavior are out of the scope. This paper is organized as follows: Next section gives summary about the pros and cons of different methods. Then a conclusion remark is given.

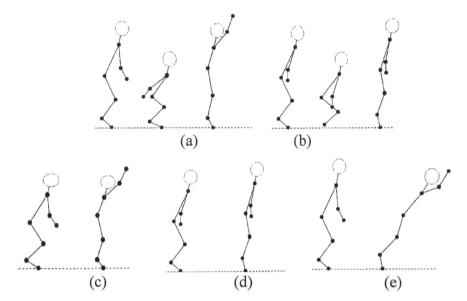

Fig. 1 Human jumping: (a) Countermovement Vertical Jump with arm swing (CMJAS); (b) Countermovement vertical jump with restricted arms (CMJNAS); (c) Squat Vertical Jump with arm swing (SJAS); (d) Squat vertical jump with restricted arms (SJNAS); (e) Long Jump with arm swing (LJAS)

LITERATURE OVERVIEW

In the past several decades, studies have been performed to understand the human jumping. The methods of the studies are mainly divided into two groups: Experimental method and simulation-based method. This section summarizes the methods.

EXPERIMENT-BASED STUDIES

In the literature, experiments have been performed with different aims and different human movements. In table 1, these studies were summarized among the aims, requested movements and finally key findings.

Table 1 Summary of previous experimental studies

Study	Aim	Requested Movement	Key Findings
Luthanen et al. (1978)	To investigate segmental contribution to the performance	2 maximal jumps, plantar flexion, knee and trunk extension, head swing, straight, 90° and 45° arm swing	The contribution to take-off velocity from body segments: knee extension 56% plantar flexion 22% trunk extension 10% arm swing 10% head swing 2%
Luthanen et al. (1979)	To investigate the difference of force produced, force time characteristics between good and less skilled athletes	Long jump with running	Good athletes showed similar characteristics Less skilled athletes showed various characteristics. The main difference was in timing of body segments movements.
Ae et al. (1980)	To investigate the contribution of segments in vertical jumping	CMJAS	Momentum generated by segments with ratios to whole momentum in former and latter stages. Trunk: 22.9%, 20.4% Legs : 71.7%, 76.1% Arms: 5.4%, 3.5%
Fukashiro et al. (1987)	To examine joint moment and mechanical power in lower extremities	CMJNAS SJNAS Hopping	the take-off vertical velocity is greater in countermovement jumping Ground reaction force during the hopping was the greatest.
Shetty et al. (1989)	To investigate effect of arm swing to kinetic ant kinematic results	SJAS SJNAS	Contribution of arm swing to Maximum force: 6% Work done: 14% Power: 15% Release velocity: 6 Impact force was decreased 12%
Khalid et al. (1989)	To investigate contribution of arm swing to performance	CMJAS CMJNAS SJAS SJNAS	Contribution of arm swing to Countermovement by 10% Squat jump by 11%
Oddsson (1989)	To investigate a relationship between force-time curve and GRF and maximal jumping height	CMJAS CMJNAS Hopping	A relation established between force-time curve and ground reaction force. Best jumpers showed two peaks in vertical reaction force profile
Harman et al. (1990)	To investigate contribution of arm swing and countermovement to performance	CMJAS CMJNAS SJAS SJNAS	Arm swing increased take off velocity by 10% Arm swing's contribution is larger than countermovement Countermovement increased pre-take-off jump duration by 71%-76%

Ashby et al. (2002)	To investigate the effect of arm swing in long jump	LJAS LJNAS	Performance increased 21.2% with arm swing. Velocity at take-off increased 12.7%
Lees et al. (2004)	examined transmission of force, joint torque augmentation, pull theories	CMJAS CMJNAS	72% improvement of performance is due to the increased velocity 28% is due to the increased center of mass at take-off
Hara et al. (2006)	To examine arm swing effect to the lower extremity work	SJAS SJNAS	The increase caused by arm swing due to the increased total body work with 34.1% and 65.9% from the upper and the lower extremities.
Hara et al. (2008)	comparison of the mechanical effect of arm swing and countermovement on the lower extremities	CMJAS CMJNAS SJAS SJNAS	Torque was increased by arm swing at the hip joint at the latter half of the propulsive phase. Countermovement and arm swing have independent effects on the lower extremities
Hara et al. (2008)	Effect of arm swing direction on forward and backward jump performance	Forward and backward long squat jump with forward and backward arm swing	Performance was increased mostly when the arm swing was in the same direction with the jump. The contribution of the hip joint to the total work done was the greatest portion

SIMULATION-BASED METHOD

Another method used for studies is simulation-based method. In simulation-based method generally optimization process takes part. For the simulations an effective body model to represent the body segments is necessary. In the studies where forward dynamics is used for analysis, inputs are muscle forces which create torques at the joints. In order to generate torques at joints, muscles are modeled with Hill-type muscle model (Hill et al., 1975). In Table 2, body models used in the literature for jumping are listed. In Table 3 (see Appendix), where forward and inverse dynamics are used a summary is given about the simulation-based studies with their jump type, objective, constraints, validations and key findings.

Table 2 Body models used in literature

Study	Year	Dimension	# Segments	# d.o.f
Pandy et. al.	1990	2D	4	4
Bobbert	1996	2D	4	4
Anderson et. al	1999	3D	10	23
Spägele et.al.	1999	2D	3	3
Bobbert et. al.	2001	2D	4	4
Meghdari et. al.	2003	2D	7	10
Nagano	2005	3D	9	20
Bobbert et. al.	2005	2D	4	4
Ashby et. al.	2006	2D	5	7
Vanrenterghem et. al	2008	2D	4	4
Blajer et. al.	2007	2D	9	9
Babič et. al.	2009	2D	4	4
Cheng et. al.	2008	2D	4-5	4-5

DISCUSSION

Jumping is a complex series event of segments which also involve collision at the landing phase. Therefore it is prone to the injuries. To get an insight about the parameters for the different jumping styles such as: vertical jumping, long jumping, several studies performed in the literature experimentally and numerically. Experimental method can only give result for present conditions in the experiment. Each human being may jump in different styles. Therefore the experimental studies can just give a result for subject's performance. However the simulation-based method can predict the optimum jump scenario for maximum performance without any possible injuries with constraints.

The body model has a significant importance for a full analysis. The effect of arm swing, cannot be neglected which enhances the performance by %10, also head swing has a small contribution to the performance. As the jumping movements were considered in sagittal plane, 2D analysis performed. Since the human is a biped, for the balance a 3D analysis should be performed for more accurate findings.

Also the interaction of upper extremities with the lower extremities cannot be overlooked to understand the contribution of each segment to the performance which can be understand if the work done by each segment is studied.

In our final overall consideration in the different approaches, simulation-based method has more advantages than experimental method. Basically, being able to predict the motion is the main advantage. Also changing the parameter and running the simulation again in a short time makes the simulation-based method more effective than experimental method. In simulation-based methods, according to our opinion maximizing the jump height with minimum performance by finding the optimum muscle activation levels and timings with constraining the problem with zero ground reaction force at jumping, joint limits is the best approach. Also modeling the jumping with propulsive, flying and landing phases has importance to predict or preventing the possible injuries. Finally modeling the human body in 3-D will make the results more accurate.

REFERENCES

Ae, M., Shibukawa, K. (1980), "A biomechanical method for the analysis of the contribution of the body segments with an example of vertical jump takeoff." Japanese Journal of Physical Education, vol. 25, 233–243.

Anderson, F.C., Pandy, M.G. (1999), "A dynamic optimization solution for vertical jumping in three dimensions." Computer Methods in Biomechanics and Biomedical Engineering 2, 201–231.

Ashby, M. B., Heegaard, J. H. (2002), "Role of arm motion in standing long jump." Journal of Biomechanics, vol.35, p1631-1637.

Ashby B.M., Delp S.L. (2006), "Optimal control simulations reveal mechanisms by which arm movement improves standing long jump performance." Journal of Biomechanics, vol. 39, Issue 9,1726-1734.

Babič J., Bokman L., Omrcen D., Lenarcic J., Park F. C. (2009), "A Biarticulated Robotic Leg for Jumping Movements: Theory and Experiments." Journal of Mechanisms and Robotics, vol. 1, 011013.1-011013.9.

Blajer W., Dziewiecki K., Mazur Z. (2007). "Multibody modeling of human body for the inverse dynamics analysis of saggital plane movements." Multibody System Dynamics, vol. 18, 217-232.

Bobbert M.F., Gerritsen K.G., Litjens M.C., Van Soest A.J. (1996), "Why is countermovement jump height greater than squat jump height?" Medicine & Science in Sports & Exercise, vol. 28, 1402-1412.

Bobbert M.F., van Soest A.J. (2001), "Why do people jump the way they do?" Exercise and Sport Sciences Reviews, vol. 29, 95-102.

Bobbert M.F., Richard, C.L.J. (2005) "Is the Effect of a Countermovement on Jump Height due to Active State Development?" Medicine and Science in Sports and Exercise, vol. 37, 440-446.

Cheng K.B., Wang C.H., Chen H.C., Wu C.D., Chiu H.T (2008). "The mechanisms that enable arm motion to enhance vertical jump performance--A simulation study." Journal of Biomechanics, vol. 41, Issue 9, 1847-1854.

Fukashiro, S. and Komi, P.V. (1987), "Joint moment and mechanical power flow of the lower limb during vertical jump." International Journal of Sports Medicine 8 Suppl 1, pp. 15–21

Hara M., Shibayama A., Takeshita D., Fukashiro S. (2006), "The effect of arm swing on lower extremities in vertical jumping." Journal of Biomechanics, vol.39, 2503-2511.

Hara M., Shibayama A., Takeshita D., Hay D.C., Fukashiro S. (2008), "A comparison of the mechanical effect of arm swing and countermovement on the lower extremities in vertical jumping." Human Movement Science, vol. 27, 636-648.

Hara M., Shibayama A., Arakawa H., Fukashiro S. (2008), "Effect of arm swing direction on forward and backward jump performance." Journal of Biomechanics, vol. 41, Issue 13, 2806-2815.

Harman, E.A., Rosenstein, M.T., Frykman, P.N., Rosenstein, R.M. (1990), "The effects of arms and countermovement on vertical jumping." Medicine and Science in Sports and Exercise, vol. 22, 825–833.

Hill, T.L., Eisenberg, E., Chen, Y., Podolsky, R.J. (1975) "Some self consistent two-state sliding filament models of muscle contraction." Journal of Biophysiology, 15, 335-372.

Khalid, W., Amin, M. and Bober, T. (1989), "The influence of upper extremities movement on take-off in vertical jump." Biomechanics in Sports, pp. 375- 379.

Lees, A., Vanrenterghem J., Clercq D.D. (2004), "Understanding how an arm swing enhances performance in the vertical jump." Journal of Biomechanics, vol.37, 1929-1940.

Luhtanen, P., Komi, P.V. (1978), "Segmental contribution to force in vertical jump." European Journal of Applied Physiology, vol. 38, 181–188.

Luhtanen, P., Komi, P.V. (1979), "Mechanical power and segmental contribution to force impulses in long jump take-off" European Journal of Applied Physiology, vol. 41, 267–274.

Meghdari, A., Aryanpour M. (2003), "Dynamic Modeling and Analysis of the Human Jumping Process." Journal of Intelligent and Robotic Systems, vol. 37, 97-115

Nagano A., Komura T., Fukashiro S., Himeno R. (2005), "Force, work and power output of lower limb muscles during human maximal-effort countermovement jumping" Journal of Electromyography and Kinesiology, vol.15, 367–376.

Oddsson. L. (1989), "What factors determine vertical jumping height?" International Society of Biomechanics in Sport 5th Congress, Athens Greece., p393-401

Pandy, M.G., Zajac, F.E., Sim, E., Levine, W.S. (1990), "An optimal control model for maximum-height human jumping." Journal of Biomechanics, vol. 23, 1185–1198.

Seyfarth, A., Friedrichs, A., Wank V, Blickhan R. (1999) "Dynamics of the long jump," Journal of Biomechanics, vol. 32, 1259-1267

Shetty, A.B., Etnyre, B.R. (1989), "Contribution of arm movement to the force components of a maximum vertical jump." Journal of Orthopedic and Sports Therapy, vol.11, 198–201.

Spägele, T., Kistner, A., Gollhofer, A. (1999), "Modeling, simulation and optimization of a human vertical jump." Journal of Biomechanics, vol. 32, pp. 521–530.

Vanrenterghem J., Bobbert M.F., Casius L.J., Clercq D.D. (2008), "Is energy expenditure taken into account in human sub-maximal jumping? –A simulation study." Journal of electromyography and kinesiology." vol. 18, 108-115.

Table 3 Summary of previous simulation-based studies

Study	Aim	Jump Type	Objective	Constraints	Validation	Keynotes
Pandy et al. (1990)	To maximize jump height	SJNAS	Maximum jump height	Bilateral symmetry, zero ground reaction force at take-off, neural control signal boundary.	Previous studies	Average jump height was found 33cm and take-off time was 0.5 s.
Bobbert et al. (1996)	To investigate the effect of countermovement	CMJNAS SJNAS	None	None	None	Jump height was 3.4 cm more at countermovement.
Spägele et al. (1999)	To simulate a measured jump consisting of all stages	SJNAS	Minimize differences between the measured and calculated kinematic data,	Trajectory of hip point, activation signal boundary	Kinematics, EMG	All stages included. One legged jump. Predicted and measured muscle excitations have a close relationship.
Seyfarth et al. (1999)	To develop a mechanical model for	Long Jump	None	None	Experiment, GRF	Spring mass system for leg There is a minimum stiffness for optimum performance.
Anderson et al. (1999)	To predict the pattern of muscle excitations for maximum jump	SJNAS	Maximum height Minimum sum of squared torques at joints	Bilateral symmetry, activation signal boundary	Kinematics, EMG, GRF from the experiment	Only propulsive stage was modeled. Initial optimization problem solved for initial static posture of human body Interpolation between control nodes. Most detailed human body model.
Meghdari et al. (2003)	To develop a dynamical model	SJAS	None	None	Kinetics from experiment, GRF	Hand motion can lead to a small improvement. Inverse dynamics was used.

Study	Objective	Model	Optimization criterion	Constraints	Outputs	Remarks
Nagano et al. (2005)	To search muscle activation pattern	CMJNAS	Maximum jump height	Bilateral symmetry, activation signal boundary	Kinematics	Muscle activations controlled with a series of step function with constant duration.
Bobbert et. al. (2005)	To search for switch times of activations	CMJNAS SJNAS	Maximize jump height	Bilateral symmetry, pre-specified minimum jump height	None	Optimization took part to find optimum switch times.
Babič and Lenarčič (2009)	To determine optimum timing of the biarticular link activation	SJNAS	Maximize jump height	Passive constraints in the hip, knee and ankle	None	Timing of the biarticular link activation and stiffness influence height of jump.
Ashby et al. (2006)	To search the muscle activation pattern	LJAS LJNAS	Maximize jump distance	Bilateral symmetry, activation signal boundary	Kinematics	Case 1: to maximize the distance without considering the landing configuration. Case 2 to maximize the horizontal position of toe with an acceptable landing configuration.
Blajer et al. (2007)	To find muscle forces and joint reaction forces in lower extremities	CMJAS LJAS	None	None	GRF	Inverse dynamics was used Redundancy for lower extremities was solved by pseudoinverse method.
Cheng et al (2008)	to find effective activations	SJAS SJNAS	Maximize jump height	Zero ground reaction force at take-off, joint ranges	Kinematics, Torque at joints	Arm swing increased the ground contact duration. total work done increased with the arm swing Torque/work augmentation theory was valid only in the hip joints Transmission theory was questionable Pull/impart theory was valid.

Chapter 45

Mixture-of-Behaviors and Levels-of-Expertise in a Bayesian Autonomous Driver Model

Claus Möbus[1], Mark Eilers[2]

Learning and Cognitive Systems / Transportation Systems
C.v.O University / OFFIS, Oldenburg, Germany
http://www.lks.uni-oldenburg.de/
claus.moebus@uni-oldenburg.de, mark.eilers@offis.de

ABSTRACT

Traffic scenario simulations and risk-based design require Digital Human Models (DHMs) of human control strategies. Furthermore, it is tempting to prototype assistance systems on the basis of a human driver model cloning an expert driver. We present the model architecture for embedding probabilistic models of human driver expertise with sharing of *behaviors* in different driving *maneuvers*. These models implement the sensory-motor system of human drivers in a *mixture-of-behaviors (MoB)* architecture with *autonomous and goal-based attention allocation* processes. A Bayesian MoB model is able to decompose complex skills (*maneuvers*) into basic skills (*behaviors*) and vice versa. The Bayesian-MoB-Model defines a probability distribution over driver-vehicle trajectories so that it has the ability to *predict* agent's behavior, to *abduct* hazardous situations, to *generate* anticipatory plans and control, and to *plan* counteractive measures by *simulating*

[1]project Integrated Modeling for Safe Transportation (IMOST) sponsored by the Government of Lower Saxony, Germany under contracts ZN2245, ZN2253, ZN2366
[2]project ISi-PADAS funded by the European Commission in the 7th Framework Program, Theme 7 Transport FP7-218552

counterfactual behaviors or actions *preventing* hazardous situations.

Keywords: Bayesian models of human driver behavior and cognition, probabilistic driver model, Bayesian autonomous driver models, mixture-of-behavior model, visual attention allocation, prediction and abduction of behavior, anticipatory plans and control, counteractive measures, risk and hazardous prevention

INTRODUCTION

The Human or Cognitive Centered Design (HCD) of intelligent transport systems requires digital Models of Human Behavior and Cognition (MHBC) which are *embedded, context aware, personalized, adaptive,* and *anticipatory.* A special kind of MHBC is the *driver model* which is used mainly in traffic scenario simulations and risk-based design (Cacciabue, 2007).

Modeling drivers is a challenging topic because no well established psychological theory about driving is at hand. Even simple maneuvers like *braking* are not well understood empirically. With the need for smarter assistance the *problem of transferring human skills* (Xu, 2005) without having a well-founded skill theory becomes more and more apparent.

The conventional approach for driver modeling is the *handcrafting* of MHBC. An *ex post* evaluation of their human likeness or empirical validity and revision-evaluation cycles is obligatory. We propose as a *machine-learning* alternative the estimation of Bayesian MHBCs from human behavior traces. The learnt models are empirical valid by construction. An *ex post* evaluation of *Bayesian Autonomous Driver (BAD)* models is in principle not necessary when the statistical relations and conditional independencies between the pertinent variables in the data are mapped into the model.

The advantage of probabilistic models is their robustness facing the *irreducible incompleteness of knowledge* about the environment and the underlying psychological mechanisms (Bessiere, 2008).

A BAYESIAN MIXTURE OF BEHAVIORS MODEL

BAYESIAN AUTONOMOUS DRIVER MODELS

BAD models (Möbus et al., 2008; 2009a; 2009b, 2009c) are developed in the tradition of Bayesian expert systems (Pearl, 2009) and Bayesian (robot) Programming (Lebeltel et al., 2004, Bessiere et al., 2003, 2008). They describe phenomena on the basis of the joint probability distribution (JPD) and their factorization into conditional probability distributions (CPDs) of the observable pertinent variables. This is in contrast to models in cognitive architectures (e.g. ACT-R) which try to simulate latent or hidden cognitive algorithms and processes on a finer granular basis.

A BAD Mixture-of-Behaviors (BAD-MoB) model is a *Bayesian Program (BP)*, which is able to decompose complex skills (scenarios, maneuvers) into basic skills (= behaviors, actions) and vice versa. The basic *behaviors* or sensory-motor schemas could be shared and reused in different *maneuvers*. Context dependent complex driver behavior will be generated by mixing the pure basic *behaviors*. The BAD-MoB-Model is embedded in a dynamic Bayesian network (DBN). If its template (Fig. 5) is rolled out (Fig. 6, 7) it defines a probability distribution over driver-vehicle trajectories so that it has the ability to *predict* agent's behavior, to *abduct* hazardous situations (what could have been the initial situation), to *generate anticipatory plans and control*, and *to plan counteractive measures* by *simulating* counterfactual behaviors or actions *preventing* hazardous situations.

BAYESIAN PROGRAMS AND DESCRIPTION COMBINATION

A *BP* is defined as a mean of specifying a family of probability distributions (Bessiere et al., 2003, 2008; Lebeltel et al., 2004). On the basis of a BP it is possible to construct a BAD-MoB-model, which can effectively control a (virtual) vehicle.

As Bessiere (2008) points out it is possible to combine or select single descriptions (= BPs) by a *probabilistic if-then-else*. "*Description combination* appears to implement naturally a mechanism similar to *Hierarchical Mixture of Experts* (Jordan and Jacobs, 1994) and is also closely related to mixture models ... From a programming point of view, description combination can be seen as a *probabilistic if-then-else* construction. H is the condition. If H is known with certainty, then we have a normal branching structure. If H is known with some uncertainty through a probability distribution, then the two possible consequences are automatically combined using weights proportional to this distribution." (Bessiere, 2008).

We embedded description combination (Lebeltel et al., 1999, 2004; Bessiere et al. 2003) in our DBN-based BAD-MoB-model. The *condition* variable H is a generalized *case-statement* like a Lisp *cond* and *one* of the root variables in our template model (Fig. 5), especially the variable *Behaviors*. The marginal probability distribution $P(Behaviors^0)$ or $P(Behaviors^{t-1})$ corresponds to the weighting or mixing coefficients of the *description combination*. The number of CPDs *P(Action | behavior, States, Percepts)* equals to the cardinality *Behaviors* variable. For each *behavior* we have to establish a local CPD *P(Action | behavior, States, Percepts)*. The collection of these local CPDs is the envisioned behavior library summarized in the total CPD *P(Action | Behaviors, States, Percepts)*.

LEVELS OF EXPERTISE AND MIXTURES OF BEHAVIORS

BAD-MoB-models are *dynamic Bayesian networks (DBNs)* which can be considered as a subtype of a *Bayesian Program (BP)* (Bessiere, 2003). Under the assumption of stationarity their *template models* are specified as *2-time-slice*

Bayesian networks (2-TBNs). The template model can be unrolled so that their *interface variables* (Koller and Friedman, 2009) *Behaviors* and *State* are glued together producing an unrolled DBN over T time slices (T-TBN) like the 3-TBN in Fig. 6, 7. Learning data are time series of the pertinent domain-specific variables *goals, behaviors, actions,* observable *states,* and *actions* combined with *posthoc annotations* of maneuvers and scenarios. Information can be propagated within the T-TBN in various directions. When working *top-down,* goals emitted by higher cognitive layers of the agent activate a corresponding *behavior* which propagates *actions,* relevant *areas of interest* (AoIs), and expected *perceptions.* When working *bottom-up,* percepts trigger *AoIs, actions, behaviors,* and *goals.* When the task or goal is defined and there are percepts, evidence can be propagated *simultaneously* top-down and bottom-up, and the appropriate *behavior* can be activated.

Our DBN-based MoB model is influenced by the *visual attention allocation* model of Horrey et al. (2006) and the *Bayesian filter and action selection* model of Koike (2008). The BAD-MoB-model we present here is tailored to a virtual highway scenario assuming a hierarchy of driving skills or expertise.

A Virtual Highway Scenario

For the proof of concept we developed a 2-TBN for a simple scenario with three areas of interest (AoIs) and maneuvers (Fig. 1-3) (Möbus, et al., 2009c). The driver is sitting in the *ego* vehicle (ev). Sometimes an *alter* vehicle (av) or the *roadside* is occupying the AoIs depending on the *state* of the car (State = left, middle, or right lane).

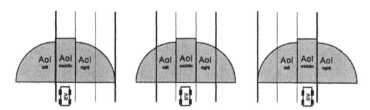

FIGURE 1. Areas of Interest (AoIs) and Ego Vehicle Positions

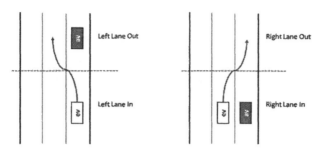

driving actions with focus on the left: left check lane, left signal, left turn
driving actions with focus in the middle: acceleration, deceleration, look forward
driving actions with focus on the right : right check lane right signal, right turn

FIGURE 2. Driving Maneuvers LeftLaneChange LLC (left) and RightLaneChange RLC (right) with two sequences of Driving Behaviors each (above, below)

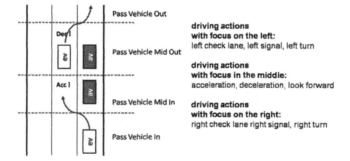

driving actions
with focus on the left:
left check lane, left signal, left turn

driving actions
with focus in the middle:
acceleration, deceleration, look forward

driving actions
with focus on the right:
right check lane right signal, right turn

FIGURE 3 Pass Vehicle Driving Maneuver with a sequence of 4 Driving Behaviors

The levels of expertise, the model components (layer, sequence) and a partial grammar of expertise are shown in Fig. 4.

Levels of Expertise	Model Component	Hierarchy of Skills, Levels of Expertise
Skills		Skills = {..., drivingScenarioSkills, ...}
Scenario Skills		DrivingScenarioSkills = { highway, countryRoad, city }
Driving Maneuver Skills	Driving **Maneuver** *Sequence* (horizontally distributed)	highway.Maneuvers = { leftLaneChange (lLC), rightLaneChange (rLC), passVehicle (pV), newManeuver }
Driving Behavior Skills	Driving **Behavior** *Layer*	Behaviors = { leftLaneIn (lLI), leftLaneOut (lLO), passIn (pI), passMidIn (pMI), passMidOut (pMO), passOut (pO), rightLaneIn (rLI), rightLaneOut (rO), newBehavior } e.g. leftLaneChange.Behaviors = {leftLaneIn, leftLaneOut }
Driving Action Skills	Driving **Actions** *Layer*	Actions = { leftCheckLane (lCL), leftSignal (lS), leftTurn (lT), middleAcceleration (mA), middleDeceleration (mD), middleLookForward (mLF), rightCheckLane (rCL), rightSignal (rS), rightTurn (rT) } e.g.. leftLaneIn.Actions = {lCL, mD, mLF, lS. lT, mA}

FIGURE 4 Levels of Expertise, Model Components, and part of Expertise Grammar

Dynamic Reactive BAD-MoB-model

For our BAD-MoB-model we propose *partially inverted* dynamic Bayesian networks (DBNs) of the 2-TBN-type (Fig. 5). We call the model *Dynamic Reactive MoB Model*. The model is *reactive* because AoIs *directly* influence actions. The model embeds two naïve Bayesian classifiers: One for the *Behaviors* and one for the *States*. This simplifies the structure of the architecture. Time slices are selected so that in each new time slice a new *behavior* is active. A *sequence* of behaviors implements a single *maneuver*. When we replace the reactive submodel for the *Actions* variable in Fig. 5 by a *third* classifier we can simplify the model and reduce the number of parameters by 79%.

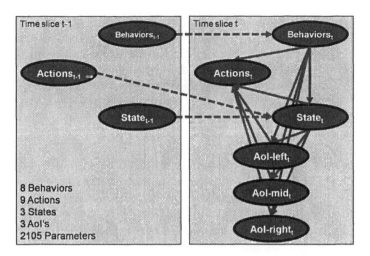

FIGURE 5 *Dynamic Reactive* BAD-MoB-model with *Behavior* and *State* Classifiers

The top layer consists of *behavior* nodes. There are behaviors for each main part of a *maneuver* (Fig. 2-4): *left_lane_in, ...* The next layer describes the *actions* the model is able to generate: *left_check_lane, ...* Below that appears the node *state* of the vehicle (*is_in_left_lane, ...*). Then there are three bottom layers contain nodes for the three *AoIs* with values *is_occupied and is_empty.*

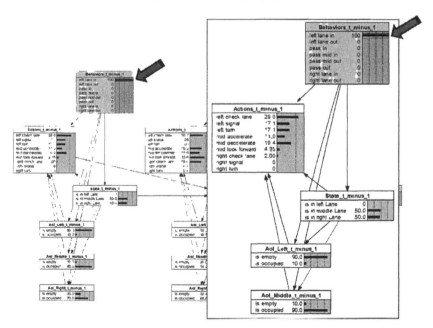

FIGURE 6 Expected behavior of the 3-TBN model with the goal behavior *left_lane_in* (the left-upper part of time-slice t-1 is expanded on the right of the figure)

An implementation in NETICA with artificial but plausible data is shown in Fig 6. When the model is urged to be in the *left_lane_in behavior* by e.g. goal setting from a higher cognitive layer, we expect in the *same* time-slice primarily that the *left lane is checked* and that the driver *decelerates the vehicle*. For the AoIs we expect that the middle AoI *is occupied* and the left AoI *is empty*. For the *this* time slice we expect the vehicle *in the right* or *middle lane*. The expected *behavior* changes between the time-slices. So the expected behavior in time-slice t is the *left_lane_out behavior*. We have higher beliefs in *acceleration, attention forward* and in *checking the left and right lane*.

When the state is known (e.g. *State = is_in_middle_lane*) we can include this as a single evidence in the model and infer the appropriate expectations (e.g. *left and right lane check, looking forward*, and both *(ac|de)celerations*).

When the model perceives a combination of AoI evidence, we can infer the *behaviors*. For instance, when the left AoI *is empty* and the middle and right AoI *is occupied*. We expect that the vehicle *is in the middle* or *right lane*, that the *behaviors left_lane_in* and *pass_in* are ambiguous, and that their appropriate *mixed* behavior (*left_lane_check, deceleration*) is activated. In the case, when all AoIs are occupied the model *is decelerating* with main attention to the middle AoI (*middle_straight_look*).

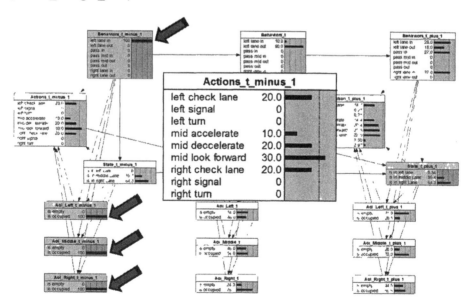

FIGURE 7 Conditional distributions in *Dynamic Reactive MoB Model* when receiving a combination of *Behavior* (goal) and *blocking AoI* evidence (Action-node expanded)

What will happen, if a goal is blocked? In Fig. 7 this is modeled by the appropriate evidence. The *lane-in behavior* is activated as a goal and at the same time the perception in the *left* and *middle* AoIs is set to *is_occupied*. This situation blocks the *left lane in* and the *pass vehicle in behaviors*. The expected actions are *looking forward, checking left and right lanes*, and *deceleration*. These are typical behavioral indicators for helplessness and stress.

This architecture has the ability to *predict* agent's behavior, to *abduct* hazardous situations (what could have been the initial situation), to *generate anticipatory plans and control*, and *to plan counteractive measures* by *simulating* counterfactual behaviors or actions *preventing* hazardous situations. For these applications we have to provide the model with appropriate evidence and questions.

For instance when *planning counteractive measures* by *simulating* counterfactual behaviors or actions *preventing* hazardous situations we need a 3-step procedure (Pearl, 2009): (1) *abduction* of a hazardous situation backwards with the *full* state-based BAD-MoB-model, (2) *mutilate* the full model to a *reduced* model, that is able to *predict intervention* effects, (3) experiment with counterfactual actions (= countermeasures) by *providing action* evidence in the *reduced* model and *predicting* the action effects.

FIRST MODELING RESULTS WITH REAL DATA

BAD-models with Mixed Behaviors are expressive enough to describe and predict a wide range of phenomena. In Möbus & Eilers (2009a) we presented a BAD model for lateral and longitudinal control *without* behavior mixing. The model showed nearly perfect behavior on the Aalborg course in the racing simulation game TORCS, though some suboptimal driving maneuvers could be observed. This is due to the fact that we used a *fixed* set of parameters in our model on a track with different segments like hair-needle curve, straight line segments etc. We modified the BAD-model architecture introducing concepts of the theory of ambient vision (Horrey et al., 2006). This led to a slightly simplified version of a BAD-MoB-model with two *behavior* and *steer-action* classifiers (Fig. 8).

The results are very promising as can be seen from Figs. 9 and 10. In Figure 9 the driver is driving in a right bended curve. His ambient vision field is sampled by 20 sensors (Fig. 9, left). Provided this perceptional evidence the conditional distribution for the *action* variable *Steer* (= steering angle) and the *behavior* variable *Behaviors* (= Experts) are inferred (Fig. 9, middle, right). As can be seen only the right-turn *behavior* (expert) is recognized and the corresponding angle of the steering-*action* is inferred. Sampling a concrete steering *action* from this conditional probability distribution gives the generated *action* of the BAD-MoB-model. Leaving the right-bended curve (Fig. 10) activates actions which are a mixture of the two *behaviors* (experts) straight and right (Fig. 10, right).

CONCLUSION AND OUTLOOK

We demonstrated that the DBN-based BAD-MoB-model has the ability to *predict* agent's behavior, to *abduct* hazardous situations (what could have been the

434

initial situation), to *generate* anticipatory plans and control, and to *plan* counteractive measures by *simulating* counterfactual behaviors or actions *preventing* hazardous situations. In Eilers and Möbus (2010) we present an efficient implementation. The next research steps will work on the vertical refinement of models interfacing single *actions* with more concrete *behaviors*.

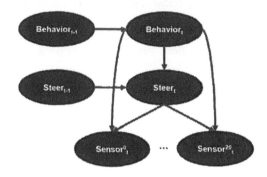

FIGURE 8 *Dynamic* BAD-MoB-model with Bayesian Classifiers *Behavior* and *Steer*

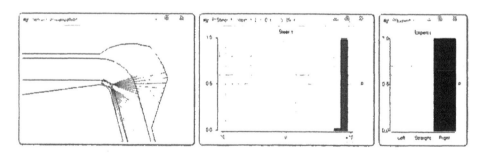

Fig. 9 Ambient perceptional evidence (left) and conditional distributions (middle, right) in *Dynamic Partial Inverted* BAD-MoB-model with 2 Bayesian Classifiers when driving in a right bended curve

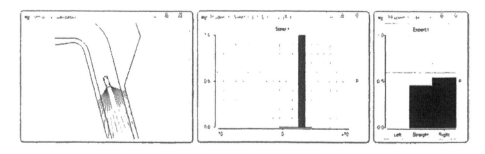

Fig. 10 Ambient perceptional evidence (left) and conditional distributions (middle, right) in *Dynamic Partial Inverted* BAD-MoB-model with 2 Bayesian Classifiers when leaving a right bended curve

REFERENCES

Bessiere, P. (2003). Survey: Probabilistic Methodology and Techniques for Artifact Conception and Development, Rapport de Recherche, No. 4730, INRIA

Bessiere, P. Laugier, Ch. and Siegwart, R. (eds.) (2008). *Probabilistic Reasoning and Decision Making in Sensory-Motor Systems*, Berlin: Springer, ISBN 978-3-540-79006-8

Cacciabue, P.C. (ed). (2007). *Modeling Driver Behaviour in Automotive Environments*, London: Springer, ISBN-10: 1-84628-617-4

Eilers, M. and Möbus, C. (2010). Learning of a Bayesian Autonomous Driver Mixture-of-Behaviors (BAD MoB) Model, (this proceedings), invited Special Interest Group: Möbus, C. & Bessiere, P., Models of Human Behavior and Cognition in the Bayesian Programming Framework, 1st International Conference On Applied Digital Human Modeling, 17-20 July, 2010, Intercontinental, Miami Florida, USA

Horrey, W.J. et al. (2006). Modeling Driver's Visual Attention Allocation While Interacting With In-Vehicle Technologies, J. Exp. Psych., 12, 67-78

Koike, C.C. Bessiere, P., and Mazer, E. (2008). Bayesian Approach to Action Selection and Attention Focusing, in Bessiere et al., (Eds.), Probabilistic Reasoning and Decision Making in Sensory-Motor Systems, Berlin: Springer, 177- 201

Koller, D., Friedman, N. (2009). Probabilistic Graphical Models, Cambridge, Mass.: MIT Press, ISBN 978-0-262-01319-2

Lebeltel, O. Bessiere, P. Diard, J. and E. Mazer (2004). Bayesian Robot Programming, Advanced Robotics, 16 (1), 49-79

Möbus, C. and Eilers, M. (2008). First Steps Towards Driver Modeling According to the Bayesian Programming Approach, Symposium Cognitive Modeling, p.59, in: L. Urbas, Th. Goschke & B. Velichkovsky (eds) *KogWis 2008*. Christoph Hille, Dresden, ISBN 978-3-939025-14-6

Möbus, C. Eilers, M. (2009a). Further Steps Towards Driver Modeling according to the Bayesian Programming Approach, in: *Conference Proceedings, HCII 2009, Digital Human Modeling*, pp. 413-422, LNCS (LNAI), Springer, San Diego, ISBN 978-3-642-02808-3

Möbus, C. Eilers, M. Garbe, H., and Zilinski, M. (2009b). Probabilistic and Empirical Grounded Modeling of Agents in (Partial) Cooperative Traffic Scenarios, in: *Conference Proceedings, HCII 2009, Digital Human Modeling*, pp. 423-432, LNCS (LNAI), Springer, San Diego, ISBN 978-3-642-02808-3

Möbus, C. Eilers, M. Zilinski, M. Garbe, H. (2009c). Mixture of Behaviors in a Bayesian Driver Model, in: Lichtenstein, A. et al. (eds), Der Mensch im Mittelpunkt technischer Systeme, p.96 and p.221-226 (CD), Düsseldorf: VDI Verlag, ISBN 978-3-18-302922-8, ISSN 1439-958X

Pearl, J. (2009). Causality – Models, Reasoning, and Inference, 2nd ed., Cambridge University Press, ISBN 978-0-521-89560-6

Yangsheng Xu, Ka Keung Caramon Lee, and Ka Keung C. Lee, *Human Behavior Learning and Transfer*, CRC Press Inc., (2005)

CHAPTER 46

Learning of a Bayesian Autonomous Driver Mixture-of-Behaviors (BAD MoB) Model

Mark Eilers[1], Claus Moebus[2]

C.v.O. University / OFFIS, Oldenburg
D-26111 Oldenburg, Germany
http://www.lks.uni-oldenburg.de/
{FirstName.LastName}@uni-oldenburg.de

ABSTRACT

The Human or Cognitive Centered Design (HCD) of intelligent transport systems requires computational *Models of Human Behavior and Cognition* (MHBC). They are developed and used as *driver models* in traffic scenario simulations and risk-based design.

The *conventional approach* is first to develop *handcrafted* control-theoretic or artificial intelligence based prototypes and then to evaluate *ex post* their learnability, usability, and human likeness. We propose a machine-learning alternative: The *Bayesian estimation* of *MHBCs* from behavior traces. The learnt *Bayesian Autonomous Driver (BAD)* models are empirical valid by construction. An ex post evaluation of BAD models is not necessary.

BAD models can be built so that they decompose or compose skills into or from

---

[1]project ISi-PADAS funded by the European Commission in the 7th Framework Program, Theme 7 Transport FP7-218552

[2]project Integrated Modeling for Safe Transportation (IMOST) sponsored by the Government of Lower Saxony, Germany under contracts ZN2245, ZN2253, ZN2366

basic skills: BAD Mixture-of-Behaviors (BAD MoB) models. We present an efficient implementation which is able to control a simulated vehicle in real-time. It is able to generate complex behaviors of several layers of expertise by mixing and sequencing simpler behavior models.

Keywords: Bayesian Autonomous Driver Models, Mixture of Behavior, Mixture of Experts, Bayesian Real-Time-Control, Levels of Expertise

INTRODUCTION

The skills and the skill acquisition process of human (traffic) agents can be described by a three-stage model consisting of a *cognitive*, an *associative*, and an *autonomous* stage or layer (Fitts, 1967; Anderson, 2002). For each stage, various modeling approaches have emerged: production-system models for the *cognitive and associative* stage, control-theoretic, or probabilistic models for the *autonomous* stage.

Due to the variability of human cognition and behavior, the *irreducible lack of knowledge* about underlying cognitive mechanisms and *irreducible incompleteness* of knowledge about the environment (Bessière, 2008) we conceptualize, estimate and implement probabilistic human traffic agent models. We described first steps to model lateral and longitudinal control behavior of single and groups of drivers with simple *reactive* Bayesian sensory-motor models (Möbus and Eilers, 2008). Then we included the time domain and reported work in progress with dynamic Bayesian sensory-motor models (Möbus and Eilers, 2009a; 2009b). In this paper we propose a dynamic BAD MoB model which is able to decompose complex *maneuvers* into basic *behaviors* and vice versa. The model facilitates the management of sensory-motor schemas (= *behaviors*) in a library. Context dependent driver behavior can then be generated by mixing pure basic *behaviors*.

BASIC CONCEPTS OF BAYESIAN PROGRAMS

BAD MoB models are developed in the tradition of Bayesian expert systems (Pearl, 2009) and Bayesian (Robot) Programming (Bessière et al., 2003, 2008). A Bayesian Program (BP) (Bessiere et al., 2003, 2008, Lebeltel et al., 2004) is defined as a mean of specifying a family of probability distributions. By using such a specification it is possible to construct a driver model, which can effectively control a (virtual) vehicle. The components of a BP are presented in Fig. 1.

An *application* consists of a (behavior model) description and a question. A *description* is constructed from preliminary knowledge π and a data set δ. *Preliminary knowledge* is constructed from a set of pertinent variables, a decomposition of their joint probability distribution (JPD) and a set of forms. *Forms* are either parametric forms or questions in other BPs.

The purpose of a *description* is to specify an effective method to compute a JPD on a set of variables given a set of (experimental) *data* and preliminary knowledge. To specify *preliminary knowledge* the modeler must *define the set of relevant variables* on which the JPD is defined, *decompose the JPD* into factors of (conditional) probability distributions (CPDs) according to conditional independency hypothesis (CIHs), and *define their forms*. Each CPD in the decomposition is a form. Either this is a *parametric form* whose parameters are estimated from batch data (behavior traces) or a *question* to another *application*. Parameter estimation from batch data is the conventional way of estimating the parameters in a BAD model. The Bayesian estimation procedure uses only a small fraction of the data (cases) for updating the model parameters.

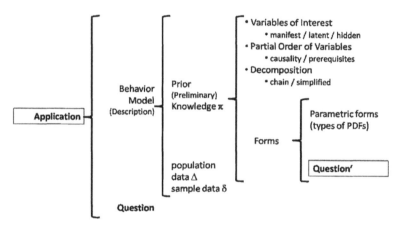

FIGURE 1. Structure of a Bayesian Program (adapted from Bessiere et al., (2003, 2008), Lebeltel et al., (2004)).

Given a description a *question* is obtained by partitioning the variables into *searched*, *known*, and *unknown* variables. A question is defined as the CPD $P(Searched|known, \pi, \delta)$. Various *policies* (Draw, Best, and Expectation) are possible whether the concrete *action* is *drawn* at random, chosen as the *best* action with highest probability, or as the *expected* action.

BAYESIAN-AUTONOMOUS-DRIVER MIXTURE-OF-BEHAVIOR MODELS

We presented a probabilistic model architecture for embedding layered models of human driver expertise which allow sharing of *behaviors* in different driving *maneuvers* (Möbus and Eilers, 2010). These models implement the sensory-motor system of human drivers in a psychological motivated *mixture-of-behaviors (MoB)* architecture with *autonomous and goal-based attention allocation* processes. A Bayesian MoB model is able to decompose complex skills into basic skills and to compose the expertise to drive complex *maneuvers* from basic *behaviors*.

We gave a proof of concept with plausible but artificial data and first modeling results with real data. We demonstrated that the Dynamic Bayesian Network (DBN)-based BAD MoB model has the ability to *predict* agent's behavior, to *abduct* hazardous situations (what could have been the initial situation), to *generate* anticipatory plans and control, and to *plan* counteractive measures by *simulating* counterfactual behaviors or actions *preventing* hazardous situations.

With an increasing number of observable action- or percept-variables and especially latent state- or behavior-variables, inferences in a BAD MoB model can soon become too complex to be computable for real-time-control. Therefore we propose an effective implementation of BAD MoB models, based on the concept of *behavior-combination* (Bessière et al., 2003), that allows to realize DBN-based BAD MoB model by several simpler BPs.

BASIC CONCEPTS OF IMPLEMENTATION

A BAD MoB model as proposed in Möbus and Eilers (2010) intends to model n *behaviors*. It contains a set of *action*-variables A, a set of *percept*-variables $P = P^1, \ldots, P^m$ and a single *behavior*-variable $B = \{1, \ldots, n\}$ with n values for the n behaviors[3]. This BAD MoB model can efficiently be implemented by BPs with three different purposes which we will call: *Action-*, *behavior-classification-* and *gating-models*.

Each *behavior* b_i $i \in \{1, \ldots, n\}$ has to be defined by an *action*-model, with preliminary knowledge π_i and sample data δ_i, consisting of the set of *action*-variables A and an own set of *percept*-variables $P_i \subseteq P$. An *action*-model defines the JPD $P(A, P_i | \pi_i \delta_i)$ that will be used to answer the question $P(A | P_i, \pi_i, \delta_i)$.

Identification of proper behaviors for a given situation is achieved using a *behavior-classification*-model. It consists of the behavior-variable B and a set of *percept*-variables $P_B \subseteq P$. They define the JPD $P(B, P_B | \pi_B \delta_B)$ and will be used to answer the question $P(B | P_B, \pi_B, \delta_B)$.

The *action*-models and *behavior-classification*-model are combined by the *gating*-model, which consists of the *action*-variables A, the *percept*-variables P and the *behavior*-variable B. Whereas the JPDs of *action-* and *behavior-classification*-models may be decomposed into simpler terms according to CIHs, the JPD of a *gating*-model is decomposed as follows:

$$P(A, P, B | \pi, \delta)$$

$$= P(P | \pi, \delta) \cdot P(B | P, \pi, \delta) \cdot P(A | P, B, \pi, \delta).$$

The decomposition of a *gating-model* consists of three terms: $P(P | \pi, \delta)$ is the prior distribution of all *percept*-variables and can be derived from (experimental)

---

[3] The implementation of BAD MoB models we propose is not restricted to static components, which may be implemented as DBNs. Here we use a static example for reasons of clarity.

440

data or assumed to be uniform. The term $P(B|P, \pi, \delta)$ denotes the probability of each *behavior* for the given percepts and will be defined as a question to the *behavior-classification*-model:

$$P(B|P, \pi, \delta) \equiv P(B|P_B, \pi_B, \delta_B).$$

For each possible behavior $B = b_i, 1 \leq i \leq n$ the term $P(A|P, B = b_i, \pi, \delta)$ is defined as a question to the corresponding i-th *action*-model:

$$P(A|P, B = b_i, \pi, \delta) \equiv P(A|P_i, \pi_i, \delta_i).$$

The question to be answered by a BAD-MoB model is $P(A|P, \pi, \delta)$. By asking this question to the *gating-model* we obtain the weighted sum over all behaviors:

$$P(A|P, \pi, \delta)$$
$$= \sum_{i=1}^{n} [P(B = b_i|P, \pi, \delta) \cdot P(A|P, B = b_i, \pi, \delta)]$$
$$= \sum_{i=1}^{n} [P(B = b_i|P_B, \pi_B, \delta_B) \cdot P(A|P_i, \pi_i, \delta_i)].$$

This structure of a BAD MoB model can be seen as a template. A BAD MoB model can be extended to *hierarchical* BAD MoB model by replacing some of its *action*-models with further BAD MoB models. An example is shown in Fig. 2

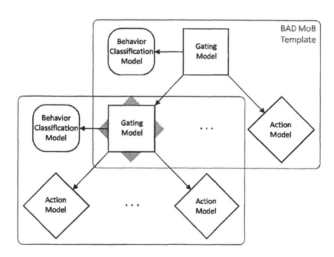

FIGURE 2. Graphical representation of a hierarchical BAD MoB model constructed by BAD MoB templates, where an *action*-model was replaced by a further BAD MoB model. Rectangle nodes represent *gating*-, rounded rectangles represent *behavior-classification*-, and diamond nodes represent *action*-models (notation adapted from

Bishop and Svensen (2003)). Directed connections represent that CPDs of the parent-model are defined to be questions of the child-model.

IMPLEMENTATION

Using the racing simulation TORCS[4] we implemented a BAD MoB model intended to master a complex *driving scenario*. The scenario covers the ability to drive on a racing track together with two other slow vehicles. When approaching a slower car, they should be followed until given the possibility for overtaking.

LEVELS OF EXPERTISE

In reference to (Möbus and Eilers, 2010), this intended *driving scenario* was split up into *driving maneuvers,* namely *Lane-Following, Car-Following* and *Overtaking. Lane-Following,* a complex maneuver by itself (Möbus and Eilers, 2009a), was supposed to be created by mixing and sequencing the *lane-following.behaviors* for driving through a left curve (*Left*), along a straight road (*Straight*) and through a right curve (*Right*). Accordingly, the maneuver *Car-Following* consists of *car-following.behaviors* for following a car through a left curve (*FollowLeft*), on a straight road (*FollowStraight*) and through a right curve (*FollowRight*). The third maneuver *Overtaking* is composed by the three *overtaking.behaviors* of veering to the left lane (*PassOut*), passing the car (*Pass Car*) and go back to the lane *(PassIn)*. Each *action*-model will infer concrete actions for steering wheel angle and a combined acceleration-braking-pedal, which refers to the *driving action* level of expertise. The referring BAD MoB model therefore consists of four *gating-*, four *behavior-classification-* and nine *action*-models on three hierarchical layers, covering four levels of expertise. The structure of the model is shown in Fig. 3.

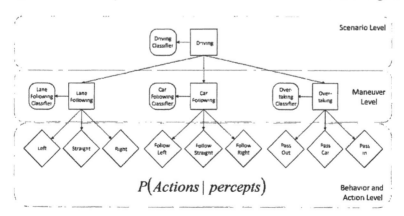

$$P(Actions \mid percepts)$$

FIGURE 3. Hierarchical structure of BAD MoB model constructed by four *gating-*,

---

[4] http://torcs.sourceforge.net/ (last retrieved 2010-02-26)

four *behavior-classification-*, and nine *action*-models, covering four levels of expertise.

MODELING PURE BEHAVIORS BY ACTION-MODELS

Each of the nine *action*-models was implemented as a DBN. The *action*-models *Left*, *Straight*, and *Right* share the same preliminary knowledge, specify the same variables and define the same decompositions. They only differ in the experimental data set used for parameter estimation. The same applies for the *FollowLeft*, *FollowStraight*, and *FollowRight action*-models, and for the *action*-models *Pass-Out*, *Pass-Car*, and *PassIn*. Their structure is shown in Fig. 4.

For each time slice, variable $Steer^t$ represents the current steering wheel angle, Acc^t represents the position of a combined acceleration-braking-pedal. $Speed^t$ denotes the longitudinal velocity. A variable $Mid\angle_i^t$ represents the angle between heading vector of the car and the vector to the middle of the right lane in a distance of i meters. In contrast to this, a variable $Cou\angle_i^t$ represents the angle between heading of the car and the course of the road in a distance of i meters. The variables Dis^t and $Car\angle^t$ represent distance and angle to the nearest other vehicle. All pertinent variables were chosen as a tradeoff between computation speed and model performance, guided by statistical methods (i.e. *likelihood maximization*).

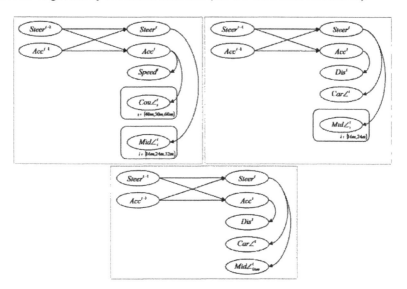

FIGURE 4. **Upper Left**: DBN of *Left, Straight*, and *Right action*-models. The boxes, called plates, denote copies of the nodes shown inside the box. **Upper Right**: DBN of *FollowLeft, FollowStraight*, and *FollowRight action*-models. **Lower Middle**: DBN of *PassOut, PassCar*, and *PassIn action*-models.

BEHAVIOR-IDENTIFICATION BY BEHAVIOR-CLASSIFICATION-MODELS

For behavior identification each *behavior-classification*-model was implemented in form of a DBN. In each time slice, the *behavior-classification*-models define a single *behavior*-variable representing the current *driving maneuver* or *behavior*, namely DM^t for the *Driving-Maneuver-Classification* model, LFB^t for the *Lane-Following-Behavior-Classification* model, CFB^t for the *Car-Following-Behavior-Classification* model, and OB^t for the *Overtaking-Behavior-Classification* model. For all *behavior-classification*-models each time slice is implemented as *naïve Bayesian classifier*. The pertinent variables were chosen as a tradeoff between computation speed and model performance, guided by statistical methods (i.e. *likelihood maximization*). The structure of the *behavior-classification*-models is shown in Fig. 5.

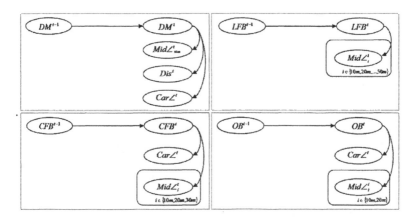

FIGURE 5. **Upper Left**: DBN of *Driving-Maneuver-Classification* model. **Upper Right**: DBN of *Lane-Following-Behavior-Classification* model. **Lower Left**: DBN of *Car-Following-Behavior-Classification* model. **Lower Right**: DBN of *Overtaking-Behavior-Classification* model.

BEHAVIOR-COMBINATION BY GATING-MODELS

Following the structure shown in Fig. 3, the *action*-models were combined by the *Lane-Following-Maneuver-*, *Car-Following-Maneuver-*, and *Overtaking-Maneuver-gating* model using their corresponding *behavior-classification*-models for behavior identification. These three *gating-models* were then combined by the *Driving-Scenario-Gating* model using the *DMC* model for maneuver identification. Considering the defined decomposition of *gating*-models, we will relinquish to show their structure.

LEARNING BY DATA COLLECTION AND BEHAVIOR ANNOTATION

For the purpose of data collection four laps were driven by a single driver, two laps at a time on two different TORCS racing tracks, containing several complex chicanes like s-shaped curves and hairpins. Instructions were given to drive sensual, stay on the right side of the road and observe a speed limit of approximately 110 km/h (70 mph). When approaching a slower car, it should be followed in short distance until a longer straight road segment would allow an overtaking-maneuver. Experimental data for parameter estimation was then obtained by recording time series of all current variable values. As values of *behavior*-variables were unknown during recording, the time series were annotated offline, manually setting the appropriated behaviors.

RESULTS

First results are very promising. With the recorded experimental data the BAD MoB model is able to accomplish the racing tracks used for data collection and other tracks of comparable complexity. The model successfully performs *Car-Following* and *Overtaking* maneuvers (an example of model-ability is shown in Fig. 6, videos are available at http://www.lks.uni-oldenburg.de/46350.html). Compared to former BAD models (Möbus and Eilers, 2008, 2009a) the driving performance was considerably improved: the BAD MoB model now stays on the right lane, sticks to the intended high speed and does not collide with roadsides anymore. In addition, the use of the proposed BAD MoB model structure significantly improved performance towards implementation of combined BAD MoB models.

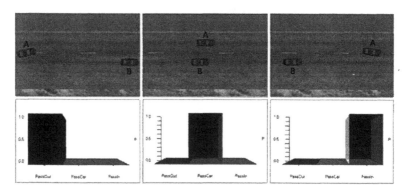

FIGURE 6. Sequencing of behaviors during *Overtaking* maneuver. Upper row shows snapshots of BAD MoB model (A) in TORCS simulation overtaking slower vehicle (B), lower row shows corresponding CPD of *overtaking.behavior* variable OB^t.

CONLUSION AND OUTLOOK

We believe that the proof of concept is convincing: Bayesian Autonomous Driver Models with Mixture-of-Behavior are expressive enough to describe and predict a wide range of phenomena. Next we have to implement further models creating a library of behaviors of various levels of expertise. To that end a careful selected taxonomy of scenarios, maneuvers, behaviors, and control actions without and with alter agents has to be defined and studied.

REFERENCES

Anderson, J.R. (2002): *Learning and Memory*, John Wiley

Bessière, P. and the BIBA INRIA Research Group (2003): Survey: Probabilistic Methodology and Techniques for Artefact Conception and Development, Technical Report RR-4730, INRIA.

Bessière, P., Laugier, Ch. and Siegwart, R. (eds.) (2008) Probabilistic Reasoning and Decision Making in Sensory-Motor Systems, Berlin: Springer, ISBN 978-3-540-79006-8

Bishop, C. M. and Svensén, M. (2003): Bayesian hierarchical mixtures of experts. In: Kjaerulff, U. and C. Meek (Ed.): Proceedings of the Nineteenth Conference on Uncertainty in Artificial Intelligence, pp. 57-64.

Fitts, P.M. and Posner, M.I., Human Performance, Belmont, CA: Brooks/Cole, ISBN 0-13-445247-X, (1967)

Möbus, C. and Eilers, M. (2008): First Steps Towards Driver Modeling According to the Bayesian Programming Approach, p.59, in L. Urbas, et al. (eds) KogWis. Christoph Hille, Dresden, ISBN 978-3-939025-14-6

Möbus, C. and Eilers, M. (2009a): Further Steps Towards Driver Modeling according to the Bayesian Programming Approach, Conference Proceedings, HCI 2009, Digital Human Modeling, San Diego, Springer: Lecture Notes in Computer Science (LNCS) and Lecture Notes in Artificial Intelligence (LNAI)

Möbus, C., Eilers, M., Garbe, H., and Zilinski, M. (2009b): Probabilistic, and Empirical Grounded Modeling of Agents in Partial Cooperative (Traffic) Scenarios, Conference Proceedings, HCI 2009, Digital Human Modeling, San Diego, Springer: LNCS and LNAI)

Möbus, C., Eilers, M., Zilinski, M., and Garbe, H. (2009c): Mixture of Behaviors in a Bayesian Driver Model, 8. Berliner Werkstatt, Mensch-Maschine-Systeme - Der Mensch im Mittelpunkt technischer Systeme, 7.-9. Oktober 2009, Berlin-Brandenburgische Akademie der Wissenschaften, VDO-Verlag

Möbus, C. and Eilers, M. (2010): Mixture-of-Behaviors and Levels-of-Expertise in a Bayesian Driver Model, (this proceedings), invited Special Interest Group Möbus, C. & Bessiere, P., Models of Human Behavior and Cognition in the Baycsian Programming Framework, 1st Intern.Conf. On Applied Digital Human Modeling, 17-20 July, 2010, Intercontinental, Miami Florida, USA

Pearl, J. Causality – Models, Reasoning, and Inference, 2nd ed., Cambridge University Press, 2009, ISBN 978-0-521-89560-6

Chapter 47

Modeling Lateral and Longitudinal Control of Human Drivers with Multiple Linear Regression Models

Jan Charles Lenk, Claus Möbus

University of Oldenburg
jan.lenk@uni-oldenburg.de
claus.moebus@uni-oldenburg.de

ABSTRACT

In this paper, we describe results to *model lateral and longitudinal control behavior* of drivers with simple *linear multiple regression models*. This approach fits into the *Bayesian Programming* (BP) approach (Bessière, 2008) because the linear multiple regression model suggests an action selection strategy which is an alternative to the BP action selection strategies *draw* and *best*. Furthermore, the inference process provided by a linear multiple regression model is a kind of *short cut inference* compared to the inference approach used in Bayesian networks or Bayesian Programming.

Keywords: digital human modeling, driver modeling, lateral and longitudinal control, linear multiple regression model, Bayesian Programming

INTRODUCTION

Modeling driver behavior is essential for developing error-compensating assistance systems (Cacciabue, 2007). The Human Centered Design of Partial Autonomous Driver Assistance Systems (PADAS) requires Digital Human Models (DHMs) of human control strategies for simulating traffic scenarios (Möbus et al., 2009). There are a number of control-theoretical driver models (Jürgensohn, 2007; Weir and Chao, 2007) available. Salvucci and Gray (2004) and Salvucci (2007) proposed an integrated model (S&G-model), which has been implemented as a production system within the wide-spread cognitive architecture ACT-R (Anderson et al., 2004). However, only lateral control in this integrated model has been achieved by using a control model, which uses visual signals as input for steering actions. Longitudinal control is missing.

Using linear multiple regression models, we estimated the optimal coefficients as well as parameters for the *lateral* control model from single human test drives. Additionally, the steering model has been reformulated in order to achieve a better fit with human data. The same techniques were applied to *longitudinal* control, i.e. acceleration and deceleration.

This approach fits into the Bayesian Programming (BP) approach (Bessière et al., 2008). The multiple regression model *E(Action | Percepts)* uses the conditional probability distribution *P(Action | Percepts)*. This is a *form* in the BP framework. The linear multiple regression model suggests the action selection strategy *selection of expected conditional action* which is an alternative to the BP action selection strategies *draw* and *best*.

The regression models are learnt from multivariate time series of driving episodes generated by a single driver. The variables of the time series describe phenomena and processes of perception, cognition, and action control of drivers according to the S&G-Model. The real-time control of virtual vehicles is achieved by inferring the appropriate actions under the evidence of sensory percepts. This is a slightly different but more efficient action selection strategy than those used in the BP framework. Here the action selection strategies are $draw(P(Action \mid Percepts))$ or $best(P(Action \mid Percepts))$. According to the *draw* strategy the concrete action is randomly selected from the conditional probability distribution $P(Action \mid Percepts)$, while under the *best* strategy the concrete action with the highest probability or density is selected from the conditional probability distribution $P(Action \mid Percepts)$. This differs from our approach, where the selected action is the *conditional expected value* $E(Action \mid Percepts)$ under the constraint of a linear model.

THE TWO-POINT VISUAL CONTROL MODEL OF STEERING

In the Two-Point Visual Control Model (S&G-Model; Salvucci and Gray, 2004), steering actions are controlled by points which are obtained from the road. It is based on the experiments of Land and Horwood (Land and Horwood, 1995; Land,

1998), where participants were shown only small visual segments of the road. This resulted in the hypothesis that the quality of driving improved with the horizontal angle of *two* visual *segments*. This hypothesis was adapted by the S&G-Model and now states that these visual signals are in fact *two points*: The Near Point (N) is defined by its distance d_N to the vehicle. The Far Point's location (F) is dependent on the situation (Figure 1.1): On straight road strips, it is defined by the *escape point*, while on bent strips it is defined by the *tangential point*. A third situation is defined by a leading vehicle, but shall not be of further interest in this work.

The respective angles between F and N and the car's longitudinal axis, θ_F and θ_N, constitute the errors for two parallel-connected controllers. Thus, the steering angle φ is computed by a Proportional (P) controller for F and a Proportional-Integral (PI) controller for N in the original S&G-Model (Salvucci, 2004):

$$\varphi = k_N \theta_N + k_F \theta_F + k_I \int \theta_N dt$$

Thus, the coefficients k_N, k_F, and k_I are unknown. In the work of Salvucci and Gray, the distance between N and the car's location is given as 6.2m. However, it should be noted that all original experiments were conducted at the relatively low constant speed of 60.84 km/h. Thus, no longitudinal control was needed in their experimental setting.

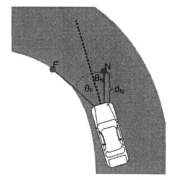

FIGURE 1.1.   Straight (left) and Bend (right) situations in the S&G Model

This leads to the question whether the parameters are dependent on the vehicle speed. In order to achieve this, test drives can be used to identify the coefficients as well as the N distance parameter. A less coercive model should be able to cope with high and variable velocities as well.

REIMPLEMENTATION OF THE S&G-MODEL

We already implemented the S&G-Model before (Möbus et al., 2007). We had some doubts whether the autonomous control of a vehicle is dependent on foveal control as is hypothesized by Salvucci and Gray (2004). There is some evidence that ambient vision is sufficient for real-time control in routine driving situations (Horrey et al., 2006). A Bayesian model for longitudinal and lateral control which rests on the assumption of ambient vision has been presented by Möbus and Eilers (2009).

The current implementation uses the open-source TORCS[1] racing simulator as simulation environment, which has been augmented with a development environment for all kinds of control-theory based driver models (Lenk, 2008). In this implementation, the selection rules for the situation-dependent Far Point calculation had to be redefined, as the original ACT-R model could not be reused. The front-vehicle situation has not been considered, thus leaving the escape and tangential point situations. These were distinguished by introducing another parameter, the distance d_F (Figure 1.1) to the Far Point. Thus, if the road strip in the given distance ahead is curved, the tangential point is calculated, otherwise the escape point serves as F.

ESTIMATING LATERAL CONTROL

A simulated test drive has been conducted. The track chosen for this drive features many different types of curves with varying radii. The drive lasted around 14 minutes, during which a multivariate time series with 95522 episodes of data had been sampled. These included the car state, such as position, velocity, acceleration, and orientation, as well as the steering, braking, and accelerating actions of the driver. The human driver had been instructed to drive fast, but careful enough to stay on the road at all times, although cutting curves was permitted. Average speed was 30.65039 m/s (around 110 km/h) with a standard deviation of 5.475582 m/s. This may be a significantly higher average speed than any one encountered in day-to-day traffic, however we felt the model should work under extreme conditions.

REGRESSION FOR THE S&G MODEL

The locations of the control points N and F are hypothetical constructs. We had to infer the distances as a second set of parameters. In order to find the optimal distance parameters, we conducted a grid search over a set of regressions on the same test data with varying distances d_N and d_F. The controller's coefficients were estimated for any distance between $10m$ and $40m$ for d_N and $10m$ to $80m$ for d_F using a multiple linear regression model.

$$y = \beta_1 x_1 + \beta_2 x_2 + \beta_3 x_3$$

The dependent variable y corresponds to the steering angle φ, while the independent variables are x_1 (θ_N), x_2 (θ_F), and x_3 ($\int \theta_N \, dt$). The parameters $\beta_1, \beta_2,$

---

[1] http://torcs.sourceforge.net/ (last retrieved: 2/25/2010)

and β_3 are estimated. The model does not include an intercept term. The determination coefficient R^2 of the regressions is very high in general (Figure 2.1). Subsequently, the parameter estimations are used as controller coefficients k_N, k_F, and k_I, respectively. The d_N^{opt}, d_F^{opt}-tuple with the highest coefficient of determination R^2 determines two optimal angles θ_N^{opt} and θ_F^{opt}, which best explain the actions of the human driver. Thus, the controller calculates the conditional expected value $E\left(\varphi|\theta_N^{opt},\theta_F^{opt}\right)$ using the sum of the products of the estimated coefficients with their respective angles.

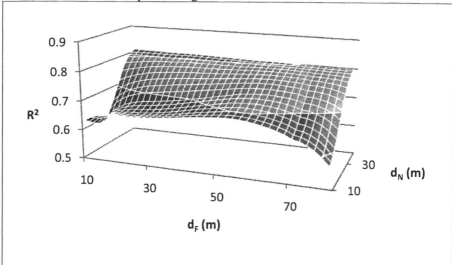

FIGURE 2.1.   Coefficient of determination R^2 for multiple regressions on varying distances d_N and d_F using the original unmodified 3-parameter S&G lateral control model.

EXTENDING THE S&G MODEL

We adapted the S&G-Model by introducing a segmentation of the road into *bent* and *straight* strips. This approach effects in a doubling of the number of controller coefficients. Thus, k_{NS}, k_{FS}, and k_{IS} guide the model on straights, while k_{NB}, k_{FB}, and k_{IB} perform the same function on bends.

$$\varphi = \begin{cases} k_{NS}\theta_N + k_{FS}\theta_F + k_{IS}\int \theta_N dt, & \text{on straight segments} \\ k_{NB}\theta_N + k_{FB}\theta_F + k_{IB}\int \theta_N dt, & \text{on bent segments} \end{cases}$$

Accordingly, the data matrix for the predictors in the regression features six columns, which are either filled with the actual values for the angles in one condition or set to zero when the opposite condition applies. The type of Far Point calculation serves as discriminator in order to determine whether the straight or bent strip condition is in place. If the tangential point is calculated, the columns in the data matrix corresponding to θ_{*B} are filled with the actual values, or vice versa if the escape point is calculated.

The results of the regressions (Figure 2.2) generally show a higher coefficient of determination for all reviewed distances d_N and d_F. The maximum $R^2 = 0.809$ can be found at $d_N^{opt} = 40$ and $d_F^{opt} = 10$. These extreme values may seem surprising at first, but the estimated controller coefficients (Table 2.1) perform well. The resulting behavior is similar to that of the actual test drive, so that while the cutting corner behavior is reproduced, the car is kept stable in the middle of the road.

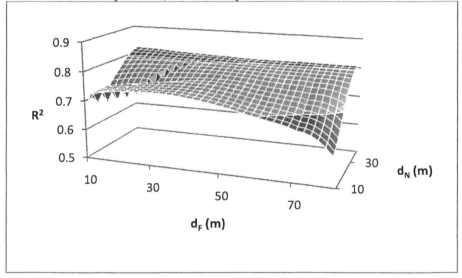

FIGURE 2.2.   Coefficient of determination R^2 for multiple regressions on varying distances d_N and d_F using the modified lateral control model with segmentation.

Due to the fact that d_F^{opt} is actually lower than d_N^{opt}, the name characterization as a "Far Point" might be disputed. However, this is only the case for the straight condition, since the distance of the tangential point in the bent condition is usually higher than the distance to the Near Point. This could be an indicator that a single-point model could be preferable on straight roads.

Table 2.1 Estimated controller coefficients for $d_N^{opt} = 40$ and $d_F^{opt} = 10$.

k_{NS}	k_{FS}	k_{IS}	k_{NB}	k_{FB}	k_{IB}
0.166	0.0038	6.16×10^{-5}	0.1295	0.0461	-0.0002

ESTIMATING LONGITUDINAL CONTROL

In order to achieve longitudinal control emitting acceleration and deceleration actions, a naïve control model has been chosen first. Thus, braking and accelerating actions u are numeric values on the same axis with opposite signs. A PID controller adjusts this value using the difference between the actual velocity v and a set-point velocity v_d (Coller, 2007).

$$u = -\left(c_P(v - v_d) + c_I \int (v - v_d)dt + c_D \frac{d(v - v_d)}{dt}\right)$$

However, while the actual velocity v is known from the experiment data, the set-point velocity v_d might be considered an internal state of the driver, thus not being observable. However, it might be approximated by a heuristic using two parameters: a braking rate b and a maximum velocity v_{top}.

For each road segment s, a maximum velocity v_s^{max} may be determined. For a bent road segment, it is defined by the radius r_s, and a friction constant f_s.

$$v_s^{max} = \begin{cases} \min\left(\sqrt{f_s \cdot G \cdot r_s}, v_{top}\right) & \text{on bends} \\ v_{top} & \text{on straights} \end{cases}$$

All road segments within a velocity-dependent look-ahead distance $d_l = v^2/(2 \cdot b)$ are examined for their maximum velocity. Thus, the set-point velocity is approximated by $v_d = \min(v_s^{max} \mid \forall s \text{ with } d_s < d_l)$.

With this approximation of v_d, it is possible to conduct yet another grid search over the parameters b and v^{top}, using multiple linear regressions to estimate c_P, c_I, and c_D (Figure 3.1). Once again, the coefficient of determination is used to select the optimal parameters b^{opt} and v_{top}^{opt}, which effect an optimal set-point speed v_d^{opt}. Thus, the controller output is the conditional expected value $E\left(u \mid v_d^{opt}\right)$.

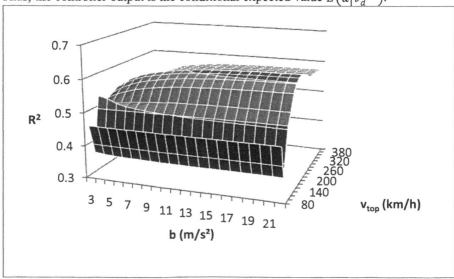

FIGURE 3.1.   Coefficient of determination R^2 for multiple regressions on varying distances b and v_{top} using the longitudinal control model.

Again, the values $b^{opt} = 22$ and $v_{top}^{opt} = 220$ may seem extreme, but provide the best fit with human data with $R^2 = 0.612$. The estimated coefficients for this configuration (Table 3.1) provide adequate acceleration and deceleration during the model run. It should be noted that the longitudinal controller ignores the current gear state altogether, even though the fuel pedal state depends on it. Thus, the estimation for the controller would have an even better fit with human data if it were adapted to accommodate this variable.

Table 3.1 Estimated longitudinal controller coefficients for $b^{opt} = 22$ and $v_{top}^{opt} = 220$

c_P	c_I	c_D
0.0069	2.59×10^{-6}	-5.35×10^{-5}

DEPENDENCY OF LONGITUDINAL CONTROL ON VISUAL PERCEPTS

Clearly, the above controller is not an entirely plausible model of a human driver, since the set-point speed v_d^{opt} cannot be readily established from human data. However, it may be derived from the absolute values of the visual percepts θ_N^{opt} and θ_F^{opt}. Thus, a nested controller may be embedded in the above longitudinal controller, which estimates the expected value $E(v_d^{opt}|\theta_N^{opt}, \theta_F^{opt})$. Again, discrimination between straight and bent segments takes place. A constant k is needed to provide a positive value when both angles converge to zero.

$$v_d^{opt} = \begin{cases} c_{NS}|\theta_N^{opt}| + c_{FS}|\theta_F^{opt}| + k & \text{on straights} \\ c_{NB}|\theta_N^{opt}| + c_{FB}|\theta_F^{opt}| + k & \text{on bends} \end{cases}$$

Using a single regression, values for the coefficients may be estimated (Table 3.2). The coefficient of determination for this regression is $R^2 = 0.5716$. In a way, both regressions for the longitudinal controller effect a role change for v and v_d^{opt}. As the velocity v is provided by human data, it can be considered the set-point velocity, while its difference to the model-provided v_d^{opt} is minimized by the nested regressions.

Table 3.2 Estimated controller coefficients v_d^{opt}.

c_{NS}	c_{FS}	c_{NB}	c_{FB}	k

-43.1718	-12.7844	-105.2268	-9.0415	60.6528

MODEL RUN

If the integrated model is run on the same track as the test drive using the estimated coefficients (Tables 2.1, 3.2, and 3.3) the model performance shows similar behavior for lateral as well as longitudinal control. The achieved average velocity of $29.98395\ m/s$ is slightly lower than in the original human data and the model does not achieve the same high velocities as the original human driver (Figure 3.2) with a standard deviation of $5.123619\ m/s$. Nevertheless, it performs well on the road. Generally spoken, the model actions tend to be more temperate than those of the human driver.

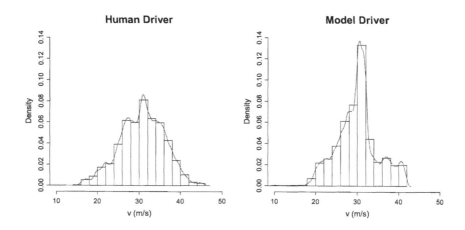

FIGURE 3.2. Density of achieved velocities by Human Driver (left) and integrated model (right) while driving.

CONCLUSION

We reformulate the S&G Model of steering as a linear multiple regression model and extend this model with a second linear multiple regression model for the purpose of longitudinal control. The inference process provided by a linear multiple regression model is a kind of *short cut inference* compared to the inference approach used in Bayesian networks or Bayesian Programming. From this perspective the percept-based inference or selection of actions on the basis of linear multiple regression models is suited for reactive agents and seems to be more efficient than the usual inference process of the BP approach. But there is an efficiency-flexibility trade-off. Linear models are more restrictive than the flexible

Bayesian Programs. They only allow one direction of inference.

The proceedings introduced in this work are reproducible for other human test data, even though the concrete estimations may vary. Further study is required to determine the relationships between lateral and longitudinal control.

REFERENCES

Anderson, J.R., Bothell, D., Byrne, M. D., Douglass, S., Lebiere, C., and Qin, Y. (2004). "An integrated theory of the mind", *Psychological Review*, Vol. 4, pp. 1036-1060

Bessière, P., Laugier, Ch., and Siegwart, R., (eds.) (2008). *Probabilistic Reasoning and Decision Making in Sensory-Motor Systems*, Springer, Berlin

Cacciabue, P.C. (2007). *Modelling Driver Behaviour in Automotive Environments*, Springer, London

Coller, B.D. (2007). "Implementing a video game to teach principles of mechanical engineering", *Proceedings of the 2007 American Society for Engineering Education Annual Conference*

Horrey, W.J., Wickens, Ch.D., and Consalus, K.P.: "Modeling Driver's Visual Attention Allocation While Interacting With In-Vehicle Technologies", *J. Exp. Psych.*, 2006, 12, 67-78

Jürgensohn, T. (2007). "Control Theory Models of the Driver", in Cacciabue, 2007, pp. 277-292

Land, M. (1998). *The Visual Control of Steering, Vision and Action*, editors Harris, L.R. and Jenkin, M., Cambridge University Press, Cambridge, pp. 163-180

Land, M. and Horwood, J. (1995). "Which Parts of the Road Guide Steering?", *Nature*, Vol. 377, pp. 339-340

Lenk, J. C. (2008). *Zum Rapid Prototyping von Fahrermodellen*, B.Sc.-Thesis, Department of Computing Science, University of Oldenburg

Möbus, C. and Eilers, M., "Further Steps Towards Driver Modeling according to the Bayesian Programming Approach", in: Vincent G. Duffy (Ed.), *Digital Human Modeling, HCI 2009, San Diego, CA, USA*, Springer: Lecture Notes in Computer Science (LNCS 5620) and Lecture Notes in Artificial Intelligence (LNAI), ISBN 978-3-642-02808-3, p. 413 - 422

Möbus, C., Eilers, M., Garbe, H., and Zilinski, M. (2009). "Probabilistic and Empirical Grounded Modeling of Agents in (Partial) Cooperative Traffic Scenarios", in: *Conference Proceedings, HCI 2009, Digital Human Modeling*, pp. 423-432, LNCS (LNAI), Springer, San Diego

Möbus, C., Hübner, S., and Garbe, H., "Driver Modelling: Two-Point- or Inverted Gaze-Beam-Steering", in M. Rötting, G. Wozny, A. Klostermann und J. Huss (eds), *Prospektive Gestaltung von Mensch-Technik-Interaktion, Fortschritt-Berichte VDI-Reihe 22*, Nr. 25, 483 – 488, Düsseldorf: VDI Verlag, 2007, ISBN 978-3-18-302522-0

Salvucci, D.D. (2007). "Integrated Models of Driver Behavior", in *Integrated models of cognitive systems*, editor Gray, W.D., Oxford University Press, New York, pp. 356-367

Salvucci, D.D. and Gray, W.D. (2004). "A Two-Point Visual Control Model of

Steering", *Perception*, Vol. 33, pp. 1233-1248

Weir, D.H. and Chao, K.C. (2007). "Review of Control Theory Models for Directional and Speed Control", in Cacciabue, 2007, pp. 293-311

CHAPTER **48**

A Unified Theoretical Bayesian Model of Speech Communication

Clément Moulin-Frier[1], Jean-Luc Schwartz[1], Julien Diard[2], Pierre Bessière[3]

[1]GIPSA-Lab, Speech and Cognition Department (ex-ICP), UMR 5216
[2]Laboratoire de Psychologie et NeuroCognition (LPNC), UMR 5105
[3]Laboratoire d'Informatique de Grenoble (LIG-Lab), UMR 5217
[1, 2, 3]CNRS – Grenoble University

ABSTRACT

Based on a review of models and theories in speech communication, this paper proposes an original Bayesian framework able to express each of them in a unified way. This framework allows to selectively incorporate motor processes in perception or auditory representations in production, thus implementing components of a perceptuo-motor link in speech communication processes. This provides a basis for future computational works on the joint study of perception, production and their coupling in speech communication.

Keywords: Speech Communication, Cognitive Bayesian Modeling, Sensory-Motor interaction

INTRODUCTION: MODELS AND THEORIES IN SPEECH COMMUNICATION

Speech communication involves a set of actuators for producing speech stimuli (enabling to control the orofacial system: lungs, glottis, jaw, tongue, lips, velum) and a set of sensors for perceiving them (audition of course, but also vision for lip-reading, and haptics and proprioception for sensing the state of the vocal tract). This enables the speaker to control the task in speech production that is achieving the

458

correct gestures for uttering the adequate sounds. Hence, speech production can be conceived as a typical robotics problem, involving proximal control in reference to given distal objectives, together with learning, adaptability, or any other problem related to cognitive robotics. But the special issue in speech communication is that the task IS communicative. The speaker is also a listener, and has probably a model of the listener incorporated in the production task itself. Production and perception are closely related in communication and probably also in the human's brain. This intimate link between production and perception in speech communication has been largely discussed by phoneticians and cognitive (neuro)psychology, but seldom addressed from a modeling perspective. This is the focus of the present paper.

A key question in speech science concerns the nature of the content of communication, with three major frameworks that are motor, auditory, and sensory-motor theories of speech communication. We shall describe here how each of these theories considers both the speech production and perception processes. Then we will propose a Bayesian formal framework able to express each of them in a unified way, and discuss the possible interpretations of this model. Finally, we will conclude on the possible functional roles for the perceptuo-motor link in speech communication.

MOTOR THEORIES

Motor theories consider the reference frames of speech communication as gestures. In the Articulatory Phonology framework (Browman and Goldstein, 1989), production is modeled as scores of overlapping gestures (Fig. 1.1), able to express the context-dependent variability of speech, without taking explicitly into account the auditory consequence of a motor event.

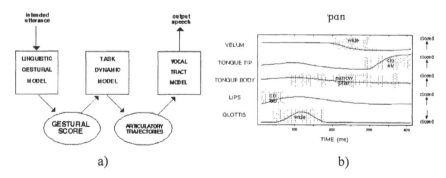

a)   b)

FIGURE 1.1   Articulatory Phonology (from Browman and Goldstein, 1989). a) Gestural computational model. b) Gestural score for the utterance 'pan'.

In line with the idea that the frames of speech communication are motor events, the Motor Theory of Speech Perception (Liberman and Mattingly, 1985) proposes that perceiving speech amounts to perceiving gestures. A main argument is coarticulation-driven signal variability, which makes the auditory content of a given phoneme dependent on the phonetic context (e.g. /d/ does not produce the same sound in /da/ vs. /du/, see Fig. 1.2), whereas the intended gesture is invariant. The interest for the Motor Theory of Speech Perception was recently renewed by the discovery of mirror neurons (located in the premotor cortex of the macaque, active both when the macaque performs a transitive action or observes another individual performing the same action, Rizzolatti and Arbib, 1998).

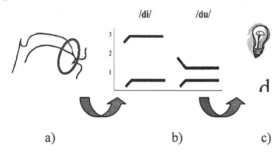

a)   b)   c)

FIGURE 1.2. Illustration of the core argument for the Motor Theory of Speech Perception, regarding the phoneme /d/ in /di/ vs /du/: a) the gesture is the same, that is closing the front of the vocal tract by putting the tongue against the teeth, b) the signal is different and c) the percept is the same. Therefore, the invariant would be the gesture.

AUDITORY THEORIES

Auditory Theories consider that the reference frame for speech is auditory. In the case of speech production, the target would be a region in the auditory space (Guenther et al., 1998). The main argument is motor equivalence, showing that various articulatory configurations are used for achieving the same auditory goal, as shown in perturbation experiments: if the articulatory apparatus is constrained, e.g. by inserting a tube between the lips (Savariaux et al., 1999), speakers reorganize their motor configuration to achieve the same auditory region.

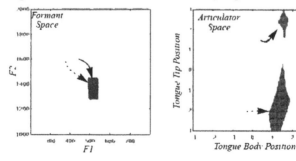

FIGURE 1.3. Motor equivalence for /r/ in English in the DIVA model (from Guenther et al., 1998). Relationship between a simple convex region corresponding to /r/ in the acoustic space (left) and the corresponding regions in the articulatory space (right). Arrows indicate model trajectories when producing /r/ starting from a /d/ configuration (solid lines) and from a /g/ configuration (dashed lines).

460

Speech production would exploit this adaptability, as in the case of /r/ in English, pronounced in the DIVA articulatory model (Guenther et al., 1998) with different configurations (bunched vs. retroflex) depending on the previous consonant (Fig. 1.3).

In the case of speech perception, proponents of auditory theories consider that speech perception involves auditory or multisensory representations and processing, with no reference to speech production (Diehl et al., 2004).

SENSORY-MOTOR THEORIES

Sensory-motor theories have recently emerged for both speech perception and production. They generally consider auditory frames as the core for communication, but they include the sensory-motor link inside the global architecture. They claim that in normal conditions, production involves cortical motor (frontal) areas and perception involves cortical (temporal) auditory ones, but that the perceptuo-motor link, necessary for speech acquisition, could also play a role in adverse conditions.

Regarding speech production, the DIVA model (Guenther, 2006) combines a feedforward control sub-system for on-line production, and a feedback control sub-system when the auditory consequence of a gesture is not congruent (Fig. 1.4).

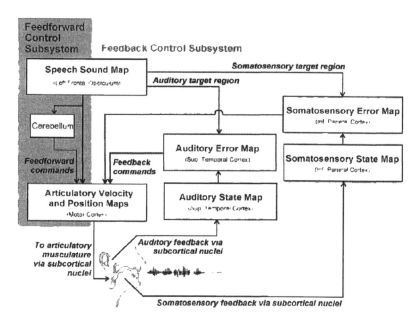

FIGURE 1.4. The DIVA model (from Guenther, 2006): a feedforward control subsystem located in the motor cortex is coupled to a parieto-temporal feedback control subsystem.

In a similar way, sensory-motor theories of speech perception argue for a core auditory (or audio-visual) system for speech perception, enhanced by motor processes in complex conditions such as noise, through "binding" (Schwartz et al., 2010, see Fig. 1.5) or "prediction" (Skipper et al., 2007).

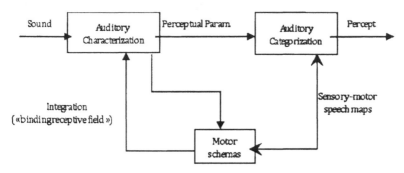

FIGURE 1.5. Perception for Action Control Theory (PACT): motor schemas are involved both in extracting relevant auditory information through binding, and in improving categorization through sensory-motor maps.

TAXONOMY

This review of speech production and perception theories and models shows that both fields share the same debates about the nature of the reference frame and the functionality of the perceptuo-motor link. Table 1.1 thus proposes an original classification of all these theories in a unified way. The next section will formalize this classification in a Bayesian framework. In other words, we aim at filling Table 1.1 with probabilistic expressions.

Table 1.1 Taxonomy of speech production and perception theories and models

Task Theory	Production	Perception
Motor	Articulatory Phonology (Browman and Goldstein, 1989)	Motor Theory (Liberman and Mattingly, 1985)
Auditory	Auditory reference frames for speech planning (Guenther, 1998)	Auditory theories (Diehl et al., 2004)
Sensory-motor	DIVA model (Guenther et al., 2006)	Perception for Action Control Theory (Schwartz et al., 2010)

A UNIFIED BAYESIAN FRAMEWORK

In this section, we define in probabilistic terms what are a motor subsystem, a sensory subsystem, and a sensory-motor link, and how they can be combined in a general system for communication. We then show how selectively disabling subsystems leads to a unified expression of the six categories of Table 1.1 in probabilistic terms. We limit our analysis to an abstract model from which we extract possible interpretations. The reliability of this model in realistic simulations of language emergence has been studied in Moulin-Frier et al. (2008, 2010).

BAYESIAN MODELING

Bayesian Robot Programming (BRP, Lebeltel et al., 2004) specifies the knowledge of a sensory-motor agent as a joint probability distribution over variables of interest (typically motor, sensory and internal variables). This joint distribution is generally expressed as a product of simpler distributions, using Bayes rule and conditional independence hypotheses. Using this knowledge, a sensory-motor behavior is then defined as a conditional probability distribution computed from the joint distribution (for example: "given the values of some sensory variables, what is the probability distribution over the motor variables"), called a question to the model.

Our modeling of a general communication system is based on four variables (but we will discuss in more details the possible interpretations):

- M: the speaker's motor gesture,
- S: the listener's sensory percept,
- O_S, O_L: the object of communication (in a very general sense, hereafter the object), respectively from the speaker and the listener point of view.

We then define the three subsystems as follow:

- The motor subsystem is defined as a conditional probability distribution $P(M|O_S)$: given an object to communicate, what is the probability distribution over the speaker's motor gesture?
- The sensory subsystem is defined as a conditional probability distribution $P(O_L|S)$: given a sensory percept, what is the probability distribution over the objects that can be inferred by the listener by the listener?
- The sensory-motor subsystem is defined as a conditional probability distribution $P(S|M)$: given a motor gesture, what is the probability distribution over the sensory percepts?

FIGURE 2.1. Model structure.

Finally, considering the *a priori* knowledge on the speaker's object $P(O_S)$ as uniform, the general communication system is a joint probability distribution:

$$P(O_S \wedge M \wedge S \wedge O_L) = P(O_S)P(M \mid O_S)\,P(S \mid M)\,P(O_L \mid S)$$

In this framework, a successful communication corresponds to equality between O_S and O_L (the object inferred by the listener must be the same as the object intended by the speaker). Therefore, each question asked to the model will be under the constraint $O_S = O_L$.

Speech production and perception are defined as the following probabilistic questions to the general model $P(O_S{}^{\wedge}M{}^{\wedge}S{}^{\wedge}O_L)$:

- Production: $P(M \mid O_S = O_L)$. For a given object in the speaker's mind, and knowing that O_L is equal to O_S to evoke the same object in the listener's mind, what is the probability distribution over motor gestures?
- Perception: $P(O_S = O_L \mid S)$. Knowing the sensory input perceived by the listener, what is the probability distribution over objects, inferred by the listener and likely to have been in the speaker's head at the input?

In probabilistic terms, motor and sensory subsystems can be deactivated by setting the corresponding distribution as uniform (that is, without explicit knowledge about the corresponding link). Motor, auditory and sensory-motor theories of speech communication will thus be expressed as the following:

- Motor theories correspond to a deactivation of the sensory subsystem, defining $P(O_L \mid S)$ as a uniform distribution,
- Auditory theories correspond to a deactivation of the motor subsystem, defining $P(M \mid O_S)$ as a uniform distribution,
- Sensory-motor theories let both the motor and sensory subsystem active, each distribution being considered as informative.

Table 2.1 Model Taxonomy

Theory \ Task	Production $P(M \mid O_S = O_L)$	Perception $P(O_S = O_L \mid S)$
Motor $P(O_L \mid S) =$ Uniform	$P(M \mid O_S)$	$\sum_M P(M \mid O_s)P(S \mid M)$
Auditory $P(M \mid O_S) =$ Uniform	$\sum_S P(S \mid M)P(O_L \mid S)$	$P(O_L \mid S)$
Sensory-motor	$P(M \mid O_s)\sum_S P(S \mid M)P(O_L \mid S)$	$P(O_L \mid S)\sum_M P(M \mid O_s)P(S \mid M)$

Finally, using these definitions and rules of Bayesian inference (typically Bayes and normalization rules), we can now assign a probabilistic expression to each type of theory (Table 2.1).

INTERPRETATIONS

General Model

The general model in Figure 2.1 can be interpreted from different points of view:

- As an objective model of communication, where the motor model is a model of the speaker, the sensory one a model of the listener, and the perceptuo-motor link is a model of the environment.
- As a subjective neurolinguistic model, where the motor and sensory models would correspond respectively to the motor and auditory cortices, and the perceptuo-motor link as the neural connections between them.
- As a subjective model of the Theory of Mind, where the motor model (resp. the sensory model) would be an internal representation of the speaker (resp. the listener) in the brain of the listener (resp. the speaker).

Let us focus on the computational interpretation of the behaviors defined by the probabilistic questions in Table 2.1.

Behaviors: Motor theories

In our Bayesian framework, motor theories of speech communication correspond to a deactivation of the sensory subsystem, setting $P(O_L|S)$ as uniform. Speech production thus leads to select a motor gesture M for a given object to communicate O_S according to the distribution $P(M|O_S)$, considering that the sensory subsystem does not provide any information. This is in line with Articulatory Phonology (Browman and Goldstein, 1989), which considers speech production as motor gestures scores, not influenced by the auditory consequence of those gestures.

Conversely, motor theories of speech perception are defined by the probabilistic question:

$$\sum_M P(M|O_S)P(S|M)$$

For a given auditory percept S heard by the listener, the inferred object thus corresponds to an object for which the speaker would have produced a gesture with the same auditory consequence. This is in line with the Motor Theory of speech perception (Liberman and Mattingly, 1985), which considers that perceiving speech actually amounts to perceiving the intended gestures of the speaker.

Behaviors: Auditory theories

Auditory theories of speech communication correspond in our framework to a deactivation of the motor subsystem, setting $P(M|O_S)$ as uniform. Speech production is then defined by the probabilistic question:

$$\sum_{S} P(S|M)P(O_L|S)$$

For a given object to communicate by the speaker, the selected motor gesture thus corresponds to a gesture for which the auditory consequence would allow the speaker to correctly infer the object using his/her sensory subsystem. This is in line with models considering that speech production targets are defined by regions in the acoustic/auditory space (Guenther et al., 1998).

Regarding speech perception, the probabilistic question is simply $P(O_L|S)$, without any information from the motor subsystem. This is in line with the claim that speech perception does not incorporate any input from speech production mechanisms (Diehl et al., 2004).

Behaviors: Sensory-motor theories

Finally, sensory-motor theories correspond in our framework to activating both the motor and the sensory subsystems, leading to distributions which are the products of those for motor and auditory theories. Speech production is thus defined by the probabilistic question:

$$P(M|O_S)\sum_{S} P(S|M)P(O_L|S)$$

For a given object to communicate, the selected motor gesture is then a compromise between an often-used gesture, and a gesture for which the auditory consequence would allow the speaker to correctly infer the object. This is in line with models like DIVA (Guenther, 2006) with its two components, feedforward for on-line production (the first factor) and feedback for correction (the second factor).

Regarding speech perception, the corresponding probabilistic question is:

$$P(O_L|S)\sum_{M} P(M|O_S)P(S|M)$$

For a given sound heard by the listener, the inferred object has both to satisfy the sensory subsystem and to correspond to an object for which the listener would have produced a motor gesture with the same auditory consequence. This is in line with the Perception for Action Control Theory (Schwartz et al., 2010), which considers the cues of speech perception as essentially auditory (the first factor), but possibly helped by access to motor knowledge (the second factor).

Conclusion and Perspectives

In this paper, we proposed a unified formal framework for speech production and perception, based on a Bayesian model able to express the major theories in the field.

In further works, our aim is to computationally study the possible functional role of the perceptuo-motor link in speech communication. Previous works (Moulin-Frier et al., 2008, 2010) already showed that this link is necessary in production for realistic simulations of language emergence (backed by data showing that human phonological systems are optimized for perceptual distinctiveness).

Regarding speech perception, we are planning simulations showing that in simple cases like vowel categorization, the sensory subsystem is better than the motor one to infer the corresponding object (in favor of auditory theories of speech perception) but, in more complex cases like syllable categorizations, the motor subsystem can add reliable information (in favor of sensory-motor theories).

REFERENCES

Browman, C.P., & Goldstein, L. (1989). "Articulatory Gestures as Phonological Units." *Phonology*, 6, 201–251.

Diehl, R.L., Lotto, A.J., & Holt, L.L. (2004). "Speech Perception." *Annual Review of Psychology*, 55, 149-179.

Guenther, F.H., Hampson, M., & Johnson, D. (1998). "A Theoretical Investigation of Reference Frames for the Planning of Speech Movements." *Psychological Review*, 105, 611–633.

Guenther, F.H., (2006). "Cortical interactions underlying the production of speech sounds." *Journal of Communication Disorders*, 39, 350–365.

Hickok, G., & Poeppel, D. (2007). "The cortical organization of speech processing." *Nature Reviews Neuroscience*, 8, 393–402.

Lebeltel, O., Bessière, P., Diard, J., & Mazer, E. (2004). "Bayesian Robot Programming." *Autonomous Robots*, 16, 49-79.

Liberman, A.M., & Mattingly, I.G. (1985). "The Motor Theory of Speech Perception Revised." *Cognition*, 21, 1–36.

Moulin-Frier, C., Schwartz, J. L., Diard, J., & Bessiere, P. (2008). *Emergence of a language through deictic games within a society of sensori-motor agents in interaction.* 8th International Seminar on Speech Production, ISSP'08, Strasbourg, France.

Moulin-Frier, C., Schwartz, J. L., Diard, J., & Bessiere, P. (2010). "Emergence of phonology through deictic games within a society of sensori-motor agents in interaction". Book chapter in *Vocalization, Communication, Imitation and Deixis*, to appear.

Rizzolatti, G., & Arbib, M. A. (1998). "Language within our grasp". *Trends in Neurosciences*, 21, 188–194.

Savariaux, C., Perrier, P., Orliaguet, J.-P., & Schwartz, J.-L. (1999). "Compensation strategies for the perturbation of French [u] using lip-tube. II. Perceptual analysis." *Journal of Acoustical Society of America*, 106, 381–393.

Schwartz, J.-L. Basirat, A., Menard, L., Sato, M., (2010). "The Perception-for-Action-Control Theory (PACT): A perceptuo-motor theory of speech perception." *Journal of Neurolinguistics (in press)*.

Skipper, J. I., Van Wassenhove, V., Nusbaum, H. C., & Small, S. L. (2007). "Hearing lips and seeing voices: how cortical areas supporting speech production mediate audiovisual speech perception." *Cerebral Cortex*, 17, 2387–2399.

CHAPTER 49

Bayesian Modeling of Human Performance in a Visual Processing Training Software

Julien Diard, Muriel Lobier, Sylviane Valdois

Laboratoire de Psychologie et NeuroCognition - CNRS
Université Pierre-Mendès-France
BSHM, BP 47, 38040 Grenoble, France

ABSTRACT

Dyslexia is a deficit of the identification of words, which is thought to be a consequence of different possible cognitive impairments. Recent data suggest that one of these might be a specific deficit of the visual attention span (VAS). We are developing a remediation software for dyslexic children that focuses on the VAS and its training. A central component of this software is the estimation of the performance of a given participant for all possible exercises.

We describe a preliminary probabilistic model of participant performance, based on Bayesian modeling and inference. We mathematically define the model, making explicit underlying generalization hypotheses. The model yields a computation of the most probable predicted performance space, and, as a direct extension, an exercise selection strategy.

Keywords: Bayesian modeling, dyslexia, user modeling, human performance.

INTRODUCTION

Developmental dyslexia is a specific learning disability characterized by a deficit in word identification. A dyslexic child is unable to acquire basic reading skills despite adequate intelligence, education, and sensory abilities. Developmental dyslexia is considered to be a cognitive disorder, this cognitive disorder being a consequence of an underlying neurobiological dysfunction.

For many years, the only recognized theory explaining developmental dyslexia was the phonological theory. It states that dyslexic children have a specific impairment of phonological skills: because of such a deficit, dyslexic children are unable to appropriately segment a word into single sounds and to link these sounds to the appropriate letters (Vellutino et al. 2004). However, some cases of developmental dyslexia are clearly not phonological and there is emerging evidence that visual attention might play a critical role in this disorder (Boden & Giaschi, 2007; Vidyasagar & Pammer, in press).

A key cognitive skill to fluent reading is the ability to recognize and process several letters in the same fixation. Data has shown that a sub-group of dyslexic children has significant difficulties in identifying a sufficient amount of letters in the short time frame of a fixation. This finding, together with the Multi-Trace Memory (MTM) reading model (Ans et al., 1998), has led to the visual attention span deficit theory of dyslexia (Bosse et al., 2007; Valdois et al., 2004). The visual attention span (VAS) is defined as the number of visual elements that can be processed simultaneously. It is measured using a global report task. In this task, a 5-letter consonant string is displayed during 200 ms. Subjects need to verbally report all the letters they have identified. Performance in this task is both reliably correlated with reading performance across all primary grades (Bosse et al., 2009) and significantly lower for a sub-group of dyslexic children than for age-matched controls (Bosse et al., 2007). Based on Bundesen (1990)'s theory of visual attention, a recent case study (Dubois et al., in press) of two dyslexic children has linked two potential cognitive sources to this VAS deficit: a reduced visual information processing rate or a limited number of items that can be stored in visual short term memory.

This insight on the specific cognitive components that are linked to the VAS is central in order to develop a targeted training regimen. We are developing evidence-based software that aims to train the deficit in visual information processing rate. The goal is to improve the reading performance of dyslexic children.

In this paper, we describe a preliminary probabilistic model of participant performance using Bayesian modeling and Bayesian model comparison. We provide the mathematical definition of the model that makes explicit all the underlying generalization hypotheses, contrary to previous approaches. We show that the model, for a given set of experimental observations, yields a computation of the most probable predicted performance space. It also yields, as a direct extension, a computation of the probability distribution over exercises to propose to participants. In other words, it naturally provides an exercise selection strategy that is optimal with respect to current observations.

VISUAL ATTENTION PROCESSING TRAINING SOFTWARE

The training software is designed to be used daily on a home computer. The typical training regimen calls for six training sessions per week, each session lasting 20 min. During these sessions, the subject has to perform visual processing tasks on multi-element arrays. Up to 150 "trials" will thus be performed in each training session. The two main characteristics of this software are the specificity of its tasks and the adaptability of the presented exercises. The tasks were designed to tap specifically visual processing of multi-element arrays. Three different exercise parameters are used to modulate the difficulty of each trial presented to the subject. For the purpose of this paper, we do not delve in depth on task specificity but concentrate on the topic of trial adaptability.

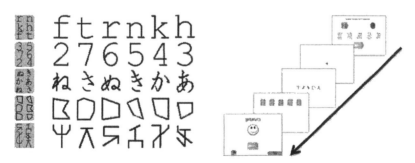

Figure 1. Left: families of characters used in the software. Right: succession of screens presented in any given trial, from the instruction screen to the feedback screen.

The training tasks all involve the visual categorization of both alphanumeric and non-alphanumeric character arrays. Five different character categories or "families" are thus defined: letters, digits, hiragana characters, polygons and pseudo-letters (see Fig. 1, left). All tasks involve the identification of the visual category of one or more characters. We now describe the sequence of events of a single trial.

The following screens are successively displayed during a single trial: instruction screen, fixation screen, stimulus screen, answer screen and feedback screen (see Fig. 1, right). The instruction screen displays the specific categorization task to be carried out in the current trial. On the stimulus screen an array of 2 to 7 characters belonging to one or two visual categories is displayed during 120 to 420 ms.

Three difficulty parameters characterize exercises, each of which can take 6 values. The character array can hold 2, 3, 4, 5, 6 or 7 elements. The display duration can be of 420, 360, 300, 240, 180 or 120 ms. The 6 different tasks are: 1- Is there an element of this family in the display? 2- How many families are there in the display? 3- How many elements of this family are there in the display? 4- Which two families are present in this display? 5- How many elements of these two families are present in the display? 6- Which two families are present in the display and how many elements of each are there? These difficulty parameters are ordered by increasing difficulty, i.e. task number 3 is easier than task number 6 but harder than task number 1.

These three difficulty parameters, numbered from 1 to 6, are the three dimensions of a matrix (number of elements, display duration, task), which we call the performance space. We call "exercise" a particular triplet in this space: for example, the coordinates (2, 2, 3) denote an exercise in which the subject will be asked how many elements of a given category are displayed (task = 3) and 3 characters will be displayed (number of elements = 2) during 360 ms (duration = 2). Exercises with small coordinates are easy whilst exercises with large coordinates are harder.

BAYESIAN MODELING OF HUMAN PERFORMANCE

A central component of the remediation software is the estimation of the performance of a given participant for all possible exercises. Its use is two-fold. Firstly, it allows us to define the exercise selection strategy, that is to say, the strategy used to automatically select the next exercise to propose to the participant. An exercise with a predicted success rate of 75% is assumed to be optimal. Indeed, it is both easy enough to maintain motivation and hard enough to drive learning effectively. Secondly, a correct estimation of the performance of a subject allows us to track, over time, the overall increase of performance and thus to assess the quality of the remediation process.

A previous approach, in the context of dyscalculia (Wilson et al., 2006), was based on heuristic estimation of human performance. In other words, an algorithmic solution was developed that tracked the subjects' performance. This solution was unable to correctly converge to the assumed representation: in simulation, the actual performance space was a cuboid but the estimation would not converge to a cuboid.

We now present a probabilistic model, based on Bayesian Programming (Lebeltel et al., 2004; Bessière et al., 2008), that solves this issue. More precisely, the model includes an explicit hypothesis about the performance space of participants. This hypothesis constrains the recognition algorithm: if the model computes the most probable cuboid, the output is guaranteed to be a cuboid.

BAYESIAN MODEL DEFINITION

The model relates the performance space of a participant at a point in time with his/her successes or failures for presented exercises.

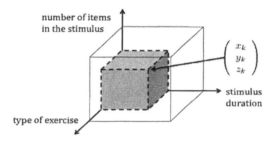

Figure 2. The performance space of a participant is a cuboid inscribed in the cube of all possible exercises, described by the point $K = (x_k, y_k, z_k)$.

We firstly assume that performance spaces are represented by a single point of the performance matrix M (see Fig. 2). At this point $K = (x_k, y_k, z_k)$, the success rate is midway between 100% and chance level (denoted c). We further assume that exercises that are easier than K (inside the cuboid) have higher success rates, and exercises harder than K have lower success rates: more precisely, we assume that success rates vary continuously, following 3 sigmoid functions of (unique) parameter α that are successively applied to each dimension (following Wilson et al. (2006)):

$$M_{x_k,y_k,z_k}(x,y,z) = 1/\sqrt[3]{e^{\alpha(x-x_k)}e^{\alpha(y-y_k)}e^{\alpha(z-z_k)}}$$

In probabilistic terms, this translates into the π_P model. Let $(x^{0:T}, y^{0:T}, z^{0:T})$ be the trials from time 0 to time T, and $S^{0:T}$ be a set of boolean variables that describe whether these trials resulted in successes or failures. Assuming that trials are independent and identically distributed over time, the resulting decomposition of the joint probability distribution is:

$$P(x_k \ y_k \ z_k \ x^{0:T} \ y^{0:T} \ z^{0:T} \ S^{0:T} \mid \pi_P)$$

$$= \ P(x_k \ y_k \ z_k \mid \pi_P) \prod_{i=0}^{T} P(x^i \ y^i \ z^i \mid \pi_P) P(S^i \mid x^i \ y^i \ z^i \ x_k \ y_k \ z_k \ \pi_P)$$

In this decomposition, $P(x_k \ y_k \ z_k \mid \pi_P)$ and $P(x^i \ y^i \ z^i \mid \pi_P)$ are assumed to follow uniform probability distributions. The last term, $P(S^i \mid x^i \ y^i \ z^i \ x_k \ y_k \ z_k \ \pi_P)$ is the prediction term, in the sense that given a supposed performance space and an exercise, it predicts the corresponding success rate. It is computed by applying the 3-dimensional sigmoid function of parameter α, according to the distance between $K = (x_k \ y_k \ z_k)$ and the presented exercise $x^i \ y^i \ z^i$.

HUMAN PERFORMANCE EVALUATION

The model being defined, it can now be used to recognize the performance space of a participant given observations of trial results. This is done by computing:

$$P(x_k \ y_k \ z_k \mid x^{0:T} \ y^{0:T} \ z^{0:T} \ S^{0:T} \ \pi_P)$$

$$\propto \ P(x_k \ y_k \ z_k \mid \pi_P) \prod_{i=0}^{T} P(x^i \ y^i \ z^i \mid \pi_P) P(S^i \mid x^i \ y^i \ z^i \ x_k \ y_k \ z_k \ \pi_P)$$

$$\propto \ \prod_{i=0}^{T} P(S^i \mid x^i \ y^i \ z^i \ x_k \ y_k \ z_k \ \pi_P)$$

In other words, maximizing the probability over performance spaces K is reduced to maximizing the likelihood of the observed data $S^{0:T}$. In order to avoid that the product of probabilities degenerates to numerical zeroes, the log of the likelihood is evaluated and maximized. This transforms the product of probabilities into a sum of log probabilities.

EXPERIMENTAL RESULTS

We evaluate our recognition algorithm in simulation: we define a "true" performance space K and use it to simulate the results $S^{0:T}$ of T trials. We then use these observations first to compute the probability distribution over K, as described above, and afterwards to maximize this probability in order to output an estimate $\hat{K}$ of K. To measure the quality of our algorithm, we compute the recognition error as the Manhattan distance between $\hat{K}$ and K. Fig. 3 (left) shows an example of recognition errors as the number of simulated trials T increases: obviously, adding observations reduces the recognition error. Using the actual software and dyslexic children, in a typical twenty-minute remediation session, around 100 exercises can be presented. This would yield, on average, and assuming that π_P is an adequate model, an error around 1. This is fairly acceptable, as it amounts to correct recognition along two dimensions and an error of 1 in the last dimension.

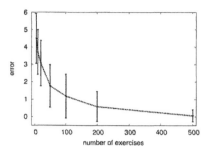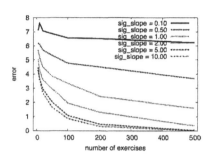

Figure 3. Left: error in the performance space recognition as a function of the number of trials (with sigmoid slope $\alpha = 5.0$; mean and standard-deviations over 100 simulations). Right: errors for different values of the sigmoid slope α.

However, Fig. 3 (right) shows the impact of the α parameter on the difficulty of the problem: when α is high, predicted success rates vary quickly. The problem is easy and the recognition algorithm's error is low with few trials. On the other hand, when α is low, predicted success rates vary slowly, which makes the problem harder and limits the convergence of the algorithm.

USING PERFORMANCE SPACE RECOGNITION FOR EXERCISE SELECTION

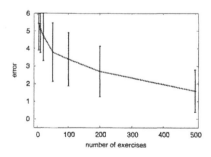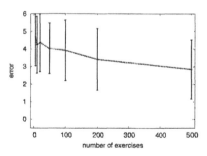

Figure 4. Left: error in the participant recognition as a function of the number of trials (sigmoid slope $\alpha = 1.0$; 100 simulations), with exercises chosen at random. Right: Same configuration, but exercises are chosen, at each time step, to be at the current most probable (x_k, y_k, z_k).

Another major aspect of the simulation has an impact on the quality of the recognition algorithm. It is the strategy for selecting exercises. Indeed, whereas previous simulations used purely random selection, the goal is to use the recognition algorithm in order to present exercises with a predicted success rate around the one at K (easy enough to maintain motivation and hard enough to drive learning effectively). However, it can be experimentally observed that such an exercise selection makes the recognition more difficult (see Fig. 4). Indeed, a random selection strategy "explores" the space of possible exercises, whereas using exercises around the estimated K concentrates the trials in a narrow portion of the space, possibly slowing convergence in case of erroneous initial estimates.

This effect is most obvious for α parameter values that correspond to difficult configurations. Indeed, in easy cases ($\alpha = 5.0$ for instance), the difference in convergence rate is only marginal.

There is a Bayesian answer to this issue. The exercise selection strategy can be added to the probabilistic model, in the form of a $P(x^i \, y^i \, z^i \mid x_k \, y_k \, z_k \, \pi_P)$ term. Then, instead of explicitly having to compute an estimated K value, the uncertainties of $P(x_k \, y_k \, z_k \mid x^{0:T} \, y^{0:T} \, z^{0:T} \, S^{0:T} \, \pi_P)$ could be propagated by summing over K:

$$
\begin{aligned}
&P(x^{T+1} \, y^{T+1} \, z^{T+1} \mid x^{0:T} \, y^{0:T} \, z^{0:T} \, S^{0:T} \, \pi_P) \\
&= \sum_{x_k, y_k, z_k} \left(\begin{array}{l} P(x^{T+1} \, y^{T+1} \, z^{T+1} \mid x_k \, y_k \, z_k \, \pi_P) \\ \times P(x_k \, y_k \, z_k \mid x^{0:T} \, y^{0:T} \, z^{0:T} \, S^{0:T} \, \pi_P) \end{array} \right)
\end{aligned}
$$

Initially, after few trials (T small), the recognition over K would still yield high uncertainties in $P(x_k \, y_k \, z_k \mid x^{0:T} \, y^{0:T} \, z^{0:T} \, S^{0:T} \, \pi_P)$; the position of K would be uncertain and therefore the selected trial would be as if drawn at random. After more observations are gathered (T large), the recognition error would be low and $P(x_k \, y_k \, z_k \mid x^{0:T} \, y^{0:T} \, z^{0:T} \, S^{0:T} \, \pi_P)$ would be fairly peaked, leading to selecting exercises in the close neighborhood of this peak. As a consequence, this model would gradually shift from a random exploration of possible exercises to a selection of exercises as described by the $P(x^i \, y^i \, z^i \mid x_k \, y_k \, z_k \, \pi_P)$ term. In other words, it would automatically shift from initial calibration to an adequate training and remediation program.

MODEL EXTENSIONS

The model we have presented is a basis for possible extensions. For instance, we have so far considered the α parameter as an internal parameter to the model but it could be explicitly handled in a probabilistic manner. We could estimate its value, or, in a more Bayesian fashion, propagate the uncertainties about it in the computations in a principled manner (as shown previously for the exercise selection strategy).

We detail a similar idea in another context. Instead of the α parameter, consider the shape of the performance space we have assumed. The founding hypothesis of the π_P model is a cuboid shape describing the way trials are correctly or incorrectly answered. In order to experimentally assess the validity of this assumption, a hierarchical Bayesian model can be defined.

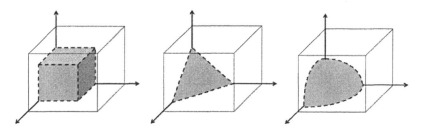

Figure 5. Alternative models of human performance that can be formally studied using Bayesian model comparison.

The first step is to define alternative models, based on different assumptions. For instance, instead of a cuboid, assume a tetrahedron or spherical section shape in the space of exercises (see Fig. 5). This is easily translated into corresponding π_T and π_S models, with different prediction terms $P(S^i \mid x^i\,y^i\,z^i\,x_k\,y_k\,z_k\,\pi_T)$ and $P(S^i \mid x^i\,y^i\,z^i\,x_k\,y_k\,z_k\,\pi_S)$ and joint probability distributions. Next, create a meta-variable, that considers all possible models: $M = \{\pi_P,\,\pi_T,\,\pi_S\}$, and encapsulate the three models in:

$$
\begin{aligned}
P(&x_k\,y_k\,z_k\,x^{0:T}\,y^{0:T}\,z^{0:T}\,S^{0:T}\,M) \\
&= \ P(M)P(x_k\,y_k\,z_k\,x^{0:T}\,y^{0:T}\,z^{0:T}\,S^{0:T} \mid M)
\end{aligned}
$$

The term $P(M)$ is a prior over models, which can be assumed uniform. Then, this model can be used to compute a probability distribution over models given experimental observations $P(M \mid x^{0:T}\,y^{0:T}\,z^{0:T}\,S^{0:T})$ in order to select the most probable model. This model would select the hypothesis that best describes the performance space of a given participant. Alternatively, as previously, uncertainties over the unknown variable M could be propagated with computations that would involve a summation over M.

CONCLUSION

We have presented a dyslexia remediation software and the component that tracks participants' performance and selects exercises. It is based on Bayesian Programming and inference. We have presented and discussed the design of the exercise selection strategy: Bayesian inference and summation over unknown variables theoretically yields a gradual shift from initial random exploration to exploitation of the optimal exercise to practice. We have outlined the extension of our model to hierarchical model comparison, in order to assess the quality of underlying assumptions.

Another natural extension would be to expand the model so as to track, over time, the displacement of the point K that describes participant performance. Technically, it would involve transforming the current model into a dynamic Bayesian filter. Instead of K, the algorithm would need to estimate the time series $K^{0:T}$. As the number of data sample is inherently limited, regularization assumptions would be required to overcome this computational challenge.

ACKNOWLEDGMENTS

This work has been supported by the VASRA ANR project.

REFERENCES

Ans, B., Carbonnel, S., & Valdois, S. (1998). A connectionist multiple-trace memory model for polysyllabic word reading. Psychological Review, 105(4), 678-723.

Bessière, P., Laugier, C., and Siegwart, R., editors (2008). Probabilistic Reasoning and Decision Making in Sensory-Motor Systems, volume 46 of Springer Tracts in Advanced Robotics.

Boden C & Giaschi D. (2007). M-stream and reading-related visual processes in developmental dyslexia. Psychological Bulletin, 133, 346-366.

Bosse, M.-L., & Valdois, S. (2009). Influence of the visual attention span on child reading performance: a cross-sectional study. Journal of Research in Reading, 32(2), 230-253.

Bosse, M. L., Tainturier, M. J., & Valdois, S. (2007). Developmental dyslexia: The visual attention span deficit hypothesis. Cognition, 104(2), 198-230.

Bundesen C. (1990) A theory of visual attention. Psychological Review, 97:523–47.

Dubois, M., Kyllingsbæk, S., Prado, C., Musca, S. C., Peiffer, E., Lassus-Sangosse, D., Valdois, S. (in press) Fractionating the multi-character processing deficit in developmental dyslexia: Evidence from two case studies. Cortex.

Shaywitz, S. E., & Shaywitz, B. A. (2005). Dyslexia (specific reading disability). Biological Psychiatry, 57(11), 1301-1309.

Lebeltel, O., Bessière, P., Diard, J., and Mazer, E. (2004). Bayesian robot programming. Autonomous Robots, 16(1):49–79.

Valdois, S., Bosse, M. L., & Tainturier, M. J. (2004). The cognitive deficits responsible for developmental dyslexia: review of evidence for a selective visual attentional disorder. Dyslexia, 10(4), 339-363.

Vellutino, F. R., Fletcher, J. M., Snowling, M. J., & Scanlon, D. M. (2004). Specific reading disability (dyslexia): what have we learned in the past four decades? Journal of Child Psychology and Psychiatry, 45(1), 2-40.

Vidyasagar, T.R. & Pammer, K (in press). Dyslexia: a deficit in visuo-spatial attention, not in phonological processing. Trends in Cognitive Science.

Wilson, A. J., Dehaene, S., Pinel, P., Revkin, S. K., Cohen, L., and Cohen, D. (2006). Principles underlying the design of "the number race", an adaptive computer game for remediation of dyscalculia. Behavioral and Brain Functions, 2(19).

Chapter 50

A Theoretical Multi-Tasking Executive Function for the Information Processing Model of the Human Brain

Joseph Kasser D.Sc., CM, CMALT

Visiting Associate Professor
Temasek Defence Systems Institute
National University of Singapore
Block E1, #05-05, 1 Engineering Drive 2
Singapore 117576

ABSTRACT

Research into holistic systems thinking needed an explanation or analogy for how the brain shifted from one task to the next. The cognitive psychology model of the human brain as an information processing system uses an executive function to control the transfer of information between short term and long term memory but seems to have little to say about how the executive function works. On the other hand, the multi-tasking operating system of a digital computer when modified provides such an analogy and is described in this paper. The key concept is that digital computer does not perform multiple tasks simultaneously. It performs one task at a time for very short periods of time, switching tasks under the control of the operating system so that it seems

478

to multi-task. The model also suggests mechanisms for why some people can multi-task, and others focus on a single task to the exclusion of other tasks. The paper summarizes the digital computer multi-tasking operating system and then discusses a conceptual theory for multi-tasking in the human brain based on adapting the digital computer multi-tasking operating system in a parallel processing environment. The paper concludes with some observations which can lead to further research.

Keywords: systems thinking, critical thinking, holistic thinking, systems engineering, multi-tasking, cognitive psychology.

PURPOSE OF PAPER

Holistic systems thinking is described as viewing a system from different perspectives (Kasser and Mackley, 2008). However, the human brain does not seem to be configured for viewing anything from different perspectives at the same time since according to Anderson we can only pay attention to one cognitively demanding task at a time (Anderson, 1995); yet the brain does seem to be able to perform a number of tasks at the same time. The cognitive psychology model of the human brain as an information processing system uses an executive function to control the transfer of information between short term and long term memory and perform other tasks, but the literature seems to have little to say about how the executive function works; see summary in (Miyake, et al., 2000).

The primary goal of this paper is to propose a hypothesis for a way of performing holistic thinking from an engineering perspective using the digital computer as an analogy. The secondary goal is to provide some speculation from different systems thinking perspectives and empirical data for future cooperation between cognitive psychologists and systems engineers to cooperatively develop a conceptual model for the operation of the executive function in the brain as a multi-tasking operating system. Consequently the focus of the paper is on holistic thinking and task switching within the context of the wider set of functions performed by the brain.

While the human brain is not configured for viewing anything from different perspectives at the same time, it can sequence through the different perspectives sequentially. Consequently when averaged over time, the brain is performing holistic systems thinking. This sequence can be considered as being similar to the manner in which a digital computer performs several applications at the same time (multi tasks). The multi-tasking operating system of a digital computer when modified to provide an analogy for multi-tasking in the human brain also suggests mechanisms for why some people can multi-task, and others focus on a single task to the exclusion of other tasks.

A DIGITAL COMPUTER MULTI-TASKING SYSTEM

The basic digital computer multi-tasking concept is broadly shown in Figure 1. Several tasks or applications are loaded in memory and represented by Tasks 1, 2, to N. Each Task contains a program that 'thinks' about something and accesses and stores data in memory. The Context Switch is the program that performs the task switching function by transferring the attention of the computer from one task to the next when it receives an interrupt signal. Task switching requires ways to save and restore the state of each task when switching occurs.

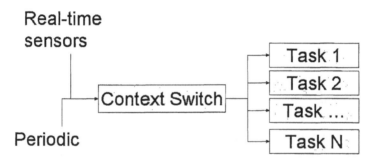

Figure 1 Digital Computer Multi-tasking Architecture

Interrupt processing

The interrupt signal may be generated periodically at fixed periods of time by a hardware signal, by a sensor in response to an event or even within the task when the program recognises the need to terminate the task.

Periodic interrupts

When a periodic interrupt is received the Context Switch responds in the following manner.

1. The state of the current task is saved.
2. The state of the next task in the sequence is retrieved.
3. The new task becomes the current task.
4. The next task is identified. If the current task is the last task, then the next task is the first task.
5. The current task is enabled to continue from where it left off in the previous task sequence cycle.

In most digital computer multi-tasking operating systems based on a periodic interrupt, the number of tasks can be large, and since the time allocated to each task is a fraction of the time available for all the tasks (including the time to save and restore the state of the task between task switches), the more tasks loaded into memory, the slower any one task seems to take[1].

Real-time interrupts

When a real time interrupt is received the Context Switch responds in a slightly different manner as follows.

1. The state of the current task is saved.
2. The state of the task associated with the specific real time interrupt is retrieved.
3. The new task becomes the current task.
4. The next task is identified. If the current task is the last task, then the next task is the first task.
5. The current task is enabled to continue from where it left off in the previous task sequence cycle.

Self-terminating tasks

When a task self terminates, the sequence of activities performed is the same as for a periodic interrupt.

Foreground and background tasks

One arrangement of tasks in a digital computer is to divide tasks between foreground and background tasks. Background tasks are those that are routine autonomic housekeeping activities such as those that monitor the state of the system, diagnostics, etc. Foreground tasks are the applications controlled by the operating system and depend on the context in which the system is deployed.

Parallel processing

The previous multi-tasking description is generally applicable to a single central processing element in a digital computer. An alternative architecture is to use more than one central processing element and split the tasks between them (using a third central processing element). Each central processing element can perform a single task or several tasks in a multi-tasking mode.

HOLISTIC THINKING VIA MULTI-TASKING

Developing an understanding of a system, issue or problem requires an analysis of the parts of the system in the set of thoughts relating to the complete system, thinking about the system in its context and verifying that the relationships between the thoughts are valid; hence holistic thinking seems to be the way to develop an understanding of a system in accordance with (Hitchins, 1992) page 14).

---

[1] This is why reducing the number of open windows on a PC can seem to speed up the computer.

Recognizing that the tight coupling between systems thinking and critical thinking has resulted in a number of [overlapping] definitions of systems thinking, critical thinking and critical systems thinking, holistic thinking (the system) is defined as sum of analysis, systems thinking and critical thinking (the subsystems) with the relationship between systems thinking and critical thinking shown in Figure 2 (Kasser, 2010).

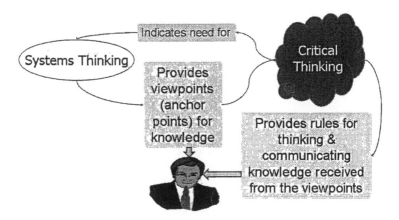

Figure 2 Holistic Thinking

Analysis

Analysis provides a white box approach for viewing a system from internal perspectives to develop an understanding of the functionality of the parts in a closed system configuration.

Systems thinking

Systems thinking is a discipline for seeing wholes (Senge, 1990), and indeed systems thinking is practiced much of the time by systems engineers but in an ad-hoc manner. The literature abounds with:

- publications advocating the use of systems thinking, e.g. (Flood and Jackson, 1991),
- philosophical and academic theories of systems thinking, e.g. (Flood and Jackson, 1991),
- the need to view problems from various perspectives, e.g. (Morgan, 1997).
- one or two publications describing how an understanding of the way things are connected together provides one with a competitive advantage over those who do not share the same understanding (Morgan, 1997; Luzatto, circa 1735),
- descriptions of the application of feedback loops (e.g. casual loops) and the claim that the use of such loops constitutes systems thinking (Senge,

1990), and

- similar descriptions of the application of systems dynamics and the claim that systems dynamics constitutes systems thinking.

Critical thinking

Critical thinking is "a unique kind of purposeful thinking in which the thinker systemically and habitually imposes criteria and intellectual standards upon the thinking, taking charge of the construction of thinking; [continually] guiding the construction of the thinking according to standards; [deliberately] assessing the effectiveness of the thinking according to the purpose, the criteria and the standards" (Paul and Willsen, 1995) page 21).

APPLICATION OF HOLISTIC THINKING

The holistic approach to the application of systems thinking was developed from a previous systematic and systemic approach to applying systems thinking (Richmond, 1993). Further research based on Richmond's work produced a set of nine viewpoints called System Thinking Perspectives (STP) (Kasser and Mackley, 2008) which have been used in teaching holistic systems thinking in postgraduate classes and workshops in Japan, Singapore, Taiwan and the UK. The systems thinking element of holistic thinking is a systemic and systematic way of viewing a system from each of the following nine viewpoints.

1. Big picture
2. Operational
3. Functional
4. Structural
5. Generic
6. Continuum
7. Temporal
8. Quantitative
9. Scientific

The first eight perspectives are descriptive, while the scientific perspective is prescriptive.

Systems engineers apply holistic thinking when using causal loops and concept maps to examine relationships and construct models. The descriptive (i.e. operational, functional and generic) perspectives provide parameters, critical thinking provides and verifies the relationships for the casual loops, the quantitative perspective provides the values for the model parameters and the model itself is a hypothesis (scientific perspective).

The perspectives perimeter

Consider the act of thinking about a problem. In general, the thinking process performs a sequence of tasks, each of which views the issue from a different perspective on the perimeter of the circle in the metaphoric representation depicted in Figure 3. Note however, that some minds seem[2]:

- To be fixed at one point on the perimeter and observe the issues from a single fixed perspective. This is akin to Wolcott and Gray's biased jumper (Wolcott and Gray, 2003).
- To only range over a limited part of the perimeter and view the issues from a limited number of perspectives.
- To range over the entire perimeter and view the issues from the set of perspectives but do not seem to do so in a systemic and systematic manner.
- To range over the entire perimeter and view the issues from the set of perspectives and seem to do so in a systemic and systematic manner.

Since there are no standard stopping points along the perspectives perimeter, each time communications between two parties takes place time is spent ensuring that both parties to the communication are viewing the issue from the same perspective (stopping point on the perspectives perimeter). This situation can be observed by the use of phrases such as "are we on the same page?" and "are we on the same wavelength?" etc. A standard set of perspectives or "anchor points" are needed to facilitate communications. One such set of anchor points is the systems thinking perspectives described above and illustrated in Figure 4.

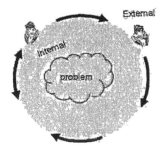

Figure 3 The perspectives perimeter

---

[2] The continuum STP suggests that this might be situational for an individual for various reasons and the same mind in different situations may view problems in different ways according to the list.

484

1. Big picture
2. Operational
3. Functional
4. Structural
5. Generic
6. Continuum
7. Temporal
8. Quantitative
9. Scientific

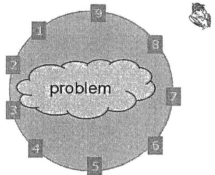

*Kasser and Mackley, INCOSE, 2008

Figure 4 System Thinking Perspective Anchor Points

Holistic thinking as multi-tasking

We can only pay attention to one demanding task at a time (Anderson, 1995). In computer terms the human information processing system, while capable of multi-tasking, can only handle one foreground or conscious task at any particular time. In holistic thinking the mind moves round the perspectives perimeter viewing the system from each of the systems thinking perspectives one perspective at a time. The approach is holistic when considered over a period of time or a number of cycles around the perspectives perimeter. The switching between perspectives may be sequential or may be driven by association of ideas where an idea from one perspective triggers a switch to a different perspective out of sequence in the manner of a self-terminating task. One focused way of switching perspectives is active brainstorming (Kasser, 2009) which uses the (Kipling, 1912) questions (who, what, where, when, which, why and how) to trigger ideas in a proactive manner.

The time spent in each perspective will depend on the attention span. While the digital computer spends a fraction of a second in each task, the attention span of the human brain (time spent on a task) seems to vary. Sometimes tasks are completed before switching to the next task; these are cases where the person is focused on that task to the exclusion of others, and at other times switching takes place before a task is completed.

MULTI-TASKING IN THE BRAIN

Multi-tasking covers autonomic and cognitive activities. Autonomic activities can be considered as the background tasks, while cognitive activities can be considered as foreground tasks. Cognitive activities include accessing, processing and storing information. The most widely used cognitive psychology information processing model of the brain based on the work of

(Atkinson and Shiffrin, 1968) cited by (Lutz and Huitt, 2003)shown in Figure 5 likens the human mind to an information processing computer. Both ingest information, process it to change its form, store it, retrieve it, and generate responses to inputs (Woolfolk, 1998).. In the multi-tasking model, the inputs from the external sensors also feed the executive as interrupts. Internal sensors for pain also feed interrupts. Some people seem to be able to set the threshold of their pain sensors (to ignore the input) at higher levels than others. From the generic perspective, some people also seem to be able to focus on a single task and set the threshold of other interrupts at higher levels which allow them to ignore the sensor inputs up to a point. Some people can set the threshold so high that they do not respond to any external stimulus and may have to be physically shaken in order to attract their attention to a different task.

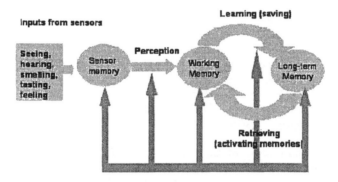

Figure 5 Human Information System

In a computer the number of tasks can be large as stated above. In the human brain, perhaps (Miller, 1956)'s rule of seven plus or minus one limits the number of anchor points on the perspectives perimeter that can be used during holistic thinking.

Observations

This section lists a number of observations to meet the secondary goal of the paper and to provide direction for future research.

- People who can multi-task may have short attention spans, and/or low thresholds on their interrupt circuits.
- Consider left brain and right brain activities as being performed by independent parallel processors. Each side of the brain can perform separate tasks but sometimes both sides of the brain process data from the same inputs. When the left brain and right brain and processes produce complimentary results they reinforce each other. However, when the processes produce contradictory results problems can be seen and might explain some of the observations in (Goleman, 1995).
- Interrupt circuit switching thresholds range along a continuum from low (boredom) to high (very interested).

- Faulty high interrupt thresholds may account for abnormal intellectual abilities which result in people being locked into one cognitive task to the exclusion of others.

SUMMARY

The paper summarized the digital computer multi-tasking operating system and then discussed a conceptual theory for multi-tasking in the human brain based on adapting the digital computer multi-tasking operating system in a parallel processing environment. The paper concluded with some observations which can lead to further research

REFERENCES

Anderson, J. R., *Cognitive psychology and its implications*. Freedman, New York, 1995.

Atkinson, R. and Shiffrin, R., "Human memory: A proposed system and its control processes," in *The psychology of learning and motivation: Advances in research and theory (Vol. 2)*, Academic Press, New York, 1968.

Flood, R. L. and Jackson, M. C., *Critical Systems Thinking Directed Readings*, John Wiley & Sons, Chichester, 1991.

Goleman, D., *Emotional Intelligence*, Bantam, 1995.

Hitchins, D. K., *Putting Systems to Work*, John Wiley & Sons, Chichester, England, 1992.

Kasser, J. E., "*Active Brainstorming: - A systemic and systematic approach for idea generation*", proceedings of the 19th International Symposium of the International Council on Systems Engineering, Singapore, 2009.

Kasser, J. E., "*Holistic Thinking and How It Can Produce Innovative Solutions to Difficult Problems*", proceedings of, 2010.

Kasser, J. E. and Mackley, T., "*Applying systems thinking and aligning it to systems engineering*", proceedings of the 18th INCOSE International Symposium, Utrecht, Holland, 2008.

Kipling, J. R., "The Elephant's Child," in *The Just So Stories*, The Country Life Press, Garden City, NY, 1912.

Lutz, S. and Huitt, W., Information processing and memory: Theory and applications. Educational Psychology Interactive, 2003, http://www.edpsycinteractive.org/brilstar/chapters/infoproc.doc, last accessed 28 February 2010

Luzatto, M. C., *The Way of God*, Feldheim Publishers, 1999, New York and Jerusalem, Israel, circa 1735.

Miller, G., "*The Magical Number Seven, Plus or Minus Two: Some Limits on Our Capacity for Processing Information.*" The Psychological Review, Vol. 63 (1956), 81-97.

Miyake, A., Friedman, N. P., Emerson, M. J., Witzki, A. H., Howerter, A. and Wager, T. D., "*The unity and diversity of executive functions and their contributions to complex "frontal lobe" tasks: A latent variable analysis*", Cognitive Psychology, Vol. 41 (2000), 49-100.

Morgan, G., *Images of Organisation*, SAGE Publications, Thousand Oaks, CA, 1997.

Paul, R. W. and Willsen, J., "Critical Thinking: Identifying the Targets," in *Critical Thinking: How to Prepare Students for a Rapidly Changing World*, J. Willsen and A. J. A. Binker (Editors), Foundation for Criical Thinking, Santa Rosa, CA, 1995.

Richmond, B., "*Systems thinking: critical thinking skills for the 1990s and beyond*", System Dynamics Review, Vol. 9 (1993), no. 2, 113-133

Senge, P. M., *The Fifth Discipline: The Art & Practice of the Learning Organization*, Doubleday, New York, 1990.

Wolcott, S. K. and Gray, C. J., Assessing and Developing Critical Thinking Skills, 2003, http://www.wolcottlynch.com/Downloadable_Files/IUPUI%20Handout_031029.pdf, last accessed

Woolfolk, A. E., "Chapter 7 Cognitive views of learning," in *Educational Psychology*, Allyn and Bacon, Boston, 1998, pp. 244-283.

CHAPTER 51

Modeling Human Response Magnitude for Electronic Message Stimulation

Katarzyna Jach, Marcin Kuliński

Institute of Organization and Management (I23)
Faculty of Computer Science and Management (W8)
Wrocław University of Technology
27 Wybrzeże Wyspiańskiego
50-370 Wrocław, Poland

ABSTRACT

By performing analysis of different visual and content attributes and their impact on email recipients it was proved that a language style, image presence and a lack of animation had an influence on improving message effectiveness. The research was conducted on a representative sample of Polish internet users with a use of eight specifically prepared email versions discriminated by five attributes like a background color, a font type, a language style, either presence or lack of animation and image.

Keywords: email, Click Through Rate, visual attributes, content attributes, effectiveness

INTRODUCTION

Electronic mail is the most frequently applied Internet service and it has been rapidly developing and evolving over the past years. Effectiveness is one of the most important parameters of the electronic message. It is essential to research which message attributes have a positive influence on an effective email. The effectiveness of email means a degree to which the sender has managed to achieve stated goals of the email transmission and provoke a desired response in the receiver. Although the studies show the decisive role of email content for the interest of recipients' email (Rettie and Chittenden, 2002), the other attributes, like a graphic form and email structure are also important (Bernard et al., 2001). While several researches into attributes determining electronic mail effectiveness have been performed (McDonald, 2004; Martin at al., 2003; Chittenden and Rettie, 2003; Zviran, 2006), all of them based their analysis on prior email campaigns. That, in turn, has impeded a precise specification of a single email attribute impact. The main goal of the study was to set out the pure effect of non content attributes on email response and to discover a power of these attributes to email recipients' reaction in comparison to content attributes.

Attributes influencing the email form had been identified based upon the Email Response Process Model, which is a modification of Email Marketing Response Process Model described by Rettie and Chittenden (2002) supplemented with elements of recipient's characteristics similarly to the holistic models of consumer's behavior (Kall, 1994). The model shows (figure 1.) that for the measure of email effectiveness, the power of recipient reaction is suitable and representative. The following analysis focused on attributes characterizing the email, with a special attention being paid to email appearance.

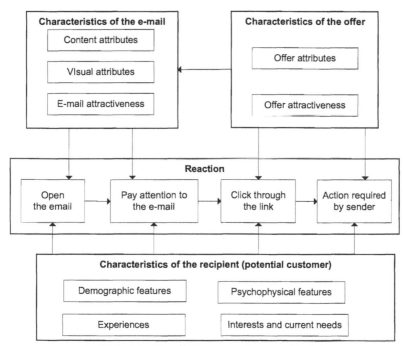

FIGURE 1. Email Response Process Model.

THE RESEARCH METHOD

A common problem for the research provided by electronic media is a misrepresentative sample, therefore it was an issue that required a special attention . Thanks to the cooperation with Wirtualna Polska (www.wp.pl), one of the biggest Polish internet portals, the research was carried out on a representative sample of fifty thousand Polish internet users. All participants gave their permission to receive emails and were selected using a stratified random sampling differentiated by age and gender. Every respondent received one of the eight prepared email versions with the request of filling the questionnaire. Five attributes were applied to differentiate the letters dispatched to the respondents, four of them concerned visual appearance of email and one conducted - its content. These were:

- Language style. Rational and emotional styles were used. The email had the same meaning but expressed in two different ways. Both styles contained the request to fill the questionnaire and some additional information about research (author, aim, etc.). The text had the same number of links and nearly the same length. As the email subject has a crucial influence on the response power (i.e. Mc Donald 2004), the subject line was the same for all the prepared email versions. This attribute was included in the research to make possible the comparison between impact of content and visual email attributes. As it was shown in the research of Lin a.o. (2006), emails that had made recipients feel positive emotions provoked a stronger response than neutral ones.
- Applied fonts. Two types of fonts were used as typical (Arial) and untypical (Txt). The main criterion of font selection was its usage frequency (Arial font is one of the most popular). Both chosen fonts were checked for their readability as well as accessibility for different software.
- Background. Three types of background were used in the experiments: default white, unsaturated green and orange with various saturation levels. As for all the email versions the black text font was applied, it was important to assure the contrast between the background and the text. The background colors were pale enough in order not to disturb the perception of the recipient as well as not to obscure text readability.
- Image. A graphic element of a mailbox was either used in the experiment or omitted.
- Animated text banner. The moving banner with the subject line text was either applied in the experiment or omitted.

All the email versions met the requirements of usability according to Nielsen (1993). As local conditions like software, screen adjustments (brightness, contrast) and its resolution are decisive for the final appearance and color of email (Pearrow, 2000; Couper, 2000), the effect of the differentiation was minimized during the test experiments made with different hardware configuration, web browsers and email client software.

The orthogonal experiment planning was applied to reduce the number of email versions from 48 to 8. The attribute levels for different email versions are listed in table. 1 below.

Table1. Attributes levels for different email versions.

No.	Language style	Font type	Background	Image	Animation
1	rational	untypical	orange	no	no
2	emotional	typical	orange	no	yes
3	emotional	untypical	green	no	yes
4	rational	typical	orange	yes	yes
5	emotional	untypical	orange	yes	no
6	rational	typical	white	no	no
7	emotional	typical	white	no	no
8	rational	untypical	green	no	yes

The stated hypotheses were based on the assumption that a greater noticeability increases a likelihood of recipient reaction. Therefore, a higher response was expected for emails with an untypical font style, a colorful background, an image and an animation applied. For the email content (the request to the research participants), the higher response was expected for email versions with an emotional language style.

ANALYSIS AND RESULTS

THE EFFECTIVENESS EVALUATION OF THE EMAIL VERSIONS

The response number was applied as the effectiveness measure of the email impact (figure 1). From the variety of email effectiveness coefficients, two options have been selected: number of unique clicks and the action ratio. The action was defined as the filling of the requested questionnaire by the recipient. The questionnaires with over 75% of missing data were deleted and not taken into account (table 2). The data were collected for two weeks after mailing action. The average Click Trough Rate (CTR) measured by click number divided by dispatched email number was 13.8%, so it was relatively high (Rettie, Chittenden 2002; McDonald 2004).

Significance of response difference measured by unique clicks for email versions was checked by chi-square test for n×m tables (Sheskin, 2000) because of abnormality of the analyzed variables. The detailed procedure was described in Jach and Grobelny (2005). Email versions bolded in table 2 had different response power measured by clicks, with statistical significance p=0.05. All the observed differences had the same trend for the action rate, however they were statistically different for version 4, 5 and 8 only. According to the analysis, it can be stated that the attributes characterizing the email really affect the reaction power. A further analysis was conducted to find the pure effect of the specified attributes.

Table 2. Response power for different email versions.

Email version	Unique Clicks	Chi^2 test[a]	p	Action	Chi^2 test[a]	p
1	839			467		
2	931			532		
3	912			508		
4	757	14.41 (6)	<0.001	445	5.14 (5)	0.08
5	1010	5.63 (4)	0.007	593	10.57 (6)	<0.001
6	803	9.06 (5)	0.009	485		
7	941			560		
8	709	40.93 (7)	<0,001	379	24.15 (7)	<0.001
Total	6902			3969		

[a] Degrees of freedom in brackets.

SIGNIFICANCE OF EMAIL ATTRIBUTES

Significance of particular attributes was specified by comparison of the reaction power for email versions on different attribute levels with the respective number of dispatched letters (table 1.). The chi-square test was applied in the analysis. The V Cramer's coefficient was accepted to measure the relation power among the different levels of the analyzed attributes (Everitt, 1977).

Table 3. Differences among various attribute levels.

Attribute	Level	Action Rate	Chi^2 test[a]	V	p
Language style	rational	12.43	59.52	0.033	<0.001
	emotional	15.18			
Font type	typical	13.73	0.18	0.002	0.67
	untypical	13.78			
Background	orange	14.15	7.65 (2)	0.012	0.021
	other	13.46			
Image	yes	13.69	1.17	0.005	0.28
	no	10.14			
Animation	yes	13.15	13.4	0.015	0.01
	no	14.45			

[a] Degrees of freedom in brackets.

The stronger and statistically significant reaction was called up by the letters with the emotional language style, the orange background and without animation (table 3), which was also confirmed by V Cramer's coefficient values. The presence of images affected the response positively, but the differences were not statistically significant, as well as no relation between the font type and the number of response was noticed. Probably the reason was a great influence of the language style. These results were confirmed by the analysis of the action ratio. In order to find the pure effects of visual email attributes (a font style, a background color, image and/or animation presence), the additional analysis was conducted for emotional and rational language styles separately.

It was expected that visual attributes had a stronger influence on response for email messages written with an emotional language style. This hypothesis was confirmed for all the investigated attributes, as it has been shown in table 4, however, not all the differences were statistically significant. For image and animation attributes in particular, the emotional language style intensified the tendencies presented in table 3. As it was stated for all email versions, a font type did not have a significant influence on the response power, however emotional emails with an untypical font applied had a slightly higher response rate.

Table 4. Differences among various attribute levels for emotional and rational language styles.

Attribute	Language style	Level	Action Rate	Chi² test[a]	V	p
Font type	rational	typical	12.48	0.04	0.001	0.84
		untypical	12.38			
	emotional	typical	14,98	0.57	0.045	0.45
		untypical	15,38			
Back-ground	rational	orange	12.77			
		green	11.34	7.12 (2)	0.016	0.02
		white	12.85			
	emotional	orange	15.53			
		green	14.59	2.16 (2)	0.009	0.34
		white	15.06			
Image	rational	yes	12.11	0.61	0.005	0.43
		no	12.54			
	emotional	yes	16.16	4.59	0.013	0.03
		no	16.85			
Animation	rational	yes	11.53	13.4	0.018	<0.001
		no	13.14			
		yes	14.74	2.67	0.009	0.10
	emotional	no	15.61			

[a] Degrees of freedom in brackets.

MODELING OF RESPONSE MAGNITUDE

In order to research a degree of influence of particular attributes on email effectiveness, a modified conjoint analysis has been applied (Vriens, 1994; Vriens et al., 1998). It was assumed that each email version was the profile which comprised a set of attributes placed on different levels and the utility function of a particular email version was the number of response for this version (unique clicks or action ratio). The variables describing the presence of an attribute at the determined level were regarded as the independent variables. A reaction power for the email version was assumed as a dependent variable. The quasi experimental encoding procedure was used for encoding the attribute levels (table 5). A multiple regression analysis was carried out.

Table 5. Attributes level encoding.

Attribute	Level	Coding Variables	Coding Values
Language style	rational	X_1	-1
	emotional		1
Font type	typical	X_2	1
	untypical		-1
Background	white		1, 0, 0
	green	X_3, X_4, X_3X_4	0, 1, 0
	orange		-1, -1, 1
Image	yes	X_5	1
	no		-1
Animation	yes	X_6	-1
	no		1

The least squares method was applied to estimate the parameters of the regression equation. It was noticed that the influence of the level of particular attributes levels on the email effectiveness could be described using statistically significant models (table 6). The application of the procedure discussed above enabled a determination of utility of attribute levels as a function of effectiveness.

The model with two independent variables indicated a high influence of the emotional language style ($X_1=1$) on the rise of the number of clicks. It also showed that the animation ($X_6= -1$) results in a decrease of response power. All the model variables proved statistically significant. The model containing three independent variables isolates the orange background ($X_3X_4=1$) from the remaining ones. The model indicates the significant influence of the orange background on the response. Similar relations were noticed in the regression models, in which the number of filled questionnaires was applied as the dependent variable. All the discussed models confirmed the chi-square test analysis.

Table 6. Selected regression models (unique clicks).

Selected attributes	Model equation	R^2	p
language style, animation	$Clicks = 862 + 85.7X_1 + {} $ $+ 35.5X_6$	0.93	0.001
language style, background, animation	$Clicks = 841 + 85.7X_1 + {} $ $+ 43X_3X_4 + 35.5X_6$	0.98	<0.001
language style, background, image, animation	$Clicks = 840 + 85.7X_1 + {} $ $+ 43X_3X_4 + 0.75X_5 + 35.5X_6$	0.98	0.005

The language style has the largest contribution to email utility, which is also confirmed by a high V Cramer's coefficient (table 3). The animation and the background color constitute the other significant attributes.

CONCLUSION

Based on the analysis, the language style had the largest influence on the message effectiveness. This attribute was related rather to the email content than to its form. As it was stated not only in this research (i.e. Martin et al., 2003), recipients paid more attention to the content elements than to the graphic form of email. Nevertheless, graphic form and email appearance played a significant role for email recipients attitudes.

It was noticed that the presence of images had a positive effect on the response power, confirming its significance as an information medium. An image has provided a very natural and readable information medium with a high precision, a large information volume (Miniard et al., 1994), natural and user-friendly reception and nearly a complete explicitness, which is greater than in a verbal communication (Młodkowski, 1998). Also, Chittenden and Rettie (2002) research confirmed emails that contained a greater quantity of images resulted in a higher click-through rate (CTR). According to the Eye Track research, more attention was paid to larger and sharper images (Outing and Ruel, 2004).

The regression models (table 4) showed that animation might reduce an email response rate, although they were very noticeable to recipients (Bayles and Chaparro, 2001) This is related to the characteristics of a human sight which enables a rapid movement perception in the peripheral vision (Młodkowski, 1998). On the other hand, many animations proved to make a focus on the text impossible and may result in a desensitization effect (Nielsen, 1993). This has been the defensive mechanism of sensorial adaptation which consists in disregarding the impulses appearing continuously. These signals cease to be received as new ones and they are transferred to the information background and consequently are disregarded by the recipient.

In the research, a positive effect of the orange background in comparison to other colors used was observed. A color attracts attention and can be the prime attention caller (Zviran et al., 2006). A color calls up the associations, which depend on individual experiences and generally accepted color symbols that are related to a particular culture (Brady, 2003). According to Chittenden and Rettie (2003) more colorful emails with images generate a greater response rate. It is also important in an email to apply a usable color set which does not disturb the text readability and the recipient's attention (Pearrow, 2000). No positive impact on the response power of the html format was stated, unlike in McDonald's (2004) survey, where CTR for html emails was twice higher than for text ones.

LIMITATIONS

The research sample was limited to Polish internet users, so the results may not be representative for other populations and cultures.

Only five email attributes determining its appearance were taken into account, and they were distributed at two or three levels only. For the future research, the survey can be expanded to other email attributes i.e. variety of a color set.

The analysis was limited to one attribute group influencing the recipient's reaction according to Email Response Process Model (figure1) and the response power was measured by the unique clicks and the action rate only, thereby in the future research, the impact of analyzed email attributes on recipients' experiences and preferences should be considered and other attributes groups should be examined.

SUMMARY

According to the analysis it can be stated that the attributes characterize the email influence on the reaction power. The applied method can be used for a preliminary evaluation of the response power of the email form. In the research, the a negative influence of animation and a positive influence of image presence on the email effectiveness has been stated. The influence of demographic features such as age or gender, or other elements of recipient's characteristics, like interests and current needs (figure 1) on the response power should be examined in the future research.

As it is seen, the email attributes play a significant role, but are not the main determining attribute in the recipients' reaction. However, as we live in a mass communication age and each email user receives daily several dozen messages on average, even little differences may /determine the recipient's reaction and response.

ACKNOWLEDGMENTS

The research was conducted at the support of Wirtualna Polska portal.

REFERENCES

Bayles M.E., Chaparro B. (2001), "Recall and Recognition of Static vs. Animated Banner Advertisements." *45th Proc. of the Human Factors and Ergonomics Society*, 1201-1204.

Bernard M.L., Liao Ch.H, Chaparro B.S., Chaparro A. (2001), "Examining Perceptions of Online Text Size and Typeface Legibility for Older Males and Females." *Proc. of the 6th Annual International Conference International Conference on Industrial Engineering– Theory, Applications, and Practice*, San Francisco, CA, USA

Brady L., Phillips C. (2003) "Aesthetics and usability: a look at color and balance." *Usability News*, 5(1), Retrieved 20 December 2006 from http://psychology.wichita.edu/surl/ usabilitynews/51/usability_news.html

Chittenden L., Rettie R. (2003), "An evaluation of email marketing and factors affecting response." *Journal of Targeting, Measurement and Analysis for Marketing*, 11(3), 203-217.

Couper M.P. (2000), "Web Surveys a Review of Issues and Approaches." *Public Opinion Quarterly*, 64 (4), 464-481.

Everitt B.S. (1977), *The analysis of contingency tables*, London: Chapman & Hall

Jach K., Grobelny J. (2005), "The e-mail form influence on the message effectiveness." *Ergonomics and work safety in information community. Education and researches.* Eds Pacholski L.M., Marcinkowski J.S., Horst W.M. Poznan: Institute of Management Engineering. Poznan University of Technology, 77-86

Kall J. (1994), *Reklama*, Warszawa: PWE

Lin T.M.Y., Wu H., Liao C., Liu T., 2006, "Why are some e-mails forwarded and others not?", *Internet Research,* 16(1), 81-93

Martin B.A.S, van Durme J., Raulas M, Merisavo M. (2003), "Email Advertising: Exploratory Insights from Finland." *Journal of Advertising Research*, 43 (3), 293-300

McDonald L. (2004), "How Message Size, Number of Links and Subject Length Affects Email Results". Retrieved 26 Oktober 2004 from http://www.emaillabs.com.

Miniard P.W., Bhatla S., Lord K.R., Dickson P.R., Unava H.R. (1991), "Picture-based persuasion processes and the moderating role of involvement." *Journal of Consumer Research*, 18 (1), 92-97

Młodkowski J. (1998), *Aktywność wizualna czlowieka.* Warszawa: Wydawnictwo Naukowe PWN

Nielsen J. (1993), *Usability Engineering*, Boston: Academic Press.

498

Outing S., Ruel L. (2004), "Eyetrack III: What We Saw When We Looked Through Their Eyes", Retrieved 26 January 2005 from http://www.poynterextra.org

Pearrow M. (2000), *Web Site Usability Handbook*, Charles River Media Inc.

Rettie R., Chittenden L., (2002), "Email Marketing: Success Factors", *Eighth Australian WWW Conference*, Retrieved 2 May 2007 from http://www.kingston.ac.uk/ ~ku03468/email_marketing.htm

Sheskin D.J. (2000). *Handbook of parametric and nonparametric statistical procedures*. Boca Raton, London, New York, Washington D.C.: Chapman & Hall/CRC.

Vriens M. (1994), "Solving marketing problems with conjoint analysis." *Journal of Marketing Management*, 10 (1-3), 105-112

Vriens M., van der Scheer H., Hoekstra J.C., Bult J.R. (1998), "Conjoint experiments for direct mail response optimization," *European Journal of Marketing*, 32 (3/4), 323-339

Zviran M.,Te'eni D., Gross Y. (2006), "Does Color in email make a difference?" *Communications of the ACM*, 49 (4), 94-99

CHAPTER 52

Modeling Online Buyers' Preferences Related to Webpages Layout: Methodology and Preliminary Results

Marcin Kuliński, Katarzyna Jach,
Rafał Michalski, Jerzy Grobelny

Institute of Organization and Management (I23)
Faculty of Computer Science and Management (W8)
Wrocław University of Technology
27 Wybrzeże Wyspiańskiego
50-370 Wrocław, Poland

ABSTRACT

The chapter presents a usability study related to online shop web page layout. Layout preferences concerning the placement of the most essential elements for such web pages, as shopping cart, item description or terms of service link, were examined among adult student subjects. Dedicated software, called microSzu, that allowed the subjects to place differently named "cards" across a virtual board, which mimics a web page of an online shop, was used during subjective layout preferences gathering. Statistical analysis and visualization of layout preferences data was performed using another specialized application, microVis. Functionality of both applications was described in details, and a preliminary review of some statistically significant differences related to online shop web page layout, based on data collected from Polish student subjects, was also included.

Keywords: Usability, e-commerce, web pages, users' preferences, statistical analysis

INTRODUCTION

The layout of a web page can affect both the speed of operations performed by its visitors and their subjective impressions based on browsing that page. For example, an improperly or unconventionally placed search functionality can make virtually impossible to find desired information, thus forcing the user to leave a web site in favor of another one with similar content, yet better layout.

Studies on layout preferences related to location of typical web interface objects, such as aforementioned search engine, homepage link or internal and external link groups, were performed by several researchers, including Bernard (2001a, 2001b) and Shaikh and Lenz (2006). In case of e-commerce centered websites, some interesting research results were published by Markum and Hall (2003), as well as Bernard (2003). In this paper an apparatus suitable for such kind of research, in a form of software applications for layout preferences data gathering, analysis and visualization, is presented in details. Some selected results from the authors' two-year study, involving data collected from nearly 700 participants, which extend previously published material (Kuliński and Jach, 2008, 2009), are also introduced and discussed.

SOFTWARE DESCRIPTION

MICROSZU – PREFERENCES DATA GATHERING

For the purpose of collecting layout preferences from the subjects a dedicated program was created, called microSzu, which runs within a web browser. The program uses a metaphor of a table that mimics an area of a webpage and cards representing items under examination. A subject literally places the cards at different fields of the table, which are preferred by him/her as the location for a given object. The main window of microSzu consists of the cards container situated on the left, the table divided into fields where the cards are placed at (the center of the window), and optional control and information elements in a form of buttons, pull-down lists and textboxes at the bottom of the window, as shown in Figure 1. The size of all visible elements automatically fits browser's window size, in which application was started. This functionality allows usage of microSzu on computers working at high as well as lower screen and desktop resolutions.

The program gives a possibility of placing the cards on the table using two different modes user can switch in between at any time during survey. The first one uses a technique called "drag and drop": a chosen card is picked from the card container, moved over the table and dropped at desired location. This mode emulates physical activity linked with real cards arrangement. The second one is called "pick and paint" and mimics paint applications to some extent. A card chosen from the container with a mouse click is surrounded by a frame and becomes a default "painting" tool, so user can quickly fill up the table with specified card by repeated mouse clicks. Any mode permits card deletion from specified field of the table using simple mouse click over desired location.

File format chosen to store data collected by the application is a variant of CSV (Comma Separated Values) standard, where every line of a text file represents a single entry, i.e. data related to a single subject, and consecutive fields separated by some special character (in this particular case a semicolon) contain all the data within an entry. While working offline, the textual data formatted this way can be displayed within a web browser for copying and preserving purposes, conducted manually by a researcher, but running microSzu form a dedicated web server (e.g. the Apache HTTP Server) allows automatic storing of the data from multiple instances of the program (i.e. a survey conducted with multiple subjects at once) in a single database file. More technical details about the software can be found in the paper by Kuliński and Michalski (2005).

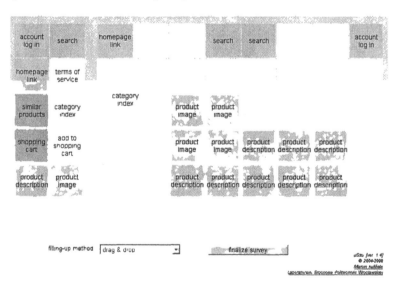

FIGURE 1. The main interface of microSzu with a sample preferences survey response

MICROVIS – DATA VISUALIZATION AND ANALYSIS

The microVis application also runs within a web browser, thus allowing data analysis to be performed remotely, i.e. using database files stored on a web server and without the need of any form of software installation, as well as locally, after downloading a database to the user's computer. It uses a form of tabbed interface,

502

where consecutive tabs (panels) reflect the analysis workflow, being activated or deactivated according to user's actions, hence guiding him and simplifying the interaction process. The *Database management* tab allows loading and parsing CSV files for further processing, *Visualization and filtering* provides tools for graphical presentation of layout preferences data, as well as for creating any desired subsets of the entries loaded, while *Statistical analysis* gives an opportunity to check the significance level of differences between selected data subsets.

The subsets creation process is based on simple, yet powerful filtering rules. Every filter applied consists of a selected special field and one or more possible values connected with it. For example, if the data contain a special field that stores subject's age, it may be selected along with some of its values, such as 20, 21 or 22 years of age. The resulting subset will contain only results from subjects within 20–22 years range. Filters can be mixed using logical *AND* operator (see Figure 2), so selecting only male subjects aged from 20 to 22 years requires an addition of another rule, this time based on a special field that stores subject's gender.

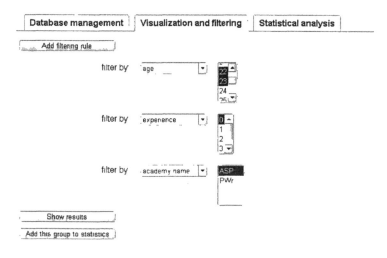

FIGURE 2. An example of applying multiple filtering rules in the microVis application

All the data, or only a given subset, can be displayed as placement density tables. For every webpage object examined during a survey, a table is generated with a percentage of selection frequency per field computed. Moreover, fields are displayed using scale of grays: the darker given field is, the more frequently it was chosen as a preferred location for analyzed object. The results are rendered within a separate window, thus can be easily saved as an HTML document for further utilization. Some additional information, including the name of a database file used, the number of total and filtered entries, the filters applied, as well as a hyperlink-based menu for convenient navigation among objects, is also provided.

The Chi-square test was implemented for statistical testing of observed differences in layout preferences between two selected data subsets. For every object under investigation the data from both groups is converted into a so called observed contingency table and examined for empty or underpopulated cells, according to formal requirements of the test (Kirkman, 1996, Hill and Lewicki, 2007). If necessary, data "rebinning" is employed to create properly populated ones, then an expected contingency table is computed and finally both tables are compared using the Chi-square formula. The results are displayed within a separate window in a form of calculated Chi-square values, degrees of freedom resulted from data rebinning, and probability values derived from the latter two, as shown in Figure 3.

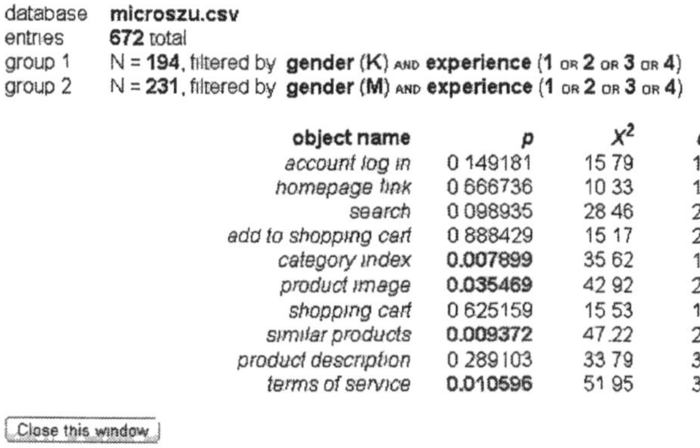

database	microszu.csv
entries	672 total
group 1	N = 194, filtered by gender (K) AND experience (1 OR 2 OR 3 OR 4)
group 2	N = 231, filtered by gender (M) AND experience (1 OR 2 OR 3 OR 4)

object name	p	x^2	df
account log in	0 149181	15 79	11
homepage link	0 666736	10 33	13
search	0 098935	28 46	20
add to shopping cart	0 888429	15 17	23
category index	**0.007899**	35 62	18
product image	**0.035469**	42 92	28
shopping cart	0 625159	15 53	18
similar products	**0.009372**	47 22	27
product description	0 289103	33 79	30
terms of service	**0.010596**	51 95	31

Close this window

FIGURE 3. Exemplary results of statistical analysis of differences performed using microVis

SURVEY DETAILS

The data presented below was collected from April, 2008 to February, 2010, and a total number of 672 subjects (292 males and 380 females) were surveyed during this period. The participants were adult undergraduate students of two Polish academies: Wrocław University of Technology and Academy of Fine Arts in Wrocław. The subjects' age ranged from 19 to 31 years, and the median was 21 years.

Online buying experience of the subjects was evaluated in terms of buying frequency, i.e. how often they perform shopping activities in an online shop. 247 individuals (37% of the sample) had no online buying experience whatsoever, while 47 subjects (7%) declared that they buy online once a month at least. Figure 4 illustrates a comprehensive experience structure of the sample, including observed and further analyzed gender differentiation. As one can see, male subjects were more experienced online buyers in general, while 49% of females stated that they never bought anything through the Internet.

A total number of 10 web objects was selected for the location preference analysis and their descriptive names presented to the subjects during the survey were as follows:

504

- *account log in,*
- *search,*
- *homepage link,*
- *terms of service,*
- *similar products* (i.e. other products in a given category),
- *category index* (i.e. a list of all categories),
- *shopping cart,*
- *add to shopping cart,*
- *product description,*
- *product image.*

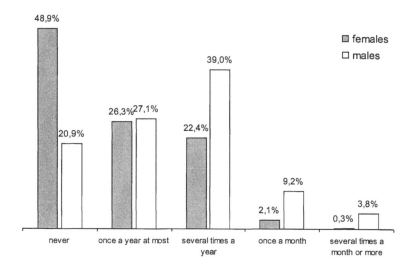

FIGURE 4. The structure of buying frequency for both genres in the sample

SELECTED RESULTS

Some of the most distinctive differences with statistically proven significances are presented and briefly discussed below, namely gender and experience related ones.

DIFFERENCES BETWEEN MALE AND FEMALE SUBJECTS

Both females and males without any online buying experience have similar preferences regarding the location of the objects under investigation (that is, no meaningful disparities were found across all the objects between both genders, except for the *account log in* object), but the situation changes when only subjects with greater than zero buying frequency are considered. In this case significant differences were statistically confirmed for *category index* (p = 0.0079), *product image* (p = 0.035), *similar products* (p = 0.0094), and *terms of service* (p = 0.011) objects, as shown in Table 1.

Table 1. Chi-square test results for differences between female (N = 194) and male (N = 231) subjects with an online buying experience

Object name	p-value	X^2	df
account log in	0.149181	15.79	11
homepage link	0.666736	10.33	13
search	0.098935	28.46	20
add to shopping cart	0.888429	15.17	23
category index	**0.007899**	35.62	18
product image	**0.035469**	42.92	28
shopping cart	0.625159	15.53	18
similar products	**0.009372**	47.22	27
product description	0.289103	33.79	30
terms of service	**0.010596**	51.95	31

Figure 5 presents the aforementioned layout preferences for the *product image* object in a form of placement density tables. The preferences of female subjects are stronger concentrated around the center of the screen and on a significantly smaller area, compared to males. Yet, note that both occupy much the same part of the screen.

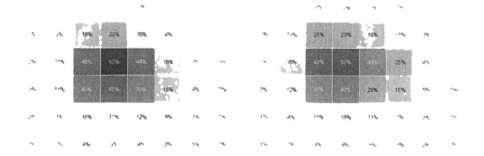

FIGURE 5. The *product image* object: layout preferences comparison of female (left) and male (right) subjects with an online buying experience

In Figure 6 differences confirmed with Chi-square test regarding preferred location of the *similar products* object are illustrated. Although both examined

506

groups pointed out the lower part of the left border being the most desired location for this object, females' preferences are clearly more straightforward and condensed. On the other hand, male subjects, along the left border, tended to choose also the lower part of the right border, as well as the bottom border of the screen (albeit the latter one to the lesser extent).

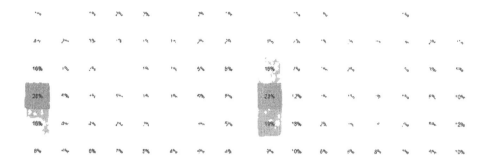

FIGURE 6. The *similar products* object: layout preferences comparison of female (left) and male (right) subjects with an online buying experience

DIFFERENCES BETWEEN EXPERIENCED AND INEXPERIENCED SUBJECTS

The most pronounced experience related differences were found and statistically confirmed between subjects, whose online buying experience was none (N = 247), and those who buy through the Internet several times a year or even more frequently (N = 246). The significance of heterogeneity was confirmed for *homepage link* (p = 0.011), *category index* (p = 0.0018), *shopping cart* (p = 0.0000000026), *similar products* (p = 0.039), and *terms of service* (p = 0.0089) objects. Table 2 presents detailed results of Chi-square tests performed.

Table 2. . Chi-square test results for differences between subjects with none
(N = 247) and moderate to high (N = 246) online buying experience

Object name	p-value	X^2	df
account log in	0.063076	17.55	10
homepage link	**0.011048**	28.82	14
search	0.540122	19.70	21
add to shopping cart	0.071274	38.43	27
category index	**0.001801**	46.31	22
product image	0.382577	32.72	31
shopping cart	**2.56e-9**	77.27	18
similar products	**0.038562**	43.77	29
product description	0.449862	29.29	29
terms of service	**0.008893**	61.68	38

Figure 7 presents differences regarding location of the *category index* object. In the case of inexperienced subjects the preferences are concentrated at the middle of the left border, but are somewhat dispersed around the rest of the screen at the same time, which is mostly evident at the upper part of it. Contrary to that, experienced subjects selected the left border as the preferred location for the object in a more consistent manner.

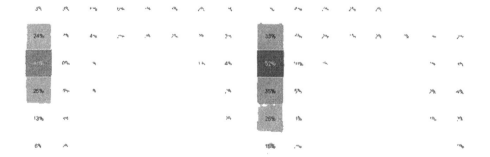

FIGURE 7. The *category index* object: layout preferences comparison of subjects with none (left) and moderate to high (right) online buying experience

Finally, Figure 8 illustrates statistically confirmed differences for the *shopping cart* object. Again, subjects with no experience placed the object in a more dispersed way, with only a slight preference towards the right border of the screen. Those who buy online several times a year or more often, strongly preferred the upper part of the same border, thus in general avoided placing the object anywhere

508

else on the screen.

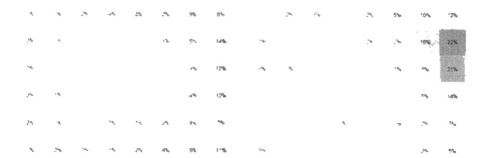

FIGURE 8. The *shopping cart* object: layout preferences comparison of subjects with none (left) and moderate to high (right) online buying experience

CONCLUSIONS

The findings described above shows that there are some gender related differences in the layout preferences analyzed. The most probably source of these are different profiles of online shops visited by women and men, respectively, as there are very little differences in case of inexperienced subjects of both genders. The data needed to confirm or discard this hypothesis was in fact collected during the study, in the form of a list containing three most often visited web shops, obtained from every subject, but the appropriate statistical analysis has yet to be done.

There is also some solid statistical evidence that the individuals, who buy online relatively often, have more crystallized expectations about the placement of certain elements of a web shop interface. Their choices during the survey were more consistent, resulting in pronounced visual patterns observed on placement density tables generated from the database, compared to subjects with none or only a minimal (i.e. those who buy once a year at most) experience. The question about the level of correlation between those expectations and actual placement of the objects, according to most frequently visited online shops registered from every subject, remains open.

The applications described in this chapter allowed us to perform a massive, long-term survey, as well as an analysis and visualization of sophisticated preferences data gathered during it. Yet, there is still room for improvements, particularly within the scope of implementing new statistical tools into the microVis software. The plans also include further development of the database management part of the program in order to enable multiple database files processing and analyzing, therefore allowing the preferences and the actual location data to be mixed and tested one against another for statistically significant differences.

REFERENCES

Bernard, M.L. (2001), "User expectations for the location of Web objects." *Proceedings of the CHI '01 Conference*, 171-172.

Bernard, M.L. (2001), "Developing schemas for the location of common Web objects." *Proceedings of the Human Factors and Ergonomics Society 45th Annual Meeting*, 1161-1165.

Bernard, M.L. (2003), "Examining user expectations for the location of common e-commerce web objects." *Proceedings of the Human Factors and Ergonomics Society 47th Annual Meeting*, 1356-1360.

Hill, T., and Lewicki, P. (2007), *STATISTICS Methods and Applications*. StatSoft, Tulsa, OK.

Kirkman, T.W. (1996), *Statistics to Use*. http://www.physics.csbsju.edu/stats/

Kuliński, M., and Jach, K. (2008), E-Commerce Websites Layout: Users' Preferences and Actual Location of Common Web Objects. [In:] Information Systems Architecture and Technology. Web Information Systems: Models, Concepts & Challenges. Borzemski, L. et al. (Eds.), Library of Informatics of University Level Schools, Wroclaw University of Technology.

Kuliński, M., and Jach, K. (2008), Layout preferences of e-commerce webpage objects. [In:] Employee wellness. Ergonomics and occupational safety. Pacholski, L.M., Marcinkowski, J.S., and Horst, W.M. (Eds.), Institute of Management Engineering, Poznan University of Technology.

Kuliński, M., and Jach, K. (2009), microVis: an application software for visualization and statistical analysis of layout preferences data. [In:] User Interface in Contemporary Ergonomics. Hankiewicz, K. (Ed.), Publishing House of Poznan University of Technology.

Kuliński, M., and Michalski, R. (2005), microSzu – a computer program for screen layout preferences analysis. [In:] Ergonomics and work safety in information community. Education and researches. Pacholski, L.M., Marcinkowski, J.S., and Horst, W.M. (Eds.), Institute of Management Engineering, Poznan University of Technology.

Markum, J., and Hall, R.H. (2003), *E-Commerce Web Objects: Importance and Expected Placement*. Laboratory for Information Technology Evaluation Technical Report, Missouri University of Science and Technology. http://lite.mst.edu/documents/LITE-2003-02.pdf

Shaikh, A.D., and Lenz, K. (2006), "Where's the Search? Re-examining User Expectations of Web Objects." *Usability News*, 8(1), http://psychology.wichita.edu/surl/usabilitynews/81/webobjects.asp

CHAPTER 53

Human Preference Modelling in Usability of Graphical Interfaces

Jerzy Grobelny, Rafał Michalski

Institute of Organization and Management (I23)
Faculty of Computer Science and Management (W8)
Wrocław University of Technology
27 Wybrzeże Wyspiańskiego
50-370 Wrocław, Poland

ABSTRACT

In this chapter the subjective assessment of the visual content of digital signage is presented as an example of applying in what way the various persons' preference modelling can be use to get the fuller picture of the attitudes towards examined stimuli. The preferences were obtained using pairwise comparisons of the designed screen formats. Then the priorities were derived by applying the Analytic Hierarchy Process (Saaty, 1977; 1980) framework. The experimentally gathered data were modelled and analysed by means of the analysis of variance, multiple regression, conjoint and factor analysis. The results show that the applied methods should be combined to acquire fuller understanding of the preference structure, which could turn out to be complex, even for the seemingly straightforward problems.

Keywords: Subjective Assessment, Digital Signage, Analytic Hierarchy Process

INTRODUCTION

The people's preferences are a crucial part of decision making process. Therefore, analysing, modelling and determining the real structure of them is essential in many areas. From the usability of graphical interfaces point of view, the users' preferences are directly connected with the user satisfaction, which in turn is one of the main dimensions of the usability concepts defined both by the HCI researchers and practitioners (Nielsen, 1993; Dix et al., 2004; Folmer and Bosch, 2004) as well as in the usability international standards e.g. ISO 9241 or ISO 9126. Because the users preferences may be quite complex and strongly depend on the context of the applications, it seems to be necessary to take into account different approaches to determine the preference structure.

In this study the focus is directed towards the comparatively new and emerging area of digital signage. Digital Signage can be defined as the set of digital devices used to present marketing information to consumers at various places (Harrison and Andrusiewicz, 2004). Generally speaking, it is a system of different screens connected into a net managed by a special-purpose computer software. The growing interest in the digital signage technologies is observed during recent years due to the great effectiveness of such tools. The big potentials of these kinds of solutions lies in the so called advertisement narrowcasting, which enables the advertisers to tailor the content of the message to the specific group of customers in given place and time. This flexibility allows, for example, to display at an airport the content targeted at business travellers on Monday mornings and family-aimed message might appear Friday afternoons.

The digital signage can be treated as not only as a means of conveying information to potential customers but also as tool for eliciting data by a user in a purchasing environment. For that reason it may be useful to treat this way of advertising as an interactive system. If this is the case, one may apply in this domain the approaches and solutions elaborated in different scientific fields including for example Human–Computer Interaction, Marketing, Decision making, Computer Science. According to widely recognized in the world international standard ISO 9241 related to the user graphical interfaces, the usability should be evaluated in three fundamental dimensions: effectiveness, efficiency and user satisfaction, taking into account the specific context of use. In the advertising domain usually the customer contentment is considered especially significant, so it seams that the third area would be of great importance to the practitioners.

The perception of digital signage communication depends on many different factors, however this specific domain is relatively new and the knowledge is mostly based on practitioners' heuristics and lacks the sound scientific background. Although in many cases it is possible to take advantage of the developments from different areas there are new challenges that arise because of the unique features of this technology. In this work the focus is given on the subjective issues related with the user preferences towards a specific graphical digital signage solution. The analysed screen variants were differentiated by two factors: the background type and the amount of the free space between different visual components of the screen layout. In the next sections of this chapter, first the experimental examination of various digital signage screen variants was described, and then the analysis of variance, multiple regression, conjoint and factor analyses were applied to obtain the broadest possible picture of subjects' preferences in this context. The chapter ends with the discussion of the obtained results and conclusions.

THE STUDY

In the sections that follow an experiment concerned with the digital signage is presented in detail. After the description of methods used, the obtained subjects' preferences were analysed by means of four different modelling techniques, namely: the analysis of variance, multiple regression, conjoint and factor analysis. The provided by these methods results seem to be rather complementary than too excessive.

METHODS

Participants

Thirty two persons participated in this research including thirteen men (41%) and nineteen men (59%). The average age of the group amounted to 23.3 (SD = 1.5) years. All the subjects were students of Wrocław University of Technology from the Computer Science and Management Faculty (17 persons, 53%) and volunteers from the Form Design Department of Wrocław Academy of Fine Arts (15 persons, 47%). The participants were quite well acquainted with new technologies spending considerable amount of time with digital devices, usually about eight and a half hours a day (SD = 2.1).

Experimental design

Because one of the most important issues related with the quality assessment of the digital signage technique is the most appropriate screen organization and design. Having this in mind, two independent factors were chosen in the current research. The first one was the width of borders that means the gap between three screen informative objects. The factor was defined at three categorial levels as small, medium, and large. The second attribute was related with the background of the digital signage screen, which was visible in the space between areas of data presentation. Three different backgrounds were designed. In the first one, vivid colours and with a gaudy facture were used, the second was more subtle and consisted of simple rectangular and colourful shapes, whereas in the last one, the uniform colour was applied.

Because there were only three attributes (factors) each of them on three levels, the full factorial design resulting in all possible full profiles amounted to (three gap sizes) × (three background types) = nine variants. Of course, in the case of the investigating more attributes on multiple levels there may be necessary to take advantage of the fractional factorial design at the cost of not being able to estimate interaction effects between factors. The experimental conditions of all studied digital signage screen profiles are illustrated in figure 1.

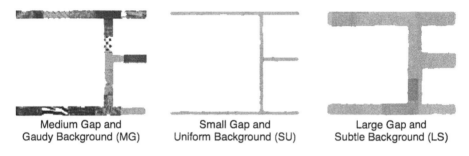

| Medium Gap and Gaudy Background (MG) | Small Gap and Uniform Background (SU) | Large Gap and Subtle Background (LS) |

FIGURE 1 Examples of studied attribute levels of digital signage screen formats

A standard within subjects model design was employed to investigate the participants' preferences towards the designed digital signage screens. That means that all combinations of the prepared digital signage profiles were assessed by every person taking part in the investigation.

Data collection

There are of course many ways of obtaining the subjects' liking concerned with various objects or phenomena. The most popular include direct ranking or pairwise comparisons. In this study, the latter method was used. This decision is rational in light of the results presented by Koczkodaj (1998), which show the extreme superiority of the pairwise comparison techniques compared to direct ranking with respect to the accuracy of stimuli estimation. However, one should also notice the main disadvantage of this approach – the number of necessary comparisons grows rapidly with the number of analysed variants.

514

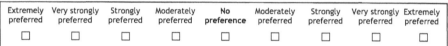

FIGURE 2 Exemplary presentation slide and a row from a questionnaire form used in AHP procedure

The research was carried out in teaching laboratories on similar personal computers equipped with 17" monitors of the LCD type. The computer screen resolution was set at 1280 by 1024 pixels. The experiments were conducted with the aid of a Microsoft® Office PowerPoint® 2003 software, and a paper version of the questionnaire including the pairwise comparison scales as well as personal questions. The exemplary slide layout along with single row of the questionnaire is given in figure 2.

The study began by informing the subjects about a purpose and a scope of the study. After answering questions regarding personal data, the participants were asked to compare each pair of screen profiles and specify which of them, she/he considers better and to what extent by means of grading the word *preferred*. Pairs were presented on the computer screen in the random order. The pairwise comparisons included all possible combinations of digital signage screen profiles.

Priorities derivation

There exist a number of methods for deriving the priorities from the pairwise comparison results. In this research the gathered data were processed within the framework of Analytic Hierarchy Process (AHP). This approach was developed at the Wharton School of Business by Thomas Saaty (1977, 1980) and has been widely used for decision making in various areas. The extensive review of its applications was provided by Zahedi (1986) and Ho (2007).

The crucial idea of this multiple criteria decision tool lies in obtaining the hierarchy of the subject' preferences from the pairwise comparisons matrix. The relative weights corresponding to the person's likings are estimated by finding the eigenvector corresponding to the maximal eigenvalue of the symmetric pairwise comparisons matrix. The idea of using the principal eigenvector was employed in the presented research to determine the relative subjective preferences towards the analysed digital signage screen profiles. The priority vectors computed individually for every subject were treated as a main dependent measure. The higher value of relative weight, the bigger preference for a given alternative. The AHP approach requires the weights to be normalized, so this dependent variable took the values between zero and one, and additionally the sum of all weights for a given participant was equal one.

PREFERENCE MODELLING AND ANALYSIS OF RESULTS

The questions related with the structure of the users preferences which influence the decision making may be answered by means o preference modelling. Conducting analyses by means of various modelling techniques allows the researcher to look at the users attitudes towards examined profiles from different points of view. The next subsections start with the consistency analysis of the obtained data, and then the condensed description of the selected methods along with the results obtained by applying them, are provided.

Consistency analysis

The application of the AHP framework allows for the consistency analysis of the gathered data for an individual subject. Therefore according to the formulas recommended in this method the consistency ratios (CR) were calculated and analysed. Higher values of this parameter correspond with the bigger inconsistencies in the pairwise comparisons. As Saaty (1980) argues, some level of inconsistency is necessary to achieve new knowledge, however the noise to signal ratio cannot be too big, and proposed to set the CR threshold at the level of 10%.

In the conducted experiment, the consistency ratio values ranged from 0.031 up to even 0.537 with the overall mean of 0.173 and standard deviation equal to 0.122. Because, in the present study, the participants did not have an opportunity of refining their initial responses, which is recommended in the AHP, the drop off value of CR was arbitrarily set at the level of 0.25. The results of the persons for whom the CRs exceeded this limit were regarded as too inconsistent and were excluded from further preference analyses. After applying this criterion, the results of 20 subjects were investigated in next subsections.

Analysis of variance

Two factorial analysis of variance was employed to assess the effects of the gap size and background type on the mean relative preferences. The results of this analysis are summarized in table 1. The graphical illustration of the mean AHP weights for the statistically significant factor is given in figure 3.

516

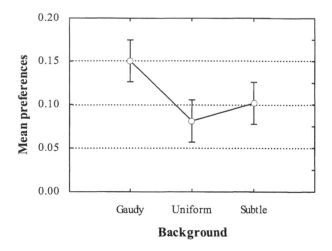

FIGURE 3 Mean relative weights depending on digital signage background type, $F(2, 171) = 8.4$, $p < 0.0005$. Vertical bars denote 0.95 confidence intervals.

Table 1 Results of the analysis of variance for the relative preferences

EFFECT	SS	DF	MSS	F	P
GAP SIZE	0.0086	2	0.0043	0.47	0.62
BACKGROUND*	0.15	2	0.076	8.4	0.00034
GAP SIZE × BACKGROUND	0.022	4	0.0056	0.62	0.65
ERROR	1.5	171	0.0091		

*$p < 0.0005$

Multiple regression

Another possibility of modelling the subjects' preferences is to take advantage of the multiple regression group of methods. In this study, all of the possible independent factors were categorical so it was necessary to use the artificial coding of the effects included in the model (Dielman, 2001). Both of the variables, the gap dimension and background type were coded using the values -1, 0, and 1. Because the AHP framework favours the use of a geometric instead of arithmetic mean while analysing the preferences across subjects, the geometric mean value of the obtained preferences were used as a dependent measure. After estimating parameters by means of the least square method, the regression model took the following form:

Mean Geometric Weights = 0.076 - 0.0041· *Gap* - 0.020 · *Background*

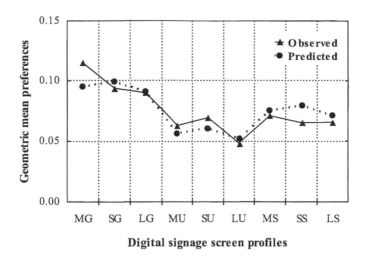

Digital signage screen profiles

FIGURE 4 Predicted versus observed geometric mean relative weights of the digital signage screen profile, $R^2 = 74\%$, $F(2, 6) = 8.7$, $p < 0.017$. The first letter denotes gap size, the second background type, e.g. MG – means medium gap and gaudy background.

The independent variables explain 74% of the data. The employed F-Snedecor test proved that R^2 considerably differs from zero $F(2, 6) = 8.7$, $p < 0.017$. The observed preferences along with values calculated from the constructed model, are illustrated in figure 4.

A t-Student test was used to verify if the model parameters differ from zero. Results of these tests, put together in table 2, demonstrated that hypotheses about insignificance of the background variable and intercept parameters should be rejected ($p < 0.01$). However the Gap variate parameter was not statistically different from zero ($p = 0.43$).

Table 2 Multiple regression of gap size and background type on relative preferences

VARIABLE	PARAMETER ESTIMATE	STANDARD ERROR	T-STATISTIC	P-LEVEL
INTERCEPT*	0.076	0.0039	19.3	0.000001
GAP	-0.0041	0.0048	-0.84	0.43
BACKGROUND**	-0.020	0.0048	-4.1	0.0064

*$p < 0.00001$
**$p < 0.01$

The degree to which the specific variable in the regression model explains the dependent variable can be assessed by calculating the squared semi partial correlations (SSC). These say by what value will R^2 be smaller if a given variable is removed from the model. In the analysed data the SSC amounted to 3% for the *Gap* variable and 71% in the case of the *Background* variate.

Conjoint analysis

The conjoint modelling of persons preferences has a decompositional nature. The main idea is to obtain the overall preferences for the given profile and then decompose the response into the partial contribution of each attribute. In this approach the possibility of estimating the so called part-worths for every attribute level along with the relative importance of particular attributes are regarded as the main merit. Since setting the foundations of this methodology in the mid sixties and early seventies (Luce and Tukey, 1964; Krantz and Tversky, 1971), the conjoint analysis have been successfully used to determine the structures of the persons' preferences in various areas (Green et al., 2001). It proved to be especially useful in the domain of the marketing and consumer research (Green and Srinivasan, 1978; 1990) thus in the context of digital signage this method seems to be particularly suitable.

In the present research, the conjoint analysis by means of the dummy variable regressions was employed. The multiple regressions were calculated for every participant, taking the AHP relative weights as the aggregate response for the individual profile of the digital signage screen format. Next, the individual part-worths together with attribute relative importances were computed. The aggregate-level results were presented in table 3. For the purpose of checking how well the regression models reflect the original ranks, the R-squared was calculated for every subject.

Table 3 Aggregate-level results of conjoint analysis.

VARIABLES	FACTOR LEVELS	RELATIVE IMPORTANCE	PART-WORTH ESTIMATES
Gap size		32.5%	
	Medium		0.00478
	Small		0.00501
	Large		-0.00978
Background type		67.5%	
	Gaudy		0.0393
	Uniform		-0.0299
	Subtle		-0.00939

The average goodness-of-fit criterion amounted to 83%, the mean F statistics for all of the regressions was equal to 7.84 and the mean significance level was $p = 0.086$. The modelled partial utilities allow now for making the decision about what version of the analysed digital signage screen profiles should be chosen. Applying the maximum utility criterion as a decision rule would result in selecting the MG profile that means a gaudy background with a medium gap. However, the BTL

(Bradley, Terry, Luce) probability choice model would recommend the SG profile (small gap with gaudy background).

Factor analysis

One of the rationales of applying factor analysis is concerned with determining the existence of the hidden common factors that can explain the covariance structure of the analysed variables. It may happen that the examined factors depend on an additional factor which was not directly measured or observed. For the extensive review of factor analysis developments from 1940 until the mid eighties, refer to the work of Mulaik (1986), and later contributions are presented for example by Steiger (1994). In this study the main purpose of the factor analysis is to further investigate the structure and nature of participants' preferences, and to check whether the obtained results depend on some other factors that were not taken into consideration during the experimental design.

A series of analyses were conducted to find possibly the best factor loading structure. The reasonable results were obtained by applying the maximum likelihood method of factors' extraction followed by the normalized varimax rotation. The scree plot analysis and the criterion of the lambda bigger than the unity resulted in selecting three factors. The obtained factor loadings are presented in table 4.

Table 4 Factor loadings obtained by factor analysis with maximum likelihood method of extraction and normalized varimax rotation

PROFILE CODE	FACTOR 1	FACTOR 2	FACTOR 2
MEDIUM GAUDY	-0.396	0.195	*-0.750*
SMALL GAUDY	-0.357	0.131	*-0.805*
LARGE GAUDY	*-0.567*	0.465	-0.117
MEDIUM UNIFORM	*0.950*	0.260	0.120
SMALL UNIFORM	*0.929*	-0.087	0.084
LARGE UNIFORM	0.140	*0.591*	0.205
MEDIUM SUBTLE	0.122	*-0.745*	0.520
SMALL SUBTLE	0.120	*-0.827*	0.502
LARGE SUBTLE	-0.217	0.019	*0.907*
PROPORTION OF VARIANCE EXPLAINED	27.4%	21.5%	29.3%

The outlined groups of factor loading shows the possible structure of the subjects preferences towards analysed variants. The last row in the table shows the percentage of the common variance explained by the individual factors. Although the values for the first and third factor are similar while the importance of the second factor is slightly lower. The presented results suggest that obtained preferences are in contrast with the factor structure assumed in the design of the experiment.

DISCUSSION AND CONCLUDING REMARKS

The results obtained by described in previous sections methods of preference modelling may lead to making different practical decisions. The analysis of variance showed that among the examined factors only the background type was important in the sense that it differentiated the mean preferences in the analysed sample. It was also demonstrated that the interaction between the factors was not statistically meaningful.

The multiple regression approach enabled to model the AHP preference weights as a function of the independent variables. It allows, additionally, checking the significance of the factors and assessing their importance. In this research the most important factor in the context of explaining the biggest portion of the dependent variate variability occurred to be the background variable with the SSC = 71% while the second factor contributed merely by 4%. The results of these two methods seem to be coherent. They indicate that the greatest impact for explaining the variability of the response variable is the background type, whereas the gap size component is almost negligible. This outcome could suggest not taking into consideration the gap size effect during the digital signage screen design.

The results from the conjoint analysis do not fully support this view. The difference between aggregated relative importances was considerably smaller than in the case of the multiple regression, and amounted to 67.5% for the background and 32.5% for the gap size. Similarly the factor analysis revealed that the digital signage profiles presented to the subject were perceived in quite a complex way. The latent factors correlated with individual profiles in a somewhat surprising pattern, which was difficult to interpret. Such a situation may be possibly attributed to either some interactions between the examined attributes and their levels, or to some external factors that might have influenced the subjects' preferences unintentionally. For instance, the mutual influence of the digital signage screen formats and the background of slides, on which they were displayed.

As it was shown in this study, even for the relatively simple experimental set-up the preference structure may be quite troublesome. The application of variety of methods could help in modelling the people's attitudes and facilitate making correct decisions. The presented combination of different approaches showed a various faces of the obtained preferences. Each of the described methods has its advantages and limitations and they should rather be treated as a set of complementary techniques than the mutually exclusive proposals.

Taking advantage of various methods will certainly allow for better understanding of how the people perceive a given stimulus. This, in turn, may help to make appropriate decisions both practical and those concerned with the design of future research objectives. Apart from the presented approaches there are some other tools that can also be applied in the more complex preference analyses – e.g. confirmatory analysis or more generally structured equation modelling as well as cluster analysis.

REFERENCES

Dielman, T.E. (2001), Applied regression analysis for business and economics. Pacific Grove: Duxbury.

Dix, A., Finlay, J., Abowd, G.D. and Beale, R. (2004), Human-Computer Interaction, 3rd edition, Harlow: Pearson Education.

Folmer, E. and Bosch, J. (2004), "Architecting for usability: a survey." *The Journal of Systems and Software*, 70, 61–78.

Green, P.E. and Srinivasan, V. (1978), "Conjoint analysis in consumer research: Issues and outlook." *Journal of Consumer Research*, 103–123.

Green, P.E. and Srinivasan, V. (1990), "Conjoint analysis in marketing: new developments with implications for research and practice." *Journal of Marketing*, October, 3–19.

Green, P.E., Krieger, A.M. and Wind, Y., (2001) "Thirty Years of Conjoint Analysis: Reflections and Prospects." *Interfaces*, 31(2), 56–73.

Harrison, J.V. and Andrusiewicz, A. (2004). "A virtual marketplace for advertising narrowcast over digital signage networks." *Electronic Commerce Research and Applications*, 3(2), 163–175.

Ho, W. (2007), "Integrated analytic hierarchy process and its applications–A literature review." *European Journal of Operational Research*, 186(1), 211–228.

ISO 9126–1, 1998, Software product quality, Part 1: Quality model, International Standard.

ISO 9241–11, 1998, Ergonomic requirements for office work with visual display terminals (VDTs), Part 11: Guidance on usability, International Standard.

Koczkodaj, W. (1998), "Testing the accuracy enhancement of pairwise comparisons by a Monte Carlo experiment." *Journal of Statistical Planning and Inference*, 69(1), 21–31.

Krantz, D.H. and Tversky, A. (1971). "Conjoint measurement analysis of composition rules in psychology." *Psychological Review*, 78, 151–169.

Luce, D.R. and Tukey, J.W. (1964), "Simultaneous conjoint measurement: a new type of fundamental measurement." *Journal of Mathematical Psychology*, 1, 1–27.

Mulaik, S.A. (1986), "Factor analysis and Psychometrika: Major developments." *Psychometrika*, 51(1), 23–33.

Nielsen, J. (1993), Usability Engineering, New York, Academic Press.

Saaty, T.L. (1977), "A scaling method for priorities in hierarchical structures." *Journal of Mathematical Psychology*, 15, 234–281.

Saaty, T.L. (1980), The analytic hierarchy process, New York: McGraw-Hill.

Steiger, J.H. (1994), Factor Analysis in the 1980's and the 1990's: Some old debates and some new developments. In Borg, I., & Mohler, Peter Ph. (Eds.) *Trends and perspectives in empirical social research.* Berlin: DeGruyter.

Zahedi, F. (1986), "The Analytic Hierarchy Process - A Survey of the Method and its Applications." *Interfaces*, 16(4), 96–108.

CHAPTER 54

Investigating Preattentive Features for Modelling Human Behaviour During Visual Search

Rafał Michalski, Jerzy Grobelny, Marcin Kuliński, Katarzyna Jach

Institute of Organization and Management (I23)
Faculty of Computer Science and Management (W8)
Wrocław University of Technology
27 Wybrzeże Wyspiańskiego
50-370 Wrocław, Poland

ABSTRACT

The main goal of this work was to verify the general visual search theoretical models involving the stage of preattentive processing in the human-computer interaction domain. In the eye tracking experiments, the participants performed three different types of visual search tasks: simple hyperlink search, tabulated data search, and a search for a specific icon in different types of graphical toolbars. The obtained results indicate that the subjects employed two distinct general strategies for finding the target in the experimental tasks.

Keywords: Eye tracking, search strategies, visual search, toolbars

INTRODUCTION

The thorough understanding how people search information has an enormous significance not only for broadening our knowledge about the human visual information processing, but also may have profound practical implications. The experimental research in this field could provide potentially invaluable information for developing the guidelines in the field of human-computer interaction (HCI). The crucial issue related with the human visual performance in this domain, is the way people deal with the search tasks in different contexts. The great variety of tasks involved in HCI is probably connected with various visual search strategies.

One of the most interesting methods that enable the researchers to record and analyse visual activity of the human being is the eye balls tracking technology. This type of visual activity registering technique has been known for a long time. First attempts were made already in the beginning of the twentieth century. However, the relatively cheap and flexible computer-aided devices developed in the last 10 years or so, substantially increased the interest of studies involving eye tracking technologies. During the years of research the visual search mechanisms have been described more and more comprehensively. The basic approaches include the spot light and zoom lens theories. The former, introduced by Posner (1980), and assumes sequential processing of some area named spotlight and preparing at the same time, the next saccade for a new place. The latter, put forward by Eriksen and St. James (1996), considers the visual search as a process of narrowing down the explored area, just like the zoom function used in cam-coders. The feature integration theory (Treisman and Gelade, 1980) and guided search proposal (Wolfe et al., 1989; Wolfe, 1994) combine the zoom lens and spotlight hypotheses. According to both approaches, image is analysed approximately in parallel in the first phase the, which is in concordance with the zoom hypothesis. In the next stage, the processing takes place within the confines of earlier identified structures, which is in turn in agreement with the spotlight proposal. The first stage is often called preattentive since it is believed that the attention is not involved at this time. The detailed summary of the findings related with the preattentive processing was presented by Wolfe and Horowitz (2004) and the extensive review of major visual processing theories can be found, for example, in the book of Findlay and Gilchrist (2003).

Applications of the eye trackers in human-computer interaction studies aim, among other things, at examining how the scan path parameters can be used in usability assessment. Many proposals in this respect were put forward by Goldberg and Kotval (1999). A general review of eye tracking studies in the domain of HCI was provided by Jacobs and Karn (2003).

RESEARCH OBJECTIVES

The main goal of this study is to explore the visual search strategies applied by users during the search for various kinds of target objects in the context of the human-computer interaction. The focus was especially given to verify in what way the theories and models developed by visual search scientists can explain behaviour of computer users' behaviour. For this purpose, two experiments were carried out, in which the visual activity was registered by the eye tracking system. In the first examination the participants were to find a hyperlink in the web page or specified information in tabulated data. The second experiment involved the execution simple search and click tasks performed on the toolbar type panels. Moreover, in the latter case, it was interesting find out to what extent the colour preattentive factor influences the efficiency of graphical object search. In next passages of this work the aforementioned eye tracking experiments are presented and discussed.

METHOD

PARTICIPANTS

Thirty subjects participated in the examinations. All of them were students of the Wroclaw University of Technology at the age between 23 and 25 years. They were not paid for taking part in the examination. None of the participants wore eye glasses or contact lenses.

APPARATUS

The experiments were carried out in a laboratory environment by means of the Pan/Tilt version of the ASL 6000 eye tracker (Applied Science Laboratories, 2005). This system registers eye ball position with the frequency of 60 Hz and enables for locating the eye sight line with the precision of one visual angle. The visual activity registration during performing the experimental tasks was supported by GazeTracker™ software, which was also used for analyses of the gathered data.

INDEPENDENT VARIABLES

Toolbar search

For the first phase of examination, the graphical panels containing objects in the shape of typical computer buttons were designed. The panels were similar to the toolbars widely used in contemporary computer programs but instead of pictorial icons, letters and numbers were put on the target items. One toolbar consisted of 36 identical objects. All the examined structures had the same shape and dimensions. The button sizes were chosen according to the one of the default values used in popular operating systems.

The layouts were differentiated by the item background colour and by the way the colours were located. The three basic colours: red (R), green (G), and blue (B) were used. They made up three pairs of colours: red-green (RG), red-blue (BR) and green-blue (BG). There were two levels of the pattern factor. The first one was called random (R), and the second one, which is similar to the chessboard was named regular (C).

Web site search

The internet book stores were examined in the second part of the examination. The most popular in Poland web sites in this domain were chosen, namely Empik, Inbook, Lideria, and Merlin. These internet web sites had different design and layout. From many sub-pages available in those portals, three pages were chosen from Empik and Inbook web sites and two from Lideria and Merlin. The persons taking part in this examination performed two types of tasks related to two kinds of the given web site profile. The first one was a simple search for a given link on the typical multimedia page, and the second involved finding information about service costs in a web page organized in a tabular form.

DEPENDENT MEASURES

The temporal parameters of the visual activity were registered by means of the GazeTracker™ computer program, which was integrated with the eye tracking equipment. For the purposes of this research one of the most widely used definitions of fixation was adopted. The fixation occurred if the time of looking at the circular area of about 40 pixels in diameter exceeded 0.2 second. The additional requirement was that the number of gaze points should be more than three in the specified area. The dependent measures included mainly the number of fixations, fixation duration, time spend on the slide, saccade lengths, and spatial locations of the fixations and saccades.

EXPERIMENTAL DESIGN AND PROCEDURE

Before the experiment started, the subjects were informed about the goal and the scope of the examination. Next, according to the ASL eye tracking system instructions, the necessary eye tracking system calibration took place. The persons taking part in the experiments performed first the toolbar search tasks and then after a short break, the examination of the web sites took place.

Toolbar search

The independent variables for the toolbar examination resulted in six experimental conditions. There were six colourful toolbars (three pairs of colours) × (two colour patterns). The simple standard within subjects design was applied, in which every participant tested all the graphical variants. The panels were always displayed in the left upper corner of the computer screen. The examined graphical structure was visible only during the visual search. The experimental task included clicking with the computer mouse on the item to be searched for, and then find and select within the shown panel the desired target. The illustration of this task is given in figure 1. The procedure was repeated by the user twice for every tested object.

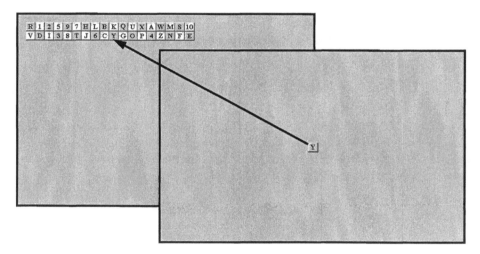

FIGURE 1 Experimental task for the graphical toolbar examination

Web site search

Two web sites represented by three sub pages and two other with further two pages produced ten experimental conditions. Two of categories of the search tasks included – the simple visual search (six trials) and the search of complex tables (four cases). The tasks were presented in a random sequence that was specified before the experiment took place. Each participant performed all ten search tasks. Every time the instruction appeared in the middle of a screen and then, after mouse click, the tested screen was exposed. The subject was to find the target object as fast as possible and click it. After that, the next instruction was displayed. The exemplary task is presented in figure 2.

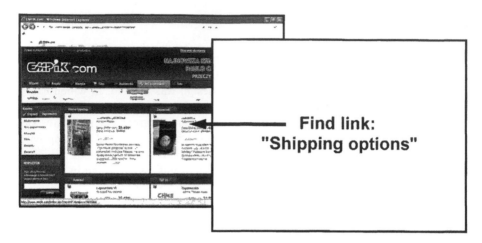

FIGURE 2 An experimental task for the website examination

RESULTS AND DISCUSSION

In the sections that follow the obtained results' analyses both for the panels and web sites are provided. The gathered data are first investigated formally by means of the analysis of variance conducted for the described in previous sections independent factors. Next, the data were qualitative analysed.

GRAPHICAL TOOLBARS EXAMINATION

ANOVA

Also in this case in order to verify the significance of discrepancies in the time on slide measure recorded for the examined toolbars, a standard two way analysis of variance was used. The calculated F statistics and respective p values for the main effects amounted to: item background colour $F(2, 384) = 0.8$, $p = 0.45$; the toolbar colour pattern $F(1, 384) = 0.26$; $p = 0.61$. The effect of interaction between item background colour and panel colour pattern was equalled $F(2, 384) = 1.25, p = 0.29$. According to the obtained results none of the effects or the interaction between the item background colour and the panel colour pattern was statistically meaningful.

Qualitative analysis

These results were quite unexpected since the similar investigation carried out by Michalski and Grobelny (2008), in which only the acquisition times were registered, the same factors were significant. To account for this considerable discrepancy an additional qualitative analysis of the recorded data was conducted. The analyses of the first fixation spatial distributions of the subjects taking part in the examination showed that the first fixations are more densely located in the right hand side of the analysed toolbars.

The further qualitative exploration revealed that the difference between the first and the second trial were possibly dependent on the specific target location within the examined graphical toolbar. This conjecture was formally verified by making an additional analysis. The toolbar was divided into eight equal areas numbered from one to eight counting from the left hand side. Next, average times spent in every panel sections were compared. The hypothesis was supported by the results of the analysis of variance: $F(7; 384) = 3.1; p < 0.01$, which showed the statistical significance. The detailed investigation of these data demonstrated that substantial number of fixations in the beginning phase of the visual search process was located on the right hand side of the examined toolbar or near the left edge of the panel. Such a distribution shows that subjects selected faster, targets located in these segments of the searched toolbar.

Apart from the fixation location analysis, also the characteristics of shifting the visual attention between the toolbar segments were explored. The obtained transition frequencies for these cells are shown in figure 3.

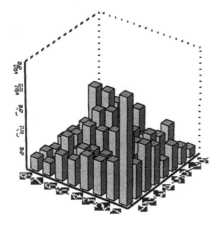

FIGURE 3 Transitions distribution between the panel segments (X and Y axis - the segment number, Z axis - the number of observations)

Saccades to the adjacent cells, which are located near the diagonal, dominate in the demonstrated pattern. This outcome was also confirmed by the average saccade length computed for the first two saccades recorded in every trial. In visual angle terms, the size amounted to about four degrees, what corresponds to the so called foveal vision. These results suggest the domination of the spotlight strategy and the covert attention mechanism during the visual search for targets within toolbars.

530

WEB SITES EXAMINATION

ANOVA

To test the significance of differences in the mean time spend on the slide recorded for individual web sites, a standard two way analysis of variance (ANOVA) was employed. The obtained results revealed that both the analysed effects were statistically meaningful. The F statistics and p values for the search type and web site type experimental factors amounted to respectively $F(1, 92) = 15$, $p < 0.0005$ and $F(3, 92) = 8.4, p < 0.0001$.

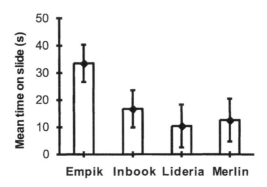

FIGURE 4 Mean times on slide depending on the web site, $F(3, 92) = 8.4$, $p < 0.0001$; whiskers denote 0.95 confidence intervals

According to the obtained results also the also the interaction effect between the way the web site was searched and the type of the web page occurred to be statistically important: web site type × search type – $F(3, 92) = 3.5, p < 0.05$. The general comparison of the total visual search efficiency results are illustrated in figures 4 and 5. These outcomes were rather expected and showed that the web site design, along with the appropriate colour use as a major preattentive feature, significantly influence the visual search efficiency. It was not also surprising that performing more complex tasks resulted in decidedly longer completion times compared with the simpler visual search task.

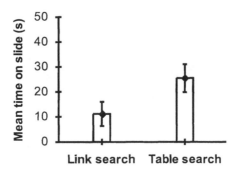

FIGURE 5 Mean times on slide depending on search type effect, $F(1, 92) = 15, p < 0.0005$; whiskers denote 0.95 confidence intervals

Quantitative analysis

To get a more wide insight into the visual search mechanisms during explored tasks, the spatial analysis of fixations and saccades was also conducted.

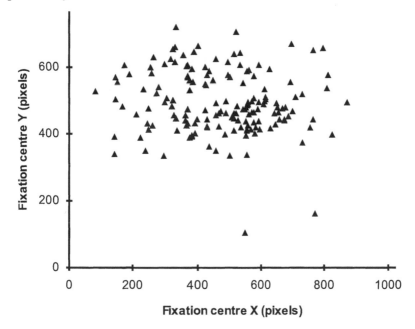

FIGURE 6 First fixations for the hyperlink search task

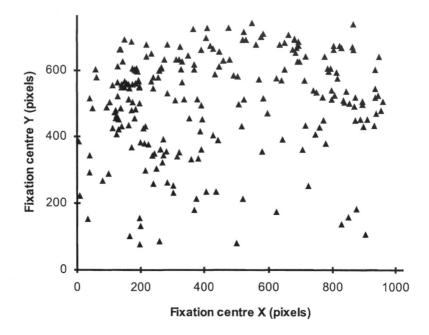

FIGURE 7 Third fixations for the simple visual search task

The spatial distribution of first fixations for the hyperlink visual search tasks for all participants is presented in figure 6. This illustration clearly shows that most of the first fixations are located a little above the centre of the screen and forms an ellipsoidal pattern. The third fixations demonstrated in figure 7, are substantially more dispersed over the whole screen. These results indicate that the search strategy applied by the subjects in this part of the examination can be explained by the visual search zoom lens model.

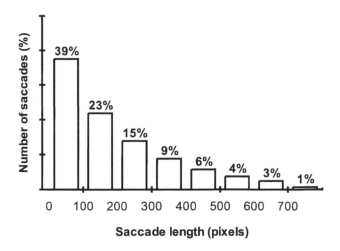

FIGURE 8 The saccade length distribution for the tabulated data in the web site examination

Similar analysis was carried out for the more complex tasks related with the search for specific information in tabulated data. However in this case the patterns were not so obvious and could not be attributed for the same strategy. To further explore the nature of the subjects' behaviour in this type of experimental tasks the characteristics of the recorded saccades were investigated. The results are demonstrated in figure 8. As can be seen from this distribution the relatively short saccades dominate in the search for very specific information in the tabulated data. The saccades not longer than 100 pixels represented the biggest proportion - 39%. Moreover, the number of saccades shorter than 200 pixels equalled to 62%. Thus, it can be said that the relatively short saccades dominated, what suggests that the spot light theory of visual activity is more appropriate in this type of search tasks.

CONCLUSIONS

The present research was designed and conducted in hope for obtaining some results showing how the general visual search hypotheses and theories work in practical HCI tasks involving the various types of target search and its selection by means of a computer mouse. The carried out experiments provided some evidence in this respect.

The main goal of the toolbar research was to verify to what extent the use of colour, which is considered as the most important factor of preattentive visual processing, helps in the search for graphical objects. The theoretical considerations and some previous research proved that grouping objects by means of colour restricts the search area and thus shortens the task completion times. Surprisingly, the colour preattentive feature did not influenced the mean time on slide in this study, which contradicted the results presented for instance by Michalski and Grobelny (2008).

The further detailed analyses suggested that this discrepancy could have been caused by the effect of target position within the given layout, which could have influenced the explicitly examined independent factors. It is also possible that target location in relation to the starting fixation point had a considerable impact on the efficiency of visual search process. In the respect of the search strategies in this part of examination, the qualitative analysis points to the spotlight model and the covert attention mechanism.

The analysis of the main factors considered in the study of web sites, demonstrated that the mean search times depend to a high degree on the way the web page was designed as well as on the complexity of the task to be accomplished. The visual activity of the subjects during the simple search for hyperlinks resembled the zoom lens mechanism, and is similar to the results published by Zelinsky (1997). The first fixation usually occurs near the web page centre while the rest of the fixations is scattered all over areas of detailed searches. This user behaviour emphasizes the significance of web page layout and the appropriate use of colours, since probably during the first fixation, the analysis of the general structure of the screen is performed. During the process of searching detailed information in the tabulated data rather short saccades dominated. That means that subjects changed their looking direction, most frequently, to the adjacent regions. Such behaviour seems to confirm the hypothesis about covert attention processing of the neighbouring areas, while the current first fixation information is analysed.

REFERENCES

Applied Science Laboratories (2005), Eye Tracking System Instructions, ASL Eye-Trac 6000 Pan/Tilt, ASL.

Eriksen, C.W. and St James, J.D. (1996), "Visual-attention within and around the field of focal attention – a zoom lens model." *Perception & Psychophysics*, 40, 225–240.

Findlay, J.M. and Gilchrist, I.D. (2003), Active vision. The psychology of looking and seeing, New York: Oxford University Press.

Goldberg, J.H. and Kotval, X.P. (1999), "Computer interface evaluation using eye movements: methods and constructs." *International Journal of Industrial Ergonomics*, 24, 631–645.

Jacob, R.J.K. and Karn, K.S. (2003), Eye tracking in human-computer interaction and usability research: Ready to deliver the promises (Section Commentary). [In:] The mind's eye: cognitive and applied aspects of eye movement research. Hyönä, J., Radach, R., Deubel, H. (Eds.), Amsterdam: Elsevier Science, 493–516.

Michalski, R. and Grobelny, J. (2008), "The role of preattentive visual information processing in human–computer interaction task efficiency: a preliminary study." *International Journal of Industrial Ergonomics*, 38, 321–332.

Posner, M.I., Snyder, C.R.R. and Davidson, B.J. (1980), "Attention and the detection of stimuli." *Journal of Experimental Psychology: General*, 109, 160–174.

Treisman, A.M. and Gelade, G. (1980), "A feature integration theory of attention."

Cognitive Psychology, 12, 97-136.

Wolfe, J.M. (1994), "Guided Search 2.0 – a Revised Model of Visual-Search." *Psychonomic Bulletin & Review*, 1, 202–238.

Wolfe, J.M., Cave, K.R. and Franzel, S.L. (1989), "Guided search – an alternative to the feature integration model for visual-search., *Journal of Experimental Psychology: Human Perception and Performance*, 15, 419–433.

Wolfe, J.M. and Horowitz, T.S. (2004), "What attributes guide the deployment of visual attention and how do they do it?" *Nature Reviews Neuroscience*, 5, 495–501.

Zelinsky, G.J., Rao, R.P.N., Hayhoe, M.M. and Ballard, D.H. (1997), "Eye movements reveal the spatiotemporal dynamics of visual search." *Psychological Science*, 8, 448–453.

CHAPTER 55

Occupant Response in Mine Blast Using Different Dummies from Two Commercial FE Codes

Kiran Irde, Ken-An Lou, Terry Wilhelm, William Perciballi

ArmorWorks
305 N. 54th Street
Chandler, AZ 85226, USA

ABSTRACT

It is commonly known that an Improvised Explosive Device (IED)/mine is one of the most dangerous threats in the U.S. operations in Iraq and Afghanistan. Death caused by an IED has increased from zero to 61 percent from 2001 to 2009 among the total hostile deaths. Design of improved IED/mine blast protection vehicles requires the use of energy absorbing seats in conjunction with vehicle underbody blast deflectors and armors.

In this paper, the occupant response from dummies of two different commercial finite element codes is compared with that from an experimental seat drop tower test. The 50th percentile Hybrid III dummy model from LS-DYNA and MADYMO software is used along with a ShockRide®'s mine blast attenuation seat. The entire seating system is modeled using LS-DYNA explicit finite element code.

There are two methods to simulate seat drop tests. In the first method, an initial velocity is applied to the seat and the occupant while a deceleration pulse is applied to the pedestal on which the seat is anchored. The second method assumes the same acceleration to the drop cage and the occupant, therefore, the mine blast acceleration pulse is applied to the occupant only. One of the drawbacks for this method is that the inertial effects of the seat are neglected. However, it allows to simulate 1 G gravity load and the impact acceleration can be offset with enough time for the occupant to load the seat cushion. In this study the first method was utilized. Data collected from Hybrid III dummy are compared to some of the acceleration time history test data.

Keywords: Mine Blast, IED, Energy Absorbing Seat, Occupant Injury, Ground Vehicle, LS-DYNA, MADYMO

INTRODUCTION

Recent increase in troop injuries due to the effects of mine blast/ Improvised Explosive Device (IED) attacks has led to a push to improve armored vehicle/seat design. The primary mode of injuries include spinal injuries due to vertical seat loading, head concussions from vehicle ceiling contact and shattering of leg bones as a consequence of floor impact. During the development process of an energy absorbing seat which provides adequate occupant protection during mine blast event, a prototype seat for each design change needs to be evaluated for its performance by conducting a drop tower test. However, these tests are expensive and lead to the destruction of the test article leaving less chance of reuse in future tests. Computer Aided Finite Element analysis has been significantly utilized by many industries during the design cycle as they provide adequate and in-depth results to the design team for analyzing a particular engineering design.

The goal of this paper is to understand the limitations of commonly used automotive simulation dummy models for pure vertical impact military ground vehicle applications. Comparing and correlating with experimental test data could give engineers confidence in the use of these dummy models. Several commercial codes with extensive dummy databases are available for determining the occupant response in a blast environment. In this paper, two such codes namely, LS-DYNA and MADYMO dummy models, have been evaluated.

CAE DUMMY MODELS

Computer simulation models using advanced, non-linear finite element code, such as LS-DYNA and MADYMO, are practical tools that complement physical seat test. Once an initial model is validated against a full-scale crash test, large parametric studies can be conducted, thus reducing the cost and time involved in the early development process.

LS-DYNA is a general purpose, explicit finite element program that can be utilized for analyzing the large deformation static and dynamic response of structures [1]. It is widely applied in automotive and aerospace industry applications including design of crashworthy energy absorbing seats, determination of normal modes and in failure analyses.

Figure 1. LS-DYNA 50th Percentile Dummy Model

The software code provides a comprehensive library of material models that includes metals, plastics, foams, fabrics and other user-defined materials to name a few. It features several robust contact algorithms and an extensive element library in addition to modeling accelerometers, sensors and seatbelts. The dummy database in LS-DYNA consists of both rigid and deformable versions of the Hybrid III 5th, 50th and 95th percentile dummies. Figure 1 shows the LSTC's Hybrid III 50th percentile dummy model.

Figure 2. MADYMO 50th Percentile Ellipsoid Dummy Model

MADYMO is a multi-body dynamics software package that provides solution to problems in crash engineering applications. It has been extensively used in both aerospace and automotive industries for occupant safety design and optimization. The program generates output that can be utilized to study the kinematics and injury parameters of the occupant, in addition to the dynamic response of the surrounding structures.

The software has an extensive library of dummy models that have been validated for frontal, side, rear and vertical impact applications. These dummy models are available as a rigid body ellipsoid, facet or finite element model types that primarily differ in the modeling techniques applied to represent the geometry and mechanical properties of the modeled components [2]. Figure 2 shows the MADYMO Hybrid III 50th percentile rigid body ellipsoid dummy model. Extensive restraint modeling such as seatbelts and airbags can be modeled using the explicit dynamic finite element code available within the software.

TEST SETUP

A typical drop tower test setup is shown in Figure 3. The seat base is rigidly mounted onto the drop tower pedestal. An instrumented Hybrid III 50th percentile dummy is placed on the seat cushion at the design H-point location and is restrained by a four point seatbelt system which is an integral part of the seat system.

Figure 3. Drop Tower Test Setup

High speed video cameras are placed at strategic locations around the drop tower test rig to monitor the overall dummy kinematics and also to capture the deformation of some specific seat components such as the Energy Absorbing (EA) links. The entire seat/drop tower pedestal assembly is raised to a height (that produces the desired impact velocity as per the design protocols) and allowed to drop onto the floor which has a layer of 12"x12"x1" Neoprene rubber foam material placed directly below the pedestal impact location. The Neoprene 70A durometer rubber foam material is used to control the pulse shape.

MODEL SETUP

The seat CAD data and the LS-DYNA seat system finite element (FE) model are shown in Figure 4. The seat FE model consisted of approximately 204,000 nodes and 202,000 elements, with SHELL elements representing a majority of the structural parts and SOLID elements representing the seat foam and EA box. The *CONTACT_TIED_NODES_TO_SURFACE card was used to represent the welding at various locations of the seat. The bolt connections were represented using *ELEMENT_BEAM card. The *MAT_LOW_DENSITY_FOAM material card was used to model the seat foam cushion.

Figure 4. CAD and Finite Element Model of Seat

The first simulation was run with a LSTC Hybrid III 50th percentile FE dummy while the second simulation was conducted with a MADYMO Hybrid III 50th percentile rigid body ellipsoid dummy model coupled to the LS-DYNA seat system. All the nodes in the seat model and the LSTC FE dummy model were provided with an initial downward velocity of 7 m/s (10.3 ft/s). In addition to this, the deceleration of the actual drop tower test pedestal was provided to the simulation pedestal part. Contacts between the various seat parts and also between the seat to the LSTC FE dummy were defined using the generic contact card *CONTACT_AUTOMATIC_SINGLE_SURFACE.

For the coupled MADYMO to LS-DYNA analysis, the MADYMO ellipsoid dummy was assigned the same initial downward velocity using the INITIAL.JOINT_VEL card. Contacts between the MADYMO dummy and LS-DYNA seat cushion were defined within MADYMO using the CONTACT.MB_FE card.

RESULTS

The main objective of this study was to evaluate the dummy head, chest, and pelvis Z acceleration performance of the LSTC FE dummy and the MADYMO ellipsoid dummy. We selected one of the ShockRide®'s mine blast attenuation seat models and test data during early seat design stage. The test to simulation model kinematics comparison at different time is shown in Figure 5.

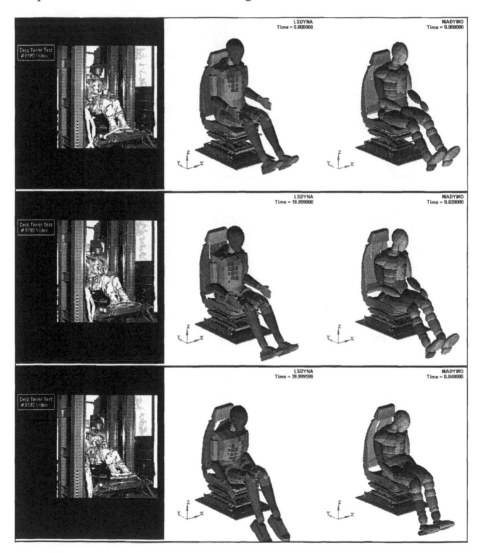

Figure 5. Test, LS-DYNA and MADYMO (coupled) Dummy Kinematics

542

Seat belts were not used in the simulation models for this study as it was noticed that they had very little effect on the simulation results.

The rear view of the seat EA box from the test and simulation is shown in Figure 6. It can be noticed that the crushing of the EA link seen in the test is very well captured by the LS-DYNA seat simulation model.

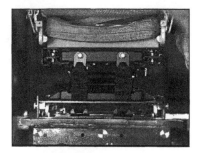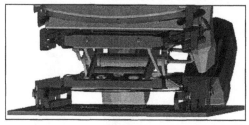

Figure 6. Crushing of EA Links in Test and Simulation

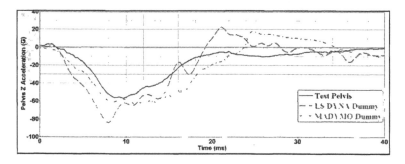

Figure 7. Dummy Pelvis Z Acceleration Comparison

Figures 7-9 show the comparison of dummy's pelvis Z, chest, and head acceleration between the experimental test, the LS-DYNA dummy and the MADYMO coupling dummy models.

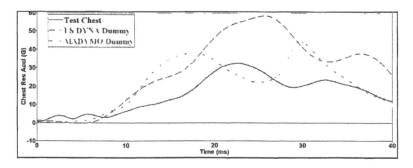

Figure 8. Dummy Chest Resultant Acceleration Comparison

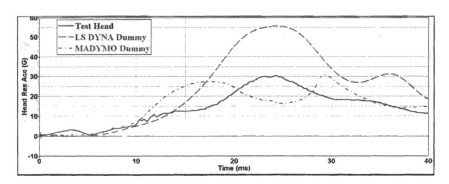

Figure 9. Dummy Head Resultant Acceleration Comparison

CONCLUSIONS

The preliminary test-analysis correlation shows that both dummy models are receiving too much load. The LS-DYNA dummy is receiving more load and the upper legs are rebounding more than the MADYMO dummy. The MADYMO dummy results are closer to the test data compared with LS-DYNA dummy. Further refinement at ArmorWorks will include improving the EA box and seat cushion model. Part of the discrepancy between the test and simulation is also attributed to the limitations of the current dummy models. Therefore, efforts must be made by both dummy development groups to either continue improving current automotive frontal crash dummies for vertical load cases or to develop a new vertical impact crash dummy model. The dummy model must be validated by occupant response in the vertical direction.

REFERENCES

1. LSTC, LS-DYNA Theory Manual, Livermore, CA, March 2006.
2. TASS, MADYMO Model Manual, Release 7.1, Delft (The Netherlands), June 2009.

CHAPTER 56

Modeling Medication Impairment and Driving Safety

Vincent G. Duffy

Schools of Industrial Engineering and
Agricultural & Biological Engineering
Purdue University
West Lafayette, IN USA 47907

ABSTRACT

Driving mishaps are still a significant contribution to years of lost life, in the U.S. and abroad - especially among younger drivers. As the rate of incidents is high among both young and older drivers, some attention could be given to ways to assist these 'at risk' drivers. As well, elderly are asked to not drive under certain medication regimens. With that, their quality of life is adversely affected. As sensor and automation technology becomes more prevalent, it does raise the potential for improved driving under distracted and possibly impaired conditions. It would be useful to know when a driver should simply just be discouraged from driving, and when they could instead be assisted by new automation that is now or will be available in the coming years. It would also useful to have this kind of information available across disciplines so that it can be considered by the 'at risk' drivers, the engineering community, and the medical and pharmacy communities.

Keywords: Medication impairment, distracted driving, driving safety, driving fatalities, automation.

INTRODUCTION

As people get older, they take more medications - by some reports, as many as 7 on average in later years. Some prescription regimens suggest that drivers should not drive, due to impairment anticipated by using these medications. Elderly drivers are considered to be at higher risk (Russell and Barkley, 2005). Community pharmacists help to convey the information to those prospective drivers. However, there really isn't much to quantify the impairment of individual medications, let alone prescription regimens with multiple medications. And if there are known impairment levels for certain medications, there certainly is not much known about medication impairment levels in contrast to something such as alcohol.

IMPAIRMENT AND DRIVING SAFETY

Some medications in fact are shown to help (eg. those with Attention Deficit disorders can be assisted in their driving through some medications - Barkley et al. 2005). Some medications may make driving worse. How much? We don't really know. One could work toward classifying the impact of medications on impairment, and in contrast to levels that can be demonstrated for alcohol impaired drivers. This, ultimately will be of interest to drivers and insurance agencies, but may take time to develop. Developing the understanding of what will help drivers to overcome impairment will also be needed.

Burns et al. (2002) have outlined, in their article in *Scientific American*, examples where new technologies are enabled in vehicles. Szczerba et al. (2006) have shown methods to determine when and how some versions of the same new technology may be more preferred than others. And some new technologies such as sensors within and outside the vehicle have some ability to go essentially undetected – and may be able to capture information that can help drivers in cooperation with other automation or information systems. Driver distraction and conditions that lead to driver crashes are of significant interest in the literature and the community-at-large (see Lee and Abdel-Aty, 2009).

There is the potential that some sensor configurations can help offset these effects of distraction or impairment, and could be ultimately justified or 'prescribed' for impaired drivers. With some further development of the underlying science, this could extend the years of quality life, and possibly could be an entry point for sensor developers in the automotive industry into the 'medical device' field. If we can develop the underlying science that explains the benefit of certain sensor configurations that include Vehicle to Vehicle (V-V), Vehicle to Infrastructure (V-I) and Human-Machine Interface considerations, the engineering community can provide a service (or you may consider it as a set of assistive technologies) to impaired drivers that could be justified from a medical expenditure or possibly occupational rehabilitation standpoint.

In order to achieve that, the new technologies would need to assist drivers in performance, or assist them in knowing/guiding when they should transfer control *to* the vehicle. And some development work would need to be done to demonstrate improvement under the impaired condition. If you consider the rate of driving mishaps over the range of ages as a U shaped curve (see Eby and Kantowitz, 2006 in G. Salvendy *Handbook of Human Factors and Ergonomics*), you could consider that some drivers would stand to benefit from some form of assistive technology when driving. New technologies such as adaptive cruise control are now available for a cost. However, some drivers still feel that they are better to control the vehicle than the newly available algorithms. They may be right. So you can consider that some drivers need assistance (perhaps a great deal of the time). And some drivers will not want assistance, particularly at the current level of control available in the new technology.

As technology improves, when should drivers be willing to relinquish control of the vehicle? And when should doctors and pharmacists consider allowing drivers under the impaired condition to still drive with assistance? The answer can be ultimately considered in cooperation with occupational therapists familiar with available technology, and a strong underlying science to explain the capabilities and limitations of such new technologies to assist drivers –impaired or otherwise.

METHODS

It becomes apparent fairly quickly that driving simulators lend themselves fairly well to this type of research – where, on behalf of driver safety, we would prefer to avoid the hazardous driving condition if at all possible. Other live or real vehicle validations may later be considered in some limited capacity with second drivers (similar to a drivers' education type car) and a test track.

As for pedestrians, Rosenbloom et al. (2008) have noted the challenges of improving pedestrian safety at street crossings, especially for children. An alternate approach that addresses both driver safety and pedestrian safety at the same time could be considered for data collection and potentially for training pedestrian crossing. Networked simulators that provide different vantage points, such as driver and pedestrian view at the same time, have not been commonly used in recent driving studies. This type of networked simulator together with specialized instrumentation such as that used for obtaining eye movement activity (outlined by Ahlstrom and Friedman-Berg, 2006) and explicit impairment regimens (as outlined in Zimmerman et al. 2008) may provide some insight into pedestrian safety as well as driver safety. Such simulators could provide an opportunity for data collection and analysis from both the driver and pedestrian vantage point in parallel.

DISCUSSION

When you consider that there are still a similar level (approx. 40,000) of fatalities each year – over many years dating at least back to the early 1990s, and that the years of lost life each year due to driving fatalities in the U.S. are comparable to those of heart disease cancer and stroke combined (Wickens et al. 2004), there certainly seems to be significant justification for additional effort to be spent on driving safety from a research standpoint. And when considering that as people in the U.S. - where the places people go regularly are typically geographically set apart by significant distance - have come to rely on their driving capabilities into their later years in life, quality of life in the context of the increased medication usage is likely to become a growing concern among the elderly. Recent research from NHTSA (Staplin et al, 2008) shows that there isn't sufficient information from the databases to draw epidemiological conclusions on multiple medications and driving. They also weren't able to get enough data to establish patterns of impairment due to the medications.

In the NHTSA report of Staplin et al. (2008), they have outlined some medication classifications that one could focus on initially, and they have noted that drivers performed differently-better when under observation of the 'occupational therapist'. This could motivate some cooperation in development of something like a 'virtual' occupational therapist that could ultimately also be part of prescription regimens, if our sensor configurations and driver models provide accurate prediction of conditions in which drivers will overcome impairment due to the medications. Those drivers considered to be impaired, but stable may be helped more than some others. For instance those who are prone to episodes of hypoglycemic stupor are more likely to end up in driving mishaps (Cox, et al. 2003) and may still need to 'ride the bus' or consider other public transportation rather than drive, even with improved vehicle control algorithms or human-system interface.

Current challenges that remain in driving safety and transportation human factors help us to see more clearly why, as noted by Wickens et al. (2004), it is considered as much as 30 times more safe to ride the bus. However, as automation technology becomes more prevalent (as outlined in Lee, 2009 in *Science*) it would certainly be good to be able to know with greater certainty which among these new automation advances have the potential to effectively improve quality of life. And it would be better if we can effectively communicate that as a scientific community, and cooperate with those of other disciplines on behalf of the community-at-large.

ACKNOWLEDGMENTS

Thanks to Akshatha Pandith for her assistance. Thanks also to Prof. Yaobin Chen, colleagues at TASI (Transportation Active Safety Institute) at IUPUI-Indianapolis for their invitation to a workshop on driving safety at IUPUI-Indianapolis and

548

NEXTRANS (USDOT Center at Purdue) for their support during the formation of this paper.

REFERENCES

Ahlstrom, U. & Friedman-Berg, F.J. (2006). Using eye movement activity as a correlate of cognitive workload, *International Journal of Industrial Ergonomics, 36 (7) July, 623-636.*

Barkley, R.A., Murphy, K.A., O'Connell, T. & Connor, D.F. (2005). Effects of two doses of methylphenidate on simulator driving performance in adults with attention deficit hyperactivity disorder, *Journal of Safety Research*, 36, 121-131.

Burns, L.D., McCormick, J.B., & Borroni-Bird, C.E. (2002). Vehicle of change. *Scientific American, 287*(4), 64-73.

Cox, D.J., Penberthy, J.K., Zrebiec, J., Weinger, K., Aikens, J.E., Frier, B., Stetson, B., DeGroot, M., Trief, P., Schaechinger, H., Hermanns, N., Gonder-Frederick, L, Clarke, W. (2003). Diabetes and Driving Mishaps, Frequency and correlations from a multinational survey. *Diabetics Care*, 26: 2329-2334.

Eby, D. W. and Kantowitz, B.H., (2006). Human factors and ergonomics in motor vehicle transportation, in G. Salvendy (Ed.) Handbook of Human Factors and Ergonomics, 3rd Ed., John Wiley & Sons, New Jersey, ch. 59. pp. 1538-1569.

Lee, C. & Abdel-Aty, M. (2008). Presence of passengers: Does it increase or reduce driver's crash potential? *Accident Analysis & Prevention, 40 (5)* 1703-1712.

Lee, J.D. (2009) Can technology get your eyes back on the road? *Science*, 324, 17, April, 344-346.

Rosenbloom, T., Ben-Eliyahu, A. & Nemrodov D. (2008). Children's crossing behavior with an accompanying adult, *Safety Science*, 46 (8) *1248-1254.*

Russell A. & Barkley, K. R. (2005). Driving Accidents in the Elderly: An Analysis of Symptoms, Diseases, and Medications. *Journal of Safety Research* , 121-131.

Staplin, L., Lococo, K.H., Gish, K.W., Martell, C. (2008). A pilot study to test multiple medication usage and driving functioning, NHTSA Report No. DOT HS 810 980 (123pp).

Szczerba, J., Duffy, V. G., Geisler, S., Rowland, Z., & Kang, J. (2007). A study in driver performance: alternative human-vehicle interface for brake actuation. *SAE 2006 Transactions Journal of Engines, 2006-01-1060*, 605-610.

Wickens, C.D., Lee, J.D., Liu, Y. and Gordon-Becker, S. (2004). An Introduction to Human Factors Engineering, Prentice-Hall, New Jersey.

Zimmermann, I.M., Mick, I., Vitvitskyi, V., Plawecki, M.H., Mann, K.F. and O'Connor, S. (2008). Development and Pilot Validation of Computer-Assisted Self-Infusion of Ethanol (CASE): A New Method to Study Alcohol Self-Administration in Humans. *Alcoholism: Clinical and Experimental Research, Wiley InterScience* , 32 (7) 1321-1328.

CHAPTER 57

Modeling Effective and Ineffective Knowledge Communication and Learning Discourses in CSCL with Hidden Markov Models

Theodor Berwe, Michael Oehl, Hans-Ruediger Pfister

Institute of Experimental Industrial Psychology
Leuphana University of Lueneburg
Germany

ABSTRACT

The grounding theory according to Clark (1996; Clark and Brennan, 1991) is a prominent approach to describe the co-construction of knowledge in computer-supported collaborative learning (CSCL) environments. Communication problems such as incoherence or inadequate coordination are common in simultaneous text-based chat tools resulting in impaired grounding processes, which again may affect learning outcomes. Thus collaboration scripts like learning protocols are implemented to increase the structure of chat discourses and to reduce communication problems in order to support grounding processes. In this study the impact of grounding itself on learning outcomes was examined by implementing a confederate in each learning group, who showed either a constructive or a destructive grounding behavior. Probabilistic models were computed for both the discourses of the present study as well as effective and ineffective discourses of a previous study. The long-term objective of this research is an adaptive script automatically detecting discourse structures in order to support grounding and thus to facilitate knowledge communication and learning processes.

Keywords: CSCL, Chat, Adaptive Scripting, Grounding, Hidden Markov Models

CHAT-BASED CSCL

Communication in computer-supported collaborative learning (*CSCL*) environments often suffers from deficiencies due to incoherence of contributions, from difficulties to maintain meaningful threads of discourse, and problems of instigating effective knowledge sharing activities (Bromme, Hesse, and Spada, 2005; Fuks, Pimentel, and Lucena, 2006; Herring, 1999; Kreijns, Kirschner, and Jochems, 2003; Oehl and Pfister, in press). However, with respect to net-based groups of collaborating learners, ample evidence shows that collaboration or cooperation scripts enhance learning performance (e.g., Kollar, Fischer, and Hesse, 2006; O'Donnell and King, 1999). Sophisticated collaboration scripts in terms of *adaptive chat tools*, which provide supporting functionalities only in circumstances, when it is necessary, might help to correct these deficiencies while maintaining the advantages of collaborative learning in general. Collaborative learning via the Internet can proceed along different ways, for example, using asynchronous discussion groups, or synchronous groups that communicate via chat or video-conferencing. In the present study, we deal with groups that communicate synchronously via text-based chat tools, enabling participants to carry out a discourse on a particular topic to be learned. In our current study we used a chat tool called 'learning protocol'.

The *learning protocol approach* in chat-based CSCL can be used on almost all levels of education, e.g., high school as well as university. Learning protocols implement a special kind of collaboration script, tailored to support efficient learning discourses (Pfister and Mühlpfordt, 2002; Pfister and Oehl, 2009). Think of a group of learners, say three or four, possibly moderated by a tutor, connected via the Internet and using a chat-based communication environment. Their task is to learn about a specified subject matter by discussing, asking questions, giving explanations, and exchanging arguments. Textual messages are produced, sent, and received by other members of the group. Learning protocols generally support these discourse activities by providing a collection of mainly three functionalities (figure 1):

(1) Explicit referencing: If a participant wants to indicate any contribution in the chat history or any part of the shared text with learning information to which his or her current contribution refers, she or he may mark this reference with the mouse and an arrow pointing from the current message to the referred to passage is automatically generated, visible for all members of the learning group. By explicit

referencing, the coherence of the discourse should be enhanced, and participants are able to track back threads by following the reference arrows.

(2) Typing of contributions: Each textual contribution, that is a single chat message, can be classified and labeled by the contributing participant with respect to its communicative type, for example, as a question, explanation or critique. The learner deliberately attaches a label to her or his message. This label precedes the message on the screen and remains visible for all participants.

(3) Moreover, a *turn-taking* rule (one-after-the-other) controls the sequencing of contributions among participants.

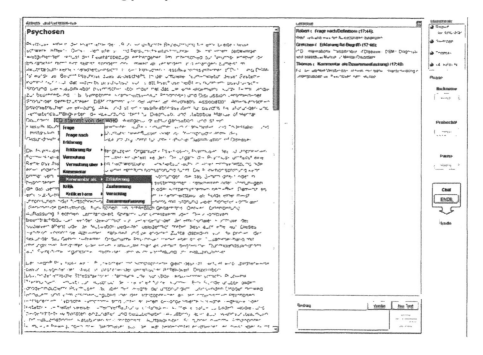

FIGURE 1. Screenshot of the learning protocol interface with the software ConcertChat (Mühlpfordt, 2006; the left pane shows the common text about the learning topic, also shown there is the pop-up menu of contribution types. The large right pane shows the chat-history with all learners' contributions to the discourse. The lower right pane depicts the input field for one's own contribution. A referencing line goes from the current contribution pane to a text fragment in the left text pane).

In this experimental study, we investigated how a manipulation of grounding processes in chat-based CSCL with learning protocols effects discourse coherence and learning outcomes. Additionally, we pursued the probabilistic modeling of grounding processes and discourse structures in order to develop adaptive learning protocols promoting and facilitating automatically grounding processes in learning groups.

GROUNDING PROCESSES IN CSCL WITH LEARNING PROTOCOLS

The *grounding theory* proposed by Clark (1996; Clark & Brennan, 1991) has received considerable attention as a framework to understand mechanisms and difficulties in computer-mediated communication and online learning (e.g., Baker, Hansen, Joiner, and Traum, 1999; Brennan, 1998; Bromme, Hesse, and Spada, 2005; Pfister and Oehl, 2009). The gist with respect to collaborative learning is that during a learning discourse, *common ground*, i.e., a shared understanding, is collaboratively constructed. For that purpose, participants' contributions need to be *grounded*. Put very simply, a contribution is considered to be grounded if it receives positive feedback from other contributors that it has been understood. In face-to-face discourses, signals for grounding often are indirect, such as nodding, eye contact, deictic gestures, and other non-verbal cues. In net-based discourses, these methods are usually not available. Hence online discourses are susceptible to insufficient grounding. Learning protocols are designed to compensate the lack of grounding possibilities and coherence typical for chat-based communication by providing and implementing tools that serve as means for grounding (Oehl and Pfister, in press; Pfister and Oehl, 2009). Referencing as well as typing indicate a relationship to previous contributions, and, in doing so, they provide feedback for other contributors that their message is considered and contemplated. The general efficiency of learning protocols could be verified in several studies (Mühlpfordt and Wessner, 2005; Pfister and Mühlpfordt, 2002; Pfister and Oehl, 2009; Stahl, 2009).

In our present study we examined the question, how different grounding behavior itself influences a chat discourse. Thereto we established a confederate in each learning group manipulating the grounding processes either in a constructive or destructive way by making use of the implemented tools to a different extent.

ADAPTIVE SCRIPTING IN CSCL

Collaboration scripts in general vary according to the degree of flexibility of the implemented tools. With respect to the optimal degree of coercion the designer is confronted with a dilemma: a lack of coercion may affect the functionality of the script, whereas a surplus could lead to a phenomenon called *over-scripting* resulting in sterile, constrained interactions and an increased cognitive load for the contributors (Dillenbourg and Tchounikine, 2007). An approach recently discussed (Dillenbourg and Tchounikine, 2007; Oehl and Pfister, in press; Soller, 2004) is the development of highly flexible and *adaptive scripts*, providing support in terms of visualizations and feedback mechanisms exclusively in situations, when it is necessary. Prospectively it might be possible to implement a computer-supported detection of structural elements and patterns characteristic for deficient communication processes, resulting in an intelligent and adaptive script to foster grounding and thus learning processes. Such collaboration scripts need mechanisms to adapt to the learners' progress and to assess the learners' evolving interactions.

A promising approach to the automatic identification of suboptimal communication and grounding processes in CSCL are probabilistic models such as *Hidden Markov Models (HMM)* used by Soller (2004). For each sequence of tokens (e.g., a specific sequence of speech acts) a model can be computed consisting of a fixed number of states, initial probabilities, transition probabilities and output probabilities for the tokens (Rabiner, 1989). Additionally, the probability of a specific sequence of tokens being generated by this model can be calculated as well.

The long-term objective of this research is an adaptive CSCL script automatically promoting and facilitating grounding processes in learning groups.

METHOD

Students (N = 56, i.e., 12 male and 44 female) at the age of M = 23.14 years (SD = 4.33) from different faculties volunteered as participants, randomly assigned to 28 small learning groups. Each group consisted of two participants and one additional confederate, who manipulated the grounding process either in a constructive or in a destructive way, i.e., in the constructive grounding condition the confederate established adjacency pairs, correct referencing as well as typing and so on. The confederate was instructed not to vary the quality of his contributions but exclusive his grounding behavior in order to avoid unintended effects on a content level. In addition to the grounding manipulation, the CSCL script was varied resulting in a *2 x 2 design* with the *first factor grounding manipulation* (constructive vs. destructive), and the *second factor CSCL script* (learning protocol vs. free chat). Participants were instructed to learn about the mechanisms and causes of the Chernobyl nuclear power plant disaster. An introductory text was provided and learners than discussed the topic using the learning protocol (vs. chat) environment as described above for ca. 60 minutes (figure 1). Each single contribution, including duration, reference, type, and text, was automatically recorded in a log file for further analyses. Finally, the learning outcome of each participant was evaluated based on both a knowledge test and a problem-solving task asking for the exact sequence of events leading to the Chernobyl disaster.

For further analyses of the *discourse structures* and the *grounding processes* with Hidden Markov Models (HMM) each contribution was rated according to a coding scheme of grounding activities based on Beers, Boshuizen, Kirschner and Gijselaers (2007). The original scheme by Beers et al. (2007) was specially developed to capture the negotiation of common ground in CSCL. It was evaluated and optimized in a validation process and achieved an inter-coder reliability (Cohen's kappa) of r = .68 (Beers et al., 2007). However, in order to fit the characteristics of our study, some modifications had to be made resulting in a coding scheme of five categories with regard to grounding activities:

- *Initiation:* A new topic in form of a statement or a question is introduced.
- *Verification:* Information is requested about a previous contribution.
- *Response:* A content-related reaction to an initiation or verification.
- *Positive feedback:* A positive reaction (e.g., expression of intelligibility or agreement).
- *Negative feedback:* A negative reaction (e.g., expression of unintelligibility or disagreement).

In addition, HMM for effective and ineffective chat discourses with learning protocols were modeled based on the data of a previous study (Pfister and Oehl, 2009). Students ($N = 118$; age: $M = 25.03$ years, $SD = 4.07$; gender: 70.10% female) volunteered as participants and were randomly assigned to 33 learning groups, consisting of three or four members. Learning domain and task were identical to the current study. In order to ensure a high comparability, only triads were considered for modeling ($n = 42$). With regard to the learning outcomes the best and worst quartile (25%) was used to generate a HMM on the on hand for effective learning discourses and on the other hand for ineffective discourses. For each discourse of the current study the probability that the specific sequence of codes could be generated by the different HMM was computed. Eventually, the arithmetic mean for both the constructive und destructive manipulation condition was determined.

We addressed four main research questions to this experimental study:

(1) Do learning protocols in contrast to free chat lead to better learning outcomes?

(2) Increases a constructive manipulation of grounding processes learning outcomes?

(3) Can different grounding behavior be modeled probabilistically?

(4) Are Hidden Markov Models an appropriate technique for the automatic detection of ineffective discourses?

RESULTS

With regard to the learning outcomes, we hypothesized that both the implementation of a learning protocol and a constructive manipulation of grounding processes leads to higher average overall post-test scores. However, an ANOVA showed no significant differences of learning outcomes between the four experimental conditions ($F_{(3,52)} = 0.189$, $p = .903$). In discourses with learning protocols the participants achieved a mean score of $M = 19.61$ ($SD = 3.53$) in comparison to $M = 19.04$ ($SD = 3.76$) in discourses with free chat. For learning groups with a constructive confederate a mean score of $M = 19.43$ ($SD = 3.70$) could be measured, whereas the mean score in groups with a destructive confederate reached $M = 19.21$ ($SD = 3.62$).

The modeling of the HMM for the learning protocol discourses revealed some noticeable differences regarding the discourse structure between the two experimental conditions of the grounding manipulation factor (constructive vs. destructive) as shown in figure 2. The different initial probabilities of the two HMM discourse models were salient, i.e., the initial probabilities in the model for the constructively manipulated discourses were equally distributed but in the model for the destructively manipulated discourses a tendency to start in state s_1 occurred. Additionally, the transition probabilities between the two states s_1 *and* s_2 were generally higher in the HMM for discourses with constructive confederate, i.e., transitions to the other state more likely occurred.

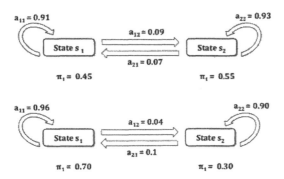

FIGURE 2. HMM for discourses with constructive (upper) and destructive (lower) grounding manipulation.

In order to specify the differences between the models, the output probabilities of grounding activities as displayed in table 1 have to be considered.

Table 1 Probabilities of grounding activities in HMM for constructive vs. destructive grounding manipulation

	Constructive manipulation		Destructive manipulation	
Code	State 1	State 2	State 1	State 2
Initiation	.363	.428	.436	.597
Verification	.015	.089	.068	.000
Response	.135	.401	.321	.172
Positive feedback	.282	.061	.082	.159
Negative feedback	.205	.021	.092	.072

The code *initiation* occurred with a high probability in all conditions. However, whereas in the HMM for discourses with constructive confederate a comparatively distinct distribution of the codes *response* and *positive* as well as *negative feedback* occurred, this difference was much less obvious in the model for destructively manipulated discourses. Furthermore, a generally higher likelihood for the feedback codes was observed in the model for constructive manipulation. Regarding the model for the discourses with constructive confederate the sequence *state s_1, state s_2, state s_1,* might be interpreted as an almost prototypical grounding process, i.e., an initiation is followed by a response and again grounded by positive or negative feedback.

It was hypothesized that the constructively manipulated discourses of our current study resemble a model based on particular effective discourses from our previous study regarding their structure, whereas the destructively manipulated discourses show a higher similarity to ineffective discourses. However, contrary to our expectations we computed a higher fit between the sequence based on discourses with a destructive confederate for both models as displayed in table 2.

Table 2 Probabilities that a given model generates a specific sequence

	Effective model	Ineffective model
Sequence for constructive manipulation	$1.05*10^{-26}$	$2.94*10^{-28}$
Sequence for destructive manipulation	$1.13*10^{-18}$	$3.83*10^{-20}$

CONCLUSION

Learning protocols, as a special version of collaboration scripts for CSCL, have been shown to be beneficial for learning performance under some conditions (Mühlpfordt and Wessner, 2005; Pfister and Mühlpfordt, 2002; Pfister and Oehl, 2009; Stahl, 2009). In this study, we investigated how a manipulation of grounding effects learning results and knowledge communication in terms of chat discourses.

Contrary to our hypothesis we found no significant differences between the learning outcomes. In a previous study (Pfister and Oehl, 2009) we found that the positive impact of learning protocols on learning performance is influenced by group size, i.e., learning outcomes improved as group size increased. Considering the small group size of two learners and one additional confederate, it should be noted that an increased group size may have enhanced the effectiveness of the learning protocol as well. Furthermore the confederate may have leveled the differences between the discourses varying not only his grounding behavior but also the quality of his contributions. In order to ensure that the confederate followed the given instructions, a list of keywords crucial for the understanding of the Chernobyl disaster was compiled and the amount of such keywords per discourse contribution was counted. Since the findings showed significant differences ($Z = -3.67, p < .001$) between constructively ($M = 0.80, SD = 1.14$) and destructively manipulated discourses ($M = 1.17, SD = 1.28$), it should be considered that the confederate might have produced more elaborated contributions in destructively manipulated discourses. Nevertheless, we observed a tendency to better learning outcomes of discourses with learning protocols and constructive manipulation, which is interpreted as an evidence for our hypothesis considering the stated restrictions.

Furthermore, we computed HMM of the discourses with constructively as well as destructively grounding behavior. The comparison of the HMM revealed some differences with regard to discourse structure of constructively vs. destructively manipulated discourses probably due to the influence of the confederate. We interpreted the states of the constructive HMM as phases of an effective grounding process. This might have a positive effect on the discourse structure as well.

In addition, the discourses of the current study were compared to an effective and an ineffective HMM and the average probabilities that these sequences were generated by the different HMM were computed. In contrast to our expectations, the sequence based on discourses with destructive confederate showed a higher likelihood for both the effective and the ineffective HMM.

However, the probabilities within a range from the 18th to the 28th decimal place are quite low, but low probabilities themselves are no problem with regard to an adaptive computer-supported identification of ineffective discourse structures or grounding processes. Nevertheless, this raises another methodical issue. Due to probabilistic characteristics of HMM, a short training sequence, i.e., short discourse threads, leads to a relatively high probability to be generated by a specific HMM while a long sequence decreases the probability. A comparison of the average number of contributions per discourse revealed that discourses with constructive grounding behavior are longer than destructively manipulated discourses ($M = 46.93$, $SD = 8.57$; vs. $M = 45.09$, $SD = 16.13$). Although this difference failed to be significant ($t_{(594)} = 1.77$, $p = .08$), this might explain the higher modeling probabilities of the destructive sequences. However, it might still be possible to use the salient differences in regard to the matching probabilities between constructive and destructive sequences for the automatic detection of ineffective learning discourses. Assumed that the discourse length can be considered as an indicator for the amount of successful grounding processes (whereas long discourses implicate a high amount of grounding activities), a comparatively high matching probability to a given HMM indicates a rather short discourse, which might be in danger of being ineffective. This again could lead to the automatic provision of supporting functionalities by the adaptive script.

Even though HMM, as applied by Soller (2004), are a promising approach with respect to adaptive scripting in CSCL, they raise methodical issues. Soller (2004) used in her studies a quite restrictive collaboration script providing sentence openers as well as selective collaboration tasks, whereas learning protocols operate exclusively on the level of communication processes. The application of HMM in rather coercive CSCL scripts, which ensures discourses of a similar structure, or enhanced algorithms of HMM, might be a requirement for further use of HMM towards adaptive CSCL scripts.

REFERENCES

Baker, M.J., Hansen, T., Joiner, R., and Traum, D. (1999), The role of grounding in collaborative learning tasks. In P. Dillenbourg (Ed.), *Collaborative learning: Cognitive and computational approaches* (pp. 31-63). Elsevier Science, Oxford, UK.

Beers, P.J., Boshuizen, H.P.A., Kirschner, P.A., and Gijselaers, W.H. (2007), The analysis of negotiation of common ground in CSCL. *Learning and Instruction, 17*, 427-435.

Brennan, S. (1998), The grounding problem in conversations with and through computers. In S.R. Fusseland, and R.J. Kreuz (Eds.), *Social and cognitive psychological approaches to interpersonal communication* (pp. 201-225). Erlbaum, Mahwah, NJ.

Bromme, R., Hesse, F.W., and Spada, H. (2005), *Barriers and biases in computer-mediated knowledge communication - and how they may be overcome.* Springer, New York, NY.

Clark, H.H. (1996), *Using language.* Cambridge University Press, Cambridge, UK.

Clark, H.H., and Brennan, S.E. (1991), Grounding in communication. In L.B. Resnick, J.M. Levine, and S.D. Teasley (Eds.), *Perspectives on socially shared cognition* (pp. 127-149). APA, Washington, DC.

Dillenbourg, P., and Tchounikine, P. (2007), Flexibility in macro-scripts for computer-supported collaborative learning. *Journal of Computer Assisted Learning, 23*, 1-13.

Fuks, H., Pimentel, M.G., and de Lucena, C.J.P. (2006), R-U-Typing-2-Me? Evolving a chat tool to increase understanding in learning activities. *International Journal of Computer-Supported Collaborative Learning, 1*, 117-142.

Herring, S. (1999), Interactional coherence in CMC. *Journal of Computer-Mediated Communication, 4*, [http://www.ascusc.org/jcmc/vol4/issue4/herring.html].

Kollar, I., Fischer, F., and Hesse, F.W. (2006), Collaboration scripts – a conceptual analysis. *Educational Psychology Review, 18*, 159-185.

Kreijns, K., Kirschner, P.A., and Jochems, W. (2003), Identifying the pitfalls for social interaction in computer-supported collaborative learning environments: A review of the research. *Computers in Human Behavior, 19(3)*, 335-353.

Mühlpfordt, M. (2006). *ConcertChat* [Computer software]. Retrieved February 24, 2010, from http://www.ipsi.fraunhofer.de/concert/index_en.shtml?projects/chat

Mühlpfordt, M., and Wessner, M. (2005), Explicit Referencing In Chat Supports Collaborative Learning. In T. Koschmann, D. Suthers, and T. W. Chan (Eds.), *Proceedings of the CSCL 2005 Conference on Computer Supported Collaborative Learning. The Next 10 Years*. Erlbaum, Mahwah, NJ.

O'Donnell, A.M., and King, A. (Eds.) (1999), *Cognitive perspectives on peer learning*. Erlbaum, Mahwah, NJ.

Oehl, M., and Pfister, H.-R. (in press), E-Collaborative Knowledge Construction in Chat Environments. In B. Ertl (Ed.), *E-Collaborative Knowledge Construction: Learning from Computer-Supported and Virtual Environments*. IGI Global, Hershey, NY.

Pfister, H.-R., and Mühlpfordt, M. (2002), Supporting discourse in a synchronous learning environment: The learning protocol approach. In G. Stahl (Ed.), *Computer Support for Collaborative learning: Foundations for a CSCL Community. Proceedings of CSCL2002 - Conference on Computer Supported Collaborative Learning, Boulder, Colorado* (pp. 581-589). Erlbaum, Hillsdale, UK.

Pfister, H.-R., and Oehl, M. (2009), The Impact of Goal Focus, Task Type, and Group Size on Synchronous Net-Based Collaborative Learning Discourses. *Journal of Computer Assisted Learning, 25*, 161-176.

Rabiner, L.R. (1989), A tutorial on hidden Markov models and selected applications in speech recognition. *Proceedings of the IEEE, 77(2)*, 257-286.

Soller, A. (2004), Understanding knowledge sharing breakdowns: A meeting of the quantitative and qualitative minds. *Journal of Computer Assisted Learning, 20*, 212-223.

Stahl, G. (Ed.) (2009), *Studying virtual math teams*. Springer, New York, NY.

CHAPTER 58

IMMA – Intelligently Moving Manikin – Project Status

Lars Hanson[1,2], Dan Högberg[3], Robert Bohlin[4] and Johan S. Carlson[4]

[1] Industrial Development
Scania CV
Södertälje, Sweden

[2] Wingquist Laboratory
Chalmers University of Technology
Göteborg, Sweden

[3] School of Technology and Society
University of Skövde
Skövde, Sweden

[4] Fraunhofer-Chalmers Research Centre for Industrial Mathematics, FCC
Chalmers Science Park
Göteborg, Sweden

ABSTRACT

The overall rationale and assumption for the research project presented is this paper is that a fast, easy to use, and reliable procedure to predict and validate manual assembly tasks is of major importance in product and production development processes to ensure high and robust product quality and process performance. A basic condition for the research is the belief that tools with such functionality are currently not available for companies to utilise in their development processes. Hence more research and development is needed in the area. This paper describes the basic concepts and initial steps taken in the recently commenced research project IMMA – Intelligently Moving Manikin.

Keywords: Digital Human Modelling, Ergonomics, Simulation, Motion Planning

INTRODUCTION

Automatic path planning, i.e. the methodology to automatically find collision free motion patterns of objects being moved in space, is a well established research area and there are a number of distinguished groups in the world. Complete algorithms are of little industrial relevance due to the complexity of the problem (Canny, 1988). Instead, sampling based techniques, trading completeness for speed and simplicity, are typically the methods of choice. Common for these methods are the needs for efficient collision detection, nearest neighbour searching, graph searching and graph representation. The two most popular methods are: Probabilistic Roadmap Methods (PRM) (Bohlin and Kavraki, 2000) and Rapidly-Exploring Random Trees (RRT) (LaValle and Kuffner, 1999). There are commercial available path planner tools, e.g. from Fraunhofer-Chalmers Centre (FCC) and KineoCam, on the market today that investigate if a part without any external support can find a collision free way from a starting position outside the assembly to its end position in the assembly structure. FCC has, inspired by both these probabilistic methods, developed a novel deterministic path planning algorithm. Path planning tools give valuable support when evaluating new product or workstation concepts and when comparing alternative solutions. Also different assembly sequences can be compared and verified.

In an actual assembly context, no part can find its way from one point to another without external support. The part needs to be manoeuvred into is correct position, e.g. by a robot or by a human. Consequently, path planner tools that find a collision free way for the part and the robot arm holding the part have been developed. An industrial robot has normally six revolute joints making it a six degrees of freedom mechanism. However, far from every part in an assembly structure is assembled by a robot. The majority of the parts are assembled by humans. In a sense, a human can be seen as a high dimensional robot; typically a virtual manikin is modelled with 120-150 degrees of freedom. Most research about motion planning of manikins has been done in the computer graphics community and is focused on animations of characters in movies and games (Yamane et al., 2004). However, these methods are not addressing industrial assembly situations where the motion is often highly constrained by surrounding geometries. The work by Laumond et al. (2005) is an exception, where assembly situations involving a manikin are investigated by solving path planning problems with 7 degrees of freedom. This work needs to be improved in several directions: (i) high dimensional path planning using more degrees of freedom will increase the ability to avoid collisions, (ii) ergonomics aspects, such as load and joint angle analysis, should be incorporated, affecting the acceptable joint configurations, (iii) several manikin configurations, e.g. related to anthropometry, should be considered at both start and end positions, including multiple grasping alternatives, (iv) allowing for re-grasping during the motion, (v) a hybrid strategy for continuously mixing path planning in forward kinematics (manikin joints) and inverse kinematics (object position), allowing the path planner to solve harder problems and consider motion constraints. Within the area of ergonomics, several approaches for manikin motion generation have been investigated. Among other techniques, neural networks, a structure that aims to imitate the human brain, have been tested for motion generation. Fuzzy logics, a logic that aims to imitate human reasoning, have also been tested to generate human like motions. These techniques were tested in the EU-project ANNIE (Application of Neural Networks to Integrated Ergonomics) (Hanson et al., 1999) but are currently not used in ergonomics simulation and visualisation tools on the market. In Siemens ergonomics simulation and visualisation tools two approaches are currently used: a task-based simulation approach with inverse kinematics (Raschke et al., 2005) and an approach that modify root motions gathered from real humans using motion capture systems (Park et al., 2008). These approaches are however rarely used in real applications in industry. Most commonly the manikin is manually manipulated joint by joint. Manual adjustment is time consuming and posture and motion results may vary, both within and between tool users (Lämkull et al., 2008). From an industrial point of view such variation within the process is not adequate. Frequently, the validity of the manually generated manikin motion is questioned. Therefore, researchers and industry sometimes use motion capture systems in combination with ergonomics simulation and visualisation tools (e.g. Stephens and Godin, 2006). Motion capture systems regularly require a laboratory like environment and physical prototypes. By using such a set-up, several of the advantages with virtual manufacturing disappear or are reduced. Therefore, there is a call for valid motion generators in digital human modelling tools in order to support proactive ergonomics (Chaffin, 2005). The development of motion prediction functionality would enable full work cycles to more easily be simulated, instead of static postures as is the most common procedure today (Lämkull et al., 2008). The possibility of simulating whole work cycles will however put pressure on having proper functionality of the ergonomics simulation tools, e.g. an easy to use programming language for describing the tasks the manikin is to perform, and ergonomics evaluation methods that consider time factors and the aggregation of ergonomic loads.

This aim of the paper is to describe the current status of the ergonomics simulation and visualisation tool IMMA – Intelligently Moving Manikin.

APPROACH

The biomechanical model is inspired by studies of anatomic and biomechanics literature as well as by benchmarking of manikins in contemporary ergonomics simulation and visualisation tools. The motion prediction functionality is inspired by motion algorithm techniques used in path planning for industrial robots. The ergonomics evaluation functionality is inspired by investigating the ergonomics assessment methods presently used by the industry partners in the project, as well as by studies of ergonomics literature and benchmarking of methods integrated in ergonomics simulation and visualisation tools. The IMMA research project is initiated and carried out in collaboration between academia and industry. The demonstrator developed so far is visually evaluated by the participating industry partners.

RESULTS

Since the project recently commenced, this section reports on initial steps and portrays in general terms what outcomes are expected in the project related to biomechanical model and motion generator, anthropometrics and ergonomics assessment functionality.

Biomechanical model and motion generator

A first version of a biomechanical model has been developed (Figure 1). The biomechanical skeleton model consists of 72 joints and 81 segments, representing 148 degrees of freedom. Each hand has 16 joints and 20 segments.

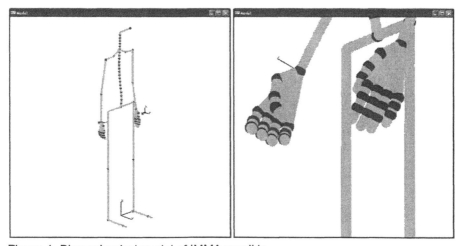

Figure 1. Biomechanical model of IMMA manikin.

The biomechanical model is a full human body replica and acts as a starting point for further refinement, and as a basis for further functionality development such as building an associated skin model. Also, a first motion controller has been developed. This controller enables the manikin motions to be controlled manually, either in single joints or links of segments. The controller algorithms can also automatically predict manikin movements in order for the manikin to reach for specific targets in space while satisfying biomechanical constraints.

Anthropometrics

Swedish anthropometric data from the study performed by Hanson et al. (2009) has been implemented in the tool. Manikin anthropometry is varied by adjusting the three key measurements: stature, corpulence and upper body/lower body ratio. Defining these three measurements is seen as a good starting point for defining the overall anthropometric characteristics for an individual, based on the assumption that the other body measurements, respectively, have high correlation with one of these measures, but lower correlation with the other two (Flügel et al., 1986; Speyer, 1996; Bubb et al., 2006).

Ergonomics assessment

The ergonomics assessment functionality has basically two purposes. The first purpose is to assist the motion prediction functionality to optimise among the number of possible motion paths, and to propose a path that is both physically possible and that cause low ergonomics load, according to the assessment method employed. The second purpose of the assessment functionality is to be able to assess already defined motions, e.g. to enable the assessment of a motion that has been generated using another ergonomics assessment method, or possibly by using a motion capture system.

Up to now, an ergonomics assessment method developed for the manufacturing engineering department within Volvo Car Corporation in Sweden has been implemented in the tool (Lämkull, 2005; Högberg et al., 2008). The method assesses joint angle values of neck, back, shoulders, elbows, hands, hips, knees and feet. The method is based on joint angle recommendations by Petrén (1968) and Munck-Ulfsfält et al. (2003). The assessment is structured according to a colour scheme where *Green* = Allowed, *Yellow* = Semi-allowed and *Red* = Not allowed. Figure 2 illustrates this assessment technique for angular values in shoulder and arm joints.

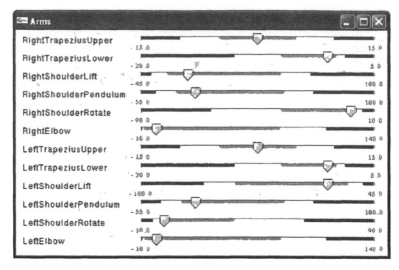

Figure 2. Assessment of angular values in shoulder and arm joints.

Evaluation of project outcome

The evaluation of the outcomes of the project so far, made by the participating industry partners, has been positive. This view is mainly based on visual evaluation of the functional demonstrator of the IMMA tool. However, the project partners ask for more functionality such as additional company specific ergonomics assessment methods. The view is that the motion generator is a strong, and long asked for, feature of the tool.

DISCUSSION

The biomechanical model of the IMMA manikin is comparable to the manikin models in Siemens/UGS *Jack* and Catia *V5 Human* (Table 1). They all have detailed spine and hand models, whereas the foot model is simpler. Compared to the other tools, the current shoulder model in the IMMA manikin is less advanced. Similarly to many other ergonomics simulation and visualisation tools, the IMMA manikin has no muscle models. The *AnyBody* manikin is one example of a manikin with a full muscle model. The IMMA biomechanical model will be further updated, with focus on refining joint and skeleton representations.

Table 1. Comparison of biomechanical model data.

	Jack	V5 Human	IMMA
No of joints	68	74	72
No of segments	69	~100	81
Degrees of freedom	135	143	148

The basic posture manipulation and motion generation method in the IMMA tool is comparable to corresponding functionality in other ergonomics simulation and visualisation tools on the market. The IMMA tool allows manual manipulation of separate joints or a series of joints, which still is the most common way to adjust manikin postures and define manikin motions. Other manikin tools such as Siemens/UGS *Jack* and Catia *V5 Human* additionally offer more advanced manipulation methods such as task simulation based modules and interfaces to motion capture systems. The corresponding functionality in the IMMA tool is based on a motion planning algorithm that today automatically can find a smooth motion between different targets in space and compute postures that consider biomechanical rules. The automatic motion generator for predicting manikin movements will be further developed in respect to the algorithms and the ergonomics assessment methods.

At present the IMMA anthropometric module include one nationality and one age group. The functionality of the IMMA anthropometric module to use three key anthropometric variables to define manikin anthropometry is comparable to the method implemented in the ergonomics and visualisation tool *Ramsis,* based on the anthropometric survey and analysis work of Flügel, Greil and Sommer (Flügel et al. 1986), further described in Speyer (1996) and Bubb et al. (2006). Other ergonomics simulation and visualisation tools on the market typically offer more nationalities, more age groups and the consideration of secular trends. Furthermore, tools available on the market offer the possibility to create unique manikins by manual input of anthropometric data and the creation of boundary manikins. The IMMA anthropometric module will be further developed in these respects.

Up to now, the IMMA tool has few ergonomics assessment methods implemented. As described earlier, IMMA includes the established Volvo Cars assessment method. This method was implemented at the outset to illustrate the option to implement specific methods. Volvo Cars have previously customised the *eM-human* tool and Saab Automobile has customised the *IGRIP* tool (Högberg et al., 2008). Biomechanical calculations are currently not possible to perform in IMMA, and no established ergonomics assessment methods, such as OWAS (Karhu et al.,

1977) and RULA (McAtamney and Corlett, 1993), are implemented in the first demonstrator. These assessment methods are common in other ergonomics simulation and visualisation tools on the market. The IMMA ergonomics assessment module will be further developed, focusing on assessment methods applied in industry and methods that consider time factors and the aggregation of ergonomic loads.

The participative collaboration method between industry and academia employed in the IMMA project is rewarding since it supports that the research will lead to meaningful results directly applicable in industry, and that appropriate science and technology will be incorporated in the tool being developed. This kind of collaboration is not unique though. Industry, preliminary the automotive industry, has sponsored and participated in the development of ergonomics simulation and visualisation tools such as *Jack* (Badler et al., 1993) and *Ramsis* (Bubb et al., 2006). Still, Sweden is believed to have an exceptionally tight relation between academia and industry that provides for a stimulating environment for developing a relevant user friendly industry tool such as the IMMA manikin tool.

ACKNOWLEDGEMENTS

This work is carried out as a collaboration within the Wingquist Laboratory at Chalmers and the other organisations involved in the IMMA project, with support from the Swedish Foundation for Strategic Research/ProViking II and the participating organisations. This support is gratefully acknowledged.

REFERENCES

Badler, N., Phillips, C. B. and Webber, B. L. (1993). Simulating humans: computer graphics animation and control. New York, Oxford University Press.

Bohlin, R. and Kavraki, L.E. (2000) Path Planning Using Lazy PRM. In Proc. IEEE Int. Conf. on Robotics and Automation, San Francisco, USA, pp. 521-528.

Bubb, H., Engstler, F., Fritzsche, F., Mergl, C., Sabbah, O., Schaefer, P. and Zacher, I. (2006) The development of RAMSIS in past and future as an example for the cooperation between industry and university. Int. J. of Human Factors Modelling and Simulation, Vol.1, No.1, pp. 140-157.

Canny, J.F. (1988) The Complexity of Robot Motion Planning. MIT Press.

Chaffin, D.B. (2005) Improving digital human modelling for proactive ergonomics in design. Ergonomics, Vol.48, No.5, pp. 478-491.

Flügel, B., Greil, H. and Sommer, K. (1986) Anthropologischer atlas: grundlagen und daten. Verlag Tribüne, Berlin (in German).

Hanson, L., Akselsson, R., Andreoni, G., Rigotti, C., Palm, R., Wienholt, W., Costa, M., Lundin, A., Rizzuto, F., Gaia, E., Engström, T., Sperling, L., Sundin, A. and Wolfer, B. (1999) ANNIE, a Tool for Integrating Ergonomics in the Design of Car Interiors. SAE Transactions. J. of Material and Manufacturing, Section 5, Vol.108, pp. 1114-1124.

Hanson, L., Sperling, L., Gard, G., Ipsen, S. and Vergara, C.O. (2009) Swedish anthropometrics for product and workplace design. Int. J. of Applied Ergonomics, Vol.40, pp. 797-806.

Högberg, D., Bäckstrand, G., Lämkull, D., Hanson, L. and Örtengren, R. (2008) Industrial customisation of digital human modelling tools. Int. J. of Services Operations and Informatics, Vol 3, No 1, pp. 53-70.

Karhu, O., Kansi, P. and Kuorinka, I. (1977) Correcting working postures in industry: a practical method for analysis. Applied Ergonomics, Vol.8, No.4, pp. 199-201.

Laumond, J.-P., Ferré, E., Arechavaleta, G. and Estevès, C. (2005) Mechanical Part Assembly Planning with Virtual Mannequins. IEEE Int. Symposium on Assembly and Task Planning, Montréal, Canada, pp. 132-137.

LaValle, S.M. and Kuffner, J.J. (1999) Randomized Kinodynamic Planning. In Proc. IEEE Int. Conf. on Robotics and Automation, Detroit, USA, pp. 473-479.

Lämkull, D. (2005) The daily use of manikins within the manufacturing department at Volvo Car Corporation – working methodology, developments and wanted improvements. In Proc. of Nordic Ergonomics Society Conference, Oslo, Norway, pp. 86-90.

Lämkull, D., Hanson, L. and Örtengren, R. (2008) Uniformity in manikin posturing: A comparison between posture prediction and manual joint manipulation. Int. J. of Human Factors Modelling and Simulation, Vol.1, No.2, pp. 225-243.

McAtamney, L. and Corlett, E.N. (1993) RULA: a survey method for the investigation of work-related upper limb disorders. Applied Ergonomics, Vol.24, No.2, pp. 91-99.

Munck-Ulfsfält, U., Falck, A., Forsberg, A., Dahlin, C. and Eriksson, A. (2003) Corporate ergonomics programme at Volvo Car Corporation. Applied Ergonomics, Vol.34, No.1, pp. 17-22.

Park, W., Chaffin, D.B., Martin, B.J. and Jonghwa, Y. (2008) Memory-Based Human Motion Simulation for Computer-Aided Ergonomic Design. IEEE Transactions on Systems, Man, and Cybernetics, Part A, Vol.38, No.3, pp. 513-527.

Petrén, T. (1968) Lärobok i anatomi: rörelseapparaten (Anatomy textbook). Nordiska Bokhandeln, Stockholm, Sweden (in Swedish).

Raschke, U., Kuhlmann, H. and Hollick, M. (2005) On the design of a task based human simulation system. Warrendale, SAE Technical Paper 2005-01-2702.

Speyer, H. (1996) On the definition and generation of optimal test samples for design problems. Kaiserslautern, Human Solutions GmbH.

Stephens, A. and Godin, C. (2006) The Truck that Jack Built: Digital Human Models and their Role in the Design of Work Cells and Product Design. Warrendale, SAE Technical paper 2006-01-2314.

Yamane, K., Kuffner, J.J. and Hodgins J.K. (2004) Synthesizing animations of human manipulation tasks. In Proc. of SIGGRAPH, Los Angeles, USA. pp. 532-539.

CHAPTER 59

Digital Human Model Module and Work Process for Considering Anthropometric Diversity

Erik Bertilsson[1,2], Dan Högberg[1] and Lars Hanson[2,3]

[1] School of Technology and Society
University of Skövde
Skövde, Sweden

[2] Department of Product and Production Development
Chalmers University of Technology
Göteborg, Sweden

[3] Industrial Development
Scania CV
Södertälje, Sweden

ABSTRACT

In digital human modelling (DHM), ergonomics evaluations are typically done with few human models. However, humans vary a lot in sizes and shapes. Therefore, few manikins can rarely ensure accommodation of an entire target population. Different approaches exist on how to consider anthropometric diversity. This paper reviews current DHM tools and clarify problems and opportunities when working with anthropometric diversity. The aim is to suggest functionality for a state of the art DHM module and work process for considering anthropometric diversity. The study is done by an analysis of some of the current DHM systems and by interviews of personnel at car companies about their way of working with anthropometric diversity. The study confirmed that critical production simulations are often done in early development stages with only one or a few human models. The reason for this is claimed to be time consuming processes, both at the creation of the human model but mainly when correctly positioning the model in the CAD environment. The development of a new method and work process for considering anthropometric diversity is suggested. Necessary features for such a module are that it shall be easy to use and not require expert knowledge about the consideration of anthropometric diversity. It shall also be configurable and transparent, in a sense that it should be possible to work with own anthropometric data and ergonomics evaluation standards. The module has to be flexible and have different entrances depending on the type of anthropometric problem being analyzed. An improved work method is expected to lead to faster and more correct analyses.

Keywords: Digital Human Modelling, Ergonomics, Work Process, Anthropometry

INTRODUCTION

In product and production development it is often necessary to study how a product, workplace or task will affect a potential user, both related to physical and cognitive ergonomics. Human-machine interaction has traditionally been evaluated relatively late in the development phase (Porter et al., 1993), and this has often been done by physical mock-ups which have been expensive and time demanding (Helander, 1999). To address anthropometric issues at early stages, problems are typically simplified and one or a few specific body dimensions are considered. The identified anthropometric measurement is then regularly collected from anthropometric data in a reference text or human factors handbook (Peacock and Karwowski, 1993; HFES 300 Committee, 2004; Salvendy, 2006; Jung et al., 2009). In order to reduce the need for physical tests there are benefits from using DHM simulation systems, which facilitates and improves simulations and analyses (Chaffin, 2001). A DHM tool is a computer program that uses a human model to create, modify, present and analyze human-machine interaction.

There are generally two methods to work with anthropometric diversity; a percentile based method and a custom-built method. With the percentile method a human model is created of the 1st, 5th, 50th, 95th or 99th percentile according to stature and weight measurements, of a certain gender, age group and nationality, generated from anthropometric data. The custom-built method means that any desired anthropometric values are defined; omitted dimensions are calculated by regression equations. Forms of custom-built methods are the boundary manikin method (Bittner, 2000; Eynard et al., 2000; Reed and Flannagan, 2000) and the distributed method (HFES 300 Committee, 2004; Jung et al., 2008) in which body dimensions are defined for a number of cases intended to cover the target population. The problem with the percentile method is that usually very few models are tested and that it is likely that that not all users are represented in the analysis (Jimmerson, 2001; Nelson, 2001; Thompson, 2001). Creating several custom-built human models takes quite long time and quickly becomes ineffective (Jung et al., 2009).

This paper will evaluate current DHM simulation systems and clarify problems, opportunities and solutions when working with human diversity in DHM systems. The aim is to review and suggest a state of the art DHM module and work process for considering anthropometric diversity.

METHODS

To review current DHM simulation systems, and their modules considering anthropometric diversity, a comparison of following DHM simulation systems was done: Siemens/UGS Jack 6.0, Pro Engineer Manikin and Catia V5 Human. Each system was evaluated by examining a number of criteria and by analyzing how well the system met the criteria. These criteria were defined based on existing functions, own ideas and the usability methods Cognitive Walkthrough (CW) and Predictive Human Error Analysis (PHEA):

- Choosing anthropometric data base

- Inserting own anthropometric data

- Creating percentile manikin

- Creating unique manikin

- Creating manikin family

- Save and load manikin data

- Other

Three qualitative interviews were conducted to get an understanding of different methods and approaches when working with anthropometric diversity in today's manufacturing industry. Six persons were in interviewed and they were all working in Swedish vehicle manufacturing industry. Their work positions varied from simulation engineers with an up to date expertise of DHM system to people with more overall responsibility for virtual manufacturing and simulation where DHM is one part. The interview questions covered topics such as previously mentioned criteria, key anthropometric variables, pros and cons as well as suggestions for improvements.

The results from benchmarking and interviews were combined with own knowledge and informal brainstorming sessions to create a model and work method for creating manikins in a DHM system. The model is presented by using the black box method where a "black box" converts certain inputs into desired outputs (Pahl and Beitz, 1996).

RESULTS

The result from the analysis of current DHM simulation systems was implemented into a matrix table that describes each criterion and how well each system meets the criteria (Table 1).

Table 1 Results from benchmarking analysis.

System	Siemens/UGS Jack 6.0	Catia V5 Human	Pro Engineer Manikin
Choosing anthropometric data base	Not possible to choose, beside the implemented data for USA (based on Ansur88).	Possible to choose between USA, Canada, France, Japan or Korea.	Depending on the library of manikins which is possible to import into the system.
Inserting own anthropometric data base	No possibility.	No possibility.	No possibility.
Creating percentile manikin	Possible to create 1st, 5th, 50th, 95th and 99th percentile for stature and weight.	Possible to continuously adjust percentile for stature.	Depending on the predefined manikins in the library.
Creating unique manikin	Possible to individually adjust 26 anthropometric measures.	Possible to individually adjust 103 anthropometric measures.	Manikins are pre defined in the library.
Creating manikin family	Principal component tool available in add on modules.	No possibility.	No possibility.
Save and load manikin data	Possible to save and load data for manikins.	Possible to save and load data for manikins.	All manikins are pre defined in the library.
Other	Possible to import 3D scan and SAE physical test manikins are integrated.	Possible to change manikins in terms of population.	Possible to change manikins in terms of population.

The interviews at the companies gave that methods and work processes are still much in a development phase, but work environment and ergonomics are something that is interesting throughout the organizations. Not meaning that DHM simulations always are given as much focus compared to other improvement methods like lean production. In a global organization the interest and understanding for ergonomics may vary a lot between production plants, and more personnel are sometimes seen as a solution to ergonomically bad planned production lines. According to the interviews, a DHM tool needs to be fast and easy to use. Using a simulation system should lead to better quality with the same work effort and the result needs to be trustworthy. An advantage of using DHM tools is the possibility to solve problems in an early phase and being able to make better decisions based on the simulations.

Today, working with DHM systems are frequently combined with qualified guessing of the final result. A simulation analysis is often done with only one or two manikins and the rest of the results are produced by imagining how the outcome would be if another person did the analysed task. There are some recurring elements which often involves poor ergonomics in assembly like engine compartment and interior ceiling. A critical area is the hands and wrists which are complicated to position correctly and this will influence analysis in a negative way. It has also been noticed that hands of the manikins are scaled proportionally to stature leading to unrealistically big hands on big manikins.

To cover all intended users the companies uses a very rough strategy involving one or two manikins based on stature percentiles. The goal can be that the biggest male (95th percentile) and the smallest female (5th percentile) should be able to do the task. Another approach is that a woman of the 50th percentile stature should be able to reach the work area. It is not unusual that even these objectives are not possible to fulfil. If that happens studies are done to expose what is possible to achieve depending on the workstation. The reason for these simplified solutions is the time-consuming processes when working with several manikins even if good features exist to assist in the positioning of a manikin. There are some alternative methods besides using a DHM system like using anthropometric data directly from tables or video analyses. A feature that the interviewed subjects would like to see in DHM simulation systems is the possibility to rapidly scale a manikin in order to see how a work position will affect a person with other body dimensions.

573

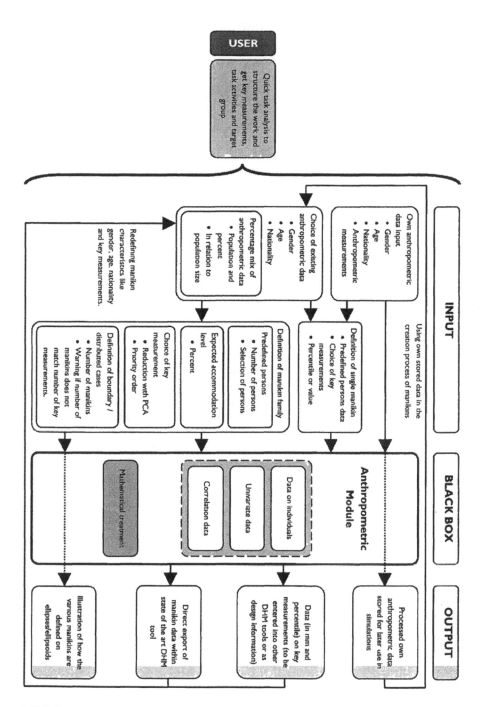

FIGURE 1 Flowchart depicting the module and work process.

DISCUSSION

Current DHM simulation systems are advanced, but many functions are not fully utilized because of time constraints in today's work. This can be compared to the fact that there are several clear requests not met by today's DHM system. The results from this study can be used when developing a new DHM simulation system or improving an existing system. The resulting work method will most likely evolve and transform during such a development.

Such a method and module needs to make it easy when working with anthropometric diversity. The module should have an easy to use graphical user interface similar to the web-based generation system created by Jung et al. (2009). It should be possible to create single manikins as well as manikin families. Possibilities to implement own anthropometric data would lead to a better connection between the simulation and the physical work station and its users. The module must be transparent, meaning that it should be easy to import and export data of anthropometric measurements and body angles. This would make it possible to communicate with manikins in other DHM simulation systems with minor modifications. The module combined with a work process inspired by Hanson et al. (2006) is illustrated in Figure 1. The process starts with the user doing task analysis. Depending on the situation it is possible to add own anthropometric data or choose data from within the system. Next step is to define single manikin or a manikin family. The choices and data are processed in the anthropometric module and depending on previous choices the result can either be inserted into a existing DHM tool or within a new DHM system in which the module are integrated.

The coming steps in the development of such a module will add focus on related areas like anthropometric measurements, correlation data, statistics and biomechanics. To be able to fully use functions like manikin families in a proper way there is a need for the DHM simulation systems to evolve and be more automatic and supportive for the tool user. Such a system should calculate a possible human motion from one position to another and repeat the simulation for a number of manikins with different body dimensions. Focus would be more on covering the target population with several manikins and finding key measurements than on visual simulations and guessing.

ACKNOWLEDGEMENTS

This work has been made possible with the support from the Swedish Foundation for Strategic Research/ProViking and by the participating organisations. This support is gratefully acknowledged.

REFERENCES

Bittner, A.C. (2000) A-CADRE: Advanced family of manikins for workstation design. In: Proceedings of the IEA 2000/HFES 2000 Congress, Human Factors and Ergonomics Society, Santa Monica, CA, pp. 774-777

Chaffin, D.B. (2001) Digital Human Modelling for Vehicle and Workplace Design. SAE International, Warrendale, PA.

Eynard, E., Fubini, E., Masali, M., Cerrone, M., Tarzia, A. (2000) Generation of Virtual Man Models Representative of Different Body Proportions and Application to Ergonomic Design of Vehicles. In: Proceedings of the IEA2000/HFES2000 Congress, Human Factors and Ergonomics Society, Santa Monica, CA, pp. 489-492.

Hanson, L., Blomé, M., Dukic, T., Högberg, D. (2006) Guide and documentation system to support digital human modelling applications. International Journal of Industrial Ergonomics, Vol. 36, pp. 17-24.

Helander, M.G. (1999) Seven common reasons to not implement ergonomics. International Journal of Industrial Ergonomics, Vol. 25, No. 1, pp 97-101.

HFES 300 Committee (2004) Guidelines for Using Anthropometric Data in Product Design, Human Factors and Ergonomics Society, Santa Monica, CA.

Jung, K., You, H., Kwon, O. (2008) A multivariate evaluation method for representative human model generation methods: application to grid method. In: Proceedings of the Human Factors and Ergonomics Society 52nd Annual Meeting, Human Factors and Ergonomics Society, Santa Monica, CA, pp. 1665-1669.

Jung, K., You, H., Kwon, O. (2009) Development of a digital human model generation method for ergonomic design in virtual environment, International Journal of Industrial Ergonomics, Vol. 39, No. 5, pp 744-748.

Jimmerson, D.G. (2001) Digital human modelling for improved product and process feasibility studies. In: Chaffin, D.B. (Ed.), Digital Human Modelling for Vehicle and Workplace Design, SAE International, Warrendale, PA, pp. 127–135.

Nelson, C. (2001) Anthropometric analyses of crew interfaces and component accessibility for the international space station. In: Chaffin, D.B. (Ed.), Digital Human Modeling for Vehicle and Workplace Design, SAE International,Warrendale, PA, pp. 17–36.

Pahl, G., Beitz, W. (1996) Engineering Design: A Systematic Approach (second edition). Bauert Springer-Verlag, ISBN 3-540-19917-9.

Porter, J.M., Case, K., Freer, M.T., Bonney, M.C. (1993) Computer aided ergonomics design of automobiles. In: Peacock, B., W. Karwowski (Eds.), Automotive Ergonomics. Taylor & Francis, London, pp. 43-77.

Peacock, B., Karwowski, W. (Eds.) (1993) Automotive Ergonomics. Taylor & Francis, London.

Reed, M.P., Flannagan, C.A.C. (2000) Anthropometric and postural variability: limitations of the boundary manikin approach. SAE Transactions: Journal of Passenger Cars-Mechanical Systems 109, pp. 2247–2252. Technical Paper No. 2000-01-2172.

Salvendy, G. (Ed.) (2006) Handbook of Human Factors and Ergonomics, third ed. John Wiley & Sons.

Thompson, D.D. (2001) The determination of the human factors/occupant packaging requirements for adjustable pedal systems. In: Chaffin, D.B. (Ed.), Digital Human Modeling for Vehicle and Workplace Design. SAE International,Warrendale, PA, pp. 101–111.

CHAPTER 60

Large-Scale Data Based Modeling in Everyday Life for Service Engineering

Yoichi Motomura

Digital Human Research Center / Center for Service Research
The National Institute of Advanced Industrial Science and Technology
Koto-ku, Tokyo 135-0064 Japan

ABSTRACT

Large-scale data has been generated in the real daily life by many kinds of electric devices. We can utilize such data to construct computational human models that predict user's behavior, consumer's preference, and so on. Probabilistic models, statistical learning and probabilistic inference method can be applied for computational human modeling. In this paper, examples of applications using Bayesian networks are shown. Bayesian networks have advantage that can represent mutual interaction for situation depend user's preference models. One good example is a contents recommendation depends on different situations and users. In order to get large-scale data corresponding to many kinds of different situations, point of act data collection scheme during service providing is necessary. We call this data collection scheme 'research as a service'. This human modeling framework is also discussed as a key technology for service engineering to improve productivity of service industry.

Keywords: Service engineering, Bayesian networks, computational modeling

INTRODUCTION

The range of applications of information processing technology is steadily increasing. At the same time, information services to aid in everyday life are increasingly in demand. Therefore, a model is necessary which describes the activities of daily life in a quantifiable way, in terms of what a person is trying to accomplish in various circumstances. Using such a quantitative theoretical model, we consider a system that predicts background requirements and expected results from the user's activities, rapidly implements them, and makes possible the development of new services that aid activities of everyday life. Additionally, by continuously implementing such cooperative operations with people during everyday life, it becomes possible to acquire meaningful data in large quantities that was not previously obtainable in a laboratory environment. Using this large-scale data, it is possible to bring about a cycle in which services continue to be used while the model is constantly being updated.

However, during these daily activities, information processing based on uncertain information (such as predictions that result in indeterminate or incomplete observational information) is of fundamental importance. What is needed is a paradigm shift from the deterministic approach, which has until now played a central role in system recording methods, to a non-deterministic approach. The non-deterministic approach is an approach to calculation in which ambiguous or uncertain information is processed as is, as far as possible. Calculations are made with the probability distribution as an object variable, along with the stochastic inference, which makes the prediction. [1]. This stochastic inference has come to be used naturally as a naïve Bayes model or a Hidden Markov Model (HHM) using, for example, a pattern recognition device that maximizes the posterior probability. Further, in order to control the system based on decision theory, express useful knowledge, and perform complex processing, calculations with high-dimensional probability distributions involving multiple variables are necessary. As the number of variables becomes enormous, calculations involving high-dimensional probability distributions become complex; therefore, one has no recourse but to approximate locally using low-dimensional probability distributions. In order to facilitate this, a graph structure is introduced which stipulates the relationship between variables. As a multidimensional probability distribution model having this type of graph structure, we have the example of a Bayesian network [2]. Bayesian networks are general models that stipulate dependencies among many variables by a conditional probability and network structure. Bayesian networks can construct a model by statistical learning from large-scale data, which in turn becomes an important feature in handling uncertainty.

In the current work, after discussing the non-deterministic approach and probabilistic modeling, together with Bayesian networks and techniques of constructing models that use them for predicting human activities in everyday life, actual cases in which they are applied are discussed. Finally, a hypothesis about "Research as a Service," the construction of which has become inevitable in the process of implementation, is proposed and discussed.

BAYESIAN NETWORKS

LARGE-SCALE DATA BASED MODELING

Mathematically, in Bayesian networks, a model is defined by a graph structure, which considers random variables to be nodes, and in which a conditional probability distribution is allotted to each node .

In the case of discrete random variables, the conditional probability distribution of each variable is given by means of a conditional probability table (CPT). Giving a table of conditional probabilities in this way allows the probability distribution to be expressed with more degrees of freedom than is the case by specifying a density function and a parameter. In other words, it is useful as a non-deterministic modeling procedure when the nature of the object is not known in advance.

Destination variables that give the conditional probability are referred to as child nodes. In this way, directed acyclic graphs defined by a conditional probability table, variables, and graph structures are constructed as Bayesian network models.

When Bayesian network models become large, it is not easy to determine the network structure or the entire conditional probability table manual. In such cases, a procedure is necessary for constructing a model from statistical studies of large amounts of data.

Utilized data sets that include cases which deal with all items in the conditional probability table are called complete data. In this case, the statistical data is counted to obtain the frequencies; and these, when normalized, become the most likely estimators of the conditional probability values. In the case of incomplete data having deficiencies, conditional probability values are presumed, compensating for various types. There are instances when it is desirable to construct the model network from data. Studies of the construction then search for the graph structure from some initial conditions. As a measurement criterion for the appropriateness of a graph structure, information criteria other than likelihood, such as AIC, BIC, or MDL, etc., are used. When the graph node number is large, the search space increases explosively, and from a computational load perspective, searching all graph structures is difficult; therefore, it is necessary to use a greedy algorithm or various types of heuristics to search for quasi-optimal structures. The K-2 algorithm [3] is a study algorithm for this type of graph structure. This search algorithm is as follows:

(i) for each node, limit the candidates that can become new nodes,
(ii) select a child node, add and graph the new candidate nodes one by one,
(iii) decide on and evaluate the parameters on which the graph is based,
(iv) only when evaluated highly, use as a new node,

(v) when there are no more new node candidates to add, or when the evaluation does not increase even if a new candidate is added, move to another child node,
(vi) repeat (i) – (v) for all child nodes.

In general, new search spaces increase combinatorically; therefore, a device is needed to avoid an increase in computational load by limiting combinations of new nodes that become ranked from the beginning to be candidates. Furthermore, we consider independently the search portion of the graph (ii), (v), and the mode evaluation portion (iii) and think about various study methods.

One can expect that the use of a Bayesian network would be an effective approach to construct a non-deterministic model from large amounts of data by means of statistical learning. However, obtaining a causal structure from only statistical data is fundamentally difficult, and the task of searching the graph structure is NP hard. In such cases, it is actually necessary to skillfully implement the variable candidates or search range limitation, or to introduce appropriate latent variables.

BAYESIAN NETWORK APPLIED INFORMATION SERVICES

The advantage of using Bayesian networks is that by performing probabilistic inferences, we can determine the probability distribution of arbitrary variables and conduct quantitative evaluations in various situations. In many conventional multivariate analysis procedures, quantitative relationships are often modeled based on a covariance relationship that assumes linearity among variables (linear independence). In the Baycsian network model, quantitative relationships are represented by a conditional probability table.

It is extremely important that these characteristics respond by recommending information or products that are acceptable matches, depending upon the user or customer activities (Web browsing history, etc.), attributes, or circumstances. In collaborative filtering, information or products desired by customers or users cannot reflect situation dependence when displayed by a portable telephone or car navigation system. Information recommendation technology for such activities that change depending upon the environment is important, even in ubiquitous computing, in which a variety of situation changes in actual space are imagined.

It sometimes happens that the driver of a car wants to stop somewhere while driving. For example, while driving for some purpose, the driver decides to stop to eat at a restaurant. In conventional car navigation systems, a category is specified, and all corresponding restaurants are listed in order by distance. The user must find the appropriate restaurant from within the list; however, the user has to operate a touch switch or remote control in order to see detailed information about restaurants, so it is not easy for the driver to locate the desired restaurant.
Therefore, if a car navigation system were to model the driver's preference of various restaurants, given various situations and criteria using a Bayesian network,

it would be an extremely practical function that the system automatically selects the appropriate destination restaurant for drivers. A person's taste depends largely upon their personality and upon the situation while driving. While driving, it is necessary to select the most appropriate choice at the time, among conditions that change moment by moment.

To illustrate this dependence on situations and personal differences, a Bayesian network can be efficiently applied that can model complex relationships among variables and uncertainty. Therefore, we test and evaluate a car navigation system that suggests content appropriate for the user [4]. This system possesses, as a Bayesian network, a user taste model within the vehicle information system. Content, such as restaurants or music, is suggested by content providers, and a score showing how appropriate it is for the user and conditions at the time is calculated as a conditional probability when the situation and user attributes are given. It then recommends items with a high score, limiting them to superior content. 300 test subjects completed a questionnaire and selected desired restaurants in six situations from 182 actual restaurant in the Shinagawa neighborhood. A model was constructed from the gathered data. Restaurants desired in six situations (scenarios) were selected from the 182 restaurants in the Shinagawa neighborhood. Concerning the selection procedure, firstly, the user was queried about desired categories, and stores corresponding to those categories were displayed. If disliked, the next genre was chosen by the same selection method as in currently existing car navigation systems. There were multiple answers for selected restaurants, and ultimately 3778 records were obtained. There were 12 situation attributes, 17 restaurant attributes, and 12 user attributes. The model in Fig. 1 was constructed as a result. There are four attribute nodes representing users, three representing situations, and six representing restaurants. The model consists of all 13 of these random variables, and the probability distribution of restaurant attributes favored by specific users in a given situation is calculated by probabilistic inference.

In the model of Fig. 1, for drivers with a light driving history, the probability is high that franchise restaurants such as family restaurants and fast food chains will be chosen; conversely, for extensive driving histories, the probability that these restaurants will be chosen is low. Franchise restaurants often provide parking areas and show a tendency to be favored by young or beginning drivers. In addition to "driving history", there is a "have plans" interaction. This reflects the tendency that even in cases wherein the driving history may be long, in situations when the driver has plans and must hurry, there is a high probability that a franchise restaurant will be used. The proper tendency is obtained intuitively for other relationships, such as that between budget level and vehicle type.

A prototype of a restaurant recommendation system was also designed (Fig. 2). Favored content attributes are forecast as probability distributions from user variables and situation variables.

Human/User Models in
personalized, situated restaurant-navigation

P(restaurant¦ driver, situation)

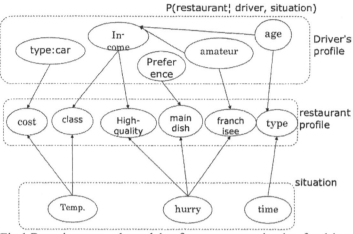

Fig.1 Bayesian network models of restaurant-navigation for drivers

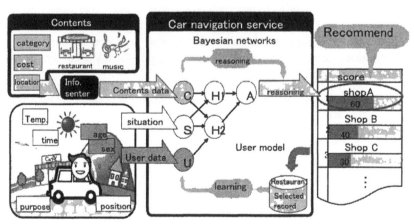

Fig.2 Restaurant recommendation service in a car navigation system

The system generates the list of appropriate restaurants according to given situation and the user profile by calculating posterior probability with the model. Upon comparing this prototype system and a conventional car navigation system, its effectiveness was confirmed by the fact that prediction results for restaurants matched the users' preferences and the situation.

The other example of application using Bayesian networks in a movie recommendation service in portable phone services have been introduced [5,6]. For approximately 1600 test subjects, their content evaluation history, user and content attributes were collected via a questionnaire that suggested movie content. Other

582

than demographic attributes such as age, gender, employment, etc., questions regarding lifestyle, appreciation frequency as attitude attributes concerning movie viewing, concern over movie selection time, the primary purpose for watching movies (seven questions about wanting to be emotionally moved), evaluation of content (good/bad), one's mood at the time (seven questions about being emotionally moved) were collected. Furthermore, for approximately 1000 people, all of the following were collected separately as free-form text: the content of each movie, what kind of feeling or situation, (theater, DVD, etc.), with who, with how many people, what time of day, was the movie appreciated. This data was input into BayoNet [7,8](Fig.3), the Bayesian network construction software developed by the author, and a Bayesian network model was constructed automatically. Through the Bayesian network constructed in this way, a prototype of the portable information system was developed that makes movie recommendations, based on situation and user tastes. If the user sends requests to services from the portable phone, together with information about the situation, the system implements the probability inference using registered user attribute and situation information from the database.

Content whose probability of being selected is judged to be high is recommended as superior. This movie recommendation system was also developed into an Internet service at auOne lab (http://labs.auone.jp) and released generally in 2007 with approximately 7000 recommendations implemented. Further, the model is being restudied from this recommendation history, and experiments are being conducted to improve recommendation precision. Using the computational model for movie selection constructed in this way, we also proposed cooperation with a movie distribution company to optimize sales strategies for DVD content for which the movie release period has passed [9].

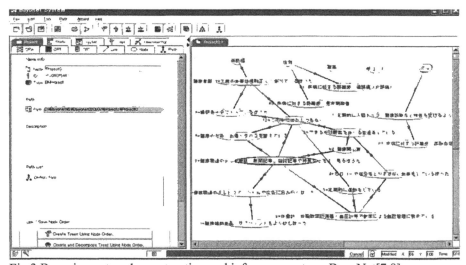

Fig.3 Bayesian network construction and inference system: BayoNet[7,8]

RESEARCH AS A SERVICE

As this information service spreads and multiple users utilize the system, the history of selection content accumulates ever-larger amounts of statistical data. Improvements in the Bayesian network model resulting from that data will increase the appropriateness and inference precision of the model, create a self-supporting feedback loop, and allow horizontal development of other services to be realized (Fig.4). Data obtained from the market through actual services becomes reusable knowledge for the calculation model; this knowledge cycle, reflected in the next service, can be called "Research as a Service". Research activities through this type of actual service can even be put into practice in a service engineering research center through construction of a calculation model from large-scale data and through implementation of an optimization design loop in the field. Such research activities are proposed as a business to improve the productivity of the service industry [10].

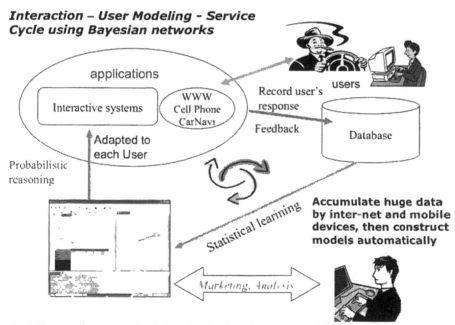

Fig.4 'Research as a service' that obtain data through providing real services

CONCLUSION

In the present research, the development of human modeling systems could be categorized as basic research; however, such applied human modeling system development, which is obviously outcome oriented service engineering, could be considered applied fundamental research. It seems that there were several conditions that implied that we rethink the criterion for application selection intuitively recommended in that process.

1. There are unresolved problems in existing procedures.
2. There were problems actualized by user requests.
3. There are stakeholders that profit from resolutions of these problems and bear the corresponding cost and risk.

The large-scale data based approach using Bayesian networks to model human behavior, forecasting customer or user activities, and achieving improvements in value and efficiency by optimization of associated services is thought to be promising tool for this direction. Client enterprises that can realize these outcomes exist in industry types that possess contact points (channels) with various customers.

Selecting the outcomes mentioned above, the appropriate fields become channels that can collect large amounts of data from customers such as the Internet, portable telephones, car navigation systems, and call centers. However, among these choices, two necessary types are: being able to adequately respond by transferring the present technology, and the development of additional technology for outcome realization. In the former, the venture responds; in the latter, the choice is made to promote cooperative research between industrial technology research institutes and enterprises

Engineering implementation and societal implementation differ. In engineering, even if the technology is already established, in order to produce societal value, participation of many more stakeholders is necessary. It is necessary to convey value to these stakeholders, which will not necessarily have an engineering background, in order to persuade them to bear the cost and risk; and it is necessary to demonstrate that the outcome has high reliability. Therefore, societal implementation through department-level cooperative research and technology transfer to industrial research institute ventures is necessary, and results need to be proven in the field. In other words, evaluation of the outcome and societal implementation occurs simultaneously.

In order to clarify the conditions under which implementation in society, a marketing research is necessary. The cost benefit analysis, which did not need consideration in the basic research, was critical for the real services. In order to smoothly advance societal implementation, reductions of cost and risk are sought while improving benefits. At this step, the outcome itself is corrected, and there is a possibility of motivating fundamental research out of necessity for a new outcome. This promotes fundamental research, becomes feedback to fledgling basic research that is called "Application driven fundamental research." It has also become possible to acquire large-scale data that includes situations and context involving the results of activities through actual services and actual users. Bayesian networks constructed from this data represent the cognitive and evaluation structures and behaviors of existing consumers and others. Being causal models rather than merely descriptive models of the statistical data, they are cognitive models with high reusability and potential for horizontal development in other services [11](Fig.5).

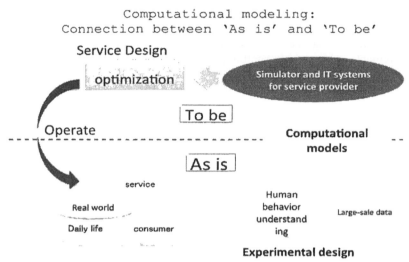

Fig.5 Concept of computational human modeling in everyday life through services

Concerning issues required for implementation in society, and from the standpoint of fundamental research, whether or not a quick response can be given is thought to be an important issue associated with establishing fundamental research on problem resolutions requested by society in the future. The very fact that speed is requested of technology in society requires that many "buds" be nurtured for the future. Such choices can only be performed by those thoroughly grounded in fundamental research; consequently, in order to perform fundamental research, views aimed at societal technology that clearly envisions the future are surely required.

REFERENCES

[1] S. Russell and P. Norvig: Artificial Intelligence, A Modern Approach, Prentice Hall Series (2002).

[2] J. Pearl: Probabilistic inference and expert systems, Morgan Kaufmann,CA, (1988).

[3] G. Cooper and E. Herskovits: A bayesian method for the induction of probabilistic networks from data. Machine Learning. 9(4), 309-347 (1992).

[4] Y. Motomura, T. Iwasaki: Bayesian network Technology, Tokyo Denki University. press (2006) in Japanese.

[5] C. Ono, M. Kurokawa, Y. Motomura and H. Asoh: A context-aware movie preference model using a bayesian network for recommendation and promotion, Proc. of User Modeling 2007, LNCS, 4511, 257-266, Springer, (2007).

[6] C.Ono, Y.Motomura, H.Asoh: Context-aware Preference Handling Technologies on Mobile Devices, Journal of Information Processing Society

586

of Japan, 48 (9), 989-994 (2007) in Japanese.

[7] Y. Motomura: BAYONET: Bayesian network on neural network, Foundation of Real-World Intelligence, 28-37, CSLI calfornia, (2001).

[8] Y.Motomura: Bayesian network software BayoNet, SICE Journal of Control, Measurement, and System Integration, 42 (8), 693-694 (2003) in Japanese.

[9] K.Ochiai, T.Shimokado, C.Ono, H.Asoh, Y.Motomura:Supporting Movie Contents Marketing using Bayesian networks, Annual Conference on Japanese Society for Artificial Intelligence (2009) in Japanese.

[10] Y.Motomura: Predictive modeling of everyday behavior from large-scale data: Learning and inference from Bayesian networks based on actual services, Synthesiology - English edition, 2 (1), 1-12 (2009) .

[11] Y.Motomura, Y.Nishida: Difficulty of Behavior Understanding Research and A Knowledge Sharing Framework, Annual Conference on Japanese Society for Artificial Intelligence, (2007) in Japanese.

CHAPTER 61

A Social System That Circulates Safety Knowledge for Injury Prevention

Yoshifumi Nishida[1,2], Yoichi Motomura[1,2],
Koji Kitamura[1,2], Tatsuhiro Yamanaka[1,2,3]

[1]Digital Human Research Center
National Institute of Advanced Industrial Science and Technology
Tokyo, Japan

[2]Core Research for Evolution Science and Technology Program
Japan Science and Technology Agency
Japan

[3]Ryokuen Children's Clinic
Yokohama, Japan

ABSTRACT

To design products safe for children, we must measure and understand children's behavior quantitatively, and design products in an evidenced-based way. This paper describes a case study on evidence-based design of playground equipment for childhood injury prevention by utilizing a behavior sensing system that we developed. Our research focused on the following case and analyzed it using a child dummy and a location sensor. At a park in Kitakyushu, Japan, a five-year-old child fell from the spiral staircase of playground equipment on which she was playing in an accident that caused a kidney injury. First, to test whether the circumstances described could actually occur, we went to the accident site and conducted an experiment to reenact how the child fell using a child dummy. From the experiment, we found that the child fell backward, her shoulder struck the edge of a stair on the spiral staircase and as a result of that collision, she fell to the ground. Second, we conducted experiments using an ultrasonic location sensor to understand how children actually use the equipment, and in particular, how they use the spiral staircase in play. The experiment reveals that younger children aged three and four frequently traversed the inner side of the equipment, while larger children aged five and six often traversed the equipment using both the inner and outer sides of the spiral staircase. Based on the above research, we devised a hand rail that keeps children outer sides of the spiral staircase as preventive measures. The preventive measures were applied to 34 equipment by a municipality. This paper discusses a necessary social system for designing safe product based on injury data for injury prevention by generalizing this case study

Keywords: Childhood Injury Prevention, Product Safety, Information Share, Behavior Understanding

INTRODUCTION

Childhood injuries are the most significant issue in child health care. Given the decreasing birthrate and the increasing proportion of elderly, there is a pressing need to address this as a social issue affecting the future of nations. WHO (World Health Organization) announces a ten-year plan of action for children injury prevention in 2006 (World Health Organization, 2006). Children injury is a real problem to be responded transdisciplinarily and globally (Hyder, 2006).

Although it can easily be assumed that without examples of accident cases we would be unable to even consider prevention measures, it is by no means clear that the collection of a large number of examples is truly beneficial for injury prevention. Although injury surveillance systems involving the collection of injury case data exist in other countries, no systematic framework for using injury information to develop prevention methods exists. To avoid abstract armchair arguments short on effectiveness and to search for clues about injury prevention, it is necessary to think concretely. From this viewpoint, the authors conducted a case study in which a series of works from data collection to implementation of preventive measure were completed.

In this report, we first relate the series of operations, from investigation to implementation of injury prevention methods, in chronological order. Then, in light of what is learnt from this series of operations, we will discuss the tasks required to build a society that can conduct injury prevention as an appropriate society.

CASE STUDY OF INJURY PREVENTION

OVERVIEW OF ACCIDENT AND INJURIES

Figure 1. The playground equipment that caused the accident.

At a park in Kitakyushu, Japan, a five-year-old child fell from a playground equipment on which she was playing in an accident that caused a kidney injury. Fortunately, the little girl who met with the injury suffered no catastrophic complications, and was released following nine days of hospitalization. Figure 1 shows the playground equipment and its photograph.

INTERVIEW INQUIRY WITH THE ACCIDENT VICTIM

First of all, we conducted an interview inquiry with the injured girl and her parents. The inquiry involved several difficulties. Most importantly, the circumstances aren't known in detail as no one was watching. In addition, the girl who fell couldn't observe her accident objectively from an outside perspective, and there were limits to her ability to explain the situation. Nevertheless, the following approximate picture of the circumstances emerged.

1. The child slipped and fell backward while leaning her shoulder against the central pole, and struck her back somewhere on the equipment.
2. The child's head did not come into contact with any part of the equipment when she fell, and in fact there was no wound to the head.
3. The child eventually fell all the way to the ground.

EXPERIMENT FOR REPRODUCING ACCIDENT AT THE ACCIDENT SITE

On the basis of the interview inquiry, to test whether the circumstances described could actually occur, we went to the accident site and conducted an experiment to reenact how the child fell. In the experiment we used the actual playground equipment on which the accident occurred and a 15-kilogram dummy of a three-year-old child. As a result of varying positions and postures and conducting several dozen iterations of a fall reenactment experiment, a fall that could nearly explain the details related in the interview survey was reenacted. Figure 2 is photographs of reproduction of the accident. From the reenactment experiment, we learned that the child's shoulder apparently slipped while she was resting it against the central pole. The child fell backward, her shoulder struck the edge of a stair on the spiral staircase, and as a result of that collision, she fell to the ground. In a fall under these circumstances, it is conceivable that extremely great pressure acted focally as a result of the stair edge directly striking the child's back, and that a kidney rupture occurred as a result of that pressure being transmitted to the child's internal organs.

Figure 2. Reproduction of accident using three-year-old dummy at accident site.

MEASUREMENT OF PLAY AT THE NATIONAL INSTITUTION OF ADVANCED INDUSTRIAL SCIENCE AND TECHNOLOGY

Although we had learned the circumstances of the fall from the on-site reproduction experiment stated in the previous section, the obtained knowledge was limited to a single accident that a young girl had on a playground equipment. To devise prevention methods that would contribute to the prevention of future accidents, it was necessary to generalize this case example. To that end, we lacked data concerning how children actually use the playground equipment in question, and in particular, how they use the spiral staircase in play. Consequently, taking into account ethical considerations, we made the same playground equipment as shown in Fig. 3 at our laboratory, in such a way as to implement safety measures while retaining the original equipment shape, and equipped it with ultrasonic location sensors that could measure the way children play. The ultrasonic location system consists of ultrasonic receivers, ultrasonic tags with a transmitter, and a radio controller. By attaching the ultrasonic tag to a child, we can detect and record the three-dimensional position data of the child. The ultrasonic location system can track the positions of the child within an error of 3 cm (Nishida, 2003). The left part of Fig. 4 shows the ultrasonic location sensor that is attached to children's back for measuring his or her position. Using this playground equipment, we observed behavior of children which consists of four age groups (three, four, five, and six year-old) as shown in the right part of Fig. 4.

Figure 3. Experimental system for observing child behavior.

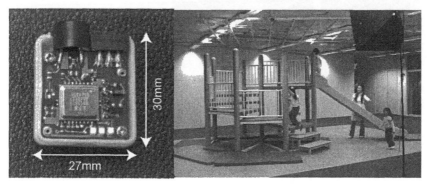

Figure 4. Ultrasonic location sensor (left) and behavior observation (right).

RESULTS OF THE PLAY MEASUREMENT EXPERIMENT IN THE LABORATORY

Figure 5 shows the results of analysis of the sections of the playground equipment that the children frequently traversed. White areas indicate high frequency of occupation by children. The figure reveals that younger children aged three and four frequently traversed the inner side of the equipment, while larger children aged five and six often traversed the equipment using both the inner and outer sides of the spiral staircase.

Spiral staircase was originally invented as an architectural structure for achieving the function of ascending and descending in limited space. In conducting the investigation we came to understand that a characteristic inherent in the structure of spiral staircases is that spiral staircases have a sharp slope when the inner side is traversed, and a gradual slope when the outer is traversed. For instance, as shown in Fig. 6, the spiral staircase used for the playground equipment studied has a slope of 14.6° when the line (r=1000mm) is traversed, a slope of 27.6°, when the line (r=500mm) is traversed, and an extremely steep slope of 52.6° when the line (r=200mm) is traversed.

As explained previously, the data from the play measurement experiment revealed that younger children aged three and four tended to traverse the inner side of the spiral staircase, and larger children aged five and six tended to traverse both the inner and outer sides of the staircase. When these tendencies and the characteristics of spiral staircases are considered together, it is evident that children age three and four use the steep part of the spiral staircase.

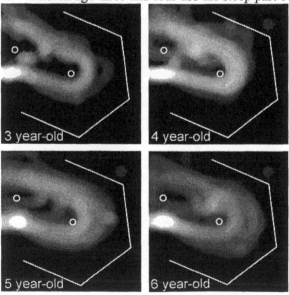

Figure 5. Measured data of the sections of the equipment used for play (existence probability distribution).

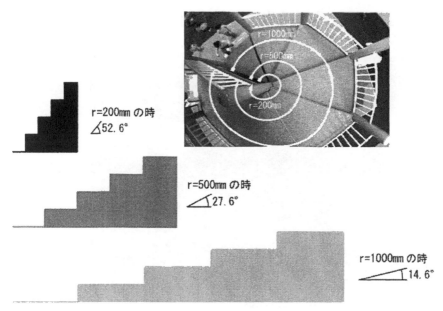

Figure 6. Characteristics of the spiral staircase.

COLLABORATION WITH PLAYGROUND EQUIPMENT MAKER

In cooperation with a playground equipment maker, we developed preventive measures based on the findings stated above. Figure 7 shows the developed preventive measures. One of developed preventive measure is a handrail. This handrail incorporates not only a handrail function, but also a "barrier handrail" function. Combining the two functions might be effective for preventing the same injury due to fall accident. The other preventive measure is a fence that prevents children from directly falling down to the ground.

594

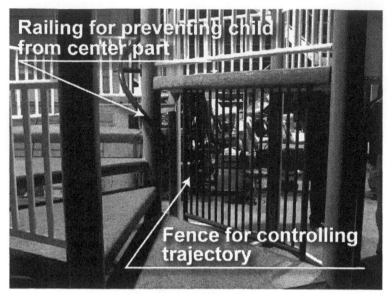

Figure 7. Developed preventive measures.

DISCUSSION AT THE KITAKYUSHU CITY OFFICE

The project members visited Kitakyushu City Office and reported the results of the investigation and the countermeasures proposed to a person who was in charge at the Construction Bureau. The Kitakyushu City Office adopted the proposed measure and improved 34 equipments. Figure 8 shows photograph of one of improved equipment.

Figure 8. Improved playground equipment.

GENERALIZATION OF CASE STUDY—PROPOSAL OF KNOWLEDGE CIRCULATING SOCIETY

The remainder of this report, in light of what is learnt from this series of operations, we discuss the tasks required to build a society that can conduct injury prevention as an appropriate society.

First of all, in the case investigated, an accident occurred in a park. Subsequently, the injured person was conveyed by a friend's car, examined and treated at a hospital, and released after nine days. In this case study, an information gathering team of project members discovered the accident information and conveyed the information to other project members. This allowed the subsequent site investigation, analysis of children's play, injury prevention measure scheme from a playground equipment manufacturer, and implementation of accident prevention measures by a municipal government.

Figure 9 shows a concept of a social system in which all necessary tasks such as injury data collection, safety knowledge creation, and safety knowledge dissemination are chained and completed as one loop. We call this concept "a knowledge circulating social system." We believe that a society that can precede injury prevention would be one where this type of injury prevention loop continues for multiple cycles.

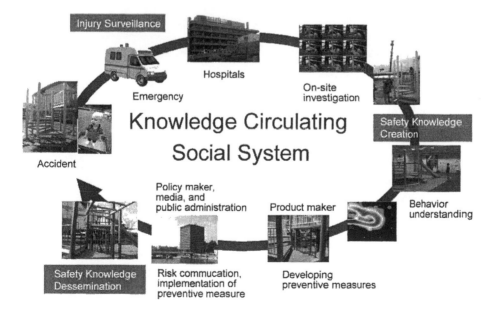

Figure 9. Knowledge circulating social system for completing injury prevention from injury data collection to dissemination of safety knowledge.

REFERENCES

Hyder, A.A.(2006), "Childhood Injuries: Defining A Global Agenda For Research and Action." *A Journal of Injury and Violence Prevention*, 4(1), 87-95.

Nishida, Y., Aizawa, H., Hori, T., Hoffman, N.H., Kanade, T., and Kakikura, M. (2003), "3D Ultrasonic Tagging System for Observing Human Activity," Proc. of IEEE International Conference on Intelligent Robots and Systems (IROS2003), 785-791.

World Health Organization. (2006), Child and Adolescent Injury Prevention -A WHO Plan of Action-.

CHAPTER 62

A System for Presenting Potential High-risk Situation by Integrating Biomechanical, Injury, and Child Behavior Model

Yoshinori Koizumi[1,2], Yoshifumi Nishida[2], Yoichi Motomura[2], Yusuke Miyazaki[3], and Hiroshi Mizoguchi[1,2]

[1]Graduate School of Science and Technology
Tokyo University of Science
2641 Yamazaki Noda-shi Chiba 278-8510, Japan

[2]Digital Human Research Center
National Institute of Advanced Industrial Science and Technology
2-3-26 Aomi Koto-ku Tokyo 135-0064, Japan

[3]Department of Mechanical Science and Engineering
Kanazawa University
Kakuma-cho Kanazawa-shi Ishikawa 920-1192, Japan

ABSTRACT

Injury prevention is one of the most important and urgent issues in children's health since the primary cause of death of children is unintentional injury. Passive approach, namely injury prevention approach by product modification is strongly needed. The risk assessment is one of the most fundamental methods to design safety products. However, the conventional risk assessment has not been able to deal with risk of childhood injury due to product use because product manufacturers have poor childhood injury data and child behavior data. Developing methodology of quantitative risk assessment for childhood injury prevention is strongly required by product manufacturers. This paper proposes a system for presenting potential high-risk situation by integrating biomechanical, injury, and child behavior model. To prove the effectiveness of the proposed system, this paper describes the application of the system to the risk analysis of a playground swing in a park. The system calculated the potential risk of head injury around the swing equipment.

Keywords: Hazard identification, Injury prevention, Injury surveillance, Impact biomechanics, Injury informatics

INTRODUCTION

Recently, childhood unintentional injuries have been recognized as a social problem. According to the world report on children injury (World Health Organization (WHO), 2008), unintentional injury is a major killer of children under the age of 18 years and is responsible for approximately 950,000 deaths. After the age of 1 year, unintentional injuries are significant contributors to the deaths of children and teenagers. Due to the present awareness of this subject, WHO has formulated a 10-year global strategy for child injury prevention by promoting research and preventive activities in all countries (World Health Organization (WHO), 2006).

Injury prevention requires making safe products as well as safe environments by evaluating the risks. In the industrial world, the International Organization for Standardization (ISO) and the International Electrotechnical Commission (IEC) prepared ISO/IEC Guide 51 to provide guidelines on the safety aspects in standards (ISO/ICE, 1999). This guide presents the process of risk assessment for reducing the risk in the use of products. This Process includes "foreseeing the unintended product usage as well as the intended" and "estimating and evaluating the risk." In fact, carrying out these steps is difficult because product manufacturers have poor behavior data and injury data, especially data pertaining to children.

In the impact biomechanics field, many studies have been carried out using a cadaver (Nahum, 1977), dummy (Shaw, 2003), multi-body model (Miyazaki, 2007), and finite element model (Mizuno, 2005) to clarify the effects of impacts and the process of injury occurrence when a person is subjected to forces and impacts. These technologies provide impact prediction data. However, it is difficult to assess the injury risk of a product by only these technologies since the relationship between impact prediction and injury is not clarified. Children's behavior is also not dealt with.

This paper proposes a new system for presenting and evaluating the potential injury risk by using a biomechanical simulation based on both bodygraphic injury data collected in hospitals and product usage data collected by sensors. To prove the effectiveness of the proposed system, this paper reports the risk analysis of a playground swing in a park as an application of the proposed system.

SYSTEM FOR PRESENTING POTENTIAL HIGH-RISK SITUATION BY INTEGRATING BIOMECHANICAL, INJURY, AND CHILD BEHAVIOR MODEL

Fig. 1 shows the configuration of the proposed system for presenting potential high-risk situation by integrating biomechanical, injury, and child behavior model. This system consists of a biomechanical model an injury model, a child behavior model, and a biomechanical simulation engine. The details are as follows.

BIOMECHANICAL MODEL

Multi-body Model

The multi-body model, composed of rigid links connected by joints, has the advantages of short calculation time and ease of posture changes. The multi-body model is useful for carrying out a large amount of simulations needed to reproduce the accident situation and present the potential injury risk. The multi-body model used in this research is constructed based on the method in Miyazaki et al. (Miyazaki, 2007). It has the dimensions of Japanese children and consists of 17 ellipsoidal segments and 16 joints.

The relationships for contact stiffness are defined by using the data of a Hybrid-III dummy. By using this model, a large amount of simulations can be carried out to clarify whole body behavior and to provide us with the Head Injury Criterion (HIC) in this research. HIC, calculated from linear acceleration responses at the head's center of gravity (COG), is the tolerance for a skull fracture or brain concussion. The tolerance is 1000.

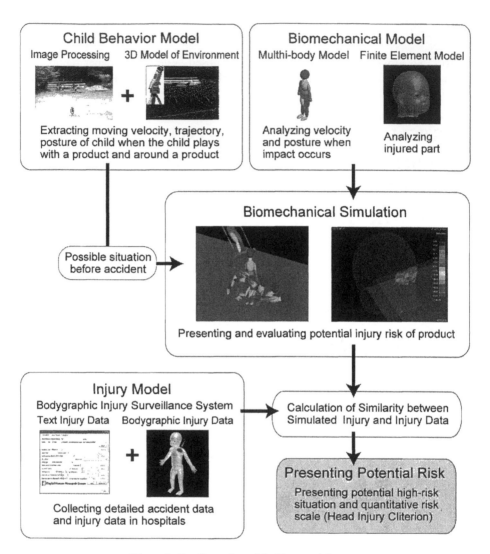

Figure 1. Configuration of the Proposed System

Finite Element Model

The finite element model allows us to analyze in more detail, although the calculation time is longer than that of the multi-body model. For example, the finite element method (FEM) can evaluate the risks for the occurrence of head injury and display the location of injury by applying the head motion calculated in the multi-body simulation as the boundary condition.

In this research, the three-dimensional human body model of BISS forms the basis for the finite element model of a head to integrate the injury model and biomechanical simulation. The mechanical attributes of the model are defined by using the data of a Hybrid-III dummy as well as the multi-body model.

The authors are also proceeding in joint research with the Autopsy Imaging (Ai) Center at Chiba University Hospital and the Forensic Medicine Class, Chiba University, to investigate the dynamical property of human body tissue to improve the technology with medical expert opinion. In the future, we plan to apply the knowledge obtained in this joint research to injury prevention.

INJURY MODEL

The authors have been developing Bodygraphic Injury Surveillance System (BISS) for collecting detailed accident information in hospitals (Tsuboi, 2008). BISS enables us to collect more than 20 kinds of text data, such as type of injury, cause of injury, degree of severity, and person(s) involved in the accident. In addition, the location, shape and attribute data of an external injury is collected by drawing on a three-dimensional human body model with a computer mouse, as shown in Fig. 2. The input injury data is converted to raster data and stored in the database system. Since the standard human body model is used, injury data is normalized.

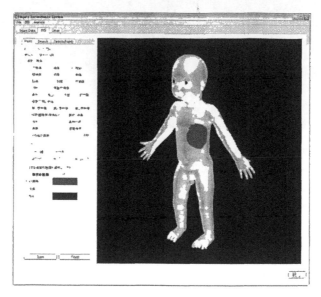

Figure 2. Input of Injury Data by Using a Computer Mouse to Draw a Shape

BISS also allows us to conduct new statistical analyses (Nishida, 2009). As one example of data visualization using BISS, we describe the bodygraphy of injury frequency. Fig. 3 shows the injury frequencies retrieved and visualized with only the necessary data on demand. Since each instance of injury data is expressed in a normalized and structured form in BISS, we can calculate the conditional probability distribution of an injury by counting injury frequencies under given conditions. In the figures shown here, a red color indicates the area with the highest frequency of external injury. Part A of the figure visualizes the bodygraphic frequency of a scald injury, Part B is the frequency of injury due to a fall by children of ages 1 to 2 years, Part C is the frequency of injury due to a playground slide, and Part D is injury due to all kinds of playground equipment.

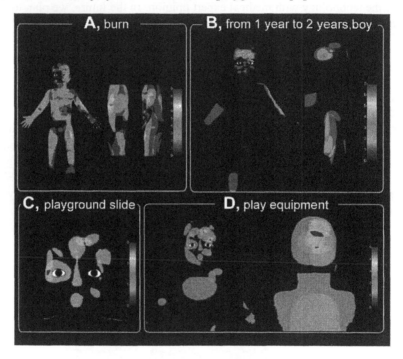

Figure 3. Examples of Injury Retrieval and Frequency Visualization
(Conditional Injury Probability Distribution)

CHILD BEHAVIOR MODEL

In this paper, product usage data means two kinds of data. One is the child behavior data when the child plays with or on the product. The other is the child behavior data when another child plays around the product. In particular, these two data are important when we analyze products such as playground equipment, in which injury stems from interactions such as colliding and pushing with the other child as well as interactions with the product itself. Product usage data is often obtained as image data by image processing. The authors are developing a computer program for extracting product usage from videos. For example, we can extract the distance between a product, a parent and a child, a child's moving velocity, trajectory, and posture. Data obtained in this way is represented on the image coordinate system (two dimensions); therefore, camera calibration is required to obtain the moving velocity, trajectory, etc., in real space (three dimensions). Three-dimensional point cloud data collected by 3D laser scanning is useful for camera calibration. Three-dimensional laser scanning is used for setting up CAD models of 3D objects by irradiating them with a laser and measuring a large number of points on their surfaces. Camera calibration is performed by finding the camera angle on the 3D laser scanner's coordinate system from the correspondence between points on the image coordinate system and points on the 3D laser scanner's coordinate system. This method allows us to obtain a children's behavior model for moving velocity, trajectory, etc., in real space. The obtained behavior model is a function of time. It is used to determine the initial condition for biomechanical simulation using the multi-body model. Thus, we can consider the children's behavior before an accident in an accident simulation.

SIMULATION PROCEDURE

In this subsection, we describe how the injury model, child's behavior model and biomechanical simulation are integrated through the simulation procedure, shown as follows.

1. Based on the child behavior model, the system produces a candidate of the situation at the time when a child uses a target product and another child plays around the target product.

2. The system simulates the possible accident process caused by the candidate situation before the accident and calculates the posture and velocity immediately before the impact occurs.

3. The system calculates the injured body part by the finite element model.

4. The system calculates the similarity between the simulated injured part and the actually injured part using the injury database by searching for the most similar injury case from the injury database. This allows us to estimate the type of injury and the degree of severity.

APPLICATION OF THE PROPOSED SYSTEM

The authors have been observing children's behavior at various sites. Fig. 4 shows a girl chasing a ball in a primary school's playground. The behavior that she cannot see anything but the ball, even near the swing, is one of the plausible behaviors. If the ball had rolled near the swing equipment, what would have happened to her? As an example in this research, we simulated this accident situation. By image

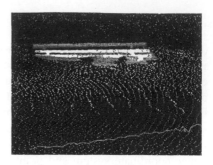

Figure 4. Children's Behavior in the
School Playground

Figure 5. Trajectory of Girl in Real
Space

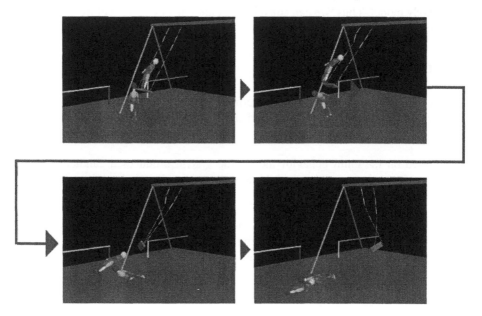

Figure 6. Possible Accident Situations

processing, we obtained her movement locus represented on the image coordinate system. By performing camera calibration based on 3D laser scanning data, her trajectory in real space was obtained as shown by the orange line in Fig. 5. The behavior data was output as a function of time.

The biomechanical simulation based on the behavior data enables us to present possible accidents, as shown in Fig. 6. An analysis of the head injury in the impact case was conducted using the finite element model of the head with the system posture and velocity data obtained from the multi-body analysis as the initial conditions. The head injury analysis allows us to obtain the location of the external injury. The coordinate system of this finite element model corresponds to that of the three-dimensional human body model of BISS. Therefore, we can calculate the similarity between the simulated injury area and the actually occurring injury area using the injury database of BISS. In this study, we calculate the distance between the centers of gravity as the similarity. In this way, we can search for corresponding case data by giving a variety of conditions for a child's behavior before an accident.

In addition, the system allows us to evaluate its risk in terms of the HIC scale. In the case shown in Fig. 6, HIC for a child moving on the ground is 749 and that for a child playing on the swing is 41. The HICs of both children did not exceed the tolerance value. This result means that the possibility of occurrence of fatal head injury is low. Fig. 7 shows the histogram of 480 calculated HICs. Repeating a large amount of simulations like this is useful for presenting the potential injury risk.

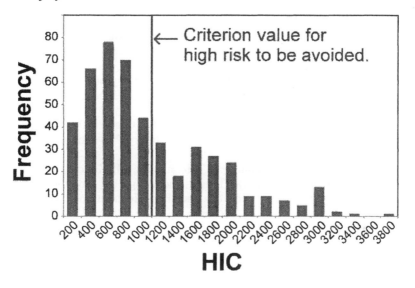

Figure 7. Histogram of HIC

CONCLUSIONS

In this paper, we proposed a system for presenting potential high-risk situation by integrating biomechanical, injury, and child behavior model. The system allows a product designer and/or park manager of a municipality to predict how children use a product and the types of injuries that can occur in a daily environment. The developed system can search the candidates of possible situations resulting in serious injuries by using the sensory data and injury data, and evaluate the risk quantitatively by using the biomechanical simulation (in terms of the head injury criterion (HIC) in our example). To prove the effectiveness of the proposed system, this paper reported the application of the system to the risk analysis of a swing in a playground where multiple children were playing.

REFERENCES

World Health Organization (WHO). (2008). World report on child injury prevention. (edited by M. Peden, K. Oyegbite, J. Ozanne-Smith, A. A. Hyder, C. Branche, A. F. Rahman, F. Rivara, and K. Bartolomeos).

World Health Organization (WHO). (2006). Child and Adolescent Injury Prevention - A Global Call to Action.

ISO/ICE. (1999). Guide 51 Safety Aspects --- Guidelines for Their Inclusion in Standards.

A.M. Nahum, R. Smith, and C. C. Ward. (1977). Intracranial pressure dynamics during head impact. In Proc. of the 21st Stapp Car Crash Conf., SAE Paper 770922, pages 339-366.

G. Shaw, D. Lessley, R. Kent, and J. Crandall. (2003). Dummy torso response to anterior quasi-static loading. Stapp Car Crash J, 47: 267-297.

Y. Miyazaki, S. Watanabe, M. Mochimaru, M. Kouchi, Y. Nishida, and S. Ujihashi. (2007). Visualization of the hazards lurking in playground equipment based on falling simulations using children multi-body models. The Impact of Technology on Sport, Vol. II: 883-888.

K. Mizuno, K. Iwata, T. Deguchi, T. Ikami, and M. Kubota. (2005). Development of a three-year-old child FE model. Traffic Injury Prevention, Volume 6(4): 361-371.

T. Tsuboi, Y. Nishida, Y. Motomura, M. Mochimaru, M. Kouchi, T. Yamanaka, and H. Mizoguchi. (2008). Injury modeling by bodygraphic injury surveillance system. In Proc. of The First International Workshop on Advanced Integrated Sensing Technologies for Safety and Security of Daily Life, pages 18-22.

Hodgson, V. R.. (1967). Tolerance of the facial bones to impact. American Journal of Anatomy, 120, pp.113-122.

Schneider, D. C. amd Nahum, A. M.. (1972). Impact studies of facial bones and skull, 16th Stapp Car Crash Conf., pp.186-203.

J. Melvin. (1980). Human Tolerance to Impact Conditions as related to Motor Vehicle Design, SAE J885 APR80.

Y. Nishida, Y. Motomura, K. Kitamura, and T. Yamanaka. (2009). Representation and statistical analysis of childhood injury by bodygraphic information system. In GeoComputation 2009.

CHAPTER 63

A Method of Evidence-Based Risk Assessment through Modeling Infant Behavior and Injury

Koji Nomori [1,2], Yoshifumi Nishida[3,4], Yoichi Motomura [3,4]
Tatsuhiro Yamanaka [3,4,5], Akinori Komatsubara [1]

[1] Graduate School of Creative Science and Engineering
Waseda University
Tokyo, Japan

[2] Research Fellow
Japan Society for the Promotion of Science
Tokyo, Japan

[3] Digital Human Research Center
National Institute of Advanced Industrial Science and Technology
Tokyo, Japan

[4] Core Research for Evolution Science and Technology Program
Japan Science and Technology Agency
Japan

[5] Ryokuen Children's Clinic
Yokohama, Japan

ABSTRACT

In Japan, childhood injury prevention is an urgent problem because unintentional injury is the leading cause of death among children, and this trend fails to improve. An injury prevention approach by product modification is especially very important. The risk assessment is one of the most fundamental methods to design safety products. However, the conventional risk assessment has been carried out subjectively because product makers have poor data on injuries. This paper proposes a new method of evidence-based risk assessment, and describes text mining of child injury data, probabilistic modeling of child behavior, and probabilistic modeling of child injury as basic methods for realizing the proposed method. Child injury data are collected from medical institutions through an injury surveillance system. Text mining of narrative parts of the injury data helps product designers learn types of potential behaviors related to a product and types of injuries due to the product use. The results of text mining allow product designers to create injury scenarios which depict injurious incidents from the time of product use to injury occurrence. Probabilistic modeling of child behavior based on behavioral observation data allows product designers to compute the probability of the occurrence of the behavior related to the product. Probabilistic modeling of child injury based on the injury data allows product designers to compute the severity of the injury that results from behavior. The risk of the injurious incident scenario is estimated by the combination of the computed probability of the behavior and the computed severity of the injury.

Keywords: Childhood Injury Prevention, Product Design, Risk Assessment, Text Mining, Foreseeing Usage of Product, Bayesian Network, Behavior Modeling, Injury Modeling, Risk Computation

INTRODUCTION

In Japan, unintentional injury is the leading cause of death among children, and the number of unintentional injuries among infants is so large that safety measures are urgently needed (Japan Ministry of Health, Labour and Welfare, 2008). Fall injuries related to consumer products and home environment are the major injuries among infants (Tsuboi et al., 2008). In general, three approaches are essential for injury prevention: education, law enforcement, and modification of product and environment. The approach of modifying product and environment prevents injuries that can be serious even if parents leave their child unsupervised. The approach has become an important approach in injury prevention, benefiting people of all ages, not just children, in the passive protection that it affords (WHO, 2008). The development of the methodology of the approach is demanded.

To prevent injuries by product modification, it is necessary to evaluate an injury risk inherent in product design and to design a safe product by controlling the risk. The risk assessment (ISO 14121, 1999) is one of the most fundamental methods to design safety products. The procedure of the risk assessment in the safe design of a product is as follows. The first is to foresee users and use types of the product. The second is to identify hazards (potential source of harm) arising in the product uses. The third is to estimate the risk: the combination of the probability of occurrence of harm and the severity of that harm. The fourth is to evaluate the risk: "Is the risk tolerable or not?" However, this risk assessment has been carried out qualitatively depending on subjective rules of thumb because product makers have poor data on injuries. In fact, carrying out evidence-based risk assessment is difficult. Quantitative risk assessment based on data is advanced in the field of environmental science such as bacteria in food (Malomy, 2008), pollution in water (Wurbs, 2000) and so on. In the field of consumer product injury, the methodology of evidence-based risk assessment using data is not established.

This study proposes a method enabling evidence-based risk assessment of consumer products by applying text mining technology and probabilistic modeling technology to large scale child injury data and child behavior data.

METHOD OF EVIDENCE-BASED RISK ASSESSMENT

This study proposes a method of evidence-based risk assessment (see Figure 1). The procedure is shown as follows.

1) Types of potential behaviors related to a product at the time of the occurrence of injuries are extracted by performing text mining to narrative part of large scale injury data.

2) An injurious incident scenario is determined based on the behavior type related to the product.

3) The probability of occurrence of the behavior causing the injury is computed based on a behavior model. The behavior model is constructed based on data on the relationship between product features and behaviors using Bayesian Network which can model the probabilistic causal relationship among factors.

4) The severity of the injury is computed based on an injury model. The injury model is constructed based on large scale injury data using Bayesian Network.

5) The risk of the injurious incident scenario is evaluated by the combination of the computed probability of the behavior and the computed severity of the injury.

6) If the risk is not tolerable, it is reduced by controlling operational variables in the behavior model and the injury model.

Generally, risk is defined as the combination of the probability of occurrence of injury and the severity of that injury. Observation of situations where injuries occur and do not occur is needed to compute the probability of injury occurrence based on evidence. It is very difficult to compute the probability because the experiment which causes injuries is never allowed. So, instead of the probability of occurrence of injury, the probability of occurrence of behavior leading to the injury is computed based on modeling of behavioral observation data. In the risk assessment proposed by this study, risk is computed with the combination of the probability of occurrence of behavior leading to an injury and the severity of that injury.

This paper presents performance of text mining of injury data to learn types of child's potential behaviors related to a product, and construction of an infant behavior model, and an injury model to compute risk. In this paper, an infant climbing behavior model and an infant fall injury model are constructed, using an injurious incident scenario: an infant climbs a product and falls from it.

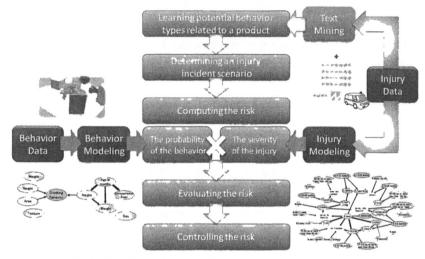

Figure 1. A method of evidence-based risk assessment.

TEXT MINING OF LARGE SCALE INJURY DATA

In this section, large scale injury data were collected and types of potential behaviors related to a product were extracted by text mining of narrative parts of the injury data. The results of the mining help product designers learn types of potential behaviors related to a product based on evidence and create an injurious incident scenario when they carry out product risk assessment. Then, a retrieval system of the relation information among products, behaviors, and injuries was developed to allow product designers to use the knowledge acquired by the data analysis.

COLLECTING LARGE SCALE CHILD INJURY DATA

Our research group is advancing an injury surveillance system continuously collecting child injury data in collaboration with medical institutions to prevent childhood injury (Motomura et al., 2006). This injury surveillance system collects data regarding injuries, such as the characteristics of the injured children (i.e., sex and age), date and time of the injuries, locations where injuries occurred, products associated with injuries, behaviors exhibited just prior to injury occurrence, types of injuries, and injured body parts. Details of the situation of an injury are recorded by free description.

TEXT MINING OF INJURY TEXT DATA

In this study, text mining was performed on narrative parts of 4,238 injury data collected by the surveillance system. The information what kind of behavior was taken on what kind of product in the situation where an injury occurs was extracted by the mining. In the text mining, morphology analysis which separates a sentence by a word was carried out, and modification relation among word was extracted. The data on the relationship between a product and a behavior when an injury occurred was collected by analyzing the modification relation between a product and a verb.

The target products in the analysis of the modification relation were 208 types of products of which the appearance frequency was five or more among products extracted by the text mining. As a result of analyzing the modification of a verb associated with these 208 types of products, 439 types of behaviors were extracted. The types of behaviors extracted by the modification analysis include verbs that mean uses of products such as "climb," "take," etc. and injuries caused by products such as "fall," "bruise," etc. For example, the types of behaviors related to a chair and the appearance frequency of the behaviors are shown in Figure 2. In the behaviors that mean injuries, there are many falling behaviors from a chair. There are also the behaviors that mean striking a body such as "strike" and "bruise." In the behaviors that mean uses, there are many behaviors to sit on a chair and some behaviors to hang or put something on it. There are also the behaviors such as "stand," "climb," "get on" and "stand up holding." These behaviors may result in the fall injuries.

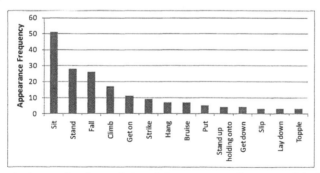

Figure 2. Behavior types related to a chair and the appearance frequency of the behaviors.

DEVELOPING A RETRIEVAL SYSTEM

A retrieval system was developed to allow product designers to use the knowledge acquired by the text mining of large scale injury text data. The system users can search the information about types of products (208 types), types of behaviors (439 types), types of injuries (27 types), age of children, and sex of children in the 4,238 injury data by setting up the condition of these 5 items. The information of types of injuries and child characteristics (age and sex) is recorded by the injury surveillance system. Further, the users can output free description texts of the injury data under the set condition. This retrieval system allows product designers to understand types of child's potential behaviors related to the product to design and types of injuries which result from the behavior and also to get details of the injuries by reading the free description texts. So, potential behavior types to a target product of risk assessment can be learned based on evidence and a scenario which depict injurious incident from the time of a product use to an injury occurrence can be created.

MODELING OF CHILD BEHAVIOR

In this section, a behavior model which can compute the probability of occurrence of child behavior from product features was constructed based on behavioral observation data to compute risk of an injurious incident scenario based on evidence. In this paper, infant climbing behavior model was constructed, using the injurious incident scenario: an infant climbs a product and falls from it.

EXPERIMENTS

To clarify the relationship among attributes of objects, characteristics of infants and infant climbing behaviors, an experiment of examining infant behaviors toward objects with different features was conducted. The objects used in this experiment were made in the forms of regular quadrangular prisms, because many objects in everyday life spaces can be expressed in these forms that are rectangular parallelepipeds. In this study, height, area, weight and texture among the object factors were selected as object variables. The levels of each variable shown in Table 1 were set, and 18 regular quadrangular prisms that combined each level of height, area, and weight were made. The "area" here is defined as the area of the top face of the object. The three levels of the textures were equally assigned at random to the 18 objects, because it was assumed that the effect of textures on climbing behaviors was small compared with the other selected variables. The effect of color was not considered in this study, so the surfaces of all objects were colored red, yellow and blue. Examples of some of the objects used in the experiment are shown in Figure 3.

The size of the experimental room was 4.7 m in length, 3.3 m in width, and 2.7 m in height. In this room, a camera that could take video of the whole room with a fish-eye lens was installed in the center of the ceiling, and two video cameras on tripods were installed in the corners of the room. During the experiment, infant behaviors were recorded by these three cameras. The experimental environment was designed to be similar to the daily living space with which infants were accustomed so that they would exhibit normal behavior. The experiment was conducted in the following procedures with 13 infants. Actual experimental situations are shown in Figure 4.

1) We explain the experiment to the infant's parent (mother) orally and with a document, and have her sign a consent form. We have her answer a questionnaire which asks sex, age, height, weight, and a developmental stage of her infant.

2) We place 6 of the 18 regular quadrangular prisms in the experiment room. These 6 objects are chosen at random. The infant freely behaves in the experiment room for 20 minutes accompanied by his or her parent.

3) We replace these 6 objects, and Step 2 is repeated 3 times for each infant.

To examine the natural behavior of each infant towards the objects, we asked each parent not to suggest that the infant interact with the objects. Each parent was instructed to only supervise the free behavior of the infant and to stop any behavior deemed to be dangerous. The experiment was approved by the ethics committee of the National Institute of Advanced Industrial Science and Technology in Japan and was conducted in an experiment room of this institute.

Table 1. Variables and levels in the objects

Variable	Height	Area	Weight	Texture
Level 1	10 cm	10 cm × 10 cm	Light	Slippery
Level 2	30 cm	20 cm × 20 cm	Heavy	Nonslippery
Level 3	50 cm	30 cm × 30 cm	–	Soft

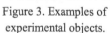
Figure 3. Examples of experimental objects.

Figure 4. Actual experimental situations.

MODELING OF INFANT CLIMBING BEHAVIOR

For the object in which an infant expressed an interest, attributes of the object, characteristics of the infant, and appearance of climbing behavior were recorded. The appearance of climbing behavior is defined as "1" if the infant climbed the object and "0" if no climbing occurred. The recorded characteristics of each infant are sex, age in months, height, weight, and a developmental stage. In the experiment, the developmental stages were three types: "Stand up holding onto something," "Walk well," and "Climb stairs easily." Climbing behavior exhibited on an object fallen sideways was not counted, because the height and width of the object changed. In total, 176 data sets were collected.

From the collected behavioral data, the probabilistic causal relationship among factors was modeled using the Bayesian Network Model (Motomura, 2001). The Bayesian Network Model is a probabilistic model that models the structure of a causal relationship among factors. In this model, the dependence among stochastic variables is shown graphically, and the quantitative relationships among the variables are decided by the conditional probability. Stochastic variables are shown as nodes and the dependence among the variables is shown by arrows. If the values of some variables are decided, the probability distributions of variables not observed can be inferred. The result of the modeling is shown in Figure 5. From the result of the modeling, it was determined that infant climbing behavior was strongly related to height of objects, area of objects, and height of the infants.

In the model shown in Figure 5, the probability of climbing behavior was able to be inferred by merely determining the values of three variables directly connected to "Climbing Behavior": "Height" of objects, "Area" of objects, and "Height" of infants. For instance, when "90 cm or more" was selected for the category of "Height" of infants and the categories of "Height" and "Area" of object changed, the probability of climbing behavior was inferred, as shown in Figure 6. From the results shown in this figure, for example, when the category of "Area" was changed from "30 cm" to "20 cm," the probability of "Climbing Behavior" decreased over 60%. Using the constructed model, infant climbing behaviors could be quantitatively predicted from object attributes and be controlled by changing these attributes.

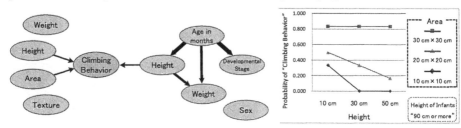

Figure 5. Infant climbing behavior model.   Figure 6. Inferred probability of
                                                    "Climbing Behavior"

MODELING OF CHILD INJURY MECHANISM

In this section, an injury model which can compute the severity of injury from the injury situation was constructed based on large scale injury data to compute risk of an injurious incident scenario based on evidence. In this paper, a fall injury model which can compute the severity of the injury when an infant fell from the height which he climbed was constructed, using the injurious incident scenario: an infant climbs a product and falls from it.

605 fall injury cases of infants (3 years or younger) which contain the information of the height the infants fell were used for constructing the fall injury model. These injury cases were compiled from 4,238 injury data collected by the injury surveillance system. The probabilistic causal relationships among six factors were modeled based on the fall injury data using Bayesian Network: infant's age, fall height, types of surface materials on which an infant fell, types of injury, injured body parts, and types of treatment after fall injury. All of the variables of these factors were set to binary variables (0 or 1). The result of the modeling is shown in

Figure 7. For example, it can be understood as follows. When the fall height was under 60 cm, injuries tend not to become serious, which require no treatment. When 6 months or younger infants fell from the height of 150 cm or more, injuries tend to become serious such as intracranial injury that requires hospitalization. This fall injury model enables us to infer what type of injuries that requires how much medical treatment occurs on what part of body from the information such as infant's age, fall height, and types of surface materials on which an infant fell.

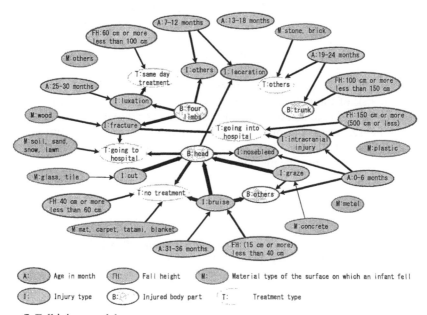

Figure 7. Fall injury model.

CONCLUSIONS

This study proposed a method of evidence-based risk assessment of consumer products by applying text mining technology and probabilistic modeling technology to large scale child injury data and child behavior data to prevent childhood injury by product modification. As basic methods for realizing the proposed method, this paper described text mining of child injury data to learn types of child's potential behaviors related to a product and constructing of a child behavior model and an injury model to compute the risk. As the behavior model and the injury model, the infant climbing behavior model and the infant fall injury model were constructed, using the injurious incident scenario: an infant climbs a product and falls from it.

In the text mining of injury data, types of potential behaviors related to a product were extracted, and the retrieval system of the relation information among types of products, types of behaviors, types of injuries, and child characteristics was developed. The retrieval system helps product designers learn types of potential behaviors related to a product and types of injuries due to the product use and create injurious incident scenarios from the time of product use to injury occurrence.

In constructing of the infant climbing behavior model, the probability of occurrence of infant climbing behavior on an object was able to be computed from the object attributes such as the height and the area. Further, the probability was able to be controlled by changing the attributes which are design values of a product. In constructing of infant fall injury model, the severity of the injury such as injury types, injured body parts and treatment types was able to be computed from the information of the situations at the time of the fall injury such as infant's age, fall height and material type of the surface on which an infant fell. By the combination of the infant climbing behavior model (see in Figure 5) and the infant fall injury model (see in Figure 7), the probability that an infant climbs a product and the severity of the injury which the infant experiences when he falls from the height he climbed were able to be computed from the information of the product and the infant. The risk of the injurious incident scenario, an infant climbs a product and falls from it, was able to be estimated based on evidence through the two models.

In the future, we plan to offer product designers and product quality managers of product makers the developed system and models, to keep receiving the feedbacks, and to improve the method.

REFERENCES

Japan Ministry of Health, Labour and Welfare (2008), "Population Survey Report" Website: http://www.mhlw.go.jp/toukei/index.html.

Tsuboi, T., Nishida, Y., Mochimaru, Y., Kouchi, M., Yamanaka, T. and Mizoguchi, H. (2008), "Injury Statistics by Bodygraphic Injury Surveillance System", Proceedings of 26th Annual Conference of the Robotics Society of Japan, pp.3G1-02(1)-(4).

WHO (World Health Organization) (2008), "World report on child injury prevention".

ISO 14121 (1999), "Safety of machinery – Principles of risk assessment"

Malorny, B., Loefstroem, C., Wagner, M., Kraemer, N., Hoorfar, J. (2008), "Enumeration of Salmonella Bacteria in Food and Feed Samples by Real-Time PCR for Quantitative Microbial Risk Assessment", Appl Environ Microbiol, Vol.74, No.5, pp.1299-1304.

Wurbs, A., Kersebaum, K. C., Merz, C. (2000), "Quantification of Leached Pollutants into the Groundwater Caused by Agricultural Land Use-Model-Based Scenario Studies as a Method for Quantitative Risk Assessment of Groundwater Pollution", Integr Watershed Manag Glob Ecosyst, pp.239-250.

Motomura, Y., Nishida, Y., Yamanaka, T., Kitamura, K., Kaneko, A., Shibata, Y. and Mizoguchi, H. (2006), "Injury Surveillance System for Preventing Children's Injury that Circulates Reusable Knowledge", Proceedings of the Institute of Statistical Mathematics, Vol.54, No.2, pp.299-314.

Motomura, Y. (2001), "BAYONET, Bayesian Network on Neural Network", Foundation of Real-World Intelligence, pp.28-37.

CHAPTER 64

Childhood Injury Modeling Based on Injury Data Collected using Bodygraphic Injury Surveillance System

Koji Kitamura, Yoshifumi Nishida, Yoichi Motomura

Digital Human Research Center
National Institute of Advanced Industrial Science and Technology (AIST)
Tokyo, Japan

ABSTRACT

This paper proposes a new technology, "a bodygraphic injury surveillance system (BISS)" that not only accumulates accident situation data but also represents injury data based on a human body coordinate system in a standardized and multilayered way. Standardized and multilayered representation of injury enables accumulation, retrieval, sharing, statistical analysis, and modeling causalities of injury across different fields such as medicine, engineering, and industry. To confirm the effectiveness of the developed system, the authors collected 3,685 children's injury data in cooperation with a hospital. As new analyses based on the developed BISS, this paper shows bodygraphically statistical analysis and childhood injury modeling using the developed BISS and Bayesian network technology.

Keywords: Surveillance System, Bodygraphic, Injury Modeling, Injury Science, Information Retrieval System, Geographical Information System

INTRODUCTION

Representing medical information of the human body in a standardized and structured form is important for advancing the science and technology of humans. However, we still do not have a good tool for this purpose. The authors have proposed a useful technology, "a bodygraphic information system" (BIS), that represents human body information by associating the information with a human body coordinate system (Tsuboi *et al.*, 2008). The concept of BIS is similar to the geographical information system (GIS) (Heywood *et al.*, 2006). BIS enables accumulation, retrieval, sharing, statistical analysis, and integration of human body information across different fields such as medicine, engineering, and industry. This paper describes the prototype of BIS integrated injury surveillance for childhood injury prevention (World Health Organization, 2006; Hyder, 2006; Peden *et al.*, 2008) and reports new statistical analyses of childhood injuries based on a large amount of injury data collected in a hospital.

BODYGRAPHIC INJURY SURVEILLANCE SYSTEM

CONFIGURATION OF BODYGRAPHIC INJURY SURVEILLANCE SYSTEM

Figure 1 shows the system configuration of the developed system in which a bodygraphic information system is applied for childhood injury surveillance. The bodygraphic injury surveillance system (BISS) consists of an injury database system, a web server, and a client software. The client part of BISS has four functions: input, retrieval and analysis, database construction, and visualization. In BISS, we can input information by typing in text data or outlining on a three-dimensional human body model with a computer mouse, and then the input data is digitized into a raster model and relevant information can be obtained. All input data is stored in a database for retrieval, analysis, and visualization according to need.

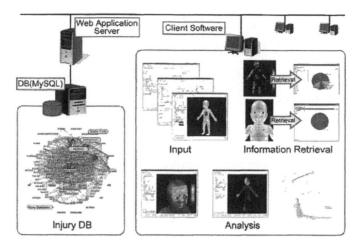

Figure 1. Bodygraphic Information System

REPRESENTATION OF INJURY DATA ON BISS

Injury information has three classifications of information: location, shape, and attribute. In BISS, this information is raster data that is structured as multi-layers based on the location on the human body, as shown in Figure 2. In this paper, raster data indicates an injury as a set of cells sectioned on the body surface. Each cell has a coordinate value in the human body coordinate system. Figure 2 shows input injury data converted to raster data. The injury data is overlaid on cells of the body surface and converted into a cell map. Each cell has information on the type of injury and the degree of severity along with the coordinates in the human body coordinate system.

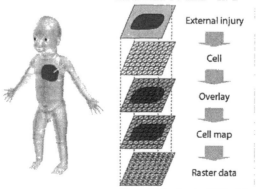

Figure 2. Representation Model by Raster Data

FUNDAMENTAL FUNCTIONS OF BODYGRAPHIC INJURY SURVEILLANCE SYSTEM

INPUT AND OUTPUT

Using the BISS, shape data of an external injury is drawn on the three-dimensional human body model with a computer mouse and other text data is typed in with a keyboard. The input injury data is converted to raster data and stored in a database system. Because the standard human body model is used, injury data is normalized.

INFORMATION RETRIEVAL FUNCTION

The BIS can retrieve target injury data by a spatial query and a text query and show the visual result. For example, by giving "1 year old" and "scald" as text queries, the BISS retrieves all scald data of 1-year-old children. By drawing a spatial query with a computer mouse, the BISS retrieves all injury data of the target body part.

OVERLAY ANALYSIS FUNCTION

Since injury data is normalized as raster data on the standard body model, we can conduct an overlay analysis. Figure 3 shows the result of injury frequency calculations. The BIS can obtain injury frequencies by summing each cell value.

AREA CALCULATION FUNCTION

According to an investigation conducted by our research group, we found that the injury area is important to judge the degree of severity of some injuries, such as burns or scalds. Therefore, we implemented a function to calculate the injury area and the ratio of injury area to the whole body surface.

ICD-10 CODE CONVERSION FUNCTION

The International Statistical Classification of Diseases and Related Health Problems (ICD-10), established by the World Health Organization (World Health Organization, 1992), is an international statistical standard of the cause of death and disease. ICD-10 codes include items concerning injuries due to external causes; these items are defined by both the type of injury and the injured body part. ICD-10 codes are widely used at medical institutions worldwide. So, we implemented a BIS function that converts input injury information into ICD-10 codes. Figure 4 shows the BISS layer defining the body parts of the ICD-10 codes. In the BIS, body parts are classified into ten areas, including the head, neck, thorax, and so on. By superimposing an input injury layer and ICD-10 code layer, input injury data is converted to the corresponding ICD-10 code.

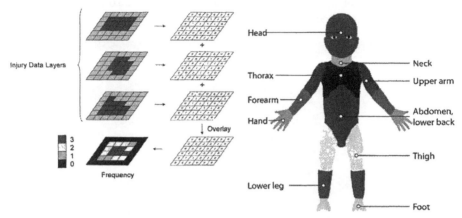

Figure 3. Overlay Analysis

Figure 4. Definition of Body Region by ICD-10

NEW STATISTICAL ANALYSIS AND INJURY MODELING BY BISS

In order to show the effectiveness of the BISS, we next present three kinds of statistical analyses. We collected 3,685 cases of injuries in children of ages 0 to 18 years in cooperation with a hospital (National Center for Child Health and Development) since 2006. In the following subsection, we describe statistical analyses using the collected data.

BODYGRAPHY OF INJURY FREQUENCY

As one example of data visualization using the BISS, Figures 5 shows injury frequencies visualized by the BISS. Since each instance of injury data is expressed in a normalized and structured form in the BISS, we can superimpose data and count frequencies by summing the data. In the figure, a red color indicates the area with the highest frequency of external injury. It is confirmed that the forehead is a significant highest frequency of external injury. Although the fact that the head part is the most frequently injured is well known in the field of child injury prevention, the BISS allows us to conduct much more extensive analyses of injured body parts. As shown in Figure 6, we can retrieve and visualize only necessary data on demand. Part A of the figure visualizes the bodygraphic frequency of scald injury, Part B is the frequency of injury due to fall by children aged from 1 to 2 year-old, Part C is the frequency of injury due to a slide type of playground equipments, and Part D is the frequency of injury due to all kinds of playground equipments. These analyses would be useful for not only initiating injury science but also designing effective protection, such as helmets.

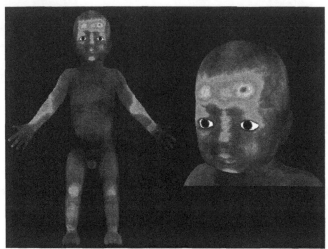

Figure 5. Injury Frequency

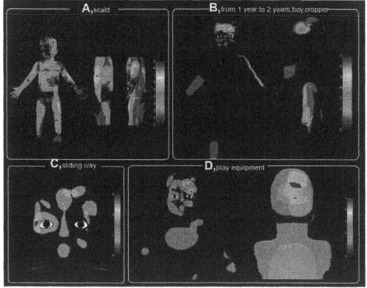

Figure 6. Examples of Injury Retrieval and Frequency Visualization

STATISTICAL TEST CONCERNING SYMMETRY OF INJURY PART

We can conduct another statistical analysis using the BISS. Here, we describe a statistical test with respect to the symmetry of injuries on a human body. To evaluate whether injury frequency is symmetric with respect to the center of the body, we conducted a chi-square test of injury data. Table 1 shows p-values of the chi-square test according to body part. This figure shows that the head and knee are significantly asymmetric at the 5% level. In this way, the BISS allows us to conduct a statistical analysis.

Table 1 Frequency and p-Value of Injury of Each Body Part

Body part	Righit	Left	Total	P–value
Head	693	792	1485	0.010
Neck	7	7	14	1.000
Thorax	18	12	30	0.273
Abdomen	20	28	48	0.248
Upper arm	31	27	58	0.599
Forearm	104	117	221	0.382
Hand	138	113	251	0.115
Upper thigh	55	32	87	0.014
Lower thigh	40	30	70	0.232
Foot	46	56	102	0.322
Total	1152	1214	2366	0.202

EXTERNAL STATISTICS: RELATIONSHIP BETWEEN INJURY AREA AND FREQUENCY

As another possible analysis, we calculate an injury area using the BISS and analysis in terms of statistics of extremes (Gutenberg, B. and Richter, C.F., 2004). The Weibull plot is widely used in various research fields such as damage kinetics, earthquake seismology, and finance since it is known that these phenomena on natural science and societal follow a Weibull distribution. Figure 7 shows the Weibull plot of injury area. Here, median rank is used for a cumulative distribution. As shown in Fig. 7, the injuries of which area is over 60 mm^2 follows a Weibull distribution. Figure 8 shows that the injury area and frequency have a power law relationship, comparable to the Gutenberg-Richter law (Gutenberg, B. and Richter, C.F., 1954) that expresses the relationship between the magnitude and total number of earthquake.

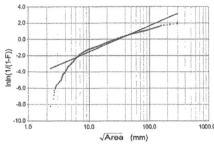

Figure 7. Weibull Plot of Injury Area   Figure 8. Power Law of Injury Area

INJURY MODELING BY BAYESIAN NETWORK MODELING PROCESS

MODELING PROCESS

Clarifying the relation among design parameters and injured body regions is useful for improving consumer products. For that purpose, we conducted injury modeling to find a causality model between the attributes of accidents and injured body parts by using the Bayesian Network. We show the procedure of injury modeling as follows.

1. The body region is automatically divided into some regions using the k-means clustering method. By giving initial cluster randomly, repeating the k-means clustering, and finding the optimal clustering, we can obtain stable division. Here we divided 20 regions as shown in Figure 9. Interestingly, automatic division indicates symmetry property of injury.

2. The user generates frequency counts in each target region defined in Step 1 by using the BISS function. If an injury covers several regions, the injury is counted in each region.

3. The user generates a cross-tabulation table from the frequency in each region and the attributes of an accident.

4. The user constructs a causal structural model from the cross-tabulation table by using the Bayesian Network. In this structural learning, Akaike's Information Criterion is used.

5. By inputting these variables of accidents, the user can estimate the injured body parts.

Figure 10 is an example of injury modeling. This figure shows the probabilistic causal structural model of the relation between body parts and accident type, such as falling and scald. Figure 11 shows an example of the results of inference by the constructed model given the conditions of A (age=0 year-old, place=dining room, object=chair), B (age=1 year-old, place=door), C (place=living room, object=liquid with high temperature such as coffee) and D (object=cutting tool).

Figure 9. Body Division by K-Means Clustering

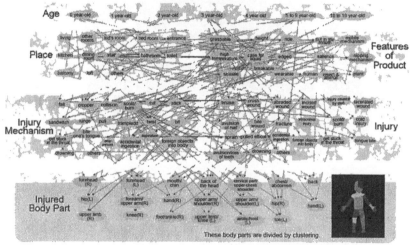

Figure 10. Probabilistic Causal Structure Model of Injury

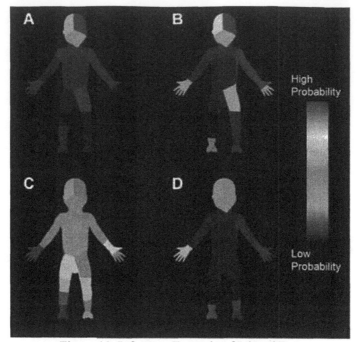

Figure 11. Inference Example of Injured Part

MODEL EVALUATION

To evaluate the performance of the constructed model, we conducted 2-fold cross-validation. As a criterion for precision of inference, we utilized a hitting ratio which is calculated as follows. First, we infer probabilities of what we want to infer by the constructed network, adopt the probability of the event corresponding to the true event as the hitting ratio, and calculate the average of the hitting ratio as a macro index. Table 2 compares the macro hitting ratio between two models when the kinds of accident, the kinds of injury, and injured body part. Model 1 indicates the result in the case of a uniform distribution, namely, all events occurs with the same probability and Model 2 indicates the result of inference by the constructed model. This table shows that the inference of the constructed model is much better than that of the uniform distribution model.

Table 2: Model Evaluation in Terms of Hitting Ratio (%); Model 1 indicates a random model and Model 2 indicates the developed model.

	Accident	Injury	Injured Part
Model 1	4.76	4.55	5.56
Model 2	41.22	41.99	38.07

As stated above, BISS allows us to construct the causal structural model that explains causality with respect to age, location where accident happens, time, and the consumer product related to the accident. If we can describe the consumer products by using control parameters that a product designer can change and then find the relationship among the parameters and injury, we can obtain new knowledge for improving consumer products. A method for representing consumer products as control parameters is one important subject of future work.

CONCLUSION

This paper proposed the concept of a bodygraphic injury surveillance system (BISS) that represents human body information in a normalized and structured form by associating the information with a human body coordinate system. As an example, we described an application of BISS by using child injury data. We presented data visualization and analyses using BISS. It is the first system to express detailed positions and shapes of external injury data and to analyze the frequency of injuries statistically based on collected data. Thus, the BISS enables us to collect and manage detailed information of external injuries that are difficult to do with conventional methods. This BISS application will open the way for injury science. Disseminating the BIS technology for injury prevention is important future work.

REFERENCES

Gutenberg, B. and Richter, C.F. (2004), *"Statistics of Extremes."* Dover Pubns.

Gutenberg, B. and Richter, C.F. (1954), *"Seismicity of the Earth and Associated Phenomena."* 2nd ed. Princeton University Press.

Heywood, I., Cornelius, S., Carver, S. (2006), *"Geographical Information Systems."* 3rd ed., Pearson Education Limited: Harlow, England.

Hyder, A. (2006), *"Childhood injuries: Defining a global agenda for research and action."* Journal of Injury and Violence Prevention, 4(1), 87.95.

Peden, M., Peden, M., Oyegbite, K., Ozanne-Smith, J., Adnan A Hyder, C. B., Rahman, A. F., et al. (2008), "World report on child injury prevention. " Switzerland: WHO Press.

Tsuboi, T., Nishida, Y., Mochimaru, M., Kouchi, M., and Mizoguchi, H. (2008), *"Bodygraphic information system."* In Proc. of the 9th world conference on injury prevention and safety promotion, p. 79.

World Health Organization. (1992), *"Icd-10: The Icd-10 Classification of Mental and Behavioural Disorders: Clinical Descriptions and Diagnostic Guidelines."* WHO Press, Switzerland.

World Health Organization. (2006), *"Child and adolescent injury prevention: A who plan of action 2006-2015."* Switzerland: WHO Press.

Chapter 65

New Capabilities for Vision-Based Posture Prediction

Lindsey Knake, Anith Mathai, Tim Marler,
Kimberly Farrell, Ross Johnson, Karim Abdel-Malek

111 Engineering Research Facility
University of Iowa
Iowa City, IA

ABSTRACT

Although field of view (FOV) is a commonly used evaluation parameter with digital human models, minimal research has involved modeling how eye motion (relative to the head and body) affects the FOV and posture of a digital human striving to see a particular target. Few models incorporate independent eye movement and the effects of obstacles, with the ability to predict human posture realistically. This work presents two new and critical components for simulating how vision affects human posture: 1) inclusion of eye movement and 2) visual obstacle avoidance. This work is conducted using Santos™, a real-time predictive physics-based virtual human with a high number of degrees-of-freedom. With optimization-based posture prediction, joint angles serve as design variables used to minimize various human performance measures that provide objective functions, subject to constraints that represent biomechanical limitations and task characteristics. Vision-based objective functions and constraints are developed and easily implemented in order to accurately predict postures. First, two new degrees of freedom were added to the Santos™ model, representing vertical and horizontal movement of the eyes. Then, functions for eye movement relative to the head and body were developed based on experimental data. The new vision-based objective function expanded on the current vision model by incorporating these new functions. Additionally, a vision-based obstacle avoidance constraint was added in order to predict postures that incorporate the tendency to look around obstacles that may be in one's line of site. Although vision alone does not govern one's posture, when combined with other performance measures, more realistic predicted postures incorporating vision were obtained. Initial subjective validation suggests the predicted postures are accurate and realistic. The consequent capabilities have proven extremely useful for ergonomic studies and analyses of automotive cab scenarios.

Keywords: Optimization, Posture Prediction, Vision, Digital Human Modeling

INTRODUCTION

The field of digital human modeling is constantly striving to provide biomechanically accurate solutions for engineering design and analysis, especially for ergonomic studies. This becomes especially apparent during ergonomic studies of automotive cabs. In order to simulate human behavior realistically, an accurate vision model is essential. However, modeling human vision and the effects it has on performance can be complex. Consequently, this work concentrates on the motion of the body, neck, and eyes, in an effort to create a clear line of sight. Nonetheless, considering such factors requires an advanced human model that includes not just head and neck motion but eye motion as well, all coupled with the ability to predict human behavior.

This paper presents a new vision model, in the context of optimization-based posture prediction. This model combines real-time posture prediction, eye movement, and the ability to look around, or to the side of objects that may interfere with one's vision.

Although much work has been completed with studying vision, most current vision models are data based. There is little work that provides a mathematical model for predicting human posture, while considering the need to see a specified target. Many studies and experiments concentrate on the coordination between the eyes and head while gazing. These studies focus on incorporating the results into robot vision (Maini, 2006; Guitton and Volle, 1987). However, there is minimal research incorporating head-eye movement ratios in human vision models. Some authors, however, have investigated eye range-of-motion (ROM) (Guitton and Volle, 1987; Huaman and Sharpe, 1993). Guitton and Volle (1987) concluded that there are neural impulses during gaze shifts that prevent the eyes from reaching these limits at all times suggesting a coordination between the head and eye movements to obtain the visual sight of the target.

Kim (2007) provides one of the first works that includes vision in a predictive virtual human. He explores the modeling of head and eye coordination and finds that vertical and horizontal head-eye movement ratios are non-linear functions dependent on the location of the target. Non-linear equations are developed from existing data in order to predict the angle of three different head degrees of freedom: horizontal rotation, vertical flexion/extension, and cyclotorsion. The calculated head and neck joint angles were contrasted with the inverse kinematics algorithms built in Jack™ for similar target location, and more natural appearances were reported for head and neck angles.

Although much work has been completed regarding predictive capabilities for virtual humans, there are few, if any computational models that incorporate eye movement in posture prediction. Marler *et al* (2009) provides extensive reviews of posture prediction capabilities, and although there are a variety of performance measures incorporating joint angles, none actually include DOFs for eyes. Nonetheless, some authors do incorporate the tendency to try to see targets, by considering body and neck motion with the eyeballs essentially looking straight ahead. In this vein, Marler *et al* (2006) describe two vision performance measures (objective functions within the context of optimization-based posture prediction): visual displacement and visual acuity. These performance measures are based solely on head position and do not include eye movement. Smith *et al* (2008) analyze posture prediction using these head-based vision performance measures and joint displacement, and discover that vision alone does not govern posture prediction. Consequently, the authors study the use of vision-based performance measures combined with other objective functions such as joint displacement, and implemented as constraints.

Most current digital human models do not include eye movement and depend primarily on head orientation to predict postures that incorporate vision. Some models include vision cones stemming from the head showing primary and periphery vision zones (Hanson, 1999).

MODEL DEVELOPMENT

The work presented in this paper uses the Santos™ human model as a platform for further development (Abdel-Malek *et al* 2006; Marler *et al*, 2008). The underlying skeletal structure for Santos™ is modeled as a series of links with each pair of links connected by one or more revolute joints. There is one joint angle for each DOF. The relationship between the joint angles and the position of points on the series of links (or on the actual avatar) is defined using the Denavit-Hartenberg (DH)-method (Denavit and Hartenberg, 1955).

Postures are predicted using an optimization-based approach detailed by Farrell et al (2005). Joint angles are the design variables, which are incorporated in various objective functions and constraints, formulated as follows:

Find: $\mathbf{q} \in R^{DOF}$

To minimize: $f(\mathbf{q})$

Subject to: $\text{Distance} = \left\| \mathbf{x}(\mathbf{q})^{\text{end-effector}} - \mathbf{x}^{\text{target point}} \right\| \le \varepsilon$

$q_i^L \le q_i \le q_i^U; \ i = 1, 2, \text{K}, DOF$

(1)

Where q is a vector of joint angles, x is the position of an end-effector or point on the avatar, and ε is a small positive number that approximates zero and *DOF* is the total number of degrees of freedom. $f(q)$ can be one of many performance measures. The primary constraint, called the *distance* constraint, requires the end-effector to contact a specified target point. q_i^U represents the upper limit, and q_i^L represents the lower limit. These limits are derived from anthropometric data.

The new eye displacement performance measure expands on the visual displacement performance measure described by Marler et al (2006), and is developed in the context of the basic optimization-based posture prediction problem formulated in (1). Visual displacement ensures that an eye vector, which emanates from the eye perpendicular to the face of Santos™ successfully, intersects the target. It essentially minimizes the absolute value of γ_{Eye_Tar} which is defined as the angle between the eye vector and the target vector (See FIGURE 2). Details of this objective function are provided by Marler et al (2006).

The new eye displacement performance measure uses visual displacement as a foundation and incorporates two new functions representing the vertical and horizontal eye displacement. In general, this performance measure ensures that literature based head-eye movement ratios are satisfied while still utilizing the old vision model to ensure that the eye vector intersects the target.

The performance measure required two new degrees of freedom to be added to Santos's™ skeleton. New axes of rotation representing vertical and horizontal DOFs for the eyes were implemented, and literature-based (Guitton and M.Volle 1987) (Huaman and Sharpe 1993) joint limits were added. The new eye displacement performance measure calculates γ_{Eye_Tar} using the new eye vector (FIGURE 1) that has a base between Santos's™ eyes and depends on the orientation of the eyes instead of solely the head orientation as with the previous eye vector.

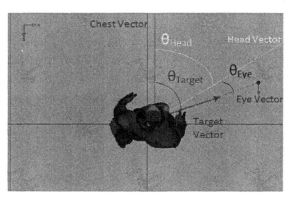

FIGURE 1 Angle and vector definitions.

All angles in FIGURE 1 are dependent on the current posture of Santos™ and thus are functions of q. The chest vector is defined as the orientation of Santos's™ upper-most spine joint and is used to represent the midsagittal plane. The head and eye vectors are defined by the orientations of the head and eyes respectively. The target vector represents the position of the target. θ_{Head} is the horizontal angle between the chest and head vectors, and θ_{Eye} is the horizontal angle between the head and eye vectors. θ_{Target} is the total horizontal target displacement from the midsagittal plane chest vector to the target vector. Using these angles and interpreting the head contribution ratio data (Kim, 2007), which depends on the θ_{Target}, eye contribution ratios were calculated and are shown in **Table 1**. θ is the horizontal component of the angles and ϕ is the vertical component of the angles. ϕ_{Target} is calculated from the horizontal plane at Santos's™ eye level (eye plane) to the target and represents vertical target displacement.

Table 1 Percentage of Eye Contribution Depending on Target Displacement

Horizontal	θ_{Eye} %	Vertical	ϕ_{Eye} %
$0° \leq \theta_{Target} \leq 10°$	100	$0° \leq \phi_{Target} \leq 19°$	84
$\theta_{Target} \geq 10°$	32	$\phi_{Target} \geq 19°$	29

From these head-eye ratios, distinct non-linear horizontal and vertical functions were developed. The literature values for θ_{Head} and ϕ_{Head} depending on θ_{Target} and ϕ_{Target} are represented by the red lines in FIGURE 2 (Kim, 2007). The y-axis represents the head angle (radians), and the x-axis represents the total target displacement (radians).

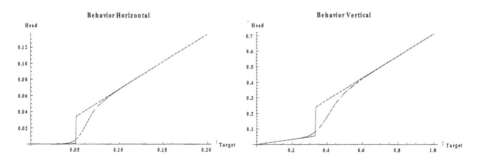

FIGURE 2 Continuous functions developed (blue line) to represent the literature model of the head angles (red line).

The literature based functions are discontinuous; therefore, two new continuous function approximations were developed, represented by the smooth blue lines in FIGURE 2. Both of these smooth functions can be represented as follows with different coefficients α, μ, β, ω, and η:

$$(2)$$

$$\theta_{Head-Desired} = f(\theta_{Target}(q)) = \alpha \frac{\theta_{Target}(q)}{(\mu + \theta_{Target}(q))^{\beta} + \eta} + \omega\theta_{Target}(q)$$

$\theta_{Head-Desired}$ is calculated using (2) with α= -4, μ=20, β=10, ω= 0.68, and η=6. $\phi_{Head-Desired}$ is also calculated using (2) but with ϕ_{Target} substituted for θ_{Target} and with α= -3.3, μ= 3, β= 10, ω= 0.71, and η= 6. $\theta_{Head-Desired}$ and $\phi_{Head-Desired}$ are the desired head angles and are used to find the $\theta_{Eye-Desired}$ and $\phi_{Eye-Desired}$ by subtracting $\theta_{Head-Desired}$ and $\phi_{Head-Desired}$ from θ_{Target} and ϕ_{Target} respectively. Therefore, the two new components of the vision model minimize the difference between the $\theta_{Eye-Desired}$ and $\phi_{Eye-Desired}$ and the measured values of $\theta_{Eye}(q)$ and $\phi_{Eye}(q)$. The complete objective function that represents the relationship between the vertical and horizontal eye movement as well as the tendency for the eye vector to coincide with the target is given as follows:

$$f_{EyeDisp}(q) = \left[\theta_{Eye}(q) - (\theta_{Target}(q) - \theta_{Head-Desired})\right]^2 + \left[\phi_{Eye}(q) - (\phi_{Target}(q) - \phi_{Head-Desired})\right]^2 + \gamma_{Eye\_Tar}(q)^2 \quad (3)$$

The first term of (3) ensures that the horizontal eye angle equates to the literature-based horizontal eye angle; the second term ensures that the vertical eye angle equates to the literature-based vertical eye angle; and γ_{Eye_Tar} utilizes the previous vision displacement objective function to ensure that the eye vector intersects the target.

The vision objective function in (3) is used as the objective function in (1) and generates posture solutions that enable the avatar to look at the target while predicting postures. However, the avatar must also able to detect obstacles obstructing the eye vector. Towards this goal, an additional vision obstacle avoidance constraint has been developed and incorporated in the predictive vision model. This constraint leverages the work of Johnson et al (2009) and ensures that the eye vector does not pass through any object in the scene.

As part of the posture prediction process, all geometry in the virtual environment is represented with sphere-based surrogate geometry (Johnson, et al. 2009). The constraint ensures that the distance between the eye vector and the eye-to-sphere vector is greater than the radius of the obstacle sphere using:

$$|x - p(q)|^2 \sin^2 \theta - r^2 > 0 \quad (4)$$

where, x is a specific obstacle sphere with radius r, $p(q)$ is the position of the eyes, and θ is the angle between the eye vector and the eye-to-sphere vector. This constraint ensures that the distance between the eye vector and the eye-to-sphere vector (defined by term 1 in (4)) minus the radius is greater than zero.

The final formulation for the new vision model includes one obstacle avoidance constraint for each sphere representing surrogate geometry, and includes (3) as the objective function.

RESULTS

In this section, we demonstrate the advantages of the new vision model, with

634

respect both to incorporation of eyes and to consideration of obstacle avoidance. Two sets of basic tests were run, followed by a test in a practical setting. First, we verified that the data used in FIGURE 2 is represented in the final predicted postures. Secondly, we compare results using the previous vision model with results using the new model that now incorporates eyes. Finally, we subjectively verify the accuracy of the predicted postures in a practical cab setting. Note that with these tests, a weighted sum of the eye displacement and joint displacement (Marler, 2009) was used as the objective functions, with weights of 0.99 and 0.01 respectively. Thus, the vision performance measure was isolated, but the use of joint displacement resulted in more realistic results for limbs.

FIGURE 3 Isolating the horizontal eye movement with a target displaced 80° (a. and b. Target located at (-714, 765, -96) in mm) and 12° (c. and d. Target located at (-37, 765, -355) in mm) from the midsagittal plane. a. and c. use the old vision model and b. and d. use the new vision model.

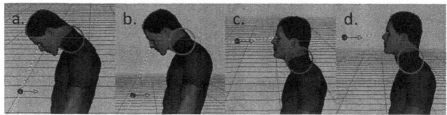

FIGURE 4 Isolating the vertical eye movement with a target displaced 71° (a., b. Target (0, 279, -317) mm) and 14° (c., d. Target (0, 715, -383) mm) from the horizontal eye plane. a., c. use the old vision model and b., d. use the new model.

Table 2 Literature Value Vadidation (*Represents Literature Values)

Figure	γ_{Target} (°)	γ_{Eye}%	γ_{Eye}%*
3b	80	32	32
3d	12	91	100
4b	71	29	29
4d	14	84	84

As Table 2 demonstrates, the predicted percentage of eye angle contribution represents closely the underlying data. One exception is shown in FIGURE 3d, where for horizontal eye displacement values from 0° to 10°, the percentage of eye movement should be 100%. This discrepancy is explained by the inherent approximation made to the original curve to remove discontinuities, as shown in FIGURE 2. However, the discrepancy translates to a difference of only one degree.

When comparing the previous vision model to the new model, FIGURES 3a and 3b provide an example of the added functionality that the new model provides. With the spine frozen and using the old vision displacement model, there is no feasible solution, and this is unrealistic. With the new eye displacement objective function (FIGURE 3b), Santos™ can easily see the target by moving his eyes.

Using only a small horizontal target displacement with the new objective function (FIGURE 3d) results in postures that are not as realistic as the previous visual displacement model (FIGURE 3c). Santos™ moves his head farther to the right and uses his eyes to look back to the left to see the target. This results from the formulation only ensuring that $\theta_{Eye\text{-}Desired}$ is equated to $\theta_{Eye}(q)$, but not ensuring that the $\theta_{Head\text{-}Desired}$ is equated to $\theta_{Head}(q)$. In most cases $\theta_{Eye\text{-}Desired}$ being equal to $\theta_{Eye}(q)$ produces solutions with $\theta_{Head}(q)$ being approximately equal to $\theta_{Head\text{-}Desired}$ as well. However, in cases with small target displacements and small $\theta_{Head\text{-}Desired}$, excess head movement where $\theta_{Head}(q)$ is greater than $\theta_{Head\text{-}Desired}$ can also result in postures with $\theta_{Eye\text{-}Desired}$ equal to $\theta_{Eye}(q)$. Since vision alone does not typically govern human posture, adding a greater weight to the joint displacement objective function may solve this problem by minimizing excess head movement.

In FIGURE 4b isolating a large vertical target displacement, using the new eye displacement objective function results in a more realistic posture that causes less movement and strain on the head and neck of Santos™. In FIGURE 4d, the resulting solutions demonstrate that with small vertical displacements, there are minimal differences between postures generated by the visual displacement or eye displacement models.

636

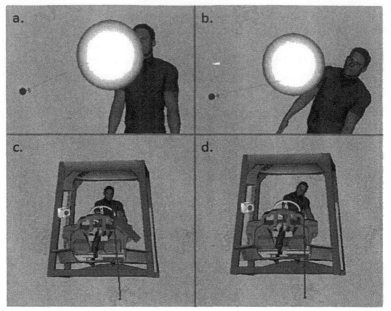

FIGURE 5 Implementation of the obstacle collision avoidance constraint with real world applications. a., c (collision shown by red arrow) use the old vision model and b., d. use the new model.

Following the above-mentioned basic tests, the new model was evaluated in the context of a cab setting. Clearly, as shown in FIGURE 6, incorporating eye movement results in more realistic postures. FIGURE 5 shows the difference in postures when using the vision obstacle avoidance constraints. Santos™ looks around obstacles that are obstructing his line of sight.

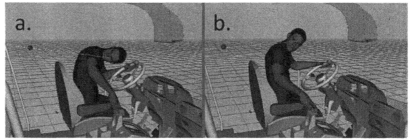

FIGURE 6 Real world application using a. the old vision model b. the new vision model

CONCLUSION

This paper demonstrates a new predictive vision model that includes eye movement and the ability to look around objects that obstruct one's view. Incorporating eye movement involved the deceptively complex task of modeling how one naturally distributes motion between the body, the neck, and the eyes. The results were tested and validated subjectively. Although we find that the proposed model is a substantial improvement over previous results, this improvement is less distinct for targets requiring minimal overall angular displacement. In a few rare cases, the results with the new model are not as realistic as those with the previous model. This detriment is attributed in part to the necessity for combing any vision objective-function with another function that represents overall body posture, using multi-objective optimization (Marler et al, 2009). That is, vision alone cannot be used as a complete posture-prediction model. It is, however, a critical component.

Although the overall approach to simulating human posture is predictive, independent of predetermined data, the model for incorporating eyes is in fact data based. Despite the initial success of the proposed model, future work will entail further investigation of mathematical models complementing the data-based aspect. To this end, the fundamental formulation for posture prediction, on which this work is based, allows us to study what governs posture (and vision) by experimenting with various performance measures and constraints. It allows us to model and test various hypotheses. In addition, various means of controlling eye and head orientation will be investigated. Initial studies suggest that combing vision with a performance measure like joint displacement have been successful in this regard. Focal views or vision cones of different types of vision tasks such as gazing, reading, or peripheral sight could be incorporated into the model. Such work would also include adding the third degree of freedom of the eyes: rotational torsion. Finally, although subjective tests and mathematical comparisons to literature values have validated initial results, a more thorough validation study including motion capture and an eye-tracking device will be conducted.

ACKNOWLEDGEMENTS

The authors gratefully acknowledge project funding from Caterpillar Inc.

REFERENCES

Abdel-Malek, K., Yang, J., Marler, T., Beck, S., Mathai, A., Zhou, X., Patrick, A., and Arora, J. (2006), "Towards a New Generation of Virtual Humans," *International Journal of Human Factors Modelling and Simulation*, 1 (1), 2-39.

Denavit, J., and R. S. Hartenberg. "A Kinematic Notation for Lower-pair Mechanisms Based on Matrices." *Journal of Applied Mechanics,* 77 (1955): 215-221.

Farrell, K., Marler, R. T., and Abdel-Malek, K. (2005), "Modeling Dual-Arm Coordination for Posture: An Optimization-Based Approach," *SAE 2005 Transactions Journal of Passenger Cars - Mechanical Systems*, 114-6, 2891, SAE paper number 2005-01-2686.

Guitton, D. and Volle, M. (1987). "Gaze Control in Humans: Eye-Head Coordination During Orienting Movements to Targets Within and Beyond the Oculomotor Range." *Journal of Neurophysiology*. 58, 427-459.

Hanson, Lars, and Akselsson, Roland. (1999). "ANNIE, a Tool for Integrating Ergonomics in the Design of Car Interiors." *SAE Souther Automotive Manufacturing Conference*, September, Birmingham, Alabama.

Huaman, Ana and Sharpe, James. (1993), "Vertical Saccades in Senescence." *Investigative Ophthalmology & Visual Science*, 34, 2588 – 2595.

Johnson, R., B. L. Smith, R. Penmatsa, T. Marler, and K. Abdel-Malek. (2009) "Real-Time Obstacle Avoidance for Posture Prediction." *3rd International Conference on Applied Human Factors and Ergonomics*. Miami: AHFE International.

Kim, Kyung Han, R. Brent Gillespie, Bernard J. Martin. (2007). "Head Movement Control in Visually Guided Tasks: Postural Goal and Optimality" *Computers in Biology and Medicine*. 37, 1009-1019.

Maini, Eliseo Stefano, Giancarlo Teti, Michele Rubino, Cecilia Laschi, Paolo. (2006) "Bio-inspired control of eye-head coordination in a roboitc anthropomorphic head." *International Conference on Biomedical Robotics and Biomechanics*. 549.

Marler, R. T., Arora, J. S., Yang, J., Kim, H. –J., and Abdel-Malek, K. (2009), "Use of Multi-objective Optimization for Digital Human Posture Prediction," *Engineering Optimization*, 41(10), 295-943.

Marler, T., Arora, J., Beck, S., Lu, J., Mathai, A., Patrick, A., Swan, C. (2008), "Computational Approaches in DHM," in *Handbook of Digital Human Modeling for Human Factors and Ergonomics*, Vincent G. Duffy, Ed., Taylor and Francis Press, London, England.

Marler, R. T., Farrell, K., Kim, J., Rahmatalla, S., Abdel-Malek, K. (2006). "Vision Performance Measures for Optimization-Based Posture Prediction." *SAE Digital Human Modeling for Design and Engineering Conference,* July, Lyon, France, Society of Automotive Engineers, Warrendale, Pennsylvania.

Smith, Brian Lewis, Marler, R. T., Abdel-Malek K. (2008). "Studying Visibility as a Constraint and as an Objective for Posture Prediction." *Digital Human Modeling for Design and Engineering Symposium,* June Pittsburgh, Pennsylvania.

CHAPTER 66

ABMiner: A Scalable Data Mining Framework to Support Human Performance Analysis

Kaizhi Tang[1], Xiong Liu[1], Yunshen Tang[1], Vikram Manikonda[1], John R. Buhrman[2], Huaining Cheng[2]

[1]Intelligent Automation, Inc.
15400 Calhoun Drive Suite 400
Rockville, MD 20855, USA

[2]Air Force Research Laboratory, Human Effectiveness Directorate
2800 Q Street, Wright-Patterson AFB, OH 45433

ABSTRACT

To address various requirements of data analysis in human performance and effectiveness such as Air Force Biodynamic Database, Navy Biodynamic Database, CAESAR database, etc., we developed a suite of flexible and scalable software tools, namely ABMiner (Agent Based data Miner). The goal of the program is to exploit large amount of bioscience sensor data, build abstract and general knowledge models from the data, and utilize those knowledge models for effective decision making. ABMiner consists of three key components. First, ABMiner provides a flexible data screening component that extracts different datasets using the similar language of domain experts, not requiring the knowledge of database and SQL. Second, ABMiner provides an optimization engine in the meta-learning level to exploit and search data mining models with best performance and efficiency among a wide range of data mining algorithms and their corresponding parameters. To speed up the optimization searching process, ABMiner utilize IAI's well-established product CybelePro for distributed computation. Finally, ABMiner provides a model base component that enables a well-trained data mining model together with its underlying data set and performance metrics to be wrapped up and accessible across various platforms.

Keywords: Data mining, meta-optimization, multi-agent systems, human effectiveness, biomechanics

INTRODUCTION

With the rapid progress of sensor, wireless, and computer technologies, enormous amounts of sensor-based data sets have been collected to study human effectiveness and performance. An imminent task is to develop highly scalable and effective data mining methods for in-depth analysis of such data. There are several databases in the public domain for human effectiveness and performance, such as AFRL Biodynamics Database (Buhrman et al, 2001), Civilian American and European Surface Anthropometry Resource (CAESAR) (Robinette et al, 2002). To systematically analyze collected data, we encounter the following challenges:

- *Data Preparation.* A data analysis task usually involves a large amount of data from various tables in the database. A domain expert will express various interests and perspectives to analyze data. It is a challenge to apply a flexible and general-purpose tool for customized data generation.
- *Model Discovery.* To discover interesting knowledge from extracted datasets, multiple data mining techniques and algorithms must be applied. How to utilize the merits of each technique and build high-quality data mining models is an interesting and challenging problem.
- *Model Deployment.* With the data mining models generated with different assumption from different aspects, how to compose these models to expose a model-driven decision making tool is also very challenging.

To overcome these challenges, we proposed and developed an innovative data mining tool to systematically analyze huge amounts of experiment and sensor data in biomechanics and other related areas. The tool, named ABMiner (where AB for Agent Based), enables users to customize their data sets and knowledge representation models based on the knowledge of domain experts.

ABMiner supports the full data mining cycle, including customized *data preparation*, large-scale *model discovery*, and convenient *model deployment*. For *data preparation*, we have developed a visual query designer which helps the user to visually build SQL queries and conveniently retrieve data from various relational databases. For *model discovery*, we have developed a meta-optimization module that enables the user to find optimal data mining models for a specific dataset. The optimization process usually involves evaluating a series of configurations of parameter values for many algorithms, which can be very time-consuming. Therefore, we have developed an agent-based computational engine to power the meta-optimization. For *model deployment*, we have developed a data mining model base, which is designed for exposing the trained knowledge models to public users via web services.

As a tool for biomechanics data analysis and knowledge discovery, ABMiner can be used in a wide range of human performance evaluation areas. A good example is guiding the design of biomechanics experiments. ABMiner will enable biomechanics engineers and scientist to conduct effective pre-studies before the physical experiments. The pre-study results will tell what type of biodynamical experiments will enrich biomechanics knowledge. Furthermore, data mining results will bring the following benefits: (a) saving the experimental cost through observing and querying the knowledge from data mining to validate the necessary of new experiments, (b) guiding the design of biodynamical experiment through a global view over the historical experiments provided by data mining, (c) extending the horizon of the validation of biodynamical simulations through the data mining models, and (d) conducting optimization design using data mining models without using simulation models.

OVERVIEW OF ABMINER

From the perspective of machine learning, data mining problems can be categorized into two main groups: *supervised learning* and *unsupervised learning*. Supervised learning can be further categorized into two subgroups, namely classification and numerical prediction. In a classification problem, the class label is discrete. While in a numerical prediction problem, the class label is continuous. Unsupervised learning does not identify a class label. A typical type of unsupervised learning is clustering analysis used to find groups of similar data records or objects.

As data mining have enjoyed great popularity and success in recent years, researchers realized the importance of a generally accepted framework. At the SIGKDD-2003 conference panel "Data Mining: The Next 10 Years" (Fayyad et al, 2003), U. Fayyad emphasizes in his position statement that "the biggest stumbling block from the scientific perspective is the lack of a fundamental theory or a clear and well-understood statement of problems and challenges". Yang and Wu (Yang and Wu, 2006) collected the opinions of a number of outstanding data mining researchers about the most challenging problems in data mining research (and presented them at ICDM-2005). Among the ten topics considered most important and worthy of further research, the development of a unifying theory is listed first. According to these interesting academic discussions, statistical analysis and data mining should be considered as a *process*, rather than individual tasks and approaches. Yang and Wu (Yang and Wu, 2006) also point out the need to support the composition of data mining operators, as well as the need to have a theory behind this.

The open source packages provide the flexibility to customize data mining solutions. The most well-known open source data mining package is Weka (Witten and Frank, 2005), a collection of Java implementations of machine learning algorithms. Rapidminer (Mierswa et al, 2006) is another open-source environment for data mining. Rapidminer seamlessly integrated all Weka algorithms. Rapidminer provides a rich variety of methods which allows rapid prototyping for new applications. By following the paradigm of visual programming, Rapidminer eases the design of processing schemes. In addition to interactive design, Rapidminer enables automated applications after the prototyping phase through the underlying XML representation.

In ABMiner, we utilized and extended the work flow design in Rapidminer to develop a fixed but reasonably flexible workflow for bioscience and biomechanics data analysis. ABMiner supports the entire cycle of data mining beginning with database selection, data extraction through data cleaning, data balancing, attribute reduction, data visualization, data mining model exploration and deployment. ABMiner contains a distributed optimization engine in the meta-learning level to select optimal algorithms and their parameters for any dataset using a flexibly composed optimization configuration. By distributing the complex tasks into a large computer cluster, the performance of different combinations can be evaluated in parallel, and one or more of the best combinations can be selected.

CUSTOMIZED DATA SCREENING AND PROCESSING

ABMiner provides a flexible data screening component for any relational database. This screening component models the database tables and their relationships using an object-oriented design. It supports a comprehensive, multi-perspective query interface that will extract different datasets using the similar language of domain experts, not requiring the knowledge of database and SQL.

This component is for amateur data mining users to prepare target data set from the data sources. Very often, the data are extracted from relational databases, as in the air force dynamics database. To interact with a relational database, the user needs to master the knowledge of SQL (Structured Query Language) to a certain level since SQL is the standard language for the retrieval and management of data in relational database management systems. For users who are not familiar with SQL, there is a need for a tool which can convert their data screening rules into the corresponding SQL queries. First, this tool should enable the user to intuitively view and operate on the objects (i.e., tables) in a relational database. Second, this tool should enable the user to define data screening rules via the graphic user interface. Last but not least, this tool should be able to correctly convert the user's data screening rules into SQL queries.

We implemented the data screening component using an open source solution, called Swing-based Wizard framework, see Figure 1. Also, we used Microsoft SQL Server JDBC APIs as our Database Programming APIs. Users need to specify the database connection parameters, such as the URL of the database server, login user name and password, and the name of the database to be connected. Once the configuration Wizard is started, it guides the user through the two-step configuration process to define a SQL Query. Users can go back and forth in these steps to modify their definitions before they choose to finish or cancel the configuration process. Once the user is done with the configuration, the data screening component automatically generates the SQL Query statement based on the user's definition.

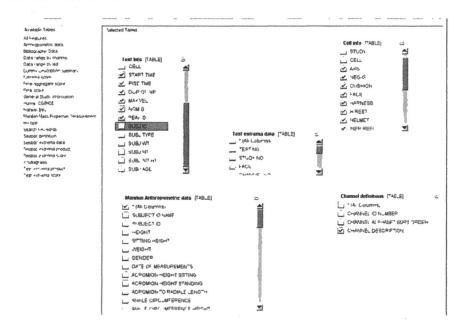

Figure 1: The Wizard Pages of the Data Screener

DATA MINING MODEL DISCOVERY THROUGH META-OPTIMIZATION

Data mining is the process of automatically searching large volumes of data for hidden patterns using algorithms such as classification, clustering and association rule mining. A key objective of data mining is to build data mining models (Liu and Tuzhilin, 2008). For example, prediction models can be built from biodynamic data to tell either the strength of human response or injury risk under different motion scenarios (Tang et al, 2007; Cheng et al, 2008).

Given a specific dataset and question in mind, there can be a number of data mining algorithms suitable for model building. Most data mining algorithms require the setting of input parameters. For example, the parameters in a support vector machine (SVM) include penalty parameters and the kernel function parameters.

These parameters usually have significant influences on the performance of the algorithm. To build a high-quality model, it is necessary to select the right algorithm with a right set of parameter values. Algorithm and parameter selection can be treated as a meta-optimization problem, i.e., finding an optimal model which has the best performance (e.g., highest prediction accuracy) for a specific dataset.

The data mining meta-optimization problem can take place at two levels: parameter level and algorithm level. The parameter-level optimization refers to finding the parameter settings that will result in optimal performance for a given algorithm. The algorithm-level optimization refers to selecting the algorithm with best performance from a list of applicable algorithms, each of which is considered for parameter optimization.

With the parameter optimization embedded, the process of algorithm-level optimization iterates within a list of feasible algorithms and finds one or multiple algorithms with highest prediction accuracy. Since there are multiple algorithms, each with multiple parameter settings, the optimization process involves expensive computation load (for both exhaustive search and heuristic search). This calls for scalable solution with distribution computation.

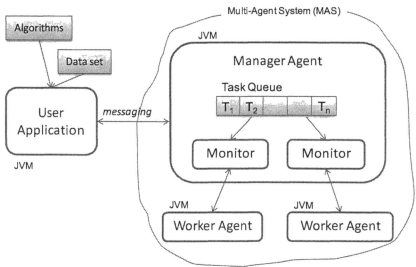

Figure 2: System architecture where user applications interact with a MAS

We have developed an agent-based framework to power the meta-optimization. This framework can evaluate the parameter settings for a number of algorithms in parallel via a multi-agent system and therefore can reduce computational time. Figure 2 shows the architecture of our framework, where user applications, e.g., Java programs running in a Java Virtual Machine (JVM), are flexibly coupled with a multiple agent system (MAS). The MAS consists of a manager agent and several worker agents. The manager agent is responsible for handling and allocating tasks, i.e., model building and cross validation. The work agents are computational engines that perform specific tasks.

There are several levels of messaging between the user application, the manager agent and the worker agents. When the MAS is started, the manager agent will send a message to all worker agents to query their processing capabilities (e.g., computation). The manager will then create an agent monitoring activity for each worker agent. When the user application (i.e., meta-optimization) is started, it will send the list of algorithms and the dataset to the manager agent. Once the manager agent receives this message, it will parse the application into a list of tasks, where each task T_i consists of a model (to be built) and a dataset. The tasks are stored in a dynamic data structure called *Task Queue*. This queue is a common object that can be accessed by the worker agents according to predefined protocols. Once the tasks are allocated, the manager agent will wait for the results from individual worker agents. When the manager agent obtains the results from all worker agents, it will deliver the combined results to the user application. Since the tasks are performed concurrently on the worker agents, the total response time is expected to be significantly reduced when compared with sequential processing. Please refer to Liu et al, 2010 for more details.

DATA MINING MODEL BASE

ABMiner provides a model base component that enables a well-trained data mining model together with its underlying data set and performance metrics to be wrapped up and accessible across various platforms. There are two major requirements when building the model base. First, the model base should be able to host the datasets and knowledge models from data mining activities, which are obtained by using IAI ABMiner IDE. Second, the model base should be able to expose these data mining outcomes to the community. These two major requirements immediately make the Client/Server model a perfect architecture to design and implement the model base, given the ubiquitous Internet nowadays. By accessing the model base, users can browse the knowledge models, and view the dataset associate with a knowledge model and query the knowledge model itself.

We developed the model base based on the technology of web services so that the data mining models can be queried and presented everywhere, including desktop, Internet and mobile devices. Figure 3 shows how Client-side applications (Web browsers and IAI ABMiner IDE) interact with the model base, as well as how the components of the model base interact with each other. The functional components of the model base are divided into two separate groups: the Business Tier module and the Web Tier module. The Business Tier module hosts the core functions, including the storage and management of the knowledge models and the associated datasets, and the computations incurred by dataset visualizations and model queries per users' requests. The computation results are consumed by the Web Tier module via the Web Services hosted in the Business Tier module. The Web Tier module generates the web pages based on the computation results from the Business Tier module and returns the pages to the users Web browsers. In addition, users can use IAI ABMiner to deploy the trained knowledge models and the associated datasets onto model base via the Web Services in the Business Tier module.

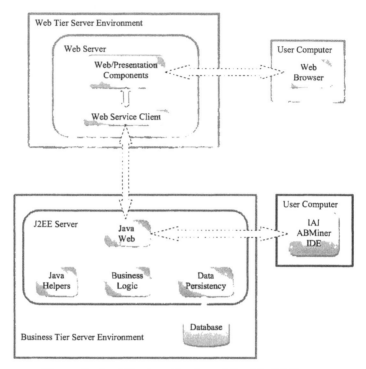

Figure 3: Architecture Design of the Model Base

CASE STUDIES

To explore the concept of parameter optimization in biodynamic data mining, we performed a case study related to Vertical Deceleration Tower (VDT) tests. The dataset has 20 independent input attributes such as NOM G (numeric), CUSHION (nominal) and HARNESS (nominal), and 1 output attribute MAXIMUM (numeric), which is the peak lumbar force in the Z direction. The problem is to predict the value of MAXIMUM using the 20 input parameters. We selected the REPtree algorithm for the prediction. REPtree is a fast decision tree learner. It builds a decision tree using information gain and prunes it using reduced-error pruning. The adjustable parameters in REPtree include:

M – Minimum number of instances per leaf. The default value is 2.

N – Number of folds for reduced error pruning. The default value is 3.

S – Seed for random data shuffling. The default value is 1.

For experimental purpose, we defined the value ranges for M and S as (1~3) and the value range for N as (2~4). That is, three possible values are allowed for each parameter. Therefore, we have $3^3 = 27$ parameter settings. We used 10-fold cross validation to build a model for each of the parameter settings. The results are shown in Table 1.

Table 1: Experiment with REPtree using different parameter settings

Set #	M	N	S	Relative Absolute Error (%)	Correlation	Solution
1	1	2	1	23.998	0.913	
2	1	2	2	24.589	0.901	
3	1	2	3	23.328	0.911	
4	1	3	1	22.484	0.923	
5	1	3	2	21.816	0.924	optimal
6	1	3	3	22.476	0.913	
7	1	4	1	21.987	0.920	
8	1	4	2	22.201	0.916	
9	1	4	3	23.662	0.894	
10	2	2	1	30.734	0.889	
11	2	2	2	29.903	0.881	
12	2	2	3	28.752	0.899	
13	2	3	1	26.785	0.911	default
14	2	3	2	26.043	0.905	
15	2	3	3	26.572	0.894	
16	2	4	1	25.946	0.909	
17	2	4	2	27.018	0.894	
18	2	4	3	26.771	0.901	
19	3	2	1	34.596	0.870	worst
20	3	2	2	32.588	0.878	
21	3	2	3	32.210	0.885	
22	3	3	1	30.912	0.903	
23	3	3	2	28.900	0.892	
24	3	3	3	31.551	0.885	
25	3	4	1	29.592	0.903	
26	3	4	2	29.626	0.883	
27	3	4	3	29.556	0.885	

As can be seen, using the default parameter setting (M=2, N=3, S=1), we obtained a relative absolute error of 26.785%. However, if using the parameter setting where M=1, N=3, and S=2, we reduced the prediction error to 21.816%. The results indicate that parameter optimization can significantly improve the prediction accuracy of a given algorithm for biodynamic data mining.

CONCLUSION

To study human effectiveness and performance, many hidden and nonlinear relationships must be studied. These relationships are usually recorded in the data which can be mined and represented as knowledge. With many years of development of data mining algorithms and systems, there are plenty of opportunities to study the data related to human effectiveness and performance. An intuitive, and customizable software tool with scalable computational capability will help the engineers and scientists in this area to discover those hidden and nonlinear relationship. ABMiner is a tool that aggregates successful experiences in data mining and provide intuitive user interface and scalable computational power to uniquely satisfy the requirements in this area. With the development of several years, ABMiner has grown into a software system that can be used for data analysis in the area of human effectiveness and performance.

REFERENCES

Buhrman, J.R., Plaga, J.A., Cheng, H., and Mosher, S.E. (2001), The AFRL Biodynamics Data Bank on the Web: A Repository of Human Impact Acceleration Response Data, SAFE Association Symposium Proceedings.

Cheng, Z., Rizer, A. L., Tang, K., Buhrman, J. R., Pellettiere, J. A. (2008), Exploration of Impact Biomechanics Using Data Mining, SAE, Detroit..

Fayyad, U., Piatetsky-Shapiro, G., Uthurusamy, R. (2003), "Summary from the KDD-2003 panel: Data Mining: The Next 10 Years." SIGKDD Explorations 5(2), 191-196.

Liu, B. and Tuzhilin, A. (2008), Managing and Analyzing Large Collections of Data Mining Models. Communications of ACM, Vol.51, No.2.

Liu, X., Tang, K., Buhrman, J.R., and Cheng, H. (2010), An Agent-based Framework for Collaborative Data Mining Optimization. IEEE International Symposium on Collaborative Technologies and Systems (CTS), Chicago, Illinois, USA, May 17 - 21, 2010.

Mierswa, I., Wurst, M., Klinkenberg, R., Scholz, M. and Euler, T. (2006), YALE: Rapid Prototyping for Complex Data Mining Tasks, in Proceedings of the 12th ACM SIGKDD International Conference on Knowledge Discovery and Data Mining (KDD-06).

Robinette, K.M., Blackwell, S., Daanen, H., Boehmer, M., Fleming, S., Brill, T., Hoeferlin, D., and Burnsides, D. (2002), Civilian American and European Surface Anthropometry Resource (CAESAR) Final Report, Volume i: Summary, Project accomplished under a Cooperative Research Agreement with SAE International, JUNE 2002.

Tang, K., Buhrman, J. R., Phadke A. (2007), Biomechanics Data Mining to Support Human Performance and Injury Prevention, INFORMS Annual Meeting, WAID Workshop, Seattle, November 2007.

Witten, I.H. and Frank, E. (2005), Data Mining: Practical machine learning tools

and techniques, 2nd Edition. Morgan Kaufmann, San Francisco.
Yang, Q., Wu, X. (2006), 10 Challenging problems in data mining research. Intl. Jrnl. of Information Technology & Decision Making 5(4), 597-604.